FIELD COLUMBIAN MUSEUM

PUBLICATION 95

ZOÖLOGICAL SERIES   VOL. IV, PART I.

THE
LAND AND SEA MAMMALS

OF

MIDDLE AMERICA AND THE
WEST INDIES

BY

DANIEL GIRAUD ELLIOT, F.R.S.E., ETC.

Curator of Department

CHICAGO, U. S. A.

1904

PUBLICATION

OF THE

FIELD COLUMBIAN MUSEUM

ZOÖLOGICAL SERIES

Vol. IV. Part I.

Chicago, U. S. A.

1904

THE

LAND AND SEA MAMMALS

OF

MIDDLE AMERICA AND THE WEST INDIES

BY

Daniel Giraud Elliot, F.R.S.E., etc.

Curator of Department.

ZOÖLOGICAL SERIES
VOL. IV. PART I.

Chicago, U. S. A.
1904

PREFACE.

In my previous volume, the "Synopsis of the Mammals of North America and the Adjacent Seas," the field covered was from the northern boundary of Mexico to and including the Arctic Ocean. The present work is supposed to contain all the Mammals of the remaining portion of the North American continent and the contiguous seas, from the northern boundary of Mexico to the Province of Cauca, South America, including the coast islands, as well as those of the Bahamas and the West Indies whose fauna is not completely related to that of South America. In the general treatment of the named forms the method adopted in the previous work has been slightly elaborated, and brief statements of the habits of the animals contained in the various families, and sometimes in the genera, have been given, together with the English name for each species or race, and keys for genera, subgenera, species, and races whenever these were sufficiently numerous to make such analytical tables desirable. For a very large number of the species and races it is well understood that no English names exist, and these had to be manufactured for the occasion, and are practically of little assistance for the recognition of the different animals; but Latin names appear to be distasteful to a small number of the laity, and only those in the vernacular are satisfactory, and it is to aid these that this departure from the previous plan has been made. Numerous named species of many genera of Mammals have so close a resemblance to each other, both in their outward covering and cranial characters, that often it is very difficult to distinguish one from the other, and for them Keys are probably less satisfactory as a means of determining the various forms than for any other class of animals; but it seems that there is a desire for such aids, which to many have become necessary, and therefore an effort has been made to meet this need, which it is hoped will serve the purpose intended. The illustrations throughout the volume comprise not only various representations of the cranium of some selected species of nearly every genus and subgenus, as in the "Synopsis," but in addition a figure is given of some species either of a family or genus, or possibly both, as the peculiarities of the animals seemed to require for a better comprehension of their appearance in life. To those unfamiliar with the diversified forms of the many mammals dwelling within the limits embraced in this volume, these figures may be of assistance, and enable them more easily to recognize the animals whose descriptions are given in the text.

The remarks made in the Preface of the "Synopsis," upon the excessive and probably unwarranted multiplications of species and races (made easy by the too liberal application of the trinominal system), may be repeated here with equal force as regards the mammalian fauna of Middle America and the various islands.

In the examination of the many specimens rendered necessary during the progress of the present work, the author has been impressed by the fact that the "characters" of a very large number of the named forms are merely comparative and not distinctive. By which is meant, characters that in themselves are not sufficient to identify the specimens, but render necessary the presence of examples of the typical form before any determination is possible, unless the locality is deemed all-sufficient to fix the status. The possession of topotypes of described forms for the majority of naturalists is impracticable except to a very limited degree, and therefore, without such aids, to accurately name specimens is, in many instances, quite impossible (for even "locality" is not always to be depended upon), and the effort often then degenerates into something very like guesswork. Every Mammalogist must at some time have been confronted with this difficulty and regretted his inability to determine his examples; and one naturally questions the value of a system that makes such a condition possible, and doubts if the giving of names to specimens on minute differences, which magnify slight comparative characters (for often there are no others, and some of these are undoubtedly due to individual variation), is scientifically warrantable or even desirable. Many specimens have been named whose cranial characters consist altogether in being "longer or shorter," "broader or narrower" than corresponding parts of some other example, and it is easily comprehended how slight is the probability that any specimen can be accurately determined whose characters are such as those given (the color of the pelage also being nearly the same), no topotypes of the forms with which these are compared by their describer being available, and in many instances no measurements of the crania having been given.

It is, of course, not to be conceived that every infinitesimal difference that an animal may possess can be intelligently demonstrated, or that the mere bestowal of a name upon a specimen would make it recognizable; and the act of naming examples that are separated from their fellows on account of these minute variations cannot fairly be regarded as an "accurate statement of the results of organic evolution." That it is desirable that all differences observed, the results of any cause whatever, should be mentioned, and in many instances dwelt upon, would not be disputed by any one, but it may well be questioned if the only wise or proper course to emphasize these slight variations is to bestow a name upon the specimen possessing them. Much confusion has been created by the multiplicity of names that burden our nomenclature, and our difficulties are by no means brushed away when one is bestowed upon some specimen, any more than are these difficulties ignored if none is given; for names are often not only no panacea for scientific woes, but on the contrary are frequently the cause of much trouble and perplexity. They are useful for the recognition of specimens possessing independent distinctive characters, but if an example has none of these its appellation is of little assistance. It is the *extreme* to which the bestowal of names has been carried that is to be deprecated, not the announcement of differences observed, however slight, and against the former custom the Author has always protested, while advocating the latter.

It is to be expected that countries like Mexico would contain a large number of animals that differ from each other in a greater or less degree; for that land probably possesses more varieties of climate within a certain number of miles square than almost any other known of an equal extent. The transition from a torrid to a temperate zone, and again to an alpine, is accomplished in a comparatively brief journey, and the several environments affect materially in certain ways the animals influenced by them. So in a small extent of country a number of forms allied to, but differentiated from each other would be expected to occur; but whether the extreme length to which the recognition of these variations has been carried by the bestowal of names is either wise or necessary, may well be doubted.

The arrangement of the Orders and Families is the same as that in the "Synopsis"; but names have in some instances been changed since that work was published, those heretofore employed having been ascertained to be either antedated by others, or previously used in this or some other branch of Zoölogy. Changes are continually occurring in Mammalian nomenclature, and it will probably be a long time before permanence in names is reached, as discoveries are being made that overturn some that are constantly employed and have become familiar by long use. But these changes will of necessity become less in time and a nomenclature that at least will approach stability may, in the distant future, be expected to be reached. By inserting the names of the described forms in this work the Author does not indorse their specific or subspecific value, and in all cases where an opinion is expressed, it will be found in a footnote on the page containing the form discussed. A critical review of all the species and races contained in this volume and their relations to each other would be a very great undertaking and cannot be properly attempted at this time. Much additional knowledge, and in many cases a greater amount of material must be acquired before any considerable success in accurately determining the proper status of the numerously named forms can be obtained.

The following is the arrangement adopted for the Orders and Families of the Mammalia comprised in this work, beginning with the lowest in degree:

The measurements of the species and races, unless otherwise stated, are given in millimeters, and usually from some selected specimen, although occasionally an average of several examples is recorded. But it must always be remembered that the dimensions of animals, even of adults belonging to the same species, vary greatly, and there is no hard and fast rule by which the exact size of any species or race of Mammals can be fixed, and allowance must be made for this variability when a comparison is instituted between the measurements given and some specimen in hand. It is really not easy to find two mammals exactly alike in all their dimensions. This fact was emphasized in the Preface of the "Synopsis," but it seems necessary to repeat it here.

The illustrations of the Crania exhibit the characteristics of every genus and subgenus contained in the work, with but few exceptions, and the Institution to which each specimen belongs and the catalogue number is given in every instance. The reason for an exception in the list is that it was not possible to obtain the cranium when desired, as no example was procurable from any collection in this country. These illustrations in half-tone of crania, with the exception of a few kindly furnished by the Director of the National Museum, were made from photographs taken by Mr. C. H. Carpenter, the Head of the Photographic Department of this Institution, and as faithful representations of the subjects exhibited with often minute and intricate details they will, it is believed, compare favorably with any heretofore published. The tooth-rows were photographed by means of an especial photomicrographic lens which causes the most minute enamel folds to be clearly visible.

The geographical distribution of many of the species and races included in this work is very imperfectly known, as a considerable number have been described only within a comparatively short time, and consequently but little information has been received regarding

them.   The extent of the dispersion of each named form has, however, been given so far as our present knowledge permits.

In the Synopsis and its Supplement 997 species and subspecies
were enumerated, and of these 789 were restricted to the regions
north of the boundary of Mexico, leaving 208 that were found on
both sides of the line.   Of genera there were 120 of which only 42
were not represented in Mexico.   In the present work the species
and subspecies number 1,018, of which 809 are restricted to Mexico
and the countries and islands embraced in the volume, leaving 209
also to be met with in the United States.   The genera number 178,
of which 78 are found north of the line, leaving 100 peculiar to the
Southern lands, the major portion of the excess over the northern
genera being found in the Chiroptera.   The Land Mammals in the
Synopsis numbered 933, and the Sea Mammals 63; in the present
volume the Land Mammals number 989 and the Sea Mammals 29,
and but two of the latter are not found in northern waters so far
as known, *Megaptera n. bellicosa* and *Prodelphinus longirostris*,
although the first named probably does go into the northern portions
of the Gulf of Mexico.

The following table exhibits the genera that are represented on
both sides of the northern boundary of Mexico, with their species and
subspecies, showing the number of those that are restricted to each
region, and also how many are common to both:

	Number of Species and Subspecies North of Mexican Line	Number of Species and Subspecies South of Mexican Line	Number of Species and Subspecies Common to Both Regions
Antilocapra	1	2?	1?
Antrozous	2	3	2
Balænoptera	8	3	3
Bassariscus	4	6	2
Blarina	7	16	1
Canis	9	11	3
Castor	4	1	1
Citellus	57	21	11
Cogia	1	1	1
Conepatus	1	8	2
Corynorhinus	3	2	2
Cratogeomys	1	8	1
Cynomys	4	4	3
Dasypterus	1	3	1
Delphinus	1	1	1
Didelphys	4	9	2
Dipodomys	14	13	6
Erethizon	5	1	1
Felis	19	24	8
Fiber	8	1	1
Geomys	15	1	1
Globicephalus	3	3	3
Heteromys	1	35	0
Lasiurus	4	5	3

	Number of Species and Sub-species North of Mexican Line	Number of Species and Sub-species South of Mexican Line	Number of Species and Sub-species Common to Both Regions
Latax	1	1	1
Lepus	56	43	19
Lutra	6	2	1
Megaptera	3	2	1
Mephitis	11	5	2
Microtus	66	7	0
Mirounga	1	1	1
Mormops	1	4	1
Mus	4	5	4
Myotis	14	20	10
Neotoma	24	29	8
Notiosorex	1	3	1
Nycticeius	1	2	1
Nyctinomus	3	5	1
Nyctinomops	1	1	1
Odontocœlus	11	17	4
Onychomys	12	7	6
Orcinus	2	1	1
Oryzomys	5	42	0
Otopterus	1	5	1
Ovis	6	2	1
Perodipus	10	5	3
Perognathus	42	33	15
Peromyscus	70	108	19
Phoca	6	1	1
Phocæna	2	1	1
Physeter	1	1	1
Pipistrellus	3	6	1
Procyon	5	5	1
Prodelphinus	4	3	2
Promops	1	5	0
Pseudorca	1	1	1
Putorius	34	7	2
Rhachianectes	1	1	1
Rhithrodontomys	16	41	4
Scapanus	7	1	0
Sciuropterus	13	1	1
Sciurus	35	51	4
Sigmodon	8	24	4
Sorex	33	12	0
Spilogale	12	7	1
Tagassu	1	9	2
Tamias	35	6	3
Tatu	1	1	1
Taxidea	3	2	1
Thomomys	30	17	4
Trichechus	1	1	1
Tursiops	2	2	2
Urocyon	8	6	3
Urus	12	2	1
Vespertilio	1	9	1
Vulpes	17	1	1
Zalophus	1	1	1
Ziphius	2	1	1

Much care has been given that all the forms that have received names before this work was sent to the press should be included, and it is hoped that few, if any, have been omitted.

Descriptions of all species and races known to the Author, which were published prior to July 1, 1904, are given in this work. No attempt has been made to add to the List after that date, as the press work was then too far advanced to permit of any additions.

In the Appendix at the end of Part II. will be found descriptions of all those Mammals that were published too late, as the pages passed through the press, to be included in their proper position in the volume.

A work like the present could not be brought to a successful issue without material assistance from various quarters, as no Museum possesses collections of such extent as to render it independent of all others, and the Author is under many obligations to his colleagues in different Institutions for the loan of material and for all other aid requested toward the successful completion of his labors. It gives him, therefore, much pleasure to name the following to whom he feels much indebted: Dr. J. A. Allen, Curator of Vertebrate Zoölogy in the American Museum of Natural History, New York, and F. M. Chapman, Esq., Assistant Curator; Dr. C. H. Merriam, Chief of the Biological Survey, Department of Agriculture, Washington, and his able assistants, Dr. A. K. Fisher, V. Bailey, W. H. Osgood, A. H. Howell, and E. A. Preble, Esqs.; R. Rathbun, Esq., Director of the National Museum, Washington; Dr. F. W. True, Curator of Biology in the National Museum, and G. S. Miller, Jr. Assistant Curator, and M. W. Lyon, Esq., of the Department of Mammals; Witmer Stone, Esq., Curator of Ornithology in the Academy of Natural Sciences of Philadelphia; Outram Bangs, Esq., of the Museum of Comparative Zoölogy, Cambridge, Mass.; Dr. E. A. Mearns, U. S. Army; and Oldfield Thomas, Esq., of the British Museum. To all these the Author desires to express his thanks for having in many instances helped to make the "rough places smooth," and the completion of this work in its present form a possibility.

D. G. E.

15th July, 1904.

CONTENTS.

VOLUME IV. PART I.

LIST OF PLATES.

VOLUME IV. PART I.

LIST OF ILLUSTRATIONS OF CRANIA IN THE TEXT.

VOLUME IV. PART I.

xvii

LIST OF FIGURES IN THE TEXT.

VOLUME IV.   PART I.

ERRATA.

VOLUME IV. PART I.

ILLUSTRATIONS.

Plates xxxvi and xxxvii, for *Tapirella dowi*, read *Tapirella bairdi*.

TEXT.

Page 74, 8th line from bottom, for 98.5 read 985.

Page 104, 12th line from top, for Hoffman's Squirrel, read Hoffmann's Squirrel.

Page 105, 9th line from top, for *S. æ. hoffmani*, read *S. æ. hoffmanni*.

Page 130, 14th line from bottom, for parieta, read parietal.

Page 152, 18th line from bottom, for announed, read announced.

*Page 177, 19th line from top, for *a.-mesomelas*, read *texensis mesomelas*.

*Page 177, 11th line from bottom, for *b.-castaneus*, read *texensis castaneus*.

Page 205, 11th line from bottom, for *fclepensis*, read *felipensis*.

Page 270, 3d line from bottom, for oranze, read orange.

Page 350, 10th line from bottom, for Forte Verde, read Fort Verde.

. Page 351, 20th line from bottom, for *hermanni*, read *heermanni*.

Page 357, 17th line from top, for *P. h. zacalecas*, read *P. h. zacatecæ*.

Page 357, 20th line from top, for *rhydinohris*, read *rhydinorhis*.

Page 369, 2d line from bottom (Footnote), for instances, read instances.

Page 414, 10th line from top, for foreman, read foramen.

---

*These were inserted after the pages were set up, and inadvertently were placed under the wrong species. They should have gone on page 188, after *e.-deserticola*.

CLASS MAMMALIA.

Order I. Marsupialia. Marsupials.

The Marsupials or Pouched Mammals have at the present time a most restricted distribution, all the families of the order but one being found in the Australian region, viz., Australia, Tasmania, New Guinea, Celebes, and smaller contiguous islands. The one family, Didelphyidæ, foreign to this portion of the world is confined to the more southern parts of North America, and to South America. Marsupials are peculiar in the majority of cases, for having a fold of skin about the milk glands which forms a pouch, and in which the undeveloped young are placed and nourished. The species vary greatly in size, from the giant kangaroo, taller than many men, to little creatures not much larger than a mouse. One, *Chironectes minimus*, an opossum from Central America, Guiana, and Brazil, is aquatic in its habits, with large webbed hind feet, and it feeds on fish and other marine creatures which it secures in the manner of the otter. Some opossums, however, are not provided with a pouch, but the young are nevertheless fastened to the teats of the mother in a similar manner as are those whose parents possess this sac, and when they are sufficiently grown to leave the teats, they are transferred to their mother's back, where they maintain their position by wrapping their tails around that of the female, which is elevated over her back and carried there for this purpose.

Fam. I. Didelphyidæ. Opossums.

Limbs rather short; feet with five distinct toes; tail prehensile. Pouch sometimes present. Habits arboreal.

O. Thomas. *Catalogue of the Marsupials and Monotremata in the collection of the British Museum*, 1888.

1

1. Chironectes.

$$I.\tfrac{5-5}{4-4}; \ C.\tfrac{1-1}{1-1}; \ P.\tfrac{3-3}{3-3}; \ M.\tfrac{4-4}{4-4} = 50.$$

Chironectes Illig., Prodr. Syst. Mamm. et Av., 1811, p. 76.   Type
Latra! minima Zimmermann.

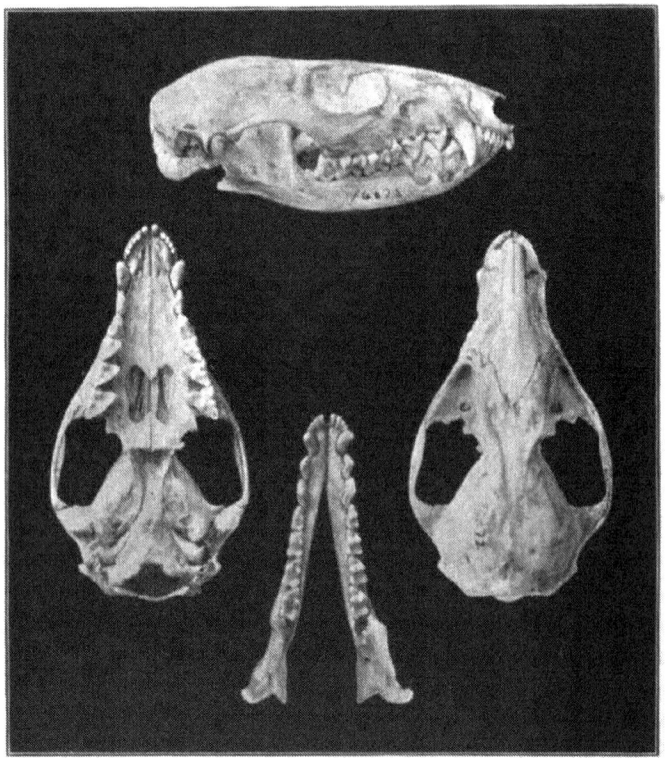

FIG. 1. CHIRONECTES MINIMUS.
No. 16072 Am. Mus. Nat. Hist. Coll.   Nat. size.

Skull similar to Didelphys; nasals expanded posteriorly; postorbital processes prominent; temporal ridges forming a crest in
adults; broad interorbital space with square edges; zygomata laterally expanded.   Single pair only of large vacuities opposite molars
on palate posteriorly, no smaller pairs.   Enlarged pisiform bone on
fore feet, forming a prominent tubercle; hind feet webbed to end of
toes; toe pads protruding beyond web.

1. **minimus** (*Latra, sic*), Zimm., Geog. Gesch., II, 1780, p. 317.
   paraguensis and *guianensis* Kerr, Linn. Anim. King., 1792, pp. 172, 174.
   memina Cuv., Tabl. Elém., 1798, p. 125.
   sarcovienna Shaw, Gen. Zoöl., I, 1800, pt. II, p. 447.
   variegatus Illig., Abh. Ak. Wiss., Berl., 1811, p. 107.
   palmata Cuv., Règn. Anim., I, 1817, p. 174.
   yapock Desm., Mamm., I, 1820, p. 261.

FIG. I. CHIRONECTES MINIMUS. WATER OPOSSUM.

WATER OPOSSUM, YAPOCK. *Zarro de Agua* in Costa Rica; *Tlacuazin de Agua* in Guatemala.

Type locality. Guiana.

Geogr. Distr. Guatemala south through Central America to southern Brazil.

Genl. Char. Size medium; ears large, rounded, metatragus very small; tufts of facial bristles above eyes, on cheeks in front of ears, and on the throat between jaws; whiskers on side of muzzle long; fur thick, woolly.

Color. Grayish white, mixed with light brown; band through eye and crown blackish brown; grayish white crescentic band between ears above the eyes; line from crown to base of tail, and transverse lines over shoulders, middle of back, loins and rump, black; ground hue between these slaty gray; chin, chest and belly white; outside of arms and legs grayish, the inner side white; tail furred at base only, remainder scaly, proximally black grading into yellowish at the tip.

Measurements. Total length, 720; tail, 395; hind foot, 72. Skull: occipito-nasal length, 53; Hensel, 48; zygomatic width, 30; interorbital constriction, 7; palatal length (palatal arch to alveoli of incisors), 31; length of upper molar series, 19; length of mandible to tips of incisors from angle, 43.

2. Marmosa.

$$I.\tfrac{5-5}{4-4};\ C.\tfrac{1-1}{1-1};\ P.\tfrac{3-3}{3-3};\ M.\tfrac{4-4}{4-4} = 50.$$

Marmosa Gray, Lond. Med. Repos., xv, 1821, p. 308. Type *Didelphis! murina*, Linnæus.

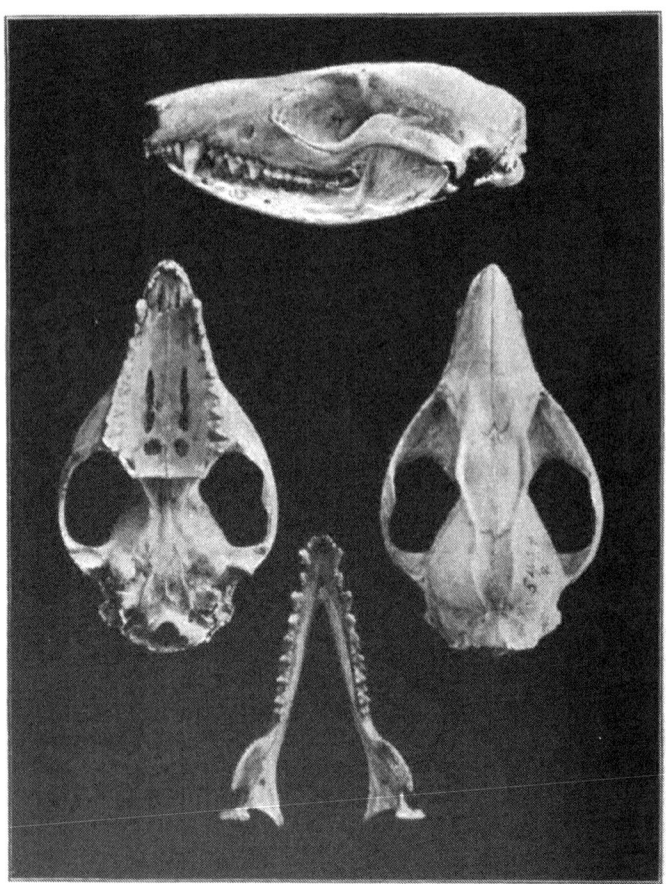

FIG. 2. MARMOSA CHAPMANI. TRINIDAD.
No. 5499 Field Columbian Mus. Coll. Enlarged ½

Micoureus Less., Tabl. Règn. Anim., 1842, p. 186.
Asagis and Notagogus Glog., Handb. Naturg., 1, 1841, p. 82.
Thylamys (sic) Gray, List Mamm. Brit. Mus., 1843, p. 101.
Grymæomys Burm., Thiere Bras., 1, 1854, p. 135.
Cuica Liais, Climats, Géol., Faun. et Geog. Botanique, Brésil,
    1872, p. 427.

Size small; pouch absent; fifth hind toe sometimes not longer than the second; tail long; body slender; teeth large, strong.

KEY TO THE SPECIES.

A. Size small; face without dark central streak.
  a. Under parts yellowish white.   PAGE

2. murina (*Didelphis!*), Linn., Syst. Nat., 1, 1758, p. 55.
MURINE OPOSSUM. *Tlacuazin Raton* in Guatemala.

Type locality. Unknown. ("In Asia; America.") Brazil?
Geogr. Distr. Central Mexico to Brazil.
Genl. Char. Tail furry at base; ear large, rounded, naked, basal projection long, pointed. Skull: nasals long, of nearly equal width throughout their length; supraorbital ridges present, tips forming postorbital processes; temporal ridges not meeting in center of braincase.

Color. General hue above deep rufous, sides paler; cheeks, throat, chin, and lips buff; orbital ring and space between eyes and nose black; under parts and inner side of limbs yellowish white or buffy; outside of limbs like back; hands and feet flesh color; tail covered with rufous hair at base, remainder pale brown.

Measurements. Total length, 395; tail vertebræ, 218; hind foot, 26; ear, 29. Skull: basal length, 33; length of nasals, 15; across postorbital processes, 7.2; palatal length, 20; length of upper molar series, 6.4.

a.—mexicana (*Marmosa*), Merr., Proc. Biol. Soc. Wash., xi, 1897, p. 44.

MEXICAN MURINE OPOSSUM.

Type locality. Juquila, State of Oaxaca, Mexico.

Geogr. Distr. States of Oaxaca and Chiapas, Mexico.

Genl. Char. Paler than *M. murina,* interparietal broader and shorter.

Color. Above cinnamon rufous, graduating into ochraceous buff on belly; orbital ring black; end of nose to between eyes buffy; tail above brown, beneath paler. Some individuals have unicolor tails.

Measurements. Total length, 330; tail vertebræ, 186; hind foot, 20.

3. canescens (*Micoureus*), Allen, Bull. Am. Mus. Nat. Hist., N. Y., 1893, p. 235.

ASHY OPOSSUM.

Type locality. Santo Domingo de Guzman, Isthmus of Tehuantepec, Mexico.

Geogr. Distr. Isthmus of Tehuantepec, northwesterly through States of Oaxaca, Guerrero and Michoacan to the Hacienda Magdalena in State of Colima, Mexico.

Genl. Char. Smaller than *M. murina.* Tail furred at base. Skull: nasals less expanded posteriorly than in *M. murina,* and the small posterior palatal vacuities absent.

Color. Above ashy brown tinged with rufous, beneath white tinged with yellow; orbital ring black, reaching nearly to the nose; sides of face, neck, and between eyes yellowish gray; ears pale brown; tail pale brown spotted with flesh color, furred at base, rest naked, terminal portion often white; feet grayish white.

Measurements. Total length, 266–288; tail vertebræ, 142–150; hind foot, 60–70; ear, 65. Skull: total length, 35.5; basal length, 33.8; zygomatic breadth, 20.8; length of nasals, 16.3; anterior border of premaxillæ to posterior border of palatal floor, 18.8; length of mandible, 26; height at condyle, 7.6; at coronoid process, 11.7.

4. sinaloæ (*Marmosa*), Allen, Bull. Am. Mus. Nat. Hist., N. Y., 1898, p. 143.

SINALOA OPOSSUM.

Type locality. Tatemales, State of Sinaloa, Mexico.

Geogr. Distr. State of Sinaloa, Mexico.

Genl. Char. Similar to *M. canescens* in color, but smaller.

Color. Above rufous brown, darkest on dorsal regions, paler on sides; beneath pale yellow washed with sooty; orbital ring black; ears and tail pale reddish brown; cheeks and throat pale yellow; feet and hands sparsely covered with yellowish hairs.

Measurements. Total length, 205-242; tail vertebræ, 115-122; hind foot, 16-18; ear, 22-25. Skull: total length, 31; basal length, 29; zygomatic width, 16.5; length of nasals, 13.7; across postorbital processes, 6.2; width of braincase, 11.2; tips of premaxillæ to palatal arch, 17; length of mandible, 22; height at condyle, 3; at coronoid process, 7.

FIG. II. MARMOSA CINEREA. GRAY OPOSSUM.
No. 7052 Field Columbian Mus. Coll.

5. cinerea (*Didelphis!*), Temm., Monog. Mamm., I, 1827, p. 46. GRAY OPOSSUM.

Type locality. Brazil.
Geogr. Distr. Costa Rica to Brazil.
Genl. Char. Size large; ear large, rounded, naked. Skull strong, with nasals expanded posteriorly; zygomata widely spread; interorbital region flat; postorbital processes conical; temporal ridges rather prominent; canines thick, short.

Color. General hue gray; sides washed with yellow, sometimes with rufous; black band inclosing the eye; under parts yellowish white; arms and legs gray; feet pale brown; tail furred at base, rest naked, scaly, slaty gray grading into yellow or yellowish white at tip; ears naked, flesh color.

Measurements. Total length, 425; tail, 248; hind foot, 24; ear, 21. Skull: basal length, 41; greatest width, 26; length of nasals, 18.5; across postorbital processes, 10; palatal length, 23.7; length of upper molar series, 7.1.

6. insularis (*Marmosa*), Merr., Proc. Biol. Soc. Wash., XII, 1898, p. 14. MARIA MADRE ISLAND OPOSSUM.

Type locality. Maria Madre Island, State of Jalisco, Mexico.

Gcogr. Distr. Tres Marias Islands, State of Jalisco, Mexico.

Genl. Char. Similar to *M. canescens*, but ears and tail longer, color more fulvous. Skull narrower and more slender.

Color. Upper parts drab brown, suffused with pale fulvous; orbital ring broad, black; median face stripe buffy fulvous; under parts buffy yellow, darkest on throat and breast; tail brown, no white.

Measurements. Total length, 270; tail vertebræ, 170; hind foot, 20.

7. oaxacæ (*Marmosa*), Merr., Proc. Biol. Soc. Wash., XI, 1897, p. 43. OAXACA OPOSSUM.

Type locality. City of Oaxaca, State of Oaxaca, Mexico.

Geogr. Distr. Range unknown. "Sonoran fauna of highlands of the State of Oaxaca, Mexico."

Genl. Char. Size small; darker than *M. canescens*; feet and ears smaller.

Color. Above dark sepia brown, reaching wrists and ankles; beneath buffy yellow; nose on top to behind eyes pale brown; orbital ring black; tail brown above, white beneath.

Measurements. Total length, 263; tail vertebræ, 144; hind foot, 18. Skull: basal length, 29; zygomatic breadth, 18.5; palatal length, 17; interorbital constriction, 4.8; breadth of frontals, 8. (ex Type.)

8. fulviventer (*Marmosa*), Bangs, Amer. Nat., XXXV, 1901, p. 632. FULVOUS-BELLIED OPOSSUM.

Type locality. San Miguel Island, Bay of Panama.

Genl. Char. Similar to *M. mitis* Bangs ex Colombia, but smaller in size and under parts fulvous instead of yellowish white

Color. Black facial markings as usual in species of this genus; upper parts cinnamon or tawny ochraceous; upper surface of arms, sides of neck and of body ochraceous rufous; under parts buff shading into ochraceous buff on lower sides, and on inner surface of arms and legs; tail dusky above, paler beneath; feet and hands grayish white.

Measurements. Total length, 325–340; tail vertebræ, 175–180; hind foot, 23–25; ear from notch, 20–22. Skull: basal length, 34.4; occipito-nasal length, 37.4; zygomatic width, 20.4; interorbital constriction, 6.2; length of nasals, 17.6; width of nasals, 5; palatal length, 18.8; upper tooth row from anterior edge of canine to posterior edge of last molar, 15.2; length of single half mandible, 27.8.

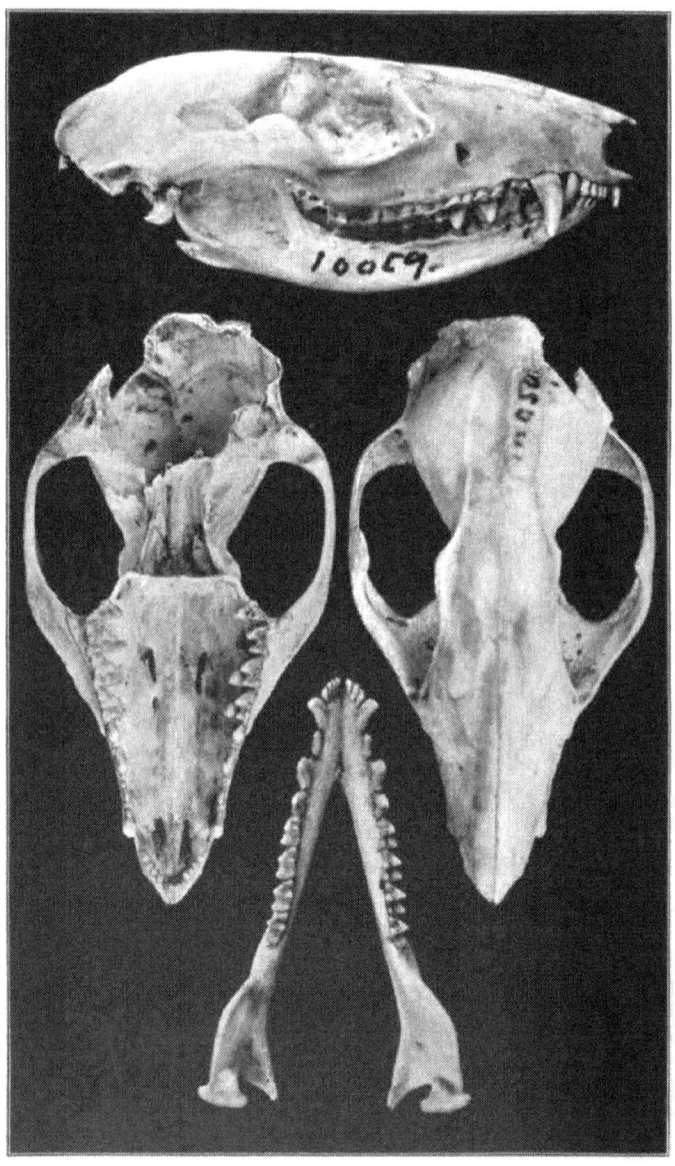

CALUROMYS ALSTONI.
No. 10059 Am. Mus. Nat. Hist. Coll. Twice nat. size.

3. Caluromys.

$$I.\frac{5-5}{4-4}; \ C.\frac{1-1}{1-1}; \ P.\frac{3-3}{3-3}; \ M.\frac{4-4}{4-4} = 50.$$

Caluromys Allen, Bull. Am. Mus. Nat. Hist., 1900, p. 189. Type *Didelphis! philander* Linnæus.

Size medium; pouch rudimentary; second hind toe shortest, fourth longest, third and fifth equal; fur thick, woolly. Skull with postorbital processes well developed; median crest absent; palate without large vacuities posteriorly.

KEY TO SPECIES AND SUBSPECIES.
PAGE
A. Under parts yellowish.........................*C. alstoni* 9
B. Under parts grayish white.
  a. Upper parts rusty........................*C. derbianus* 9
  b. Upper parts pale rufous.... ....*C. lanigera pallidus* 10

9. alstoni (*Caluromys*), Allen, Bull. Am. Mus. Nat. Hist., XIII, 1900, p. 189.
  cinerea Alston, Biol. Centr. Amer., Mamm., 1880, p. 199, pl. XXI. (nec Temm.)
ALSTON'S OPOSSUM.
  Type locality. Tres Rios, Costa Rica.
  Geogr. Distr. Costa Rica, Central America.
  Color. Above sooty, the hairs tipped with deep chestnut brown; under parts yellowish; space from nose to crown between eyes buff, inclined to reddish on top of nose; two blackish streaks on side of nose to and encircling eyes; cheeks and upper part of throat buff; ears naked, brown; hands reddish brown; feet yellowish; tail covered with hairs colored like back at base for about 37 mm., naked and yellowish for the rest of its length.
  Measurements. Total length, 430; tail, 250; hind foot to claws, 25. Skull: occipito-nasal length, 43; Hensel, 42; zygomatic width, 23.5; interorbital constriction, 7; length of nasals, 11.5.

10. *derbianus (*Didelphys*), Waterh., Jard., Nat. Libr., Mamm., XI, 1841, p. 97.
EARL OF DERBY'S OPOSSUM. Native name *Chucha Rata.*
  Type locality. Unknown.

*There is considerable variation in the markings of this species, the dorsal stripe being much more restricted in some specimens than in others, and one example. No. 11,788, Collection of the New York Museum, is almost uniform sooty above tinged with reddish, and the tips of the hairs whitish. The gray between the shoulders is indistinct and mixed with the general color, and not in a stripe, while the head above is sooty in the center, and reddish brown on sides of occiput and also on the neck. There is none of the rust color, so conspicuous on the typical style, anywhere visible.

Geogr. Distr. Nicaragua in Central America to Peru in South America.

Genl. Char. Tail longer than head and body, furred for one-third of its length.

Color. Upper parts, sides, and outer side of limbs brownish rust color; under parts soiled white; head brownish gray with a median dusky stripe from forehead to nose; orbital region brown; gray dorsal stripe from between shoulders to root of tail; gray line behind fore-arms and one on leg from knee upward; lower part of arms and

FIG. III. CALUROMYS DERBIANUS. EARL OF DERBY'S OPOSSUM.

hands white; feet dusky; tail brown on basal portion, naked part pinkish spotted with dark brown; ears pale brown, naked.

Measurements. Total length, 765; tail, 425; hind foot, 47.5; ear, 30.

laniger pallidus (*Philander*), Thomas, Ann. Mag. Nat. Hist., 7th
        Ser., IV, 1899, p. 286.
PALE WOOLLY OPOSSUM.

Type locality. Bogava, Chiriqui, Panama. Altitude, about 750 feet.

Geogr. Distr. Southeastern Central America.

Color. Above pale gray or pale rufous; face brownish; forearms, shoulders, and sides of hips pale gray; hind legs whitish or tinged with rufous; tail whitish gray, naked part mottled.

Measurements. Total length, 587; tail, 398; hind foot, 43; ear, 32. Skull: "Greatest length, 61; greatest breadth, 35; length of upper molars, 9."

4. Metachirus.

$$I.\frac{5-5}{4-4}; \ C.\frac{1-1}{1-1}; \ P.\frac{3-3}{3-3}; \ M.\frac{4-4}{4-4} = 50.$$

Metachirus Burm., Thier. Bras., I, 1854, p. 135. Type *Didelphys nudicaudata* E. Geoffroy.

Size medium; pouch rudimentary or well developed; three central hind toes subequal, longer than fifth; fur without bristles. Skull with temporal crests well developed.

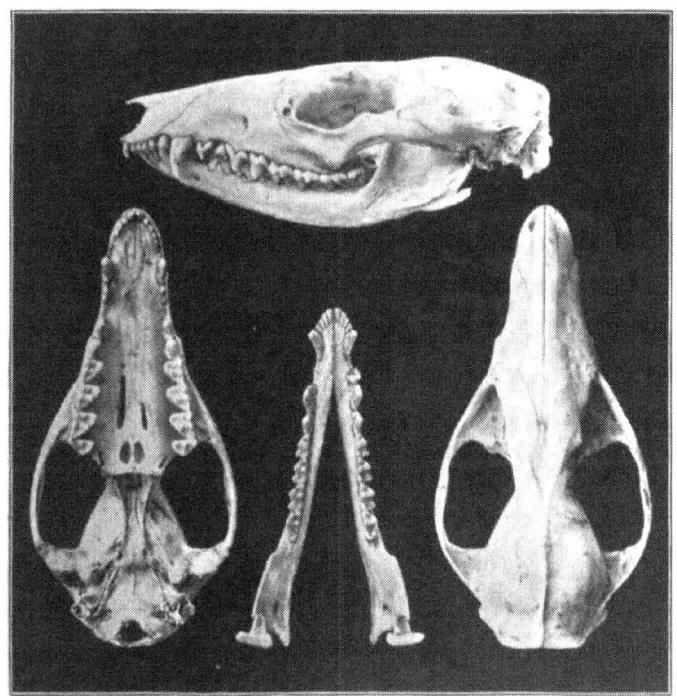

FIG. 3. METACHIRUS FUSCOGRISEUS.
No. 8252 Am. Mus. Nat. Hist. Coll. Nat. size. Type.

KEY TO THE SPECIES.

11. *nudicaudatus (*Didelphys*), E. Geoff, Cat. Mus., Paris, 1803, p. 42.
RAT-TAILED OPOSSUM.

Type locality. Cayenne, French Guiana. (Allen.)

Geogr. Distr. Costa Rica? to Brazil.

Genl. Char. Size about equaling *M. opossum* but more slender; ears large, brown, rounded, naked. Skull: postorbital processes barely apparent; interorbital space broad.

Color. Above grayish brown suffused on sides with yellowish, sometimes rufous; face rufous brown with a white or yellowish white spot over eye; under parts yellowish white; front of arms and outside of legs pale brown; indistinct yellow lateral line; tail furred at base, rest naked, scaly, brown grading to white at tip; hands and feet pale brown; ears slaty gray.

Measurements. Total length, 540; tail, 300; hind foot, 43; ear, 25. Skull: basal length, 61; greatest breadth, 36; length of nasals, 30; across postorbital processes, 11; palatal length, 36; length of upper molar series, 10.5.

FIG. IV. METACHIRUS FUSCOGRISEUS. ALLEN'S OPOSSUM.
No. 8252 Am. Mus. Nat. Hist. Coll. Type.

12. fuscogriseus (*Metachirus*), Allen, Bull. Am. Mus. Nat. Hist., N. Y., 1900, p. 194.

    quica True, Proc. U. S. Nat. Mus., VII, 1885, p. 587. (nec Temminck.) Alston, Biol. Centr. Amer., I, 1881, p. 198. (Part.)

---

*Thomas, Cat. Marsupials, p. 333, gives this species from Costa Rica.

ALLEN'S OPOSSUM.   *Tlacuazin* in Guatemala.

Type locality.   Unknown.

Geogr. Distr.   Southern Mexico to Costa Rica.

Genl. Char.   Size medium; tail longer than head and body; ears large.

Color.   Above blackish washed with gray; top of head and median line black; flanks gray; under parts yellowish white; a band above and a broader one below ears, and spots above eyes, yellowish white; outer surface of limbs paler than sides of body; inner surface yellowish white; tail dark brown at base, grading into light brown or flesh color at tip, heavily furred at base, remainder naked; feet brownish, naked; ears flesh color broadly edged with dark brown.

Measurements.   Total length, 534; tail, 283; hind foot and claws, 39.   Skull: basal length, 62; nasals, 32; canine to posterior edge of last molar, 25.5; palatal length, 38; zygomatic breadth, 32; mastoid breadth, 19; interorbital constriction, 8.5. (Type.)

a.—pallidus (*Metachirus*), Allen, Bull. Am. Mus. Nat. Hist., 1901, p. 215.

ORIZABA OPOSSUM.

Type locality.   Orizaba, State of Vera Cruz, Mexico.

Geogr. Distr.   States of Vera Cruz, Puebla, and Tabasco, south-eastern Mexico.

Genl. Char.   Similar to *M. fuscogriseus*, but lighter; orbital spots larger.

Color.   Head black above with two large spots above eyes, and a brownish white stripe on each side of nose; rest of upper parts blackish gray, darkest on dorsal line; side of body gray, paler than back; side of head, throat, hands, feet, and entire under parts yellowish white; tail at base furred like the back; naked portion black for two-thirds the length, spotted with flesh color, remaining part all flesh color.

Measurements.   Total length, 475–627; tail, 240–315; tarsus, 38–47.   Skull: total length, 77; basal length, 69; length of nasals, 37; zygomatic width, 39; across postorbital processes, 14.3; interorbital constriction, 9; mastoid width, 25; palatal length, 42; length of upper molar series, 15. (ex Type.)

5. Didelphys.

$$I.\frac{5-5}{4-4}; \; C.\frac{1-1}{1-1}; \; P.\frac{3-3}{3\;3}; \; M.\frac{4-4}{4-4} = 50.$$

J. A. Allen.   *A Preliminary study of the North American Opossums of the genus Didelphis!*, Bull. Am. Mus. Nat. Hist., 1901, p. 149.

J. A. Allen. *A Preliminary study of the South American Opossums of the genus Didelphis!* Bull. Am. Mus. Nat. Hist., 1902, p. 249.
Didelphis (*sic*) Linn., Syst. Nat., i, 1758, p. 54. Type *Didelphis!*
marsupialis Linnæus.

Size very variable; ears large; hind feet short; feet with five distinct toes, all provided with nails except the first toe of the hind foot,
which is large, opposed to the others in grasping, and is without a
nail. Tail long, prehensile, partly naked; pouch complete; long,
bristle-like hairs mingled with the fur; incisors small and pointed;
canines large; premolars with compressed pointed crowns; molars
with sharp cusps.

KEY TO THE SPECIES AND SUBSPECIES.

A. Size medium.   PAGE
    a. Tail black only at base................*D. yucatanensis*   14
    b. Tail black for two-thirds its length.......*D. y. cozumelæ*   15
B. Size large.
    a. Under fur white at base.
      a.' Under parts grayish white.
        a." Posterior end of nasals rounded....*D. mesamericana*   15
        b." Posterior end of nasals acute ....*D. m. texensis*   16
      b.' Under parts yellowish..............*D. m. tabascensis*   16
    b. Under fur orange buff at base....... ......*D. richmondi*   16
    c. Under fur yellowish white at base.
      a.' Head yellowish white to nape........*D. m. insularis*   17
      b.' Head dark, spotted with white...........*D. m. battyi*   17
      c.' Middle of head between eyes posteriorly,
        blackish...............................*D. m. etensis* . 18

13. yucatanensis (*Didelphis!*), Allen, Bull. Am. Mus. Nat. Hist., 1901,
    p. 178.
YUCATAN OPOSSUM.

    Type locality.   Chichen Itza, Yucatan, Mexico.
    Geogr. Distr.   State of Campeche, and Yucatan, Mexico.
    Genl. Char. Similar in color to *D. m. caucæ* (Allen, from southwest-
ern Columbia) but smaller. Black and gray phase equally represented.
    Color. Male, upper parts black, base of hairs white; beneath
grayish white; limbs, hands, and feet black; tail black at base;
remainder flesh color; ears black. Female similar to the male, but
covered with long yellowish white hairs, causing her to appear much
lighter, and the black at base of tail is more extensive; under parts
yellowish.
    Measurements. Total length, 634–756; tail, 312–393; tarsus,
54–60. Skull: total length, 100; basal length, 90; zygomatic width,

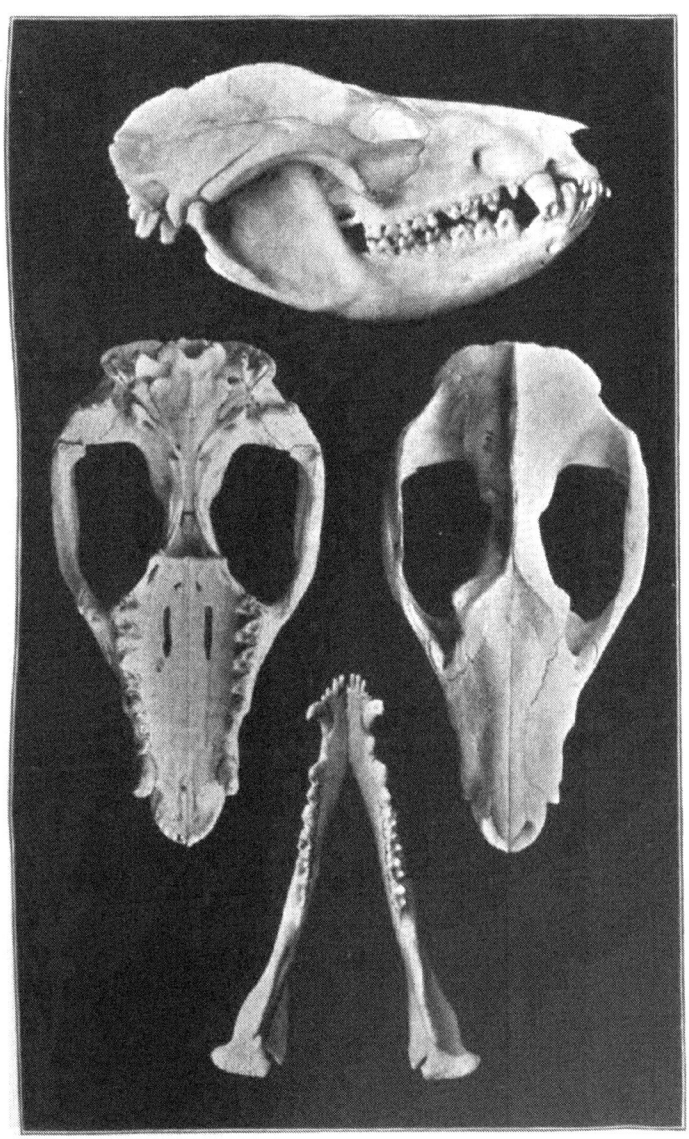

DIDELPHYS MESAMERICANA.
No. 8723 Field Columbian Mus. Coll.   ⅔ nat. size.

48; length of nasals, 46; mastoid breadth, 29.4; palatal length, 52; length of upper molar series, 19.

a - cozumelæ (*Didelphis!*), Merr., Proc. Biol. Soc. Wash., XIV, 1901, p. 101.
ISLAND OF CŎZUMEL OPOSSUM.

Type locality. Cozumel Island, Yucatan.

Genl. Char. Similar to *D. yucatanensis* but larger; tail shorter; rostrum and nasals broader; ears large, broad.

Color. Upper parts black, many long white hairs protruding; beneath dusky with white hairs intermingled; hands and feet black; tail black for two-thirds the length, remainder flesh color; ears black.

Measurements. Total length, 703; tail vertebræ, 324; hind foot, 59. (ex Type.)

FIG. V. DIDELPHYS MESAMERICANA. LINNEAN OPOSSUM.
No. 8725 Field Columbian Mus. Coll.

14. mesamericana, Oken, Lehrb. der Zoöl., II, 1816, p. 1152.
californica (*Didelphis!*) Bennett, Proc. Zoöl. Soc., 1833, p. 40.
Elliot, Syn. N. Am. Mamm., 1901, p. 3.
LINNEAN OPOSSUM.

Type locality. "Northwestern Mexico, adjacent to California."

Geogr. Distr. Oklahoma Territory, through Mexico generally into Gautemala.

Genl. Char. Black; toes reddish half way from claws.

Color. Upper parts and sides black, with occasional long white hairs, mostly on dorsal region; face and forehead whitish; around the eyes and line in the center of the crown, black; under parts white shaded with dusky; legs and feet black, digits reddish, half way from claws on hands, only at base of claws on feet; tail black at base, remainder flesh color.

Measurements. Total length, 640–940; tail vertebræ, 250–535; tarsus, 56–80. Skull: total length, 91; Hensel, 80; zygomatic width, 47; mastoid width, 27; interorbital constriction, 6; palatal length, palatal arch to alveoli of incisors, 53; length of upper molars, 20.

a. texensis (Didelphis!), Allen, Bull. Am. Mus. Nat. Hist., 1901, p. 172. TEXAS OPOSSUM.

Type locality. Brownsville, Cameron County, Texas.

Geogr. Distr. Coast region of Texas from Nueces Bay southward, and the lower Rio Grande Valley as far at least as Del Rio, Val Verde County, sporadically northward to San Antonio; Matamoros, State of Tamaulipas, Mexico.

Genl. Char. Like *D. mesamericana*, but tail relatively longer; nasals longer, usually terminating posteriorly in an acute angle.

Color. Like *D. mesamericana.*

Measurements. Total length, 698–820; tail, 255–410; tarsus, 58–73. Skull: total length, 95–128; basal length, 87–117; zygomatic width, 46–70; length of nasals, 41–57; palatal length, 53–70; length of upper molar series, 19–21.

b.—tabascensis (Didelphis!), Allen, Bull. Am. Mus. Nat. Hist., 1901, p. 173.
TABASCO OPOSSUM.

Type locality. Teapa, State of Tabasco, Mexico.

Geogr. Distr. State of Vera Cruz to that of Tabasco, and across State of Chiapas, Mexico, to northern Guatemala.

Genl. Char. Nasals long, terminating posteriorly in a pointed angle; color similar to *D. mesamericana*; tail long.

Color. Apparently not different from the typical form, except that the under parts are yellowish. A black and gray phase exists.

Measurements. Total length, 684–1017; tail, 318–463; tarsus, 57–75. Skull: total length, 90–139; basal length, 82–122; zygomatic width, 45–62.5; length of nasals, 43.5–60; across postorbital processes, 21–26; interorbital constriction, 10.5–13.3; mastoid breadth, 26–42; palatal length, 53–76.5; length of upper molar series, 19–20.6.

15. richmondi *(Didelphis!)*, Allen, Bull. Am. Mus. Nat. Hist., 1901, p. 175.

aurita (*Didelphys*), Alston, Biol. Centr. Amer. Mamm., I, 1881, p. 197. (nec Wied.)

RICHMOND'S OPOSSUM. *Zorro* in Costa Rica.

Type locality. Greytown, Nicaragua.

Geogr. Distr. Nicaragua and Costa Rica.

Genl. Char. Base of under fur orange buff; tail long. Skull: long, narrow; nasals long.

Color. Sides of head to base of ears pale buffy white; median stripe black; black band from ear to whiskers; cheeks buffy white; nape and shoulders black; the long hairs black on anterior part of body, white on posterior part; under fur with black tip, then yellowish white, and orange buff at base; under parts buffy, base of hair brownish yellow; legs, hands, and feet black, nails yellowish white; tail black on basal half, remainder flesh color; ears black.

Measurements. Total length, 948; tail, 477; tarsus, 70. Skull: total length, 114; nasals, 54; zygomatic width, 55; across postorbital processes, 25; interorbital constriction, 11.5; mastoid breadth, 32.5; palatal length, 65; length of upper molar series, 37.4.

marsupialis insularis (*Didelphis!*), Allen, Bull. Amer. Mus. Nat. Hist., 1902, p. 259.

ISLAND OPOSSUM.

Type locality. Caparo, Island of Trinidad.

Geogr. Distr. Islands of Dominica, Grenada, and St. Vincent, West Indies. Trinidad.

Genl. Char. Lighter in color than true *D. mesamericana*, and larger in size.

Color. Head yellowish white to nape, orbital ring brownish; long hairs of top of head tipped with dusky; under fur yellowish white, the coarser hair tipped with blackish; at base over nape and shoulders brownish ochraceous; long stiff overhair wholly white, or wholly black, or mixed black and white, evidently an individual peculiarity. Under parts yellow or yellowish white, tips of some hairs blackish; arms from elbows and legs from knees, blackish brown; tail naked, proximal third blackish brown, apical two-thirds flesh color or whitish; ears blackish brown.

Measurements. Total length, 810-955; tail, 425-465; hind foot, 55-66; ear, 55-65. Skull: total length, 101-110; basal length, 91-101; zygomatic breadth, 61-63.5; postorbital breadth, 21-24; occipital breadth, 30-32; length of nasals, 47-50; breadth at canines, 19-20; upper tooth row, 33-36.

marsupialis battyi (*Didelphis!*), Thomas, Novitat. Zoöl., IX, 1902, p. 137.

BATTY'S OPOSSUM.

Type locality. Coiba Island, West Coast of Panama.

Genl. Char. Similar to *D. m. caucæ*, but face dark with white spots about the supraorbital and malar tufts of bristles.

Color. Like *D. m. caucæ.* Face dark, spotted with white; tail white for less than half the length, the basal fifth being like the body; rest of pelage like that of *D. m. caucæ*, but without light dorsal bristles.

Measurements. Head and body, 430; tail, 390; hind foot, 57, to end of claws, 63; ear, 50 (skin). Skull: greatest median length, 108; basal length, 100; greatest breadth, 52.5; length of three upper molariform teeth, 18.4.

marsupialis etensis (*Didelphis!*), Allen, Bull. Am. Mus. Nat. Hist., 1902, p. 262.

carcinophaga caucæ, Bangs, Amer. Nat., xxxv, 1901, p. 633. (nec Allen.)

ETEN OPOSSUM.

Type locality. Eten, Piura, Peru. Altitude, 50 feet.

Geogr. Distr. Low coast belt of Ecuador and Peru, bordering the Gulf of Guayaquil and probably northward near the coast to Chiriqui, Panama; San Miguel Island, Bay of Panama.

Color. Similar to *D. m. caucæ*, but larger and blacker. Rostral region to the eyes, dingy brownish white, hairs tipped with blackish; whitish streak over ears meeting in front; middle of head from eyes posteriorly blackish; orbital ring blackish; ears and feet black; tail black for basal third; remainder yellowish white.

Measurements. Total length, 730–930; tail, 330–450; hind foot, 56–67; ear, 52–60; Skull: total length, 102–122; basal length, 92–118; nasals, 45–57.5; zygomatic breadth, 58–64; postorbital breadth, 23–28; postorbital constriction, 11–12; occipital breadth, 31–35.5; breadth at canines, 19–25; length of upper tooth row, 34.5–36; length of molar series, 19–20.

Order II. **Edentata. Edentates.**

The Order EDENTATA contains certain mammals of an inferior organization, and with various forms of body covering beside that of hair. The designation, *Toothless*, is not altogether correct as applied to the various species, for, while some, like the Anteaters and Pangolins, are destitute of teeth, others, as the Sloths, Armadillos, etc., are provided with them, although the incisors are wanting in all. The Sloths, so-called on account of their slow movement, have a thick covering of coarse, bristly hair, and the fingers and toes of the different species are armed with long prehensile claws, by means of which the animals maintain their position suspended from the limbs of trees. While the color of the Sloth's coat is generally some shade of gray, it is not infrequently tinged with green caused by a growth upon the hair of an algous plant whose vitality is stimulated by the dampness of the forest in which the animals dwell, and is a means of harmonizing them with the leaves and so affording conceal-ment from all enemies, as creatures without recognizable form, suspended amid the branches. Although very helpless when upon the ground, Sloths make an attempt to defend themselves by trying to seize and strangle their enemies, and sometimes they succeed in doing this, or inflict serious wounds with their sharp, hook-like claws. Their food is composed of leaves, buds, and young shoots of various trees, some species exhibiting a desire for certain kinds only. Among the branches the Sloths frequently move with considerable rapidity, but on the ground their progress is slow and performed with difficulty. These animals rarely emit any sound, but on provocation will make a curious grunting noise, or at other times disturb the stillness of the forest by a long-drawn, shrill, wail-like cry, expressive of the loneli-ness of their monotonous life.

Fam. I. **Bradypodidæ. Sloths.**

Head rounded; neck short; fore limbs very long, exceeding hind limbs in length; tail short or absent; hand with two or three claws, feet always with three claws; teeth subcylindrical; ears inconspic-uous; body clothed with long crisp hair.

6. **Choloepus.**

Teeth, $\frac{5-5}{4-4} = 18$.

Choloepus Illiger, Prodr. Syst. Mamm. et Av., 1811, p. 108. Type *Bradypus didactylus* Linnæus.

19

Two digits with strong claws on hand, three toes on foot; anterior teeth in both jaws large, separated from the others by a diastema, the upper teeth passing in front of the lower when the jaws are closed; cervical vertebrae, six; pterygoid swollen.

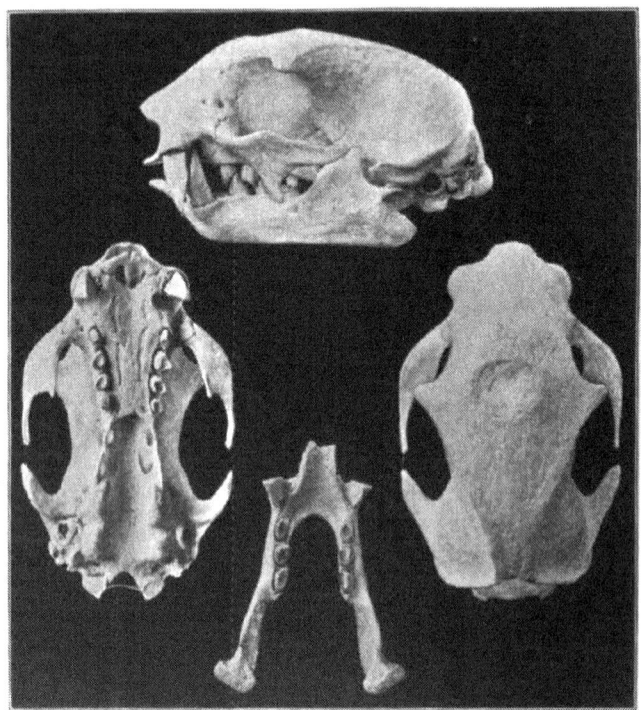

FIG. 4. CHOLOEPUS HOFFMANNI.
No 112 Field Columbian Mus. Coll. ½ nat. size.

16. hoffmanni (*Choloepus*), Peters, Monatsb. K. Preuss. Ak. Wiss.
      Berlin, 1858, p. 128.
HOFFMANN'S SLOTH.   *Perico Lijero* in Costa Rica.
   Type locality.   Costa Rica.
   Geogr. Distr.   Costa Rica, Central America.
   Genl. Char.   Those of the genus.
   Color.   Face and top of head yellowish white, grading into yellowish brown on the body above; arms and legs dark brown; under parts pale brown.

Measurements. Total length about 700 (mounted specimen). Skull: total length, 101.5; palatal arch to incisive foramina, 35; zygomatic width, 64; interorbital constriction, 34; across postorbital processes, 53; mastoid width, 45; length of nasals, 37; length of upper tooth row, 23.5; length of mandible, 77; height at condyle, 22; at coronoid process, 28; length of lower tooth row, 21.

FIG. VI. CHOLOEPUS HOFFMANNI. HOFFMANN'S SLOTH

7. Bradypus.

$$\frac{5-5}{4-4} = 18.$$

Bradypus Linn., Syst. Nat., 1, 1758, p. 34. Type *Bradypus tridactylus* Linnæus.

Ignavus Frisch, Nat. Syst. vierfüss. Thiere, in Tabellen, Tab. Gen. 1775.

Arctopithecus Gray, Proc. Zoöl. Soc., 1871, p. 446.

Arms longer than legs; bones of forearm free; three digits on hand, and three toes on foot, terminating in pointed claws forming a hook; both hands and feet are very narrow and the claws cannot be separated. Teeth rather small, anterior tooth in the upper jaw smaller than the rest, and none projecting much beyond the others. Superior outline of skull greatly arched; nasals short, broad, pointed posteriorly; interpterygoid fossa very broad and deep; palate narrowed posteriorly; arch rounded. Cervical vertebræ, nine; sometimes the eighth, and always the ninth, bear short ribs. Windpipe folded on itself before reaching the lungs. Mammæ two, pectoral.

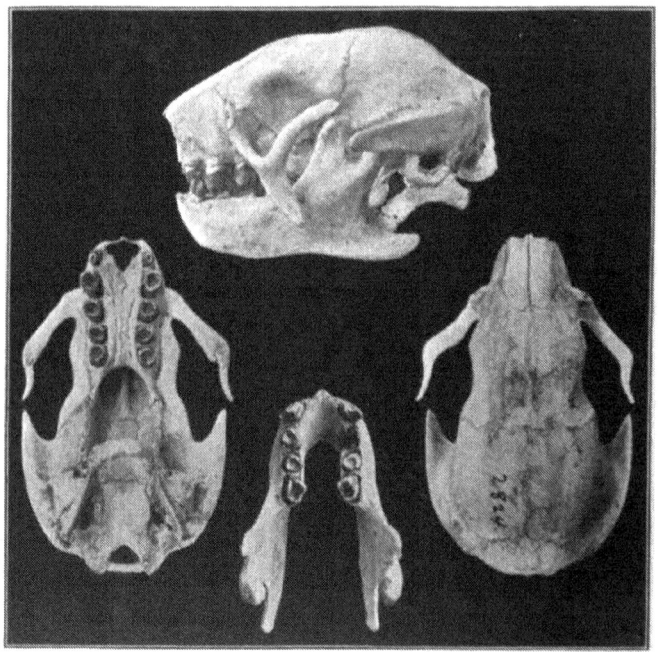

FIG. 5. BRADYPUS INFUSCATUS.
No. 2824 Am. Mus. Nat. Hist. Coll. ⅔ nat. size.

KEY TO THE SPECIES.

A. Dorsal patch yellow with black central streak.   PAGE
   a. Forehead, cheeks, and chin reddish brown .*B. castaniceps*   22
   b. Forehead, cheeks, and chin dark brown . . . . .*B. infuscatus*   23

17. castaneiceps (*Arctopithecus*), Gray. Proc. Zoöl. Soc., 1871, p. 444. CHESTNUT-HEADED SLOTH. *Camaleon* in Nicaragua.
   Type locality. Chontales, Nicaragua.
   Geogr. Distr. Nicaragua, Central America.

Genl. Char. Fur long; dorsal patch large; angle of mandible broad, rounded terminally, and projecting greatly beyond condyle.

Color. "Fur gray brown, intermixed with white hairs. Face, forehead, cheeks, and chin reddish brown; the under part of the body is pale brownish white; the sides of the neck have a long ruff of recurved dark brown hair darker than that of the face. The shoulders and hinder part of back are varied with large patches of whitish hair.

FIG. VII. BRADYPUS CASTANEICEPS. CHESTNUT-HEADED SLOTH.

The middle of the back between the shoulders has a very large patch of soft yellow hair, with a well-marked, narrow, black central streak, which commences with a triangular black spot on the upper edge of the yellow patch, and is continued into the white part of the fur on the loins." (Gray, l. c.) It is stated that when alive this sloth is a grayish green color.

Measurements. Total length, 525; tail, 25. (ex Type, Brit. Mus. Oldfield Thomas in litt.)

18. infuscatus (*Bradypus*), Wagl., Isis, 1831, p. 611.
griseus, Gray, Proc. Zoöl. Soc., 1871, p. 446.

Dusky Sloth.

Type locality.   Western Brazil.

Geogr. Distr.   Costa Rica, Central America, to Brazil and Bolivia.

Genl. Char.   Fur long; both sexes with a dark dorsal spot.

Color.   Upper parts of head dark brown; forehead, cheeks and chin white, or yellowish white; black band across forehead, and one through eye; dorsal patch pale yellow with a black central band, whitish towards edge and spotted with brown; rest of pelage grayish white; under fur white spotted with brown.

Measurements.   Head and body, 580 (cotype of *B. griscus*, Gray, in Brit. Mus., O. Thomas in litt.).   Skull: total length, 74; zygomatic width, 46; interorbital constriction, 23.5; palatal arch to end of palatal floor, 21.5; length of nasals, 18; length of upper tooth row, 25.5; length of mandible, 54; height of condyle, 27.5; at coronoid process, 31; length of lower tooth row, 21.5.

The Anteaters, as their name implies, are insectivorous, some of the species subsisting mainly if not entirely upon ants, and as they are destitute of teeth, the insects are captured by the long vermiform tongue, which is covered with a viscid secretion from the maxillary glands, that causes the ants to adhere to it.   There are three groups of Anteaters, separated by prominent and distinctive characters, and the species range in size from the Great Anteater, four feet in length without counting the huge tail, to the small arboreal species not larger than a rat.   The Great Anteater, *Myrmecophaga tridactyla*, is strictly terrestrial in its habits, and the fingers are armed with powerful claws, with which it tears apart the nests of the ants and draws the insects into its mouth by means of the flexible tongue. The species of the other genera, Cyclopes and Tamandua, are in the first, strictly, and in the latter, only partly arboreal.   When walking the toes have their points turned inwards, and the weight is supported by a pad on the fifth digit, while the soles of the hind feet are placed on the ground.   The rostrum is greatly prolonged, and the mouth is small and tubular.

Fam. II. **Myrmecophagidæ. Anteaters.**

Head conical, elongate, mouth small.   Teeth absent.   Ribs flat, dilated on outer side.   Body covered with hair.

8. **Cyclopes.**

Cyclopes Gray, Lond. Med. Repos., xv, 1821, p. 305.   Type *Myr-mecophaga didactyla* Linnæus.

Myrmydon Wagl., Nat. Syst. Amph., 1830, p. 36.

Dionyx I. Geoff., Rés. Leçons. Mamm., Mus. Paris, 1835, p. 54.

Didactyles F. Cuv., Dict. Scien. Nat., LIX, 1829, p. 501.

Myrmecolichnus Reich., K. Sächs. Naturh. Mus. Dresden, Ein Leitfaden, 1836, p. 51.

Eurypterna Glog., Hand-u. Hilfsb. Naturg., 1841, pp. XXXI, 112.

Cyclothurus Less., Nouv. Tabl. Règn. Anim., 1846, p. 152.

Didactyla Liais, Climats, Géol. Faun. Géog. Botanique Brésil, 1872, p. 356.

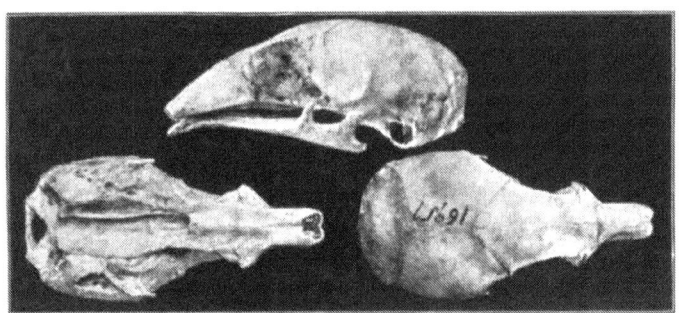

FIG. 6. CYCLOPES DORSALIS.
No. 16957 Am. Mus. Nat. Hist. Coll.   Nat. size.

Skull short and arched; palatines and pterygoids not meeting in middle line; canal for posterior nares not closed below; coronoid narrow, recurved, with angular process well developed; third digit of hand much longer than the others, the distal phalange is compressed, curved, pointed, and armed with a strong curved claw; hallux of hind foot rudimentary and contained within the skin. Ribs dilated. Habits strictly arboreal.

19. dorsalis (*Cyclothurus*), Gray, Proc. Zoöl. Soc., 1865, p. 385, pl. 19. LITTLE OR TWO-TOED ANTEATER.

Type locality.   Costa Rica.

Geogr. Distr.   Guatemala through Central America to South America.

Genl. Char.   Distinctly defined broad dorsal streak; feet and tail, yellow.

Color.   General hue golden yellow tinged in places with chestnut; dorsal stripe black or blackish chestnut; tail, hands, and feet, golden yellow; back and sides sometimes washed with black; black patch on center of breast.

Measurements. Total length, 400; tail, 205; hind foot, 34. Skull: occipito-nasal length, 50; greatest breadth of braincase, 24; median length of nasals, 13; lateral length of nasals, 14; interorbital constriction, 8.5; length of single half of mandible, 32.

FIG. VIII. CYCLOPES DORSALIS. LITTLE OR TWO-TOED ANTEATER.

9. Tamandua.

Tamandua Frisch, Nat. Syst. vierfüss. Thiere, in Tabellen, 5 Tab. Gen. 1775. Less., Nouv. Tabl. Règn. Anim., Mamm., 1842, p. 152. Gray, Proc. Zoöl. Soc., 1865, p. 383. Type *Myrmecophaga tetradactyla* Linnæus.

Tamanduas, F. Cuv., Dict. Scien. Nat., LIX, 1829, p. 501.
Uroleptes Wagl., Nat. Syst. Amphib., 1830, p. 36.

Palatine and pterygoid bones united beneath the nasal canal for the whole length. Fur of body and tail short, bristly. Tail tapering, prehensile. Skull long, slender; nose nearly as long as braincase. Habits mainly arboreal.

20. tetradactyla (*Myrmecophaga*), Linn., Syst. Nat., I, 1758, p. 35.
myosura Pall., Miscell., 1766, p. 64.
Ursine Anteater Griff., Anim. King., III, 1827, p. 304, pl.
crispus Rupp., Mus. Senck., III, 1845, p. 179.
bivittata Gray, Proc. Zoöl. Soc., 1865, p. 384.

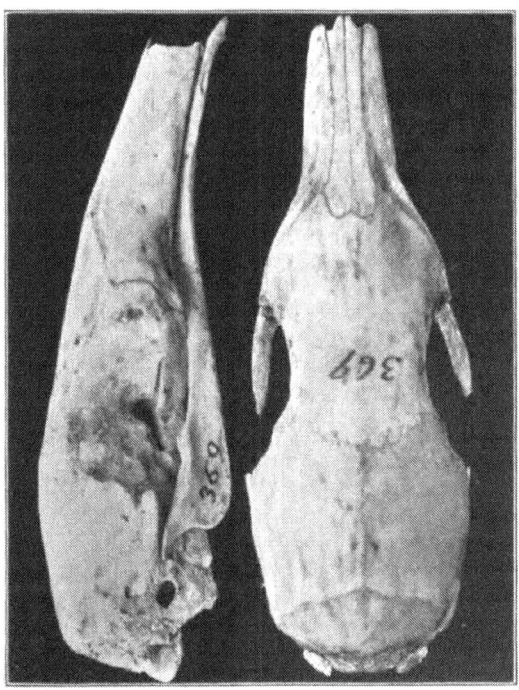

FIG. 7. TAMANDUA TETRADACTYLA.
No. 369 Am. Mus. Nat. Hist. Coll. ⅓ nat. size.

THREE-TOED ANTEATER. *Tejon, Oso Colmenero* in Costa Rica.
Type locality. "America meridionali." Brazil?
Geogr. Distr. Mexico, through Central America to Peru and Paraguay, South America.
Genl. Char. Tail long, apical half scaly; nose lengthened; claws strong.
Color. Head, neck, throat, stripe down back ending in a point on the loins, shoulders, arms and outer side of hind legs yellowish white, sometimes a deep buff; nose, broad stripe to and including

the eye, narrow bar on occiput, stripe from front of neck over shoulder, inner side of hind legs and rest of body, black; tail sparsely haired towards tip, yellowish or deep buff at base, mixed yellow and black hairs on remaining part; toes grayish.

FIG. IX. TAMANDUA TETRADACTYLA. ADULT.
THREE-TOED ANTEATER.

FIG. X. TAMANDUA TETRADACTYLA. YOUNG.
THREE-TOED ANTEATER.

Measurements. Total length, 1090; tail, 460 (mounted specimen). Skull: occipito-nasal length, 125; zygomatic width, 43; interorbital constriction, 25.5; mastoid width, 34; median length of nasals, 41; length of mandible, 107.

21. sellata (*Myrmecophaga*), Cope, Amer. Nat., XXIII, 1889, p. 133, Feb'y.
SADDLE-BACK ANTEATER.
  Type locality. Honduras.

Geogr. Distr. Honduras to French Guiana (?), South America; exact range unknown.

Genl. Char. Tail equal to head and body; hairs on extremity sparse.

Color. Band from forearm over the shoulder joining a large patch covering back and sides, black; narrow median band, thighs, rump, and tail straw color; front of eye dusky.

Measurements. Total length, 915; tail, 515.

10. *Myrmecophaga.

Myrmecophaga Linn., Syst. Nat., 1, 1758, p. 35. Type *Myrmecophaga tridactyla* Linnæus.

Falcifer Rehn., Am. Nat., XXXIV, 1900, p. 576.

Head very long; mouth tubular, small; tongue very long, vermiform; hand has third digit greatly developed, and armed with a long, falcate claw; all digits armed with claws except the fifth; foot with five unequal digits with claws; tail very long, not prehensile, equaling the body in length, covered with very long hair; ears small, oval, erect; eyes small. Skull elongate, narrow, cylindriform, and smooth on the superior surface; nares terminal; zygomatic arch incomplete.

22. tridactyla (*Myrmecophaga*) Linn., Syst. Nat., 1, 1758, p. 35.

jubata Linn., Syst. Nat., 1, 1766, p. 52.

GREAT ANTEATER. *Oso Real* in Costa Rica.

* See Thomas, Amer. Nat., XXXV, p. 143.

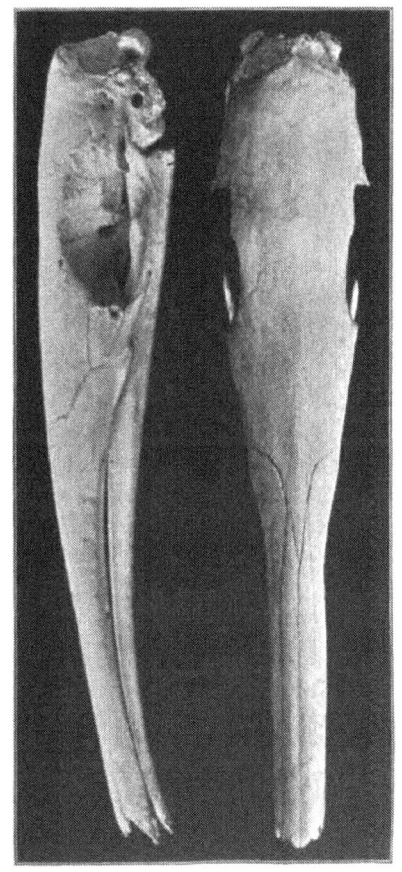

FIG. 8. MYRMECOPHAGA TRIDACTYLA.
No. 115 Field Columbian Mus. Coll. ⅓ nat. size.

Type locality.   Brazil.

Geogr. Distr.   Guatemala to Brazil.

Genl. Char.   Size large; tail very large, about as long as head and body, covered with long hairs; claws strong, curved.

Color.   Nose, head, back, loins, and tail covered with coarse hairs that are white at base, then brownish or black and tipped with buff; throat patch ending in a point on breast; stripe over shoulder ending in a point on the loins; broad band on forearms above hands; legs and under parts, black; rest of body, shoulders, breast, arms, and stripe from beneath ears to loins above the black, grayish white; white hairs on toes.

FIG. XI. MYRMECOPHAGA TRIDACTYLA. GREAT ANTEATER.

Measurements.   Total length to end of hairs on tail, 2500; tail to end of hairs, 1130. Skull: occipito-nasal length, 370; zygomatic width, 56; interorbital constriction, 43; median length of nasals, 150; lateral length of nasals, 171; length of mandible, 320.

The Armadillos are remarkable for their ossified skin, formed by the union of numerous variously shaped scales into a bony armor protecting the body, head, and limbs. In some extinct species this covering was entire, but in the living animals it is divided into three regions, the anterior, median, and posterior portions, the middle section consisting of a varying number of rings connected by a flexible skin to permit a curvature of the body. The inner surface of the limbs, and underside of the body is covered by a soft skin. Hairs often project between the bony scutes, and the skin-covered parts are more or less hairy. Fore feet with strong claws, upon the

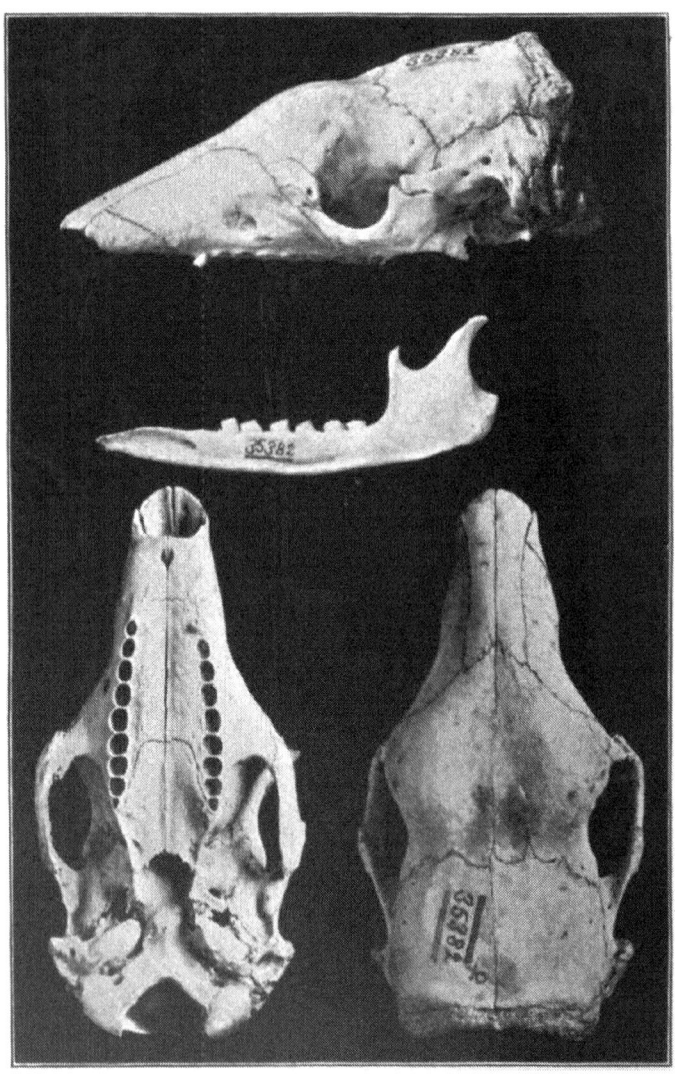

CABASSOUS CENTRALIS.
No. 35382 U. S. Nat. Mus. Coll. Nat. size. Type.

tips of which some species walk, while the soles of the hind feet are placed flat upon the ground. The tongue is long, pointed and capable of being extended.

Armadillos are harmless, nocturnal, and omnivorous, provided with numerous simple teeth that, excepting in one genus, are not shed. They are capable of running with considerable swiftness, and when frightened or attacked, they roll the body into a ball, presenting nothing but the bony armor to their enemies.

Fam. III. **Dasypodidæ. Armadillos.**

Subfam. I. **Dasypodinæ.**

Head narrow; snout long, narrow, obliquely truncate; pterygoids meeting below nasal passage; ears long, ovate, erect, placed on occiput, contiguous; bony carapace covering the elongate, narrow body, having six to twelve movable rings on the center and sides; tail long, tapering, the dermal scutes forming distinct rings. Front feet with four toes, hind feet with five, the nails strong, curved, pointed.

11. **Cabassous.**

$$\frac{8-8}{8-8} \text{ or } \frac{9-9}{9-9} = 32 \text{ or } 36.$$

Cabassous McMurtrie, Cuv., Anim. King., 1, 1831, p. 164. Type *Dasypus unicinctus* Linnæus.

Xenurus, Wagl., Nat. Syst. Amph., 1830, p. 36. (nec Boie, Aves, 1826.)

Arizostus Glog., Hand-u. Hilfsb. Naturg. 1, 1841, pp. XXII, 114.

Tatoua Gray, Proc. Zoöl. Soc., 1865, p. 378.

Intermediate bands, twelve, broader than long; fore feet with five toes; claws large, strong; tail long, tuberculate.

FIG. XII. CABASSOUS CENTRALIS. MILLER'S ARMADILLO.

23. centralis (*Tatoua*), Miller, Proc. Biol. Soc., Wash., XIII, 1899 pp. 4, 7.

cinereus hispidus True, Proc. U. S. Nat. Mus., XVIII, 1896, p. 345. MILLER'S ARMADILLO.

Type locality. Chamelicon, Honduras.

Geogr. Distr. Honduras, Central America, range unknown.

Genl. Char. Small; plates in central rings of carapace, 29–31: occipital region of skull little elevated; zygomata, as seen from above, nearly parallel with each other and main axis of skull; hamular processes of pterygoids neither thickened nor bent inward at tips. Crown shields about 38; less than a dozen small, scattered scales on cheek. Scapular shield with 7 or 8 rows, the longest with 28 plates; dorsal rings 10, the longest containing 29–31 plates. (ex Miller, l. c.)

Color. Above brownish black; lower edge of carapace yellowish; under parts light yellow; legs and face apparently flesh color; tail brownish black, tip yellowish; claws light yellow. (Skin.)

Measurements. Total length, about 505; tail, 148. Skull: occipito-nasal length, 80; zygomatic width, 43; mastoid width, 31; palatal arch to middle of fourth molar, 16; median length of nasals, 23; lateral length of nasals, 23; length of upper tooth row, 26; length of mandible, 61; length of lower tooth row, 22.

The Armadillos of the next genus are characterized by the nearly symmetrical toes on the fore feet, the second and third being longest and subequal, and the first and fourth also subequal and only slightly shorter; fifth toe obsolete. One species only, the Nine-banded Armadillo, penetrates the limits of the United States, and has a most extensive distribution from Texas to Paraguay. Several species belong to this genus, one of which, found on the Pampas of South America, from the shape of its head and the length of its ears, is known as the Mule Armadillo, or Mulita. (*T. hybrida.*)

Subfam. II. **Tatuinæ.**

12. Tatu.

$$M.\frac{\times \ \times}{\times \ \times} \text{ or } \frac{7-7}{7-7} = 32 \text{ or } 28.$$

Tatu Frisch, Nat. Syst. vierfüss. Thiere, in Tabellen, 5 Tab. Gen. 1775. *Id.* Blumenb., Handl. Naturg., 1799, p. 73. Type *Dasy-pus novemcinctus* Linnæus.

Tatusia Less., Man. de Mamm., 1827, p. 309.

Cachicamus McMurtrie, G. Cuv., Anim. Kingd., I, 1831, p. 163.

Mammæ two pair, one pectoral, one inguinal. Seven to nine movable rings in center or on sides of bony carapace. Fore toes nearly symmetrical. Milk teeth two-rooted, changed only when the animal has attained its full growth.

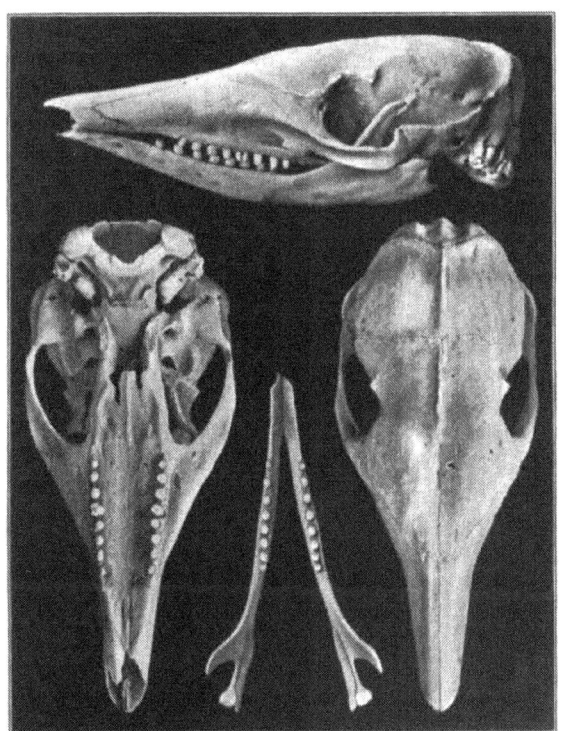

FIG. 9. TATU NOVEMCINCTUM.
Field Columbian Mus. Coll. Nat. size.

24. novemcinctum (*Dasypus*) Linn., Syst. Nat., I, 1758, p. 51. I, 1766, p. 51.
octo-cinctus (Linn.), Schrieb. Säugeth., II, 1775, p. 222, tab. LXXIII, LXXVI.
peba Desm., Mamm., 1820, p. 368.
longicaudus Wied, Breit. Naturg. Bras., II, 1825, p. 531.
mexicanus fenestratus Peters, Monatsb. Akad. Wiss. Berl., 1864, p. 180.

leptorhyncha Gray, Handl. Edent., 1873, p. 14, pl. 2, figs. 3, 4.

novemcinctus (*Cabassous*), Elliot, Syn. N. Am. Mamm., 1901, p. 4.

NINE-BANDED ARMADILLO. *Armado, Encubierto* in Mexico and Central America.

Type locality. "America meridionali." Brazil(?)

Geogr. Distr. Texas, through Mexico and Central America to Paraguay.

Genl. Char. Shield with eight movable rings in the middle, and nine on the sides; tail as long as body without head, covered by twelve rings and not enveloped in a cone; molars, 32.

FIG. XIII. TATU NOVEMCINCTUM. NINE-BANDED ARMADILLO.

Color. Bare skin on face flesh color with a few scattered yellowish hairs; head shield pale brown, that on the back black; scales on sides yellowish white; ears brown, toes yellowish with white claws; tail brownish black, the anterior half of scales yellowish white.

Measurements. Total length, 237; tail vertebræ, 90; hind foot, 31; ear, 22. Skull: total length, 70; zygomatic width, 32; across postorbital processes, 22; length of nasals, 17; length of upper tooth row, 16; length of mandible, 56; length of lower tooth row, 18.

Order III. **Sirenia. Sirenians.**

The Sirenians are mammals constituted especially for an aquatic life, and formerly were confounded with the Cetaceans, with which, however, they have no relationship. Like the members of the Order CETACEA, the Sirenians have no hind limbs, and those on the forward part of the body have been transformed into paddles, and the tail has been expanded into a flattened rudder.

The head is of the ordinary mammal type, being small for the body, with a rounded superior outline, but the nostrils are provided with flaps that open and close at the will of the animal. There are no fins. The eye is small, and the ear has no external conch. Thick lips, provided with a number of bristly hairs, cover the small mouth, and the skin of the body is thick, with sometimes hair distributed sparsely over it. The female has two pectoral mammæ. Teeth are entirely absent in some species, like Steller's Sea-Cow, but others have both incisors and molars. The bones of the skeleton are massive and dense, the skull being remarkable in this respect. Collar and nasal bones are absent and there is no sacrum, but the pelvis is represented by a pair of small bones. The two bones of the forearm are usually ankylosed at the extremities, and the digits are five in number. The lungs extend backward nearly to the last rib and are very narrow. Rough, horny plates cover the symphysis of the mandible, and the surface of the tongue is similar to these plates.

Three species of Manatee are included in the family, one of which, Steller's Sea-Cow (*Hydrodamalis gigas*), is now extinct. This animal, the largest of all, was from twenty to twenty-eight feet in length, and at the time when Steller visited Bering Sea in 1741, was very numerous around Bering and Copper Islands. The flesh, unfortunately, was found to be highly palatable, far superior to salt pork, and the sailors slaughtered the inoffensive beasts, until the last one was killed in 1768. No skin has been preserved, and a collection of bones in St. Petersburg and Washington alone remain to show what kind of animal it was. Two living species of Manatee remain in the New World, one, *T. manatus*, in southern North America; the other, *T. inunguis*, restricted to the rivers Amazon and Orinoco, in South America. In the Old World, one, *T. senegalensis*, is confined to West Africa in the district comprised between 10°-16° latitude, and 20°-27° longitude. East Africa, Australia, Ceylon, and islands in the Bay of Bengal, the Indo-Malay Archipelago and the Philippines possess the Dugong, more a marine animal than the Manatee, which is found chiefly in the rivers. Three species of Dugong have been recognized, *H. tabernaculi*, from the Red Sea, *H. dugong*, from the Indian Seas, and *H. australis* from Australia. In disposition these animals are gentle and inoffensive, feeding on water plants and grasses, and formerly, before their numbers were so greatly reduced by man, were met with in herds composed of various families, and in the case of Steller's Sea-Cow the herds were of great size.

Fam. I. **Trichechidæ. Manatees.**

13. *Trichechus.

$$I \frac{2-2}{2-2}, M \frac{1}{2-2} \text{ to } \frac{11-11}{11-11} = 32 \text{ or } 52.$$

Trichechus Linn., Syst. Nat., I. 1758. p. 34. Type *Trichechus manatus* Linnæus

Manatus Brunn., Zo. l. Fund., 1772. pp. 34. 38. 39; *Id.* Scopoli. Intr. Hist. Nat., 1777. p. 490; *H.* Storr. Prodr., Meth. Mamm., 1780. p. 41.

FIG. XIV. TRICHECHUS MANATUS. MANATEE.

* If the tenth edition, 1758, of Linn. Syst. Nat., is taken as a starting point for nomenclature then the generic term for the Manatee would be TRICHECHUS. and for the walrus, ODOBÆNUS (ODONTOBÆNUS), Briss., 1760. Should Brisson's name be rejected, as it probably ought to be, then ROSMARUS, Scopoli, 1777, would be the proper name for the Walrus. But if the twelfth edition is the starting point, then TRICHECHUS stands for the Walrus and MANATUS for the Manatees.

is found chiefly in the rivers. Three species of Dugong have been
recognized: *H. tabernaculi*, from the Red Sea, *H. dugong*, from the
Indian Seas, and *H. australis*, from Australia. In disposition these
animals are gentle and inoffensive, feeding on water plants and
grasses, and formerly, before their numbers were so greatly reduced
by man, were met with in herds composed of various families, and
in the case of Steller's Sea-Cow the herds were of great size.

Fam. I. **Trichechidæ. Manatees.**

13. *Trichechus.

$$I.\frac{2-2}{2-2}; \ M.\frac{6-6}{6-6} \text{ to } \frac{11-11}{11-11} = 32 \text{ or } 52.$$

Trichechus Linn., Syst. Nat., I, 1758, p. 34. Type *Trichechus
    manatus* Linnæus.
Manatus Brunn., Zoöl. Fund., 1772, pp. 34, 38, 39; *Id.* Scopoli,
    Intr. Hist. Nat., 1777, p. 490; *Id.* Storr, Prodr., Meth. Mamm.,
    1780, p. 41.

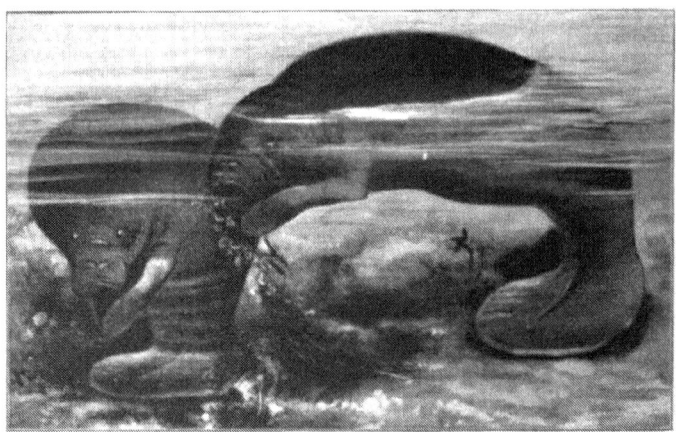

FIG. XIV. TRICHECHUS MANATUS. MANATEE.

---

    * If the tenth edition, 1758, of Linn. Syst. Nat., is taken as a starting point
for nomenclature then the generic term for the Manatee would be TRICHECHUS,
and for the walrus, ODOBÆNUS (ODONTOBÆNUS), Briss., 1760. Should Brisson's
name be rejected, as it probably ought to be, then ROSMARUS, Scopoli, 1777,
would be the proper name for the Walrus. But if the twelfth edition is the
starting point, then TRICHECHUS stands for the Walrus and MANATUS for the
Manatees.

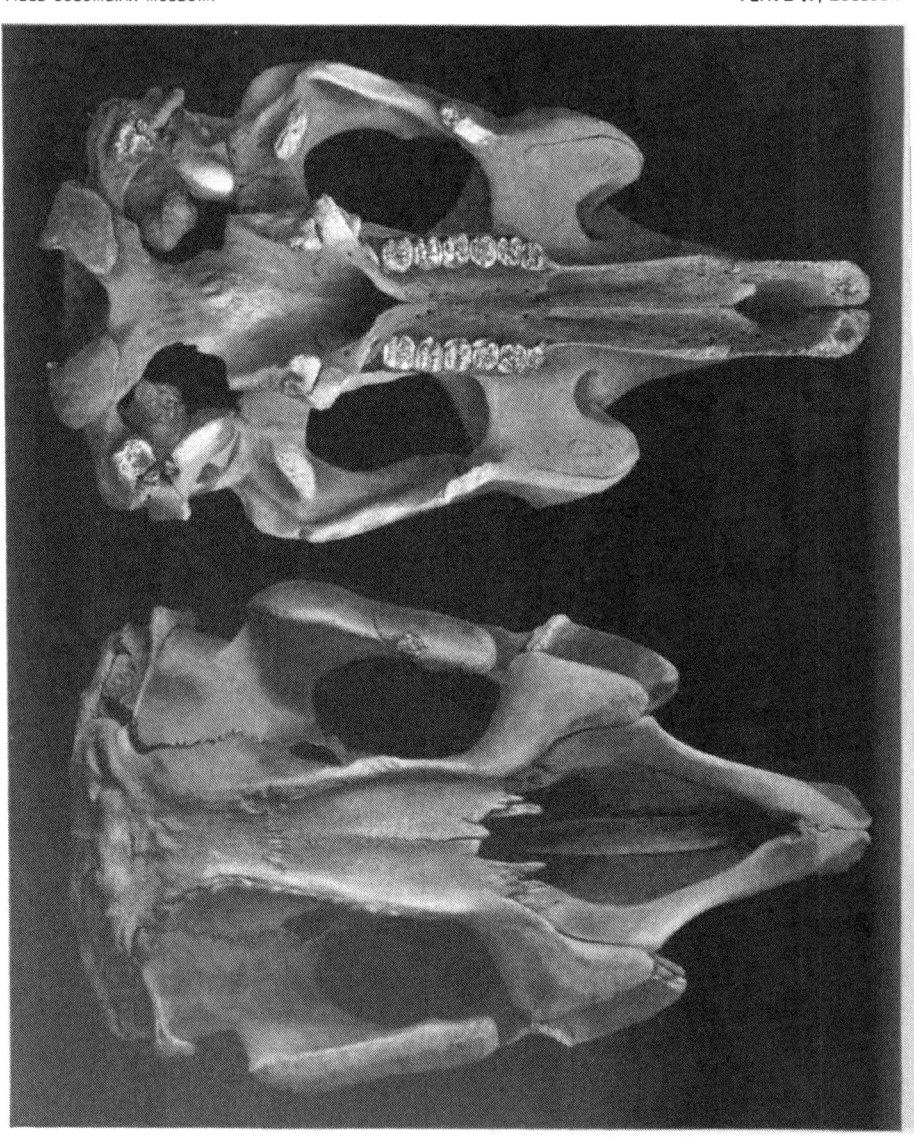

TRICHECHUS MANATUS.
No. 49 Field Columbian Mus. Coll. ⅓ nat. size.

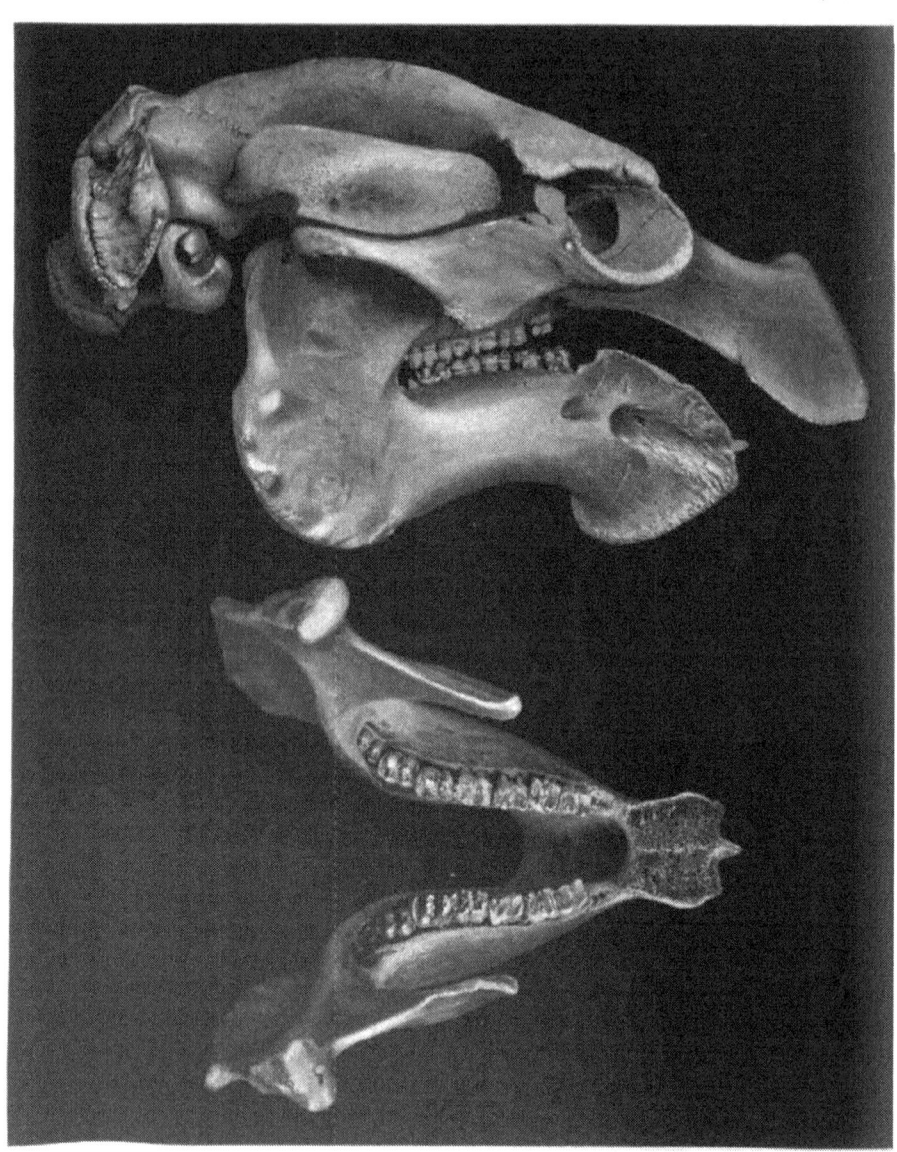

TRICHECHUS MANATUS.
No. 49 Field Columbian Mus. Coll.   ⅓ nat. size

Skull arched, descending rapidly anteriorly from frontals to nasals; zygomata large and massive; jugal greatly developed. Orbit small, prominent, almost inclosed by bone; anterior nares lozenge-shaped, extending behind orbits. The mandible is massive, with a horny plate attached anteriorly, which supplies the place of teeth. Molar teeth in both jaws similar in character, square enameled crowns elevated into transverse tuberculate ridges; those in the upper rows having two ridges and three roots, those in the lower three ridges and two roots.

25. manatus (*Trichechus*), Linn., Syst. Nat., 1, 1758, p. 34.

manatus (*Manatus*), Linn., Syst. Nat., 1, 1766, p. 49.

latirostris (*Manatus*), Harl., Journ. Acad. Nat. Scien. Phil., 1824, p. iii. Elliot, Syn. N. Am. Mamm., 1901, p. 6.

australis Tilesius, Jahrb. Naturg., 1, 1802, p. 23.

americanus Desm., Dict. Hist. Nat., 1817, p. 262, pl. 96. (Part.)

fluviatilis Schreb., Säugeth. Suppl., 1846, pl. 379.

MANATEE. *Vacca de Agua*, in Guatemala.

Type locality. East coast of Florida near the Cape.

Geogr. Distr. Florida coast to Yucatan, Mexico.

Genl. Char. Those of the genus.

Color. Grayish black over all the body.

Measurements. Total length, 2268. Skull: total length, 380; zygomatic width, 220; interorbital constriction, 70; palatal arch to end of palatal floor, 155; length of mandible, 215; height at coronoid process, 135.

Order IV. **Cetacea. Cetaceans.**

The CETACEANS, abounding in all the seas of the Globe, and also even in some of the larger rivers of both Hemispheres, live entirely in the water, in which their young are brought forth; and they never appear upon the land unless accidentally thrown upon the beach by the waves. The Order contains the largest of living mammals, the Yellow-bellied Whale, *B. sulfurea*, measuring ninety-five feet in length and weighing one hundred and forty-seven tons. Dependent upon air received into the lungs for respiration, whales are forced to rise at intervals to the surface, when the lungs are emptied with considerable force, causing a cloud of vapor to be lifted high in the air, which gives rise to the term "spouting" or "blowing," and by this act the animal betrays its presence to the whaler. Admirably adapted to a life in the water, these animals, although possessing a fish-like form, have an entire structure characteristic of the Mammalia. The Cetacea have two anterior limbs, and traces of a hinder pair. The forward pair is covered with a leathery skin, in shape like a flattened paddle, while the hinder limbs are not visible externally. Within the body there are indications of a pelvis, and two small bones that may represent the ischia. From want of use in the element in which they live, the hind limbs have become atrophied. The stomach, like that of the ruminants, is complex, and divided into several compartments, varying in number with the different genera, from three or four in PHOCÆNA to eight in ZIPHIUS, while the Sperm Whales have three, and the Whalebone Whales are stated to have four. These last, comprising the family BALÆNIDAE, are distinguished from the rest by the absence of teeth in both jaws, although, singularly enough, these are present in the early development of the embryo. The baleen, or so-called "whalebone," is a series of flattened horny plates (varying in number, amounting in some cases to as many as four hundred), which are placed on each side of the palate, leaving an open middle space. They serve as strainers, being in close proximity, and retain the small molluscs, fish and other creatures, when the water that has been taken with them into the mouth is ejected. The color of the baleen varies from jet black through different shades to creamy white. Baleen Whales are distinguished by their enormous heads, which are about one-third the total length of the animal, a curved mouth extending behind the blow-holes, apparently pigmy eyes (although they are four times the size of those of an ox); short heavy pectoral fins, and long baleen.   The largest of these animals is the Bowhead, or Arctic Right Whale, *B. mysticetus*, which yields more oil and a better quantity of whalebone than any other species.   Although huge in bulk, its gullet is not over two inches in diameter, and its food consists of microscopic organisms, millions of which are swallowed at a time. It is emphatically an ice whale, living amid floes and vast fields of ice of the Polar regions.   Various species of these whales, arranged under separate genera, frequent the different seas.

Fam. I. **Balænidæ. Baleen Whales.**

F. W. True, *On the nomenclature of the Whalebone Whales of the tenth edition of Linnæus, Systema Naturæ.*   Proc. U. S. Nat. Mus., 1898, p. 617.

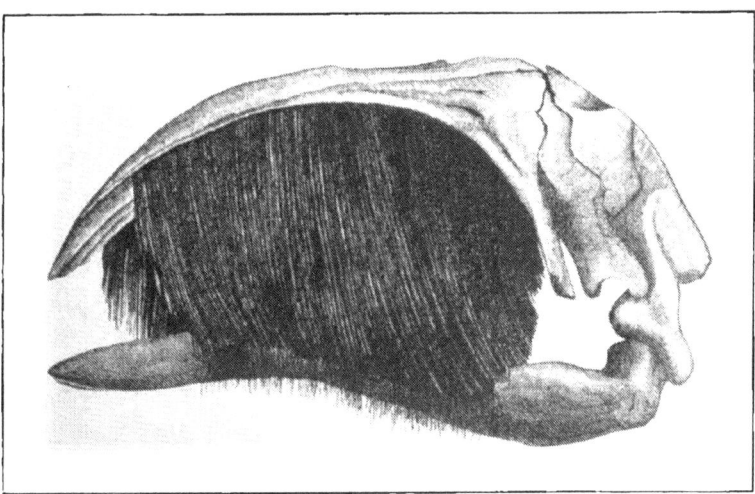

FIG. 10. BALÆNA GLACIALIS—BALEEN.
Riverside Nat. History.

Teeth absent in both jaws, present in fœtal life; palate furnished with whalebone.   Rami of mandible greatly arched outward, meeting at an angle at apex, there connected by fibrous tissues.   Skull symmetrical.   First pair of ribs alone joined to the sternum; the others are fastened to the vertebræ by ligaments.   Nasals roofing the anterior nasal passages.

FIG. XV. BALEEN WHALE ATTACKED BY KILLER WHALES.
Riverside Nat. History.

14. Rhachianectes.

Rhachianectes Cope, Proc. Acad. Nat. Scien. Phil., 1869, p. 15.
  Type *A. glaucus* Cope.

Head small, body elongate; pectoral fin narrow; no dorsal fin; skin of throat smooth; baleen short and coarse.

26. glaucus (*Agaphelus*), Cope, Proc. Acad. Nat. Scien. Phil., 1869, p. 15.
  glaucus (*Rhachianectes*) Elliot, Syn. N. Am. Mamm., 1901, p. 9.
GRAY WHALE.
  Type locality. Coast of California.
  Geogr. Distr. Coast of Lower California to north Pacific and Arctic Oceans; Okhotsk and Bering Seas.
  Genl. Char. Superior outline of head convex; size moderately large.
  Color. Mottled gray, sometimes blackish.
  Measurements. Total length, 36 to 40 feet.

15. Megaptera.

Megaptera Gray, Erebus and Terror, Zoöl., 1846, p. 16. Type
  Balæna boops Linnæus.
  Megapteron Gray, Erebus and Terror, Zoöl., 1846, p. 61.
  Kyphobalæna Eschr., Nord. Wallth., 1849, p. 56.

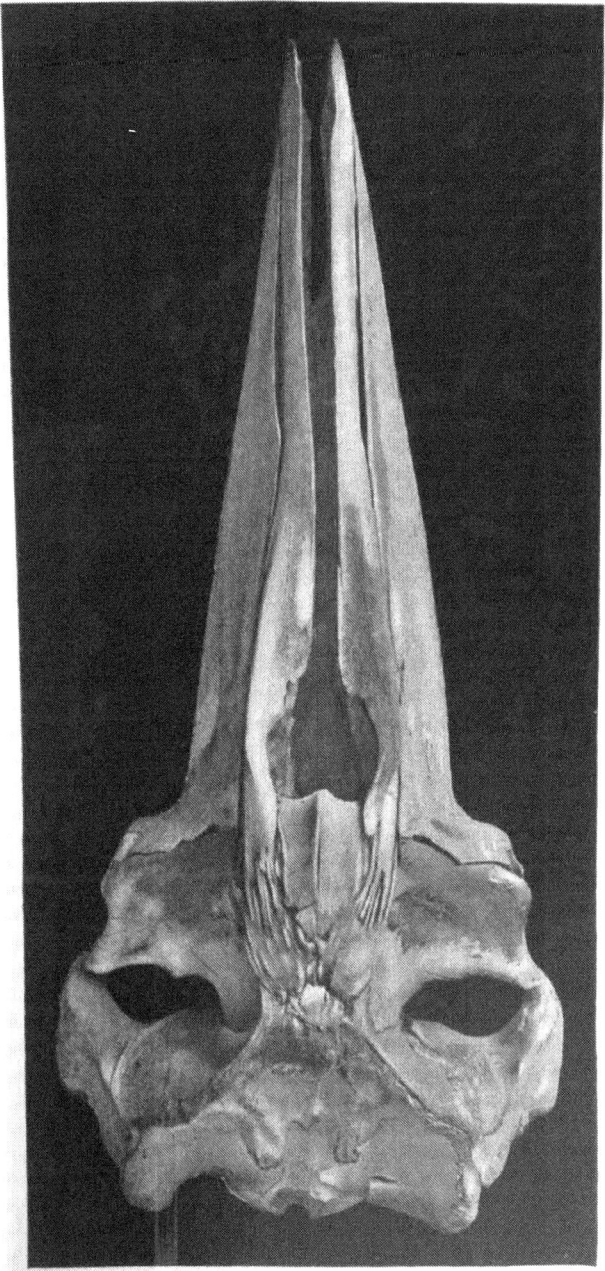

RHACHIANECTES GLAUCUS.
U. S. Nat. Mus. Coll.

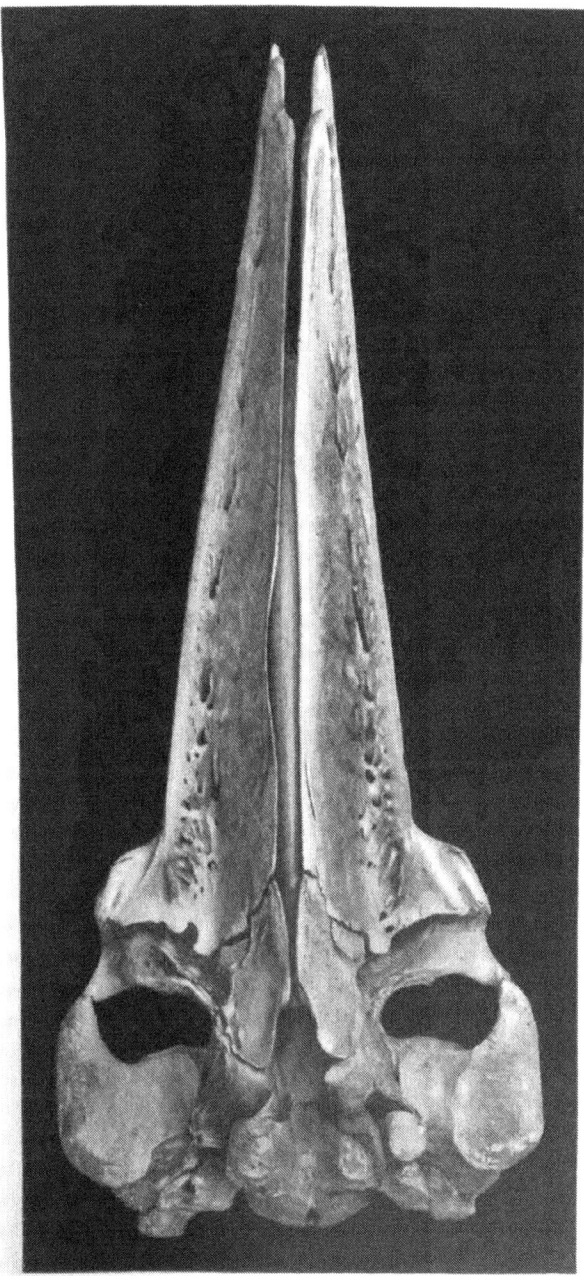

RHACHIANECTES GLAUCUS.
U. S. Nat. Mus. Coll.

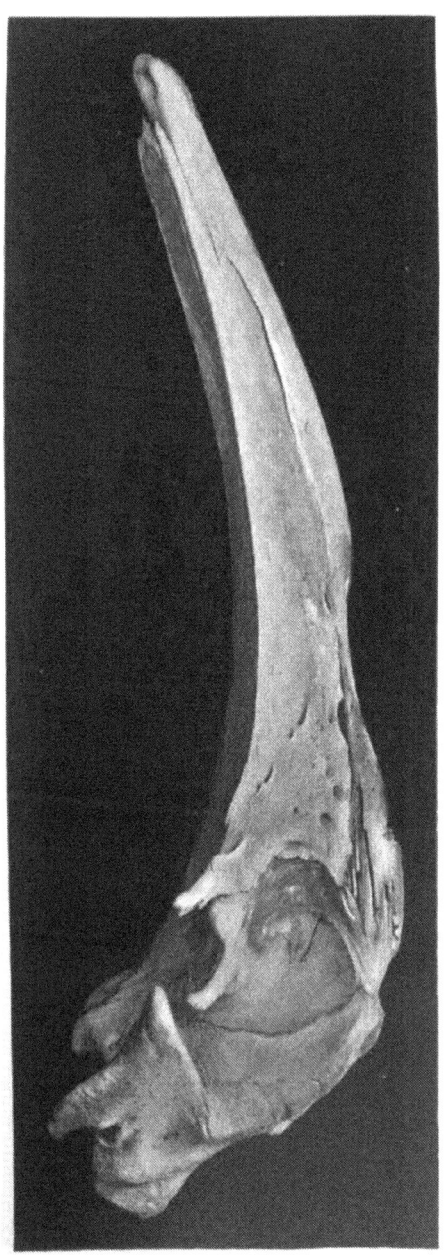

RHACHIANECTES GLAUCUS.
U. S. Nat. Mus. Coll.

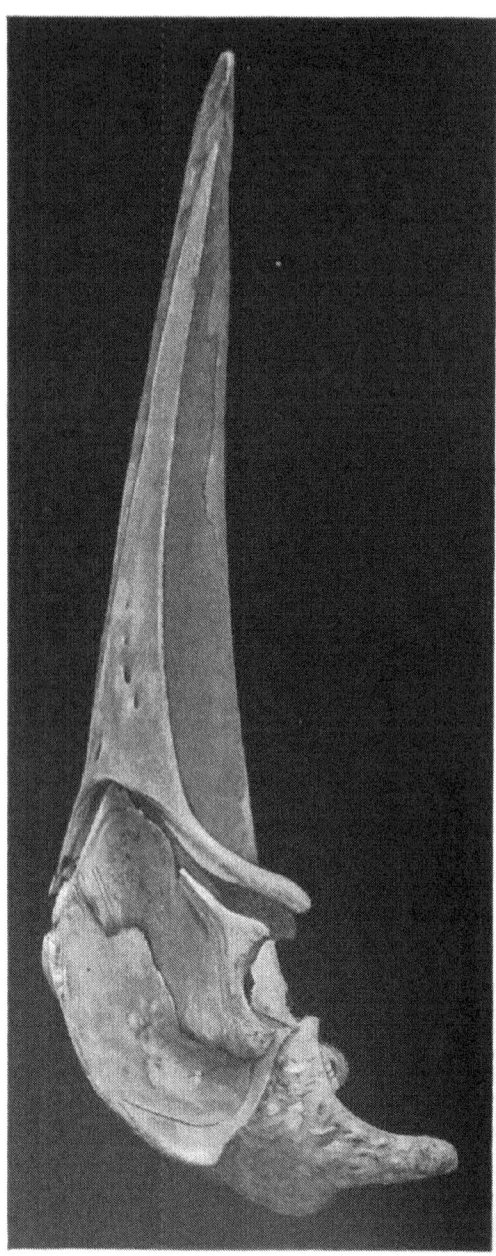

MEGAPTERA N. BELLICOSA
Acad. Nat. Sciences Coll.

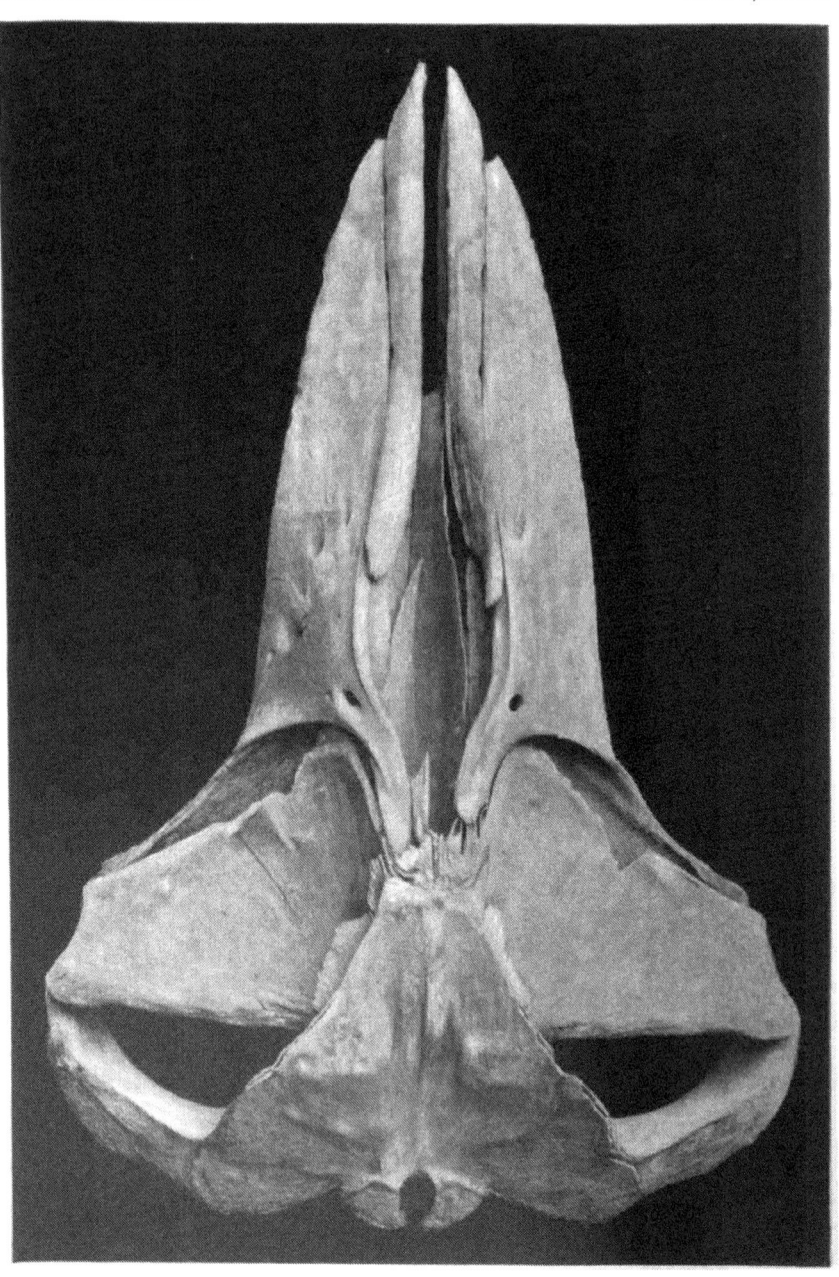

MEGAPTERA N. BELLICOSA.
Acad. Nat. Sciences Coll.

.

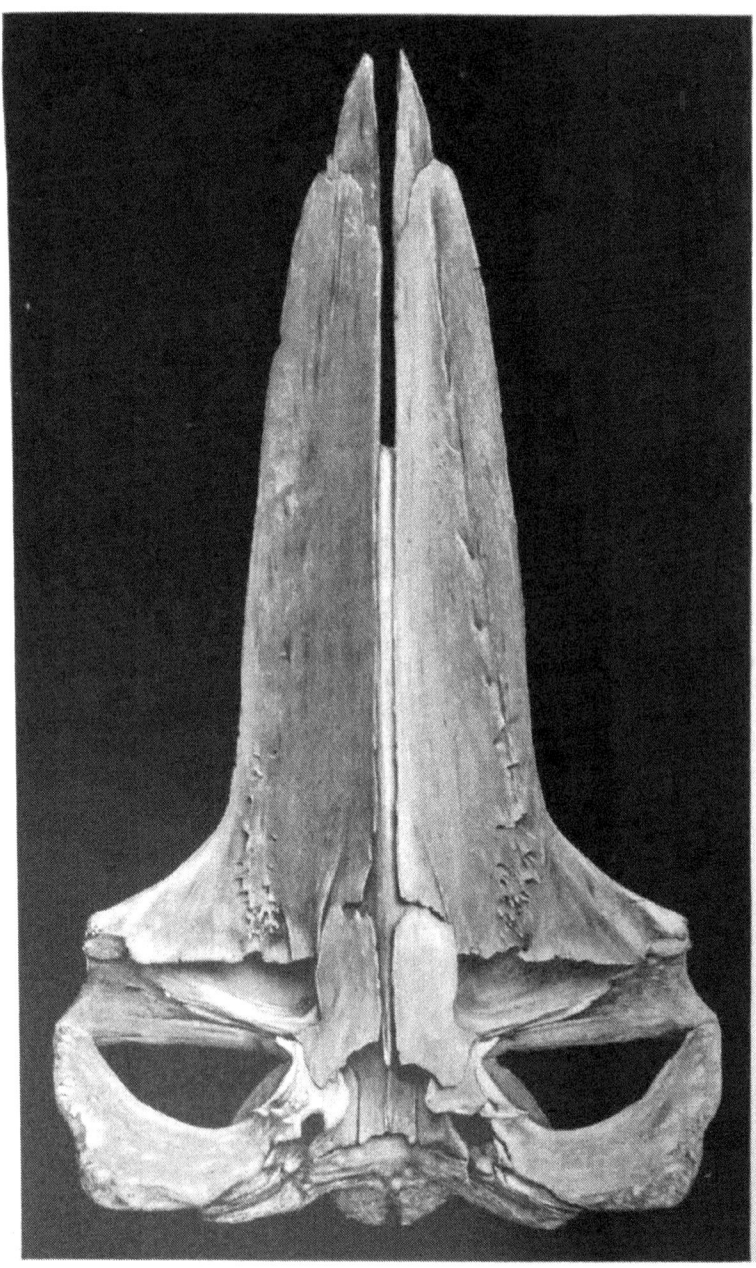

MEGAPTERA N. BELLICOSA.
Acad. Nat. Sciences Coll.

Megapteropsis Van Ben., Res. Cét. Belgique, Nouv. Mem. Acad.
    Brux., 1861, p. 38.
Pæscopia Gray, Proc. Zoöl. Soc., 1864, p. 207, fig. 3.

Head moderate; baleen short, broad; skin of throat plicate, or
folded; pectorals long and narrow, one-fourth of the entire length of
animal; dorsal low; vertebræ, 53; cervical vertebræ free, sometimes
ankylosed; nuchal arch high, subcircular; frontal broad.

KEY TO THE SPECIES AND SUBSPECIES.

A. Upper parts black.   PAGE
    a. Belly white...............................*M. n. bellicosa* 41
    b. Belly black...............................*M. versabilis* 41

nodosa bellicosa (*Megaptera*), Cope, Proc. Am. Phil. Soc., XII, 1870,
    p. 103.
FIGHTING WHALE.

Type locality. Vicinity of St. Bartholomew's Island, West Indies.
Geogr. Distr. Carribean Sea, Gulf of Mexico, South Atlantic.
Genl. Char. Cranium similar to that of *M. longimana*; supra-
occipital with a deep median groove from foramen magnum to near
superior surface, with a protuberance on each side near middle;
nasals in contact for much of their length, external beveled portion
concealed by maxillæ; otic bulla subcylindrical; ramus slender,
curved; coronoid process subtriangular, acuminate; first rib with
head simple; scapula simple.

Color. Above sooty black, under parts and pectoral fins beneath,
white spotted with black.

Measurements. Total length, 32 feet. Skull, total length, 9 feet;
nasals, 11.5 inches; width of cranium behind orbits, 5 feet, 4
inches; ramus on curve, 9 feet, 10 inches.

27. versabilis (*Megaptera*), Cope, Proc. Acad. Nat. Scien. Phil., 1869,
    p. 15. Elliot, Syn. N. Am. Mamm., 1901, p. 10.
HUMP-BACKED WHALE.

Type locality. Northwest coast of America.
Geogr. Distr. North Pacific Ocean to Alaska.
Genl. Char. Pectoral fins between one-third and one-fourth the
total length; pectoral and gular folds, 26.
Color. Above black, and in the most typical form the belly "is
said to be entirely black." (Cope, l. c.) *External face of the pectorals
black.*

The Finback Whales are numerous in species and are met with in all seas except the Polar. Their baleen is short and of inferior quality, and the amount of oil yielded by an individual is small, while their activity makes them so difficult to capture that they do not afford a sufficient recompense for the risk and labor, unless steam vessels and harpoon guns are employed. They are known by various names, such as Rorquals, Finbacks, Razor-backs, etc.

Subfam. I. **Balænopterinæ. Finback Whales.**

16. **Balænoptera.**

Balænoptera Lacép., Hist. Nat. Cét., 1804, pp. XXXVI, XXXVII, 114-141, pls. IV, V. Type *Balænoptera gibbar* Desmoulins.

Physalus Lacép., Hist. Nat. Cét., 1804, pp. XL, 219-226.

Cetoptera Rafin., Analyse Nat., Adden., 1815, p. 219.

Benedenia Gray, Proc. Zoöl. Soc., 1864, p. 211.

Sibbaldus Gray, Proc. Zoöl. Soc., 1864, p. 222.

Sibbaldius Flower, Proc. Zoöl. Soc., 1864, p. 391.

Cuvierius Gray, Cat. Seals, and Whales, 1866, p. 164, 1871, p. 54.

Head small, flat, pointed; body elongate; baleen short, broad; pectoral small, narrow, pointed; dorsal small, falcate; cervical vertebræ free; skin of throat wrinkled.

KEY TO THE SPECIES.

A. Under parts white.

   a. Width of flukes less than one-fourth total   PAGE

     length ....*B. davidsoni* 42

   b. Width of flukes one-fourth total length........*B. velifera* 43

B. Under parts yellow ....*B. sulfurea* 43

28. davidsoni (*Balænoptera*), Scamm., Proc. Calif. Acad. Scien., IV, 1872, p. 269. Elliot, Syn. N. Am. Mamm., 1901, p. 12.

DAVIDSON'S WHALE.

Type locality. Admiralty Inlet, Coast of Washington.

Geogr. Distr. West coast of North America, Mexico to Bering Straits.

Genl. Char. Dorsal small, falcate; pectorals small, narrow; baleen pure white. Laminæ 270 on each side, not exceeding 10 feet in length.

Color. Above dull black, beneath white; pectorals and caudal black above, white beneath; a white band across pectorals near their base; gular folds, 70, milky white, interspaces pinkish.

Measurements. Total length, 27 feet; pectorals, 4 feet 1 inch wide; height of dorsal, 10 inches; width of flukes, 7 feet, 6 inches.

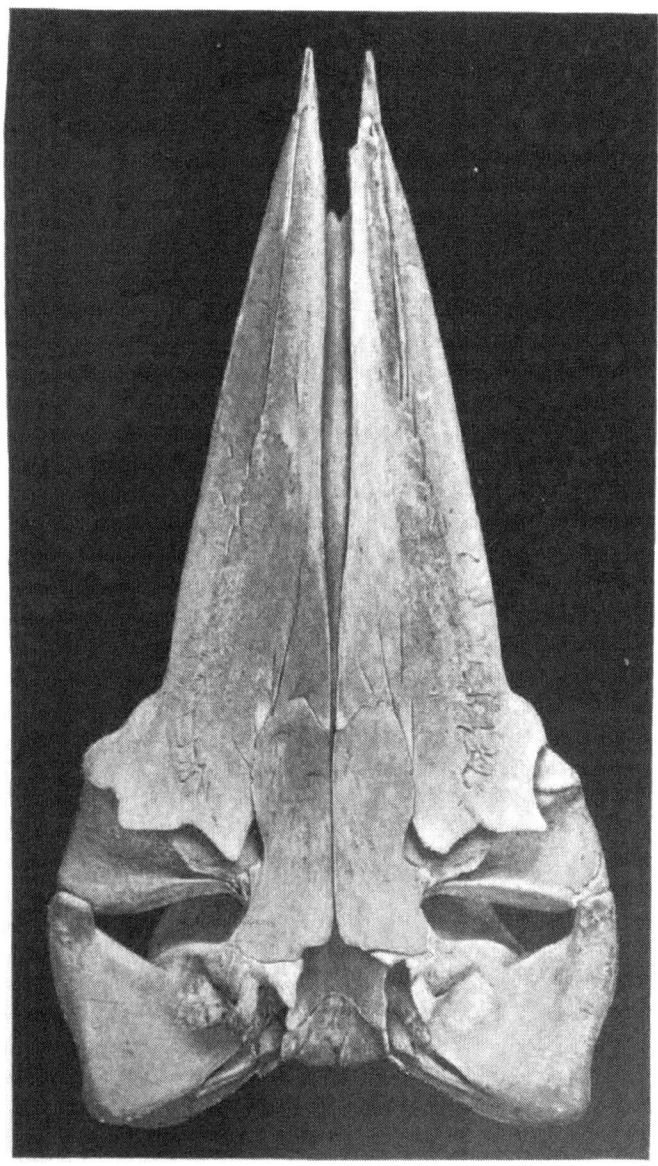

BALÆNOPTERA DAVIDSONI.
U. S. Nat. Mus. Coll.

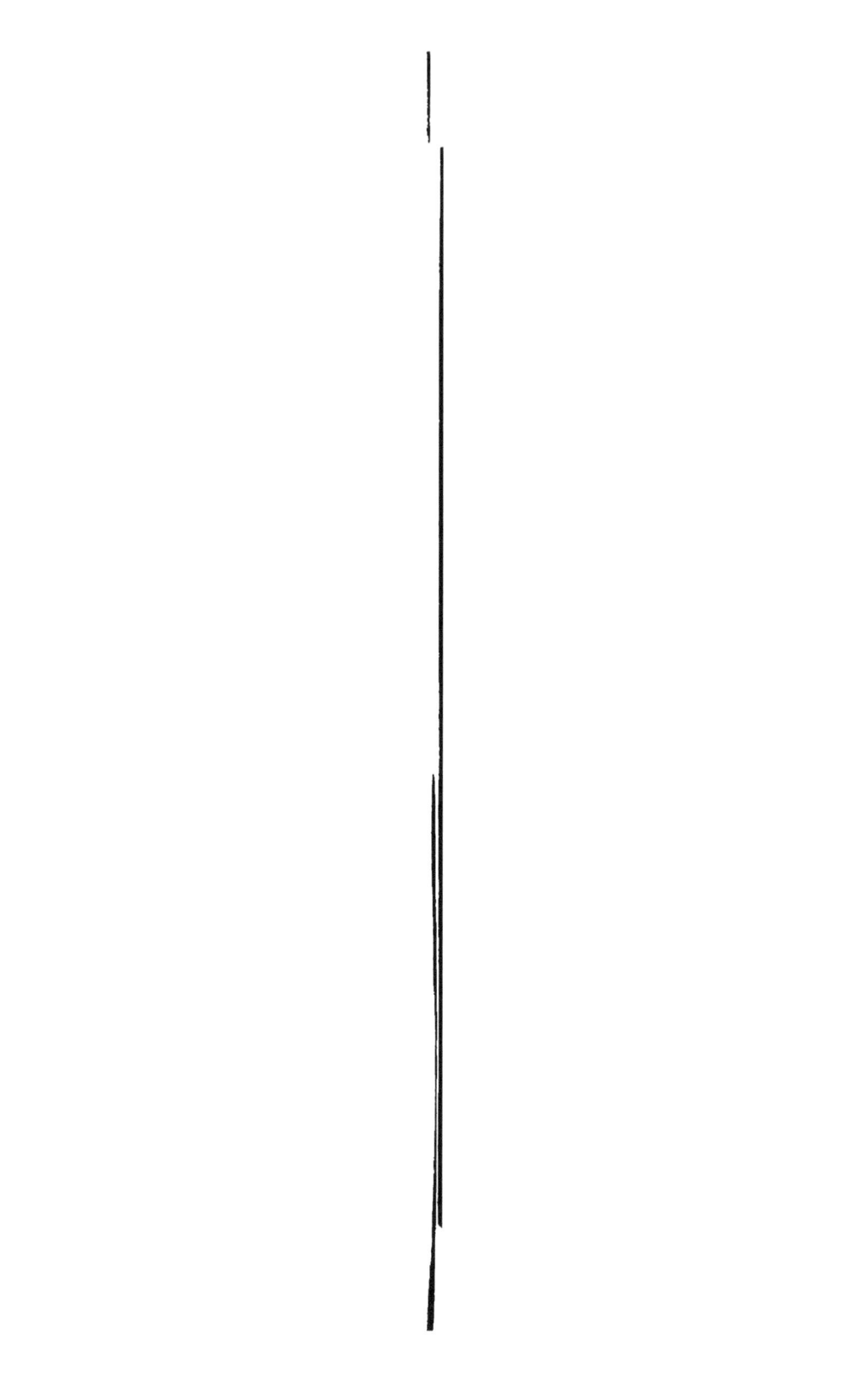

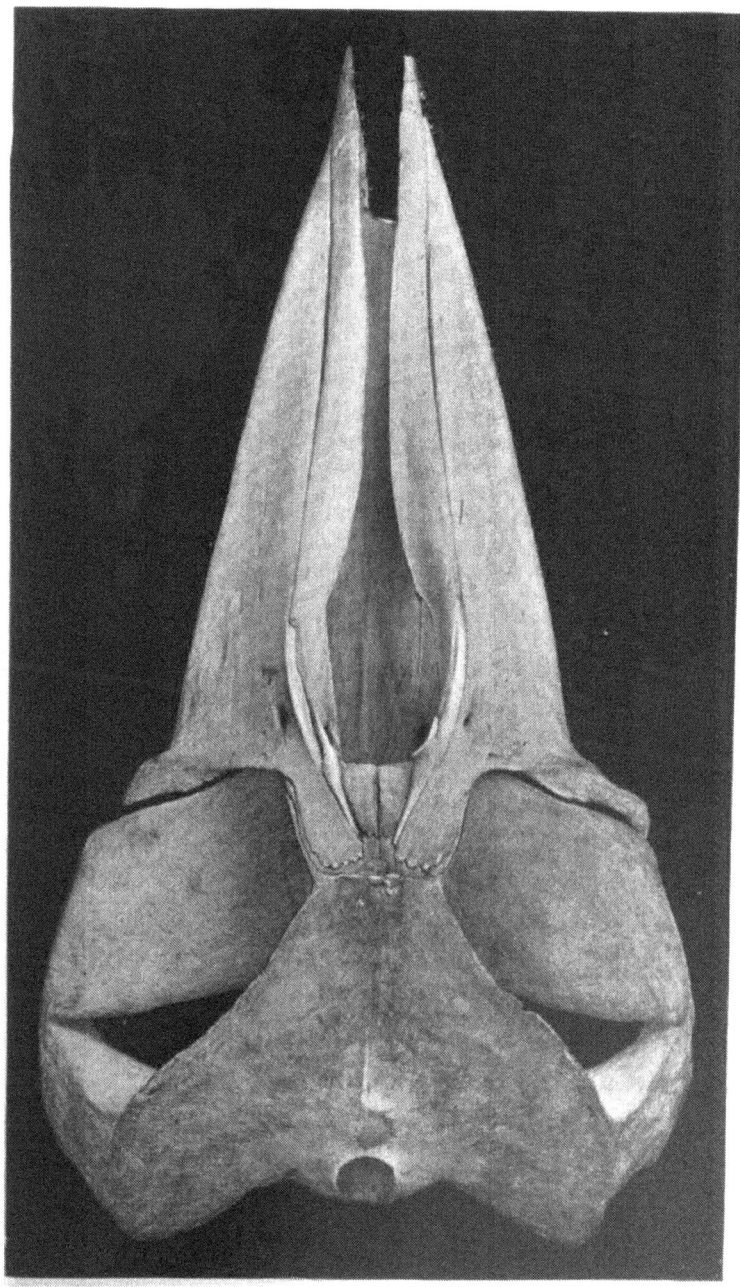

BALÆNOPTERA DAVIDSONI.
U. S. Nat. Mus. Coll.

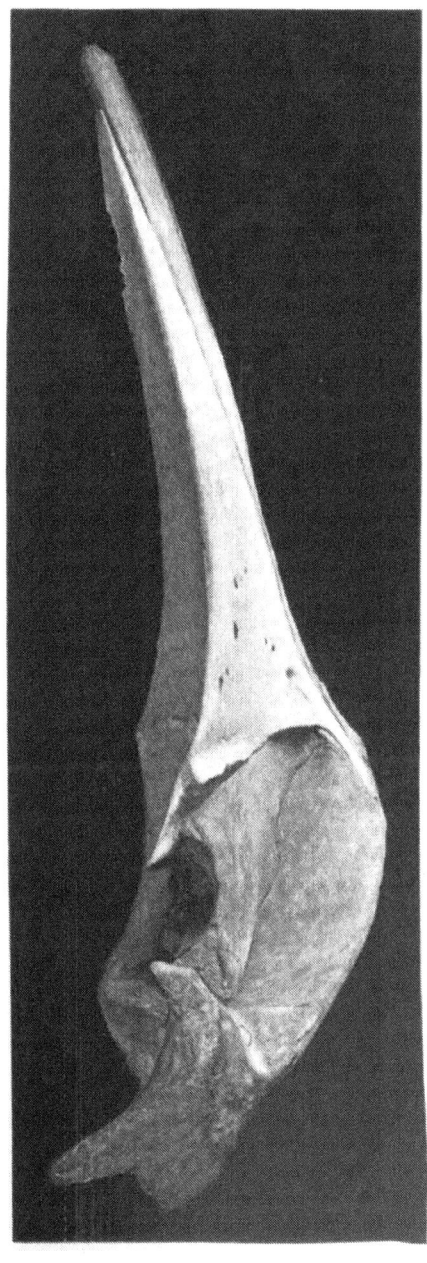

BALÆNOPTERA DAVIDSONI.
U. S. Nat. Mus. Coll.

29. velifera (*Balænoptera*), Cope, Proc. Acad. Nat. Scien. Phil., 1869, p. 16. Elliot, Syn. N. Am. Mamm., 1901, p. 12.
COPE'S WHALE.
Type locality. Shores of Oregon and California.
Geogr. Distr. North Pacific Ocean to Commander Islands.
Genl. Char. Size large; width of flukes one-fourth total length.
Color. Above black or blackish brown, beneath milky white; baleen light lead color.
Measurements. Total length, 60 feet.

30. sulfurea (*Sibbaldius*), Cope, Proc. Acad. Nat. Scien. Phil., 1869, p. 20.
sulfurea (*Balænoptera*) Elliot, Syn. N. Am. Mamm., 1901, p. 14.
YELLOW-BELLIED WHALE.
Type locality. Northwest coast of America.
Geogr. Distr. North Pacific Ocean to Bering Sea.
Genl. Char. Body slender; pectorals small, short, ends rounded; dorsal fin small, placed far back; baleen broad at base.
Color. Above light brown or brownish black, sometimes whitish; beneath yellow or a sulphur hue; baleen black or bluish black.
Measurements. Total length, 95 feet; circumference, 39 feet; length of mandible, 21 feet; longest baleen, 4 feet; weight of baleen, 800 pounds; estimated weight of animal, 147 tons. (Scammon.)

The toothed Cetacea embrace a large number of species included in several families and genera, and are known by the common names of Sperm Whales, Dolphins, Porpoises, etc. * All of them yield a certain quantity of oil, and are objects of pursuit in all the seas they inhabit. The greatest and most important species of all is, of course, the Cachalot, or Sperm Whale, which, excepting the Whalebone Whales, is the largest of living mammals, attaining at times a total length of eighty feet or more, and individuals are frequently met with over seventy feet. It is a very differently shaped animal from the Whalebone Whale, such as the Bowhead or Greenland Whale, for instance. The huge head is a high, straight-sided mass cut off square in front, and is about one-third the length of the body, and its great bulk is chiefly caused by an immense accumulation around the narial passage of an oily substance which fills the great well on top of the head and is known as spermaceti. In the intestines of this species is found the valuable commodity known as "ambergris," used in perfumery, and this substance is also met with floating on the seas this whale frequents. It is merely the "detained anal concretion of a diseased

whale, and is, therefore, composed of the refuse matter of the cepha-
lopods that form its food." Squids, cuttlefish, and octopi, large and
small, are eaten by the Sperm Whale, and the largest octopus that
ever lived, armed with its formidable beak, and long, disk-covered
arms, would be helpless when seized by the enormous jaws of this
mammal. The Sperm Whale goes in herds, at times of hundreds of
individuals, usually led by some old bulls. It has been known to
remain below after diving for more than an hour at a time, and it
requires about ten minutes to oxidize the blood after rising to the
surface, the animal respiring in that period about seventy times.
When alarmed, the Cachalot usually sinks at once, but occasionally
it will raise half of the body out of the water in the effort to see its
enemy. Hunting this whale is not without danger, and many a boat
and its crew have been destroyed by a blow from the tremendous
flukes.

Fam. II. **Physeteridæ. Sperm Whales.**

Upper jaw without functional teeth, those in the mandible
various, number often reduced. Pterygoids meeting on the median
line and hollowed on outer side. Transverse processes of the arches

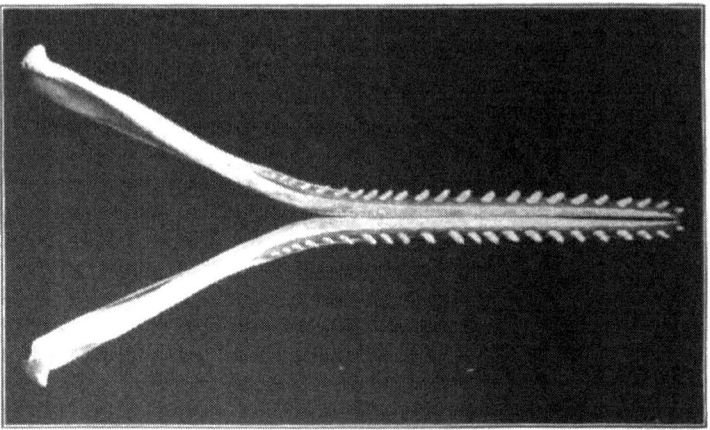

FIG. 11. PHYSETER MACROCEPHALUS—LOWER JAW.
No. 296 Field Columbian Mus. Coll.

of dorsal vertebræ cease near end of the series and are replaced at a
lower level by processes on the body. Costal cartilages not ossified.
Cranium elevated into a prominent crest behind the nares, and
asymmetrical around narial openings.

17. Physeter.

$$\frac{0-0}{20-20} \text{ to } \frac{0-0}{25-25} = 40 \text{ to } 50.$$

Physeter Linn., Syst. Nat., I, 1758, p. 76. I, 1766, p. 107. Type
  Physeter macrocephalus Linnæus.
Tursio Flem., Phil. Zoöl., II, 1822, p. 211. (nec Wagl. nec Gray.)

Upper teeth rudimentary; lower jaw with 20 to 25 on each side,
conical, pointed, and recurved; posterior and lateral edges of cranium
raised into a compressed semi-circular crest. Cranium above concave; rostrum elongate, its base broad, thence tapering to the tip;
mandible long and narrow, the symphysis being more than half the
length of the ramus; vertebræ, 50; zygomatic process of jugula thick,
massive.

FIG. XVI. PHYSETER MACROCEPHALUS. CACHALOT WHALE.

31. macrocephalus (*Physeter*), Linn., Syst. Nat., I, 1758, p. 76,
  I, 1766, p. 107. Elliot, Syn. N. Am. Mamm., 1901, p. 15.
CACHALOT. SPERM WHALE.

Type locality. North Atlantic.

Geogr. Distr. All seas.

Genl. Char. Size very large; head about one-third the length of
body, high, truncate, compressed in front; blow-hole longitudinal,
placed to the left of the median line on the upper end.

Color. Above black, shading gradually on the sides into the gray
of the under parts. Individuals sometimes are piebald.

Measurements. Total length of male, 55 to 60 feet; female much
smaller.

18. Cogia.

$$\tfrac{0-0}{9-9} \text{ to } \tfrac{0-0}{12-12} = 18 \text{ to } 24.$$

Kogia! Gray, Voy. Erebus and Terror, Zoöl., 1846, p. 22. Type *Physeter breviceps* Blainville.

"Teeth of the upper jaw absent, or reduced to a rudimentary pair in front; in the lower jaw 9 to 12 on each side, rather long, slender, pointed, and curved, with a coating of enamel. Upper surface of cranium concave, with thick, raised posterior and lateral margins, massive and rounded at their anterior terminations above the orbits. Upper edge of the methesmoid forming a prominent sinous ridge, constituting a kind of longitudinal septum to the base of the great supracranial cavity. Rostrum not longer than the cranial portion of the skull, broad at the base, and rapidly tapering to the apex. Vertebræ: C. 7, D. 13 or 14, L. and C. 30; total, 50 or 51. All the cervical vertebræ united by their bodies and arches." (Flower.)

32. breviceps (*Physeter*), Blainv., Ann. Anat. Phys., II, 1838, p. 337.
? *floweri* Gill, Amer. Nat., IV, 1871, p. 738, fig. 172.
breviceps (*Cogia*) Elliot, Syn. N. Am. Mamm., Suppl., 1901, p. 479.
PIGMY SPERM WHALE.

Type locality. Cape of Good Hope.

Geogr. Distr. Indian and Pacific Oceans, coast of southern California, possibly of Mexico, and Atlantic coast of North America.

Genl. Char. Blainville's description of a skull of this species in the Paris Museum, from an individual taken at the Cape of Good Hope, translated, is as follows: "Skull extremely wide and greatly elevated, having the frontal crests very high and consequently the nasal cavities very deep, something like those of the Cachalots, and terminate abruptly by the very short and pointed maxillæ, therefore the total length is barely an inch greater than the occipital length. The lower jaw has necessarily two branches approaching each other evenly, like a bellows, and a considerable symphysis, with a narrow extremity, but rounded termination. It is nearly certain that there are no teeth in the upper jaw, but the lower has 14 or 15 on each side, all of which are not in place, 5 only on the left side and 4 on the right remain still in their alveoli; some have been replaced by others; they are narrow, slender, conical, pointed, slightly curved interiorly, and 6 to 8 lines in length. Length of lower jaw, 13 inches; distance between condyles, 12 inches. Length of skull, 14½ inches. Another peculiarity of the skull is the inequality of the nasal cavities, the right being in nearly a rudimentary condition, and some twenty times smaller than the other."

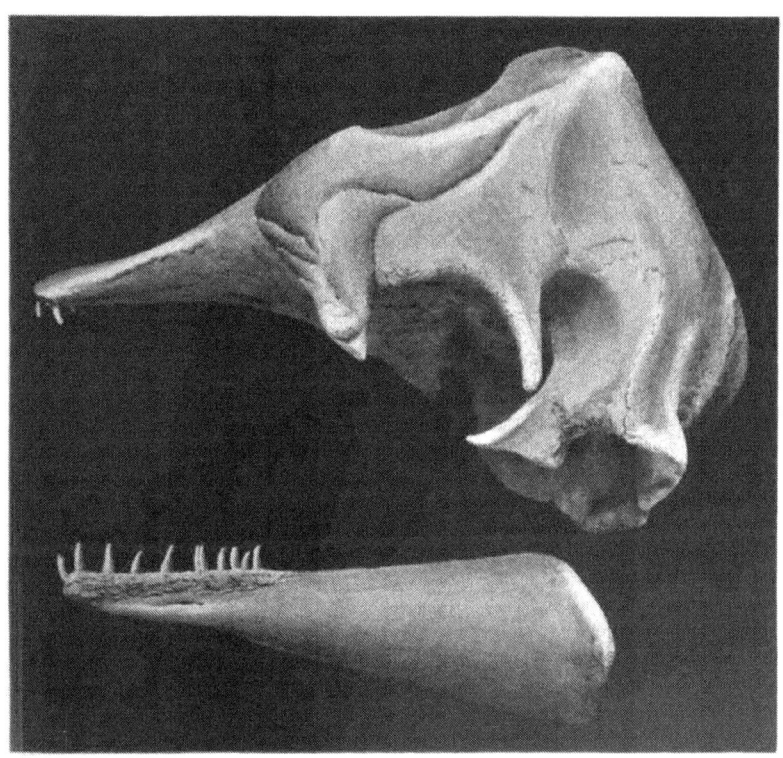

COGIA BREVICEPS.
U. S. Nat. Mus. Coll.

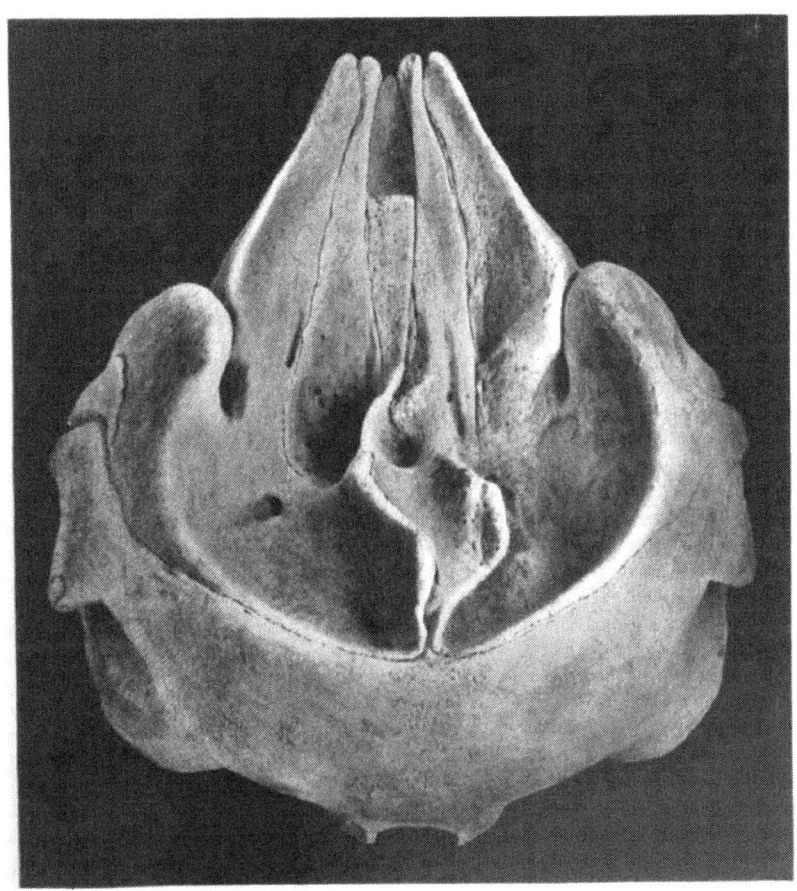

COGIA BREVICEPS.
U. S. Nat. Mus. Coll.

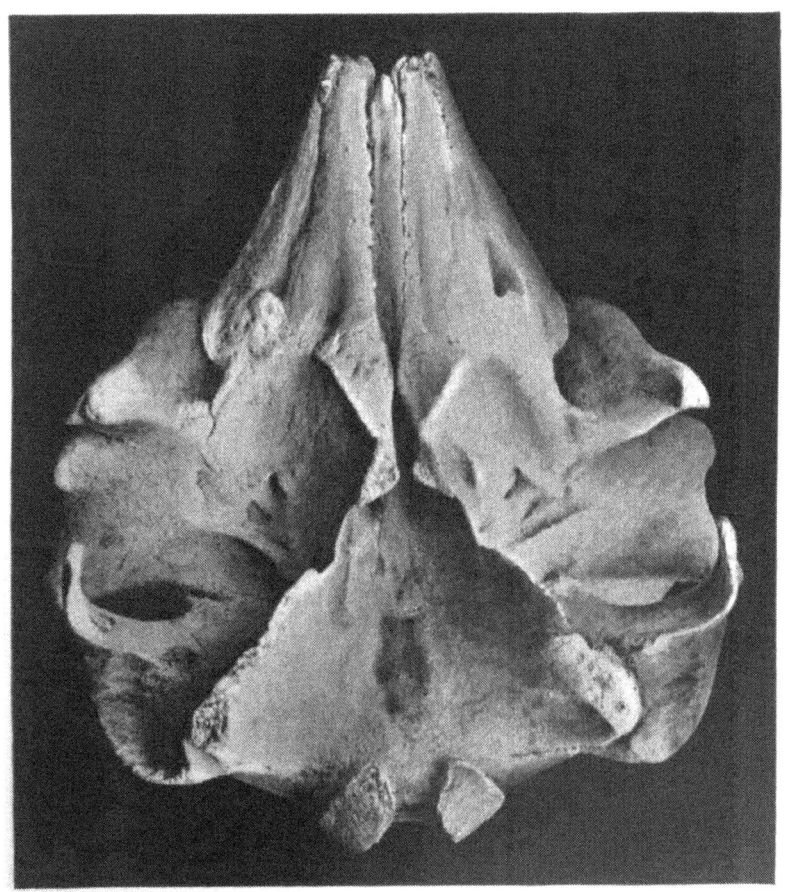

COGIA BREVICEPS.
U. S. Nat. Mus. Coll.

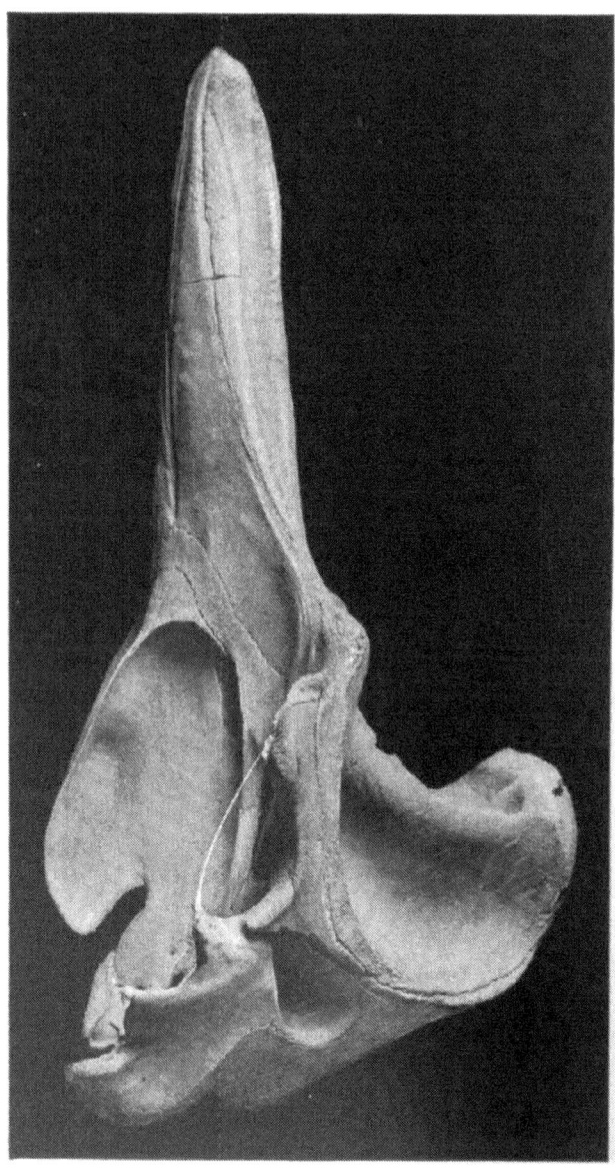

ZIPHIUS CAVIROSTRIS.
U. S. Nat. Mus. Coll.

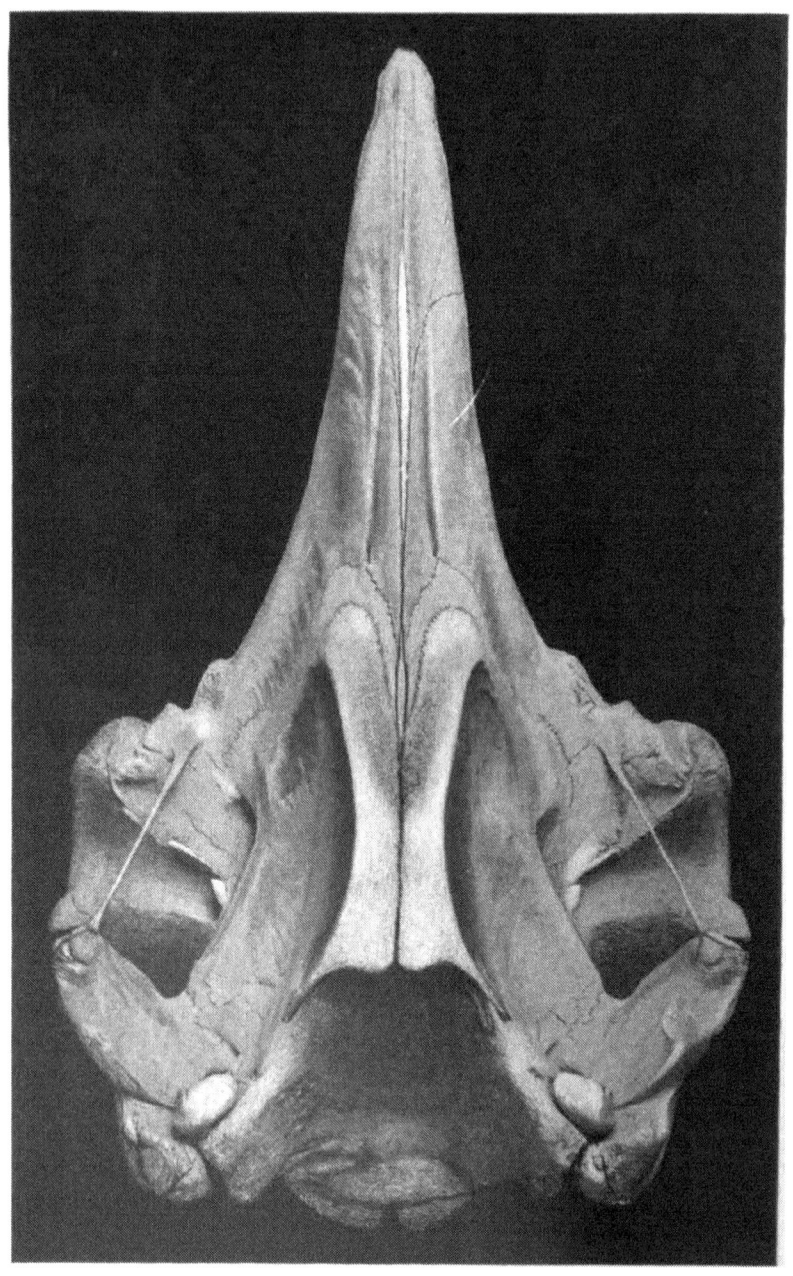

ZIPHIUS CAVIROSTRIS.
U. S. Nat. Mus. Coll.

ZIPHIUS CAVIROSTRIS.
U S. Nat. Mus. Coll.

The two-toothed Whales of the next genus were supposed to be extinct, as the imperfect skull of *Z. cavirostris*, found on the French Mediterranean coast in 1823, was described as a fossil. Various individuals have been observed since then, from as far north as the Shetland Islands, and to New Zealand in the South, and these have been separated into several species, not all probably entitled to the distinction. Specimens have occasionally drifted, or been driven ashore, and as many as twenty-five individuals were at one time stranded on the Chatham Islands east of New Zealand. This Cetacean varies in length from fourteen to twenty feet.

19. Ziphius.

$$\frac{0-0}{1-1} = 2.$$

Ziphius G. Cuv., Rech. Oss. Foss., v, 2d ed., 1823, p. 352, pl. XXVII, figs. 3, 4, 7, 9.   Type *Ziphius cavirostris* Cuvier.

Aliama Gray, Proc. Zoöl. Soc., 1864, p. 242.

Petrorhynchus Gray, Zoöl. Soc., 1865, p. 524.

Ziphiorrhynchus Burm., Revista Farmæ., Bull. Acad. Belg., 1865, *Id.* Ann. Mag. Nat. Hist., 3d Ser., XVII, 1866, p. 94.

At anterior end of the mandible on each side is a single conical tooth directed upward and forward. Rostrum triangular, tapering from base to apex; edges of maxillæ at base of rostrum raised into roughened tuberosities.

33. cavirostris (*Ziphius*), G. Cuv., Rech. Oss. Foss., v, 2d ed., 1823, p. 353.   Elliot, Syn. N. Am. Mamm., 1901, p. 16.

TWO-TOOTHED WHALE.

Geogr. Distr.   All seas.

Genl. Char.   Same as those of the genus.

Color.   Steel gray with numerous irregular white streaks; beneath white.

Total length, 16 feet.

The members of the next family, *Delphinidæ*, are many and various, and their arrangement into subfamilies, or even genera, from lack of requisite knowledge of some of the species, is not easy of accomplishment. Among the diversified forms are found, the well-known Porpoise with its many relatives; the curious Narwhal with its ivory spear, a formidable weapon both for offense as well as defense; the Cow-fish (*Tursiops gilli*); the Black-fish, or Ca'ing Whale (*Globicephalus melas*); the savage Orcas, or Killer Whales; and the numerous species of DOLPHINS, inhabitants of many seas, beside other genera and species outside the scope of this work. The beaks vary greatly, and in some species are altogether absent, while in otheis they are twice the length of the braincase.

Fam. III. **Delphinidæ—Dolphins, Porpoises, etc.**

F. W. True. *A review of the family Delphinidæ*, Bull. U. S. Nat. Mus., 1889, No. 36, pp. 1-191, pls. 1-47.

C. M. Scammon. *The Marine Mammals of Northwest North America*, 1874, p. 40.

Facial portion of skull produced usually into a beak; teeth numerous in both jaws; anterior ribs articulated to the transverse process by a tubercle; sternal ribs ossified; lacrymal not distinct from the jugal; pterygoids short, thin, and form, with a process of the palate, the outer wall of the post palatine air-sinus; mandibular symphysis short.

Porpoises, often called "Sea Pigs" or "Hog-fish," are accustomed to go in schools, sometimes in very large numbers, as they are very sociable creatures,,and are often seen in bays and harbors, as well as in the open ocean. They swim with great rapidity, and frequently play about the cutwater of a large steamer, even when the vessel is going at full speed, and these animals are better known to the majority of people than any of the other Cetaceans. The genus contains numerous species, and they are met with in all seas.

Subfam. I. **Delphinapterinæ.**

20. Phocæna. Porpoises.

$$\frac{25-25}{25-25} = 100.$$

Phocæna G. Cuv., Nouv. Dict. Hist. Nat., 2d ed., IX, 1817, p. 163, *Id.* Règn. Anim., 1829, p. 289. Type *Delphinus phocæna* Linnæus.

Head not beaked, rostrum short, broad, tapering; premaxillæ tuberculate before the nares; nasals flat; frontals elevated; mandibular symphysis short; teeth small, crowns spade-shaped, neck constricted; dorsal fin triangular, small, blunt spines often on anterior margin; pectoral fins ovate; first to sixth cervical vertebræ coalesced.

34 phocæna (*Delphinus*), Linn., Syst. Nat., I, 1758, p. 77. Elliot, Syn. N. Am. Mamm., Suppl., 1901, p. 482.

communis G. Cuv., Règn. Anim., 1817, p. 279. Elliot, Syn. N. Am. Mamm., 1901, p. 20.

vomerina Gill, Proc. Acad. Nat. Scien. Phil., 1865, p. 178.

brachycium Cope, Proc. Acad. Nat. Scien. Phil., 1865, p. 279.

? *lineata* Cope, Proc. Acad. Nat. Scien. Phil., 1876, pp. 134, 135.

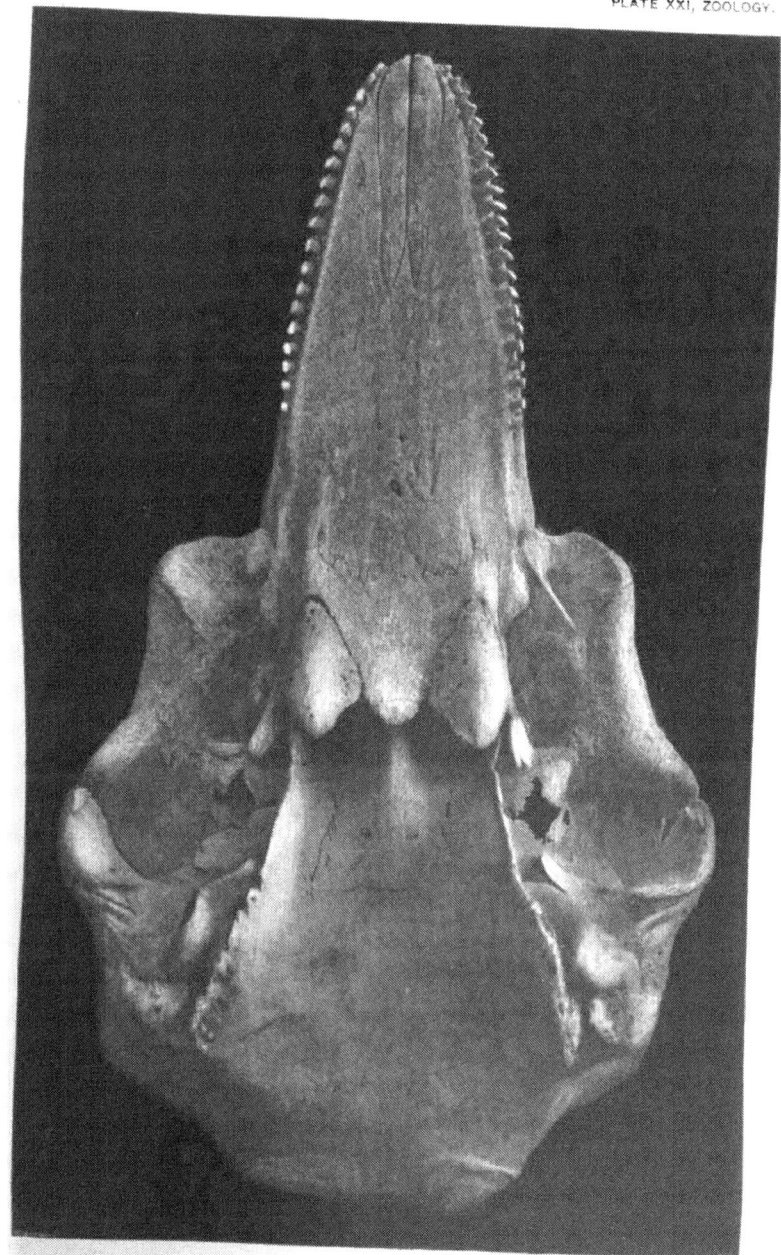

PHOCÆNA PHOCÆNA.
No. 43 Field Columbian Mus. Coll.

FIG. XVII. PHOCÆNA PHOCÆNA. PORPOISE.

COMMON PORPOISE.

Type locality. Coast of Europe.

Geogr. Distr. North Atlantic and North Pacific Oceans.

Genl. Char. Slender; dorsal fin anterior to middle of the length, triangular, posterior margin concave, anterior about straight with sometimes a row of tubercles; jaws of equal length.

Color. Upper parts slate or blackish, grading on sides into the white of lower parts; sides sometimes tinged with yellow or pink; narrow dark line from corner of mouth to anterior base of pectoral, and a broad, dark band often extends from lower jaw half-way to the pectoral.

Measurements. Total length, 1727; length of mouth, 121; end of snout to dorsal, 737; length of pectoral, 178; height of dorsal, 102; width of flukes, 317. Skull: total length, 293; length of rostrum, 137; width of beak at base, 85; at middle, 55; interorbital breadth, 137; length of temporal fossa, 6.

The "Killers" are distinguished for their great ferocity and strength, and are the wolves of the sea. They prey upon fish, and also warm-blooded animals, such as seals, and destroy a great number of the pups, and the half-grown young of other species. Banding themselves together in packs they do not hesitate to attack the Whalebone Whale, and several of them by hanging on to the lower lip, compel the huge animal, exhausted by its struggles, to open its mouth and permit the Killer to enter, when the great fleshy tongue is speedily devoured, and the unfortunate creature left to die a lingering death. Individuals of their own order are pursued and slain by these sea-wolves, and from sheer love of slaughter more creatures are killed in their forays than can be devoured. They delight in blood and rapine, and the presence of the Killers can be detected in the seas they frequent by the lofty pointed dorsal fin standing above the surface of the ocean and cutting the water like the bow of some swift vessel, as the fierce creature beneath chases its prey. Orcas do not associate together in any large numbers, a dozen being perhaps the maximum, and whenever their presence is known, or the fins are seen cleaving the surface of the ocean, all animals fly for a refuge, even the ponderous Sea Lions seeking the shore. The Killers do not possess much oil, and consequently have little or no commercial value, but some coast Indians hunt them for their flesh, which they highly esteem.   This Cetacean is usually seen in the vicinity of the Pribiloff Islands during the breeding season of the Fur Seals, and commits great destruction among the pups when these make their first attempts at swimming not far from shore, for it requires an expert in the art to be able to avoid the swift rush of this powerful mammal.

21. Orcinus. Killer Whales.

$$\tfrac{12-12}{12\ 12} = 48.$$

Orcinus Fitzin., Wiss-Popul. Naturg. Säugeth., vi, 1860, pp. 204-
217. Type *Delphinus orca* Linnæus.
Orca Gray, Erebus and Terror, Zoöl., 1846, p. 33, pls. 8-9. (nec
Wagl, 1830.)

Teeth large, stout, occupying nearly the entire length of the rostrum, which is broad, elongate, flattened above and rounded anteriorly. Pterygoids separate; premaxillæ concave before the nares, narrow in the middle and widening towards end; head depressed, no beak; dorsal large, prominent, pointed; pectorals large, ovate; first and second vertebræ, occasionally also the third, coalesced; vertebræ, 52.

FIG. XVIII. ORCINUS ORCA. KILLER WHALE.

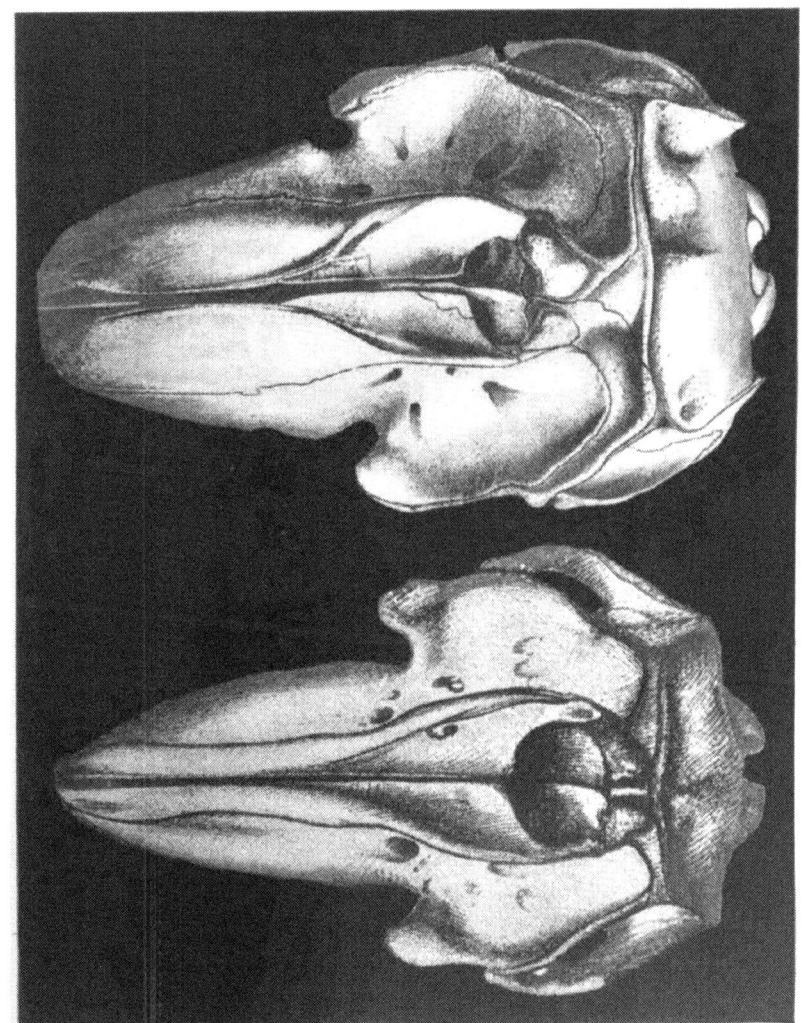

35. orca (*Delphinus*), Linn., Syst. Nat., I, 1758, p. 77.

gladiator (*Delphinus*), Bonnat., Cét., 1789, p. 23.

arcticus and europæus Gerv. & Van Ben., Ost. Cét., 1804, p. 314.

schlegelii Lilljeb., Roy. Soc., 1866, p. 235.

latirostris Gerv., Ost. Cét., 1868, p. 543.

stenorhyncha Gray, Proc. Zoöl. Soc., 1870, p. 74.

gladiator (*Orcinus*) Elliot, Syn. N. Am. Mamm., 1901, p. 22.

COMMON KILLER WHALE.

Type locality.   Coast of Europe.

Geogr. Distr.   All seas.

Genl. Char.   Size large, other characters those of the genus.

Color.   Upper parts of head and body and all the fins black; under jaw, throat, breast, and belly white; a white trident-shaped area extends back to the vent with one tine, the other two passing obliquely upward and backward on the sides; a large white patch behind the eyes; behind dorsal fin is a crescentic purple area.

Measurements.   Total length, 16 feet.

22. Pseudorca.   Killer Whales.

$$\frac{10-10}{10-10} = 40.$$

Pseudorca Reinh., Overs. K. Dan. Vidensk. Sezsk. Forh. Kjöbenh., 1862, p. 151.   Type *Pseudorca crassidens* Owen.

Rostrum short, broad; rostral portion of intermaxillæ truncate at distal end; pterygoids short, approximated on median line; teeth large, roots cylindrical; vertebræ, 50; the first to sixth or seventh coalesced; pectorals moderate, pointed; dorsal near middle of back, moderate, falcate; head elevated before blow-hole, compressed; snout truncate.

36. crassidens (*Phocæna*), Owen, Brit. Foss. Mamm., 1846, p. 516.

meridionalis Flower, Proc. Zoöl. Soc., 1864, p. 420.

destructor Cope, Proc. Acad. Nat. Scien. Phil., 1866, p. 293.

grayi Burm., Ann. Mus. Pub., Buen. Aires, I, 1864-69, p. 367, pl. XXI.

crassidens (*Pseudorca*), Elliot, Syn. N. Am. Mamm., 1901, p. 23.

LARGE-TOOTHED KILLER WHALE.

Type locality.   Coast of Lincolnshire, England.

Geogr. Distr.   All seas.

Genl. Char.   No beak; head sloping gradually from blow-hole to end of snout; dorsal in center of length, narrow, moderate; pectorals small.

Color.   All black.

Measurements. Skull: total length, 595; length of rostrum, 287; breadth at base, 208; at middle, 188; interorbital breadth, 333; length of temporal fossa, 191.

The Black-fish, the name usually given to the species of the next genus, are accustomed to go in large schools, keeping near the coast, seeking small fish, on which they subsist. These animals are inoffensive and gentle, and when alarmed can easily be driven ashore (if their pursuers are on the seaward side), as they huddle together like sheep and follow blindly any leader. They are found in all seas, and it is stated that the flesh, after having been exposed to the air and properly cooked, is not unpalatable. The Black-fish yield but little oil, not equal to that of the great whales.

23. Globicephalus. Dolphins. Black-fish.

$$\frac{8-8}{8-8} \text{ to } \frac{12-12}{12-12} = 32 \text{ to } 48.$$

Globicephala Less., Nouv. Tabl. Règn. Anim., Mamm., 1842, p. 200.

Type *Delphinus deductor* Scoresby = *Delphinus melas* Traill.

Teeth only on anterior half of rostrum and mandible, small, conical, acute, curved; rostrum short, broad; mandibular symphysis short; pterygoids in contact; skull broad, depressed; premaxillæ concave in front of nares, as wide at middle as at base; vertebræ, 57–60; first five or six cervical vertebræ coalesced; forepart of head round; dorsal low, triangular.

KEY TO THE SPECIES.

A. Head with obtuse ridge above jaw; teeth conical, persistent in both jaws, confined to anterior half of rostrum.

37. melas (*Delphinus*), Traill. Nichols, Jour., 1809, XXIII, pl. 3.
    globiceps Cuv., Ann. Mus., 1812, p. 14, pl. 1, figs. 1, 2.
    deductor Scoresby, Arct. Reg., 1, 1820, t. 13, fig. 1.
    intermedius Harl., Jour. Acad. Nat. Scien. Phil., 1829, p. 51, pl. 1, fig. 13.

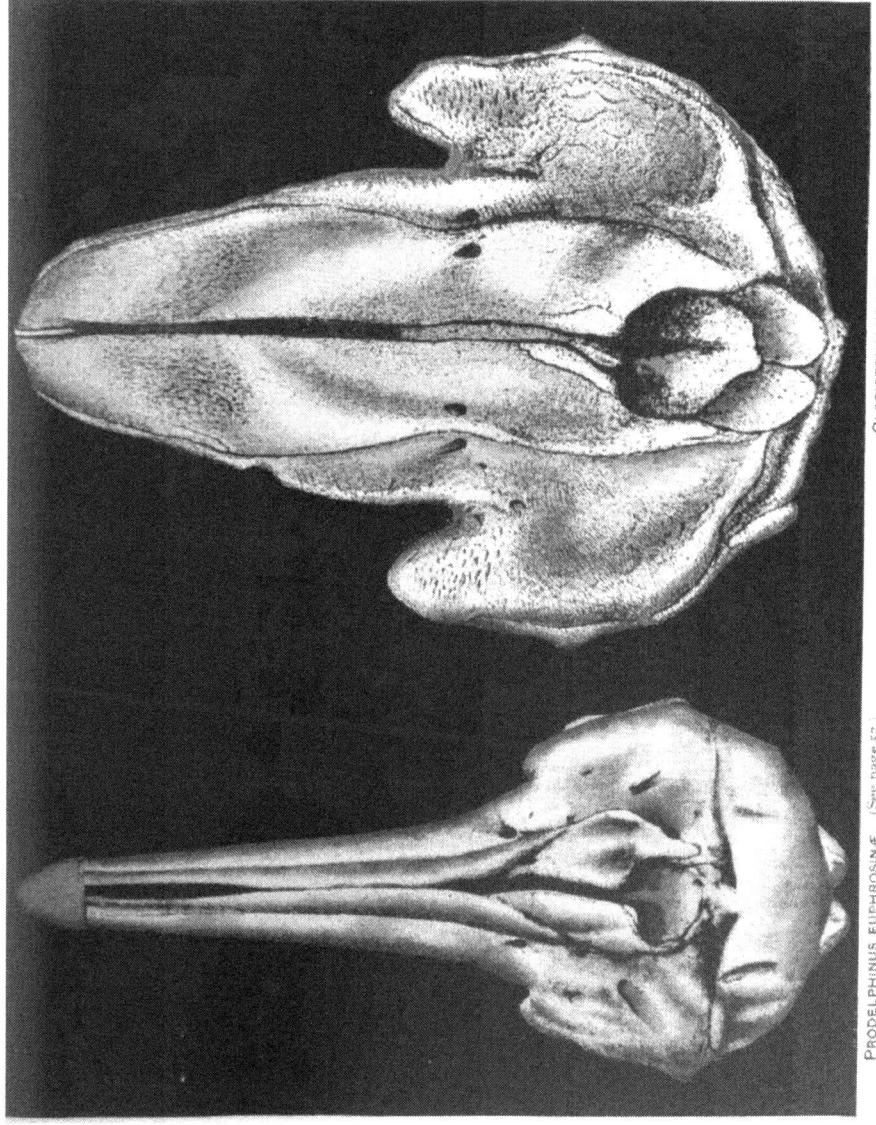

PRODELPHINUS EUPHROSINÆ. (See page 57.)
Ex True. (From Gray, Synopsis of Whales and Dolphins, 1868, pl. 22.)

GLOBICEPHALUS MELAS.
Ex True. (From Gray, Catalogue of the Whales and Dolphins, 1866, p. 316, fig. 62.)

incrassatus Gray, Proc. Zoöl. Soc., 1861, p. 309, fig. 1.

macrorhynchus Hector (nec Gray), Trans. N. Zeal. Inst., VII, 1861, pl. 16, figs. 3, 3a.

melas (*Globiocephalus!*), Elliot, Syn. N. Am. Mamm., 1901, p. 23.

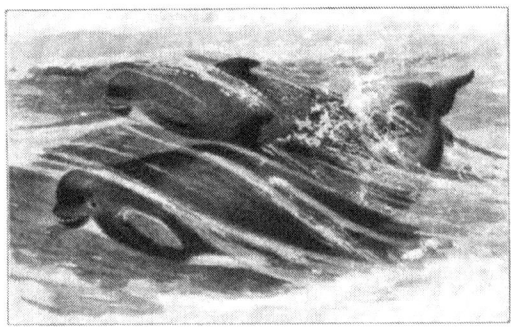

FIG. XIX. GLOBICEPHALUS MELAS. PILOT OR CA'ING WHALE.

PILOT OR CA'ING WHALE.

Type locality. Coast of England.

Geogr. Distr. South and North Atlantic Oceans, Gulf of Mexico.

Genl. Char. Vertebræ, 59–60; maxillæ and intermaxillæ rugose anteriorly; second and third vertebræ coalesced.

Color. Black with a white area beneath.

Measurements. Total length about 20 feet; length of pectoral fin, 1270 mm.; greatest breadth of pectoral, 279; from end of snout to dorsal fin, 1397.

38. brachypterus (*Globiocephalus!*), Cope, Proc. Acad. Nat. Scien. Phil., 1876, p. 129, fig. p. 131. Elliot, Syn. N. Am. Mamm., 1901, p. 24.

SHORT-FINNED BLACK-FISH.

Type locality. East coast of Delaware Bay, at the mouth of Meurice River.

Geogr. Distr. Gulf of Mexico north to New Jersey, Atlantic Ocean.

Genl. Char. Pectorals one-sixth total length of body; dorsal forward of middle length; teeth, $\frac{8}{8}$; vertebræ, 57; skull massive; rostrum broad, the basal width greater than four-fifths total length; temporal fossæ large, oval; intermaxillæ large and flat.

Color. Entirely black.

Measurements. Total length, 4648; tip of snout to dorsal, 1206; length of pectoral, 762; height of dorsal, 356; width of fluke, 1168.

Skull: total length, 662; length of rostrum, 333; breadth at base, 288; at middle, 235; interorbital breadth, 45; length of temporal fossa, 163.

39. scammoni (*Globiocephalus!*), Cope, Proc. Acad. Nat. Scien. Phil.,
      1869, p. 21.   Elliot, Syn. N. Am. Mamm., 1901, p. 24.
SCAMMON'S BLACK-FISH.
      Type locality.   Coast of Lower California.
      Geogr. Distr.   Coast of California southward; coast of South America.
      Genl. Char.   Similar to *G. brachypterus*; pectorals longer.   Skull heavy; intermaxillæ not projecting over lateral margins of rostral portion of maxillæ; superior nares broad, and bordered by narrow plates of the intermaxillæ; pterygoids short, approximate.
      Color.   Entirely black.
      Measurements.   Total length, 4724; tip of snout to dorsal, 1372; length of pectoral, 864; width of flukes, 1007.   Skull: total length, 690; length of rostrum, 340; breadth at base, 308; at middle, 252; interorbital breadth, 487; length of temporal fossa, 148.

The members of the next genus, DELPHINUS, possess prominent beaks, and the elongate rostrum is provided with a large number of teeth, which, however, is not always the same on the two sides of the jaw.   There are many species accredited to the genus, dwellers of various seas, and the one given below is common both to the Mediterranean and to the waters that wash the eastern shores of the American Continent.

24. Delphinus. Dolphins.

$$\frac{40-40}{40-40} \text{ to } \frac{60-60}{60-60} = 160 \text{ to } 240.$$

Delphinus Linn., Syst. Nat., 1, 1758, p. 77; 1, 1766, p. 108.   Type
      Delphinus delphis Linnæus.
Rhinodelphis Wagner, Schreb., Säugeth., VII, 1846, pp. 281, 316–
      349, 11 pls.
Eudelphinus Gerv., Ostéog. des Cét., 1880, p. 600.
      Teeth occupying nearly entire length of rostrum, numerous in both jaws, conical, acute, curving; rostrum twice the length of braincase; pterygoids meeting on median line for their entire length; palate deeply grooved laterally; pectorals moderate, falcate.

40. delphis (*Delphinus*), Linn., Syst. Nat., 1, 1758, p. 77; 1, 1776,
      p. 108.   Elliot, Syn. N. Am. Mamm., 1901, p. 28.
fulvo-fasciatus Wagn., Schreb. Säugeth., 1846, pl. 361, fig. 1.

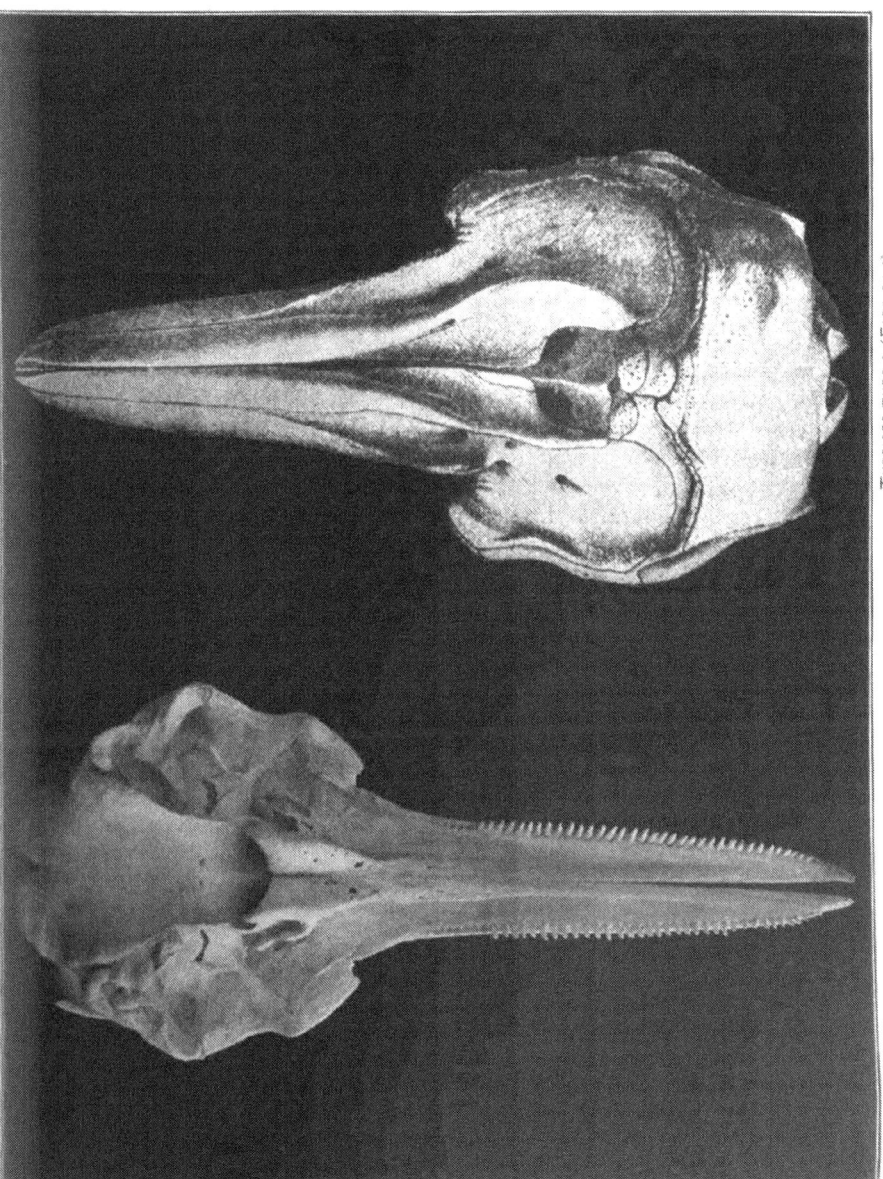

DELPHINUS DELPHIS.   TURSIOPS TURSIO (FABRICIUS).

No. 44 Field Columbian Mus. Coll. Little more than ⅓ nat. size.   Ex True.   (From Van Beneden and Gervais, Ostéographie des Cétacés, 1880-79, pl. 34, fig. 3.)

novæ-zelandiæ Quoy & Gaim., Voy. Astrolabe, Mamm., 1830,
    p. 149.
janira Gray, Erebus & Terror, Zoöl., 1846, p. 41, pl. 23.
albrinanus Peale, U. S. Expl. Exped., Mamm., v, 1848, III, p. 33.
algeriensis Loche, Rev. Mag. Zoöl., 1860, p. 474, pl. 22, fig. 1.
forsteri Gray, Cat. Cet., 1866, p. 248.
major, moorii & walkeri Gray, Cat. Cet., 1866, pp. 396, 397.
pomœgra Owen, Trans. Zoöl. Soc., VI, 1866, p. 23, pls. 6. 8.
bairdi Dall, Proc. Calif. Acad. Nat. Scien., v, 1873, p. 12.
microps Burm. (nec Gray), Desc. Phys. Argent., III, 1879, p. 534.
fuscus, sowerbianus, variegatus, batteatus, moschatus (La Font),
    Fisch., Act. Soc. Linn. Bord., v, 1881, p. 127, pls. 4, 5, 6.
marginatus, La Font., (nec Pucher.), Act. Soc. Linn. Bord., VI,
    p. 518.
curvirostris Riggio, Nat. Sicil., II, 1883, p. 157, pl. 3.
COMMON DOLPHIN.

Type locality. Coast of Europe.
Geogr. Distr. Atlantic Ocean; Gulf of Mexico.
Genl. Char. Body slender; forehead forming an angle to the beak,
which is long and slender; dorsal fin in center of back, narrow; pectorals three times longer than broad, pointed.
Color. Very variable; upper parts black and blackish gray;
beneath white or greenish white; black, gray or greenish band from
lower jaw to base of pectorals; orbital ring black, from which a black
band extends forward to the base of the beak; margin of lower jaw
black; elongate areas of light festoons of gray on sides, traversed by
two longitudinal bands of gray or greenish gray.
Measurements. Total length, 1382–2008; length of pectoral, 280–
305; width of flukes, 393–450; height of dorsal, 177–203; blow-hole to
end of nose, 330–356.

25. Tursiops. Dolphins.

$$\frac{21-21}{21-21} \text{ to } \frac{25-25}{25-25} = 84 \text{ to } 100.$$

Tursiops Gerv., Hist. Nat. Mamm., II, 1855, p. 323. Type *Del
    phinus tursio* Fabricius.
Tursio Gray (nec Wagl.), List Spec. Mamm., Brit. Mus. 1843, pp.
    XXIII, 105, *Id.* Cat. Seals and Whales, 1866, p. 254.

Rostrum with moderate taper; no groove in palate; mandibular
symphysis short; teeth stout; vertebræ, C. 7, D. 13, L. 17, C. 27=64.
Dorsal fin high, falcate.

KEY TO THE SPECIES.

A. Teeth conical, smooth; palate without lateral grooves;
pterygoids in contact.   PAGE
   a. Plumbeous gray above, white beneath......*T. truncatus* 56
   b. All black.....................................*T gilli* 56

FIG. XX. TURSIOPS TRUNCATUS. BOTTLE-NOSED PORPOISE.

41. truncatus Montagu., Mem. Wern. Soc., III, 1821, p. 73. True,
    Proc. Acad. Nat. Scien. Phil., 1903, p. 313.
    compressicauda Less., Cét., 1828, p. 199.
    communis Fitzin. (nec Cuv.), Carr, Dalm., 1846, p. 75.
    metis Gray, Ereb. & Terror, Zoöl., 1846, p. 38, pl. 17.
    cymodice Gray, Ereb. & Terror, Zoöl., 1846, p. 38, pl. 17.
    eurynome Gray, Ereb. & Terror, Zoöl., 1846, p. 38, pl. 18.
    tursio (*Tursiops*), Elliot, Syn. N. Am. Mamm., 1901, p. 29.
BOTTLED-NOSED PORPOISE OR DOLPHIN.
    Type locality. Coast of Greenland.
    Geogr. Distr. Atlantic Ocean, Mediterranean Sea, widely dis-
tributed; Atlantic coast of Atlantic States, Maine to Florida, Gulf of
Mexico.
    Genl. Char. Those of the genus. Frontal bone has no backward
extension and the parietal is broad inferiorly.
    Color. Upper part of fin is plumbeous gray, tinged with purple,
grading on sides into the pure white of the under parts.
    Measurements. Total length, 2907; of mouth, 319; height of
dorsal fin, 229; breadth of flukes, 612. Skull: total length, 432;
length of beak, 108; length of tooth row, 195; width between inter-
orbitals, 191; length of mandible, 365.

42. gilli (*Tursiops*), Dall, Proc. Calif. Acad. Scien., v, 1873, p. 13.
    Elliot, Syn. N. Am. Mamm., 1901, p. 29.
GILL'S DOLPHIN. COW-FISH.
    Type locality. North Pacific Ocean.

Geogr. Distr. Shores of southern and Lower California to northern part of Pacific Ocean.

Genl. Char. Optic canal not reaching the level of the rounded antero-internal border of the deeply concave frontal, and the lower part of the parietal is a narrow band between the anterior margin of the squamosal and the posterior margin of a backward extension of the frontal.

Color. Black, the under parts a little lighter than the upper.

Measurements. Length of beak, 29.8; breadth at base of maxillary notches, 14.1; at the middle, 8.8; length of tooth row, 25.4; length of mandible, 42.7.

26. Prodelphinus. Dolphins.

$$\frac{30-30}{30-30} \text{ to } \frac{50-50}{50-50} = 120 \text{ to } 200.$$

Prodelphinus Gerv., Ostéog. des Cétacés, 1880, p. 604, pl. XXXVIII.

Clymene Gray, Proc. Zoöl. Soc., 1864, p. 237.

Clymenia Gray, Syn. Whales & Dolphins, 1868, p. 6.

Teeth smaller than those of *Tursiops.* Rostrum long, narrow; no groove in palate; vertebræ, 73–78; symphysis of mandible short; beak elongate; dorsal and pectoral fins falcate.

FIG. XXI. PRODELPHINUS PLAGIODON. SHARP-TOOTHED DOLPHIN.

KEY TO THE SPECIES.

A. Teeth conical, small, numerous; rostral portion of intermaxillæ convex.   PAGE
    a. Purplish gray above spotted with white; breadth of rostrum at base, 109; at middle, 58; between orbits, 186...................... *P. plagiodon* 58

b. Dark gray above, spotted with light gray;   PAGE
   breadth of rostrum at base, 75; at middle,
   46; between orbits, 140..................*P. longirostris*   58
c. Black above; breadth of rostrum at middle,
   60.96; at base, 210.82.....................*P. euphrosyne*   58

43. plagiodon (*Delphinus*), Cope, Proc. Acad. Nat. Scien. Phil., 1866,
   p. 296.
plagiodon (*Prodelphinus*), Elliot, Syn. N. Am. Mamm., 1901, p. 31.
SHARP-TOOTHED DOLPHIN.
Type locality.   Eastern coast of United States.
Geogr. Distr.   Atlantic coast of the United States, Cape Hatteras
to Gulf of Mexico.
Genl. Char.   Dorsal fin high, recurved; pectoral fins broad at base;
beak stout.
Color.   Above purplish gray, shading on the sides into the white
of the under parts; upper parts and fins spotted with white or gray;
lower parts spotted with dark gray.
Measurements.   Total length, 2157; length of mouth, 280; of
pectoral fin, 304; end of beak to dorsal, 337; height of dorsal, 241;
breadth of flukes, 527.

44. longirostris (*Delphinus*), Gray, Spicil. Zoöl., 1828, p. 1.
microps Gray, Spicil. Zoöl., 1828, p. 1.
alope Gray, Erebus & Terror, Zoöl., 1846, p. 42, pl. 25.
stenorhynchus Gray, Cat. Seals & Whales, 1866, p. 396.
LONG-NOSED DOLPHIN.
Type locality.   Cape of Good Hope.
Geogr. Distr.   Southern Oceans.   Taken off Tres Marias Islands,
Mexico.
Genl. Char.   Dorsal high; nose three-fifths total length of animal;
teeth formula, $\frac{55-60}{55-60}$; vertebræ, C. 7; D. 14; L. 17-18; C. 34, =72-73.
Color.   Above dark slate gray, mottled with pale gray; beneath
white.
Measurements.   Total length, 390-420; beak, 70-80; width of
beak at base of maxillary notches, 75-86.

45. euphrosyne (*Prodelphinus*), Gray, Erebus & Terror, Zoöl., 1846,
   p. 40, pl. 22.   *Id.* Cat. Seals & Whales, 1866, p. 251.   Elliot,
   Syn. N. Am. Mamm., 1901, p. 30.
styx Gray, Erebus & Terror, Zoöl., 1846, p. 39, pl. 2.
tethyos Gerv., Bull. Soc. Agr. Herault, 1853, XL, p. 150, pl. 1.
marginatus (Duvern.), Pucher., Rev. Zoöl., 1854, p. 547.
doreides Gray, Cat. Cet., 1866, p. 400.

euphrosinoides Gray, Synops. Whales & Dolphins, 1868, p. 6.

novæ-zelandiæ Hector (nec Gray), Trans. N. Z. Inst., 1873, v,
       p. 159.

GRAY'S DOLPHIN.

Type locality.   Unknown.

Geogr. Distr.   Atlantic Ocean, Mediterranean Sea, coast of
Jamaica.

Genl. Char.   Body stout, beak long; dorsal fin high, falcate;
pectorals small.

Color.   Above black; sides blackish, beneath white; orbital ring
black; black band from eye to vent and one going downward and
backward above base of the pectorals; this black band is divided
from the dark color above by a white band which is broadest in the
middle; broad black band from eye to base of pectorals, with white
area in its center, that joins the white throat below the eye; fins
black, margined anteriorly with white.

Measurements.   Total length, 2097; end of beak to base of dorsal,
932; breadth of flukes, 420; anterior margin of pectoral fin, 305.

Order V. Ungulata. Hoofed Quadrupods.

The Order Ungulata comprises those animals formed for a terrestrial existence and whose food is mainly vegetable, although certain ones are omnivorous. The molar teeth have broad crowns and their surfaces ridged, and three pair are always present in each jaw. There are no clavicles or collar bones, and the limbs, as a rule, are only capable of a forward and backward motion, a rotary movement being impossible. The limbs generally terminate in solid bony hoofs, but in some cases the last joint of the toes is furnished with broad nails, and the number of toes varies from one to five, but in cases where the toes are numerous only two are usually of importance to the animal. In no instance are any claws present. The Ungulates vary greatly in size, from the diminutive Dik-Dik Antelope, pigmy Hog and Hyrax, to the lordly Elephant, the bulky Rhinoceros and Hippopotamus, and the lofty Giraffe. A characteristic of members of this Order is the presence of horns on some portions of the head, usually witnessed only on the male. These appendages vary greatly in size, structure, and pattern, and are most effective, in the majority of instances, as weapons of offense or defense. Many extinct species of Ungulates possessed four or five digits to each limb, but no existing species, except the elephant, has more than four, the majority indeed possessing only two, while in the horse but one remains. The Order has two divisions, those that may be termed the true Ungulates, containing the vast majority of the species, such as Hippopotamus, Swine, Llamas, Deer, Antelope, Sheep, Goats, Tapirs, Rhinoceros, Horse, etc.; and the Subungulata, with the majestic Elephant and the curious little animal, the Hyrax, very rodent-like in appearance, but entirely distinct from the other members of the Order, and which occupies quite an independent position among mammals.

The Ungulates are distributed throughout the globe excepting Australia, and are found from the Arctic regions to the Tropics, the largest number of species inhabiting the warmer portions of the earth. Some Ungulates go in herds containing many thousand individuals, like the American Bison which only a few years ago roamed the Western plains in countless numbers, but have passed away forever, save a semi-domesticated remnant; and many species of African Antelope that once covered the veldt with their mighty hosts are likewise rapidly disappearing before the rifles of the so-called sportsman and the skin hunter.

The members of the first Family of the Ungulates to be considered, the Peccaries, differ from those of the family *Suidæ* in various respects. They lack one pair of upper incisors, and the anterior premolar in each jaw. The canines, large and formidable, are directed downward, not upward, and the last premolar is complex. The stomach has three compartments, and is more complicated than that of the true pigs. These animals are fearless and pugnacious, associate in droves, sometimes of considerable numbers, and when attacked, all assume the offensive, and are capable of doing much damage with their sharp tushes, and a man in the midst of a number of enraged Peccaries is fortunate if he is able to find a tree to climb, that being about the only method of saving his life. Two species have been long known, but lately a number of others have been described, some of which may prove to possess distinctive values.

Fam. I. **Tagassuidæ. Peccaries.**

Snout elongate, truncate; flat terminal naked surface in which the nostrils are placed. Stomach complex; cæcum present. Four toes on front feet, three on hind feet. Incisors rooted; upper canines pointed downward, with cutting edges; upper outer incisors and anterior premolars of both jaws wanting. Third and fourth metapodials united at their upper ends. Ears small, erect. Body covered with bristly hair. Musk gland on the middle of the back.

27. *Tagassu.

$$I.\frac{2-2}{3-3}; \ C.\frac{1-1}{1-1}; \ P.\frac{3-3}{3-3}; \ M.\frac{3-3}{3-3} = 38.$$

T. Gill. *Note on the names of the genera of Peccaries*, Proc. Biol. Soc. Wash., 1902, p. 38.

J. A. Allen. *The Generic and Specific names of the Peccaries*, Bull. Am. Mus. Nat. Hist., 1902, p. 162.

O. Thomas. *The Generic names of the Peccaries, Northern Fur Seal and Sea Leopard*, Proc. Biol. Soc. Wash., 1902, pp. 153, 197.

Tagassu Frisch, Syst. vierfüss. Thiere, in Tabellen, 3 Tab. Gen. 1775.

Tayassu Fisch., Zoogn., III, 1814, p. 284. Type *Sus tajacu* Linnæus.

---

* Messrs. Gill, Allen, and Thomas (l. c.) have given their views regarding the proper generic names for the Peccaries, each arriving at a different conclusion, no two of them agreeing, and thus exhibiting, in a certain degree, the very unstable foundation on which nomenclature rests. Not wishing to add to the confusion by giving another opinion, even if it were necessary, I have followed Dr. Allen, whose argument is apparently the strongest of the three. Mr. Thomas in a later publication abandons his position and accepts Dr. Allen's view (l. c.).

Dicotyles G. Cuv., Règn. Anim., 1, 1817, p. 237.
Notophorus Fisch., Mem. Soc. Imp. Nat. Moscow, v, 1817, p. 418.
Adenonotus Brookes, Prodr. Syn. Anim., 11, 1828.

Size large; mane not covering the rump; sides of rostrum not flattened, and divided by zygomatic ridge; palate long, narrow, with median ridge, and extending beyond last molar; deep cavity on root of zygoma anterior to orbit, which is small and incomplete behind, the postorbital process not joining the zygoma; pterygoid plates of the alisphenoid form the interpterygoid fossa.

KEY TO THE SPECIES AND SUBSPECIES.

A. Tagassu.

46. nanus (*Tayassu*), Merr., Proc. Biol. Soc. Wash., xiv, 1901, p. 102.

DWARF PECCARY.

Type locality. Island of Cozumel, Yucatan.
Genl. Char. Small, similar to *T. angulatum*.
Color. Above grizzled black and buff; broad buffy stripe on shoulders; rest of body with black dorsal stripe from occiput to tail; ears and feet black.
Measurements. Total length, 840; tail vertebræ, 32; hind foot, 175. Skull: basal length, 176; Hensel, 168; palatal length, 120; occipito-nasal length, 189; zygomatic breadth, 100; upper tooth row, 52.

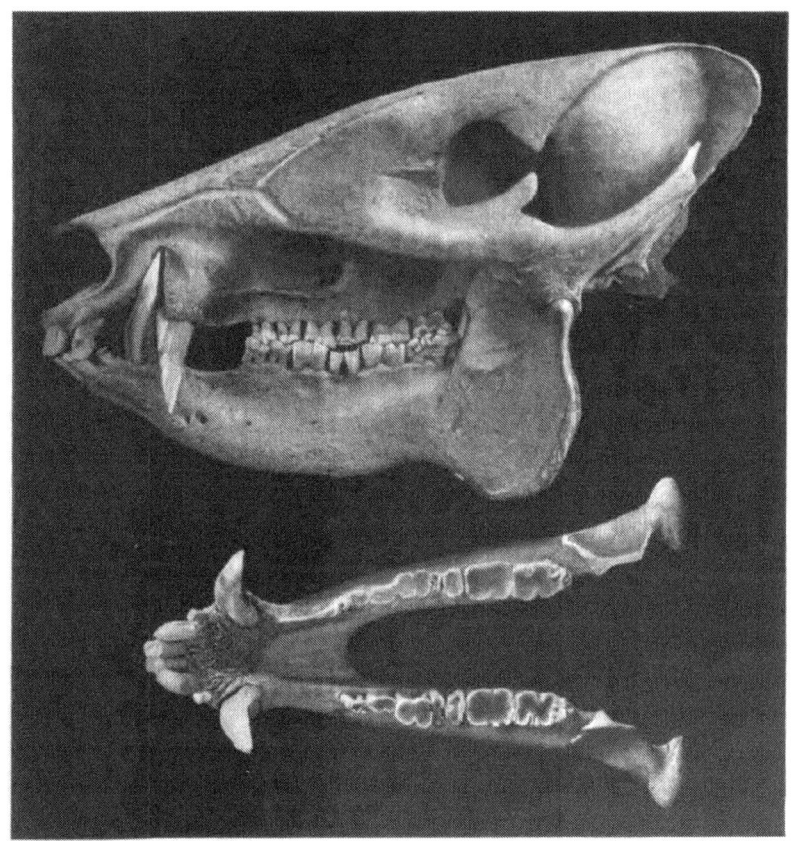

TAGASSU ANGULATUM.
No. 56 Field Columbian Mus. Coll.   ½ nat. size.

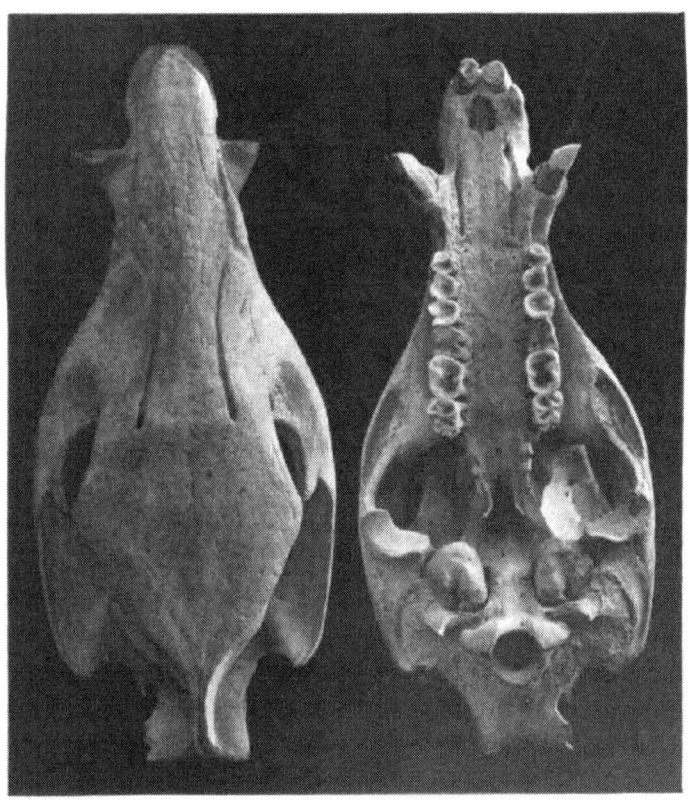

TAGASSU ANGULATUM.
No. 56 Field Columbian Mus. Coll. ½ nat. size.

47. angulatum (*Dicotyles*), Cope, Amer. Nat., xxiii, 1889, p. 147.
tajacu (*Dicotyles*), Elliot, Syn. N. Am. Mamm., 1901, p. 33 (Part).
TEXAN PECCARY. *Moran, Jabali,* in Mexico.
Type locality. Guadalupe River, Texas.
Geogr. Distr. Northeastern Mexico into Texas.
Genl. Char. "Molar crest erect, reaching base of canine alveolus; nasal bones angulate on median line; first upper molar quadri-tubercular, with intermediate tubercles, and quadrate in outline, molariform; molars wrinkled."
Color. Upper parts and sides mixed black and white, black predominating on face, mane, and along dorsal line; throat, under parts, ears and patch behind ears, nose and hoofs, black; white band over shoulders to middle of back.
Measurements. Total length, 960. (Skin from Texas.) Skull: occipito-nasal length, 206; zygomatic width, 104; Hensel, 184; inter-orbital constriction, 60; length of nasals, 80; length of upper tooth row. 62; length of mandible, 143; length of lower tooth row, 68.

a.—humerale (*Tayassu*), Merr., Proc. Biol. Soc. Wash., xiv, 1901,
  p. 122.
ARMERIA PECCARY.
Type locality. Armeria, State of Colima, Mexico.
Geogr. Distr. State of Colima to Tehuantepec, Mexico.
Genl. Char. "Similar to *T. angulatum*, but sides grayer; head yellower; dorsal band more strongly marked."
Color. Black dorsal band from behind ears to tail; general color of upper parts mixed black and white, the white predominating; top of nose and head black and ochraceous buff; sides of nose and face paler, ochraceous buff predominating; sides white and black like back; rump mostly black; stripe over shoulders to center of back straw yellow; mane black and white; chest and middle of belly black; rest of under parts dull ochraceous buff; the bristles being ringed with horn color and buff; limbs brownish or chestnut, black in front; ears blackish; hoofs brownish black.
Measurements. Total length, 960; tail, 60; hind foot, 215. Skull: basal length, 203; occipito-nasal length, 224; zygomatic breadth, 108; across squamosals posteriorly, 99; palatal length, 151; length of upper molar series, 67.

b.—yucatanense (*Tayassu*), Merr., Proc. Biol. Soc. Wash., xiv,
  1901, p. 123.
YUCATAN PECCARY.
Type locality. Tunkas, State of Yucatan.
Geogr. Distr. Arid peninsula of Yucatan.

Genl. Char. Sexes nearly alike in color and size.

Color. Similar to *T. angulatum,* but "sides decidedly whiter; shoulder stripes broader and more conspicuous and somewhat sub-triangular, broadest where they abut against the median dorsal black band, which is well developed; (shoulder stripes broadest and most striking in young); pelage coarser and scantier, the individual bristles decidedly larger and fewer in number; no black on nose or under-lip." (Merr.)

Measurements. Total length, 880; tail, 36; hind foot, 183.

c.—crassum (Tayassu), Merr., Proc. Biol. Soc. Wash., xiv, 1901,p. 124. HEAVY PECCARY.

Type locality. Metlaltoyuca, State of Puebla, Mexico.

Geogr. Distr. States of Puebla and Chiapas, Mexico.

Genl. Char. Similar to *T. angulatum,* but larger; bristles large and rigid; dorsal stripe ill defined; "anterior opening of antorbital foramen between second and third premolars."* (Merr., l. c.)

Color. General hue grizzled gray; black dorsal stripe indistinct or obsolete; hind legs grizzled black and fulvous.

Measurements. Total length, 950; tail, 54; hind foot, 203.

d.—sonoriense (Dicotyles), Mearns, Proc. U. S. Nat. Mus., xx, 1897, p. 469.
SONORA PECCARY.

Type locality. San Bernardino River, State of Sonora, Mexico, near Monument No. 77, Mexican boundary line.

Geogr. Distr. State of Sonora, Mexico, into Arizona.

Genl. Char. Larger than *T. angulatum,* with smaller and simpler molars.

Color. Above mixed grayish and yellowish white and brownish black; indistinct whitish color across neck and in front of shoulder; muzzle, cheeks and space before eyes brownish gray; under jaw yellowish, triangular black patch on chin; ears black; limbs brownish white and black, with a light band above accessory hoofs on fore legs; under parts blackish; grayish on axillar and inguinal regions; snout plumbeous; hoofs plumbeous black; mane of black-tipped bristles from crown to gland on rump. Young pale reddish brown, with black vertical stripes.

Measurements. Total length, 920; tail vertebræ, 65; ear from crown, 115; height at shoulder, 610; hind foot, 200.

---

* The position of the opening of the antorbital foramen is not a dependable character, for in a series of skulls from Texas and Mexico, this opening is found to be *between* the second and third premolar, *over* the third premolar (both specimens from the same locality), and between the third premolar and the first molar, and is thus shown to vary greatly.

48. crusnigrum (*Tayassu*), Bangs, Bull. Mus. Comp. Zoöl., xxxix, 1902, p. 20.

BOQUETE PECCARY.

Type locality. Boquete, Chiriqui, Panama, Central America. Altitude, 4,000 feet.

Genl. Char. Allied to *T. angulatum*, but shoulder stripes apparently darker (young individuals); in adults the same.

Color. "Legs, arms, central dorsal and central ventral stripes, black; rump mostly black, a few of the hairs (bristles) annulated with tawny; conspicuous shoulder stripes, tawny; sides of head and of body mixed tawny and black; all the hairs annulated with these colors; hairs on outer surface of ears mostly black; those on inner surface mostly tawny." (Type juv. ad. Bangs, l. c.)

Measurements. "Old ad. ♀. Total length, 1030; hind foot and hoof, 170; ear, 80. Skull: basal length, 197; occipito-nasal length, 222; zygomatic width, 103; greatest width across squamosals posteriorly, 98; palatal length to palatal notch, 140; breadth of basi-occipital between bullæ posteriorly, 19; length of upper molariform series, 64." (Bangs, l. c.)

B. Olidosus.

Dicotyles Gray (nec Cuv.), Proc. Zoöl. Soc., 1868, p. 45.

Olidosus Merr., Proc. Biol. Soc. Wash., 1901, p. 120. Type *Tayassu pecari* Fischer.

FIG. XXII. TAGASSU PECARI. WHITE-LIPPED PECCARY.
Ex Faun. Bor. Amer. 1, Mamm.

Size large; mane long, bristly; skull large, heavy; nasals flat on top; sides of rostrum flattened, swollen over premolars and not divided by zygomatic ridge; palate broad, flat and without ridge between first molar and inner side of canine; zygomatic ridge disappearing over secpnd premolar. Second lower molar with posterior cusp as long and high as the anterior.

KEY TO THE SPECIES AND SUBSPECIES.

49. pecari (*Tayassu*), Fischer, Zoogn., III, 1814, p. 285.
    albirostris (Sus), Illiger, Abandl, K, Preuss. Akad. Wiss. Berlin,
      1815, p. 115.
    labiatus Cuv., Règn. Anim., I, 1817, p. 238.
WHITE-LIPPED PECCARY.   *Warree, Caribbanco*, in Costa Rica.

Type locality.   Paraguay.

Geogr. Distr.   Guatemala, Central America to Paraguay in South America.

Genl. Char.   Size large; lips and breast white; cranial characters conspicuous.

Color.   Top of head and upper part of face, from above angle of mouth to a line with the ear, dark rufous brown, palest on cheeks; rest of upper parts, sides and limbs to heels, dark reddish brown and black; snout flesh color; upper lips and top of nose, chin, lower parts of cheeks, throat, breast and under parts white; limbs below heels light brown, with a blackish brown patch in front; hoofs black; ears paler than head.

Measurements.   About the same as the following subspecies. Skull: occipito-nasal length, 250; Hensel, 213; zygomatic breadth, 103; interorbital constriction, 60; length of nasals, 84; palatal length, 163; breadth of basi-occipital between bullæ, posteriorly, 20; length of upper tooth row, 72; length of mandible, exclusive of incisors, 184; height of condyle, 79; at coronoid process, 88; length of lower molar series, 79.

a.—ringens (*Tayassu*), Merr., Proc. Biol. Soc. Wash., XIV, 1901,
    p. 121.
SAVAGE PECCARY.

Type locality.   Apazote, near Yohaltun, State of Campeche, Mexico.

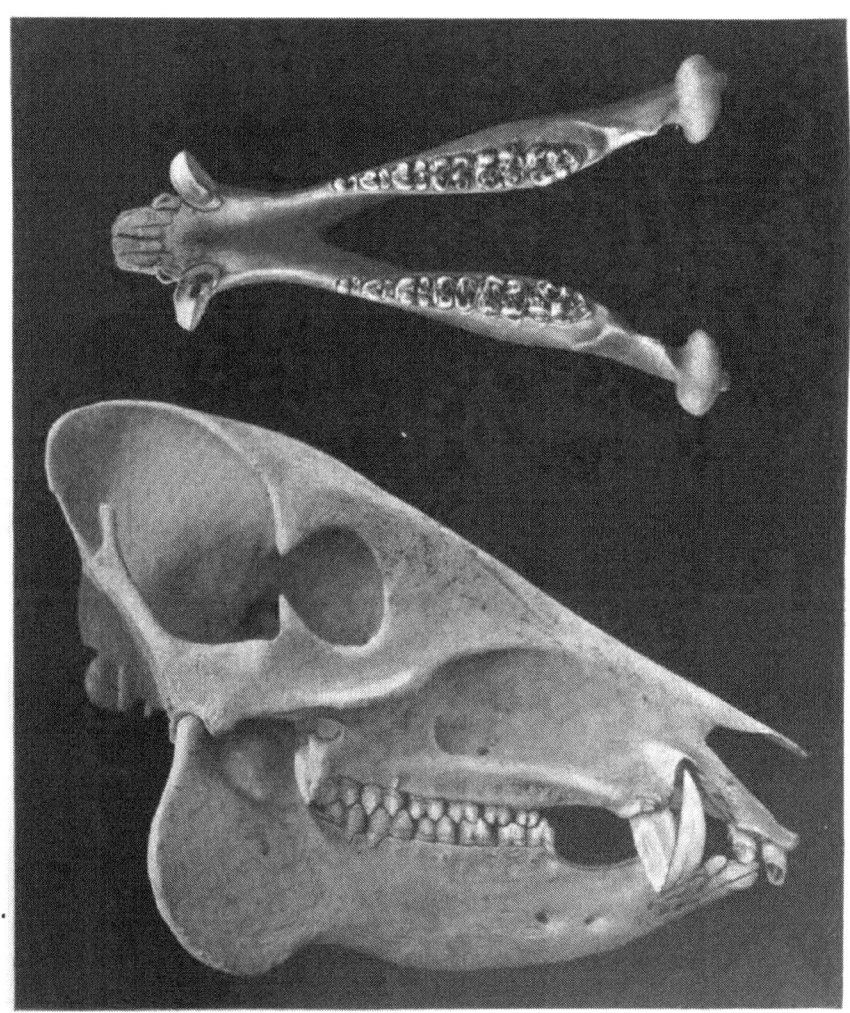

TAGASSU (OLIDOSUS) PECARI.
No. 14872 Am. Mus. Nat. Hist. Coll.   ½ nat. size.

.

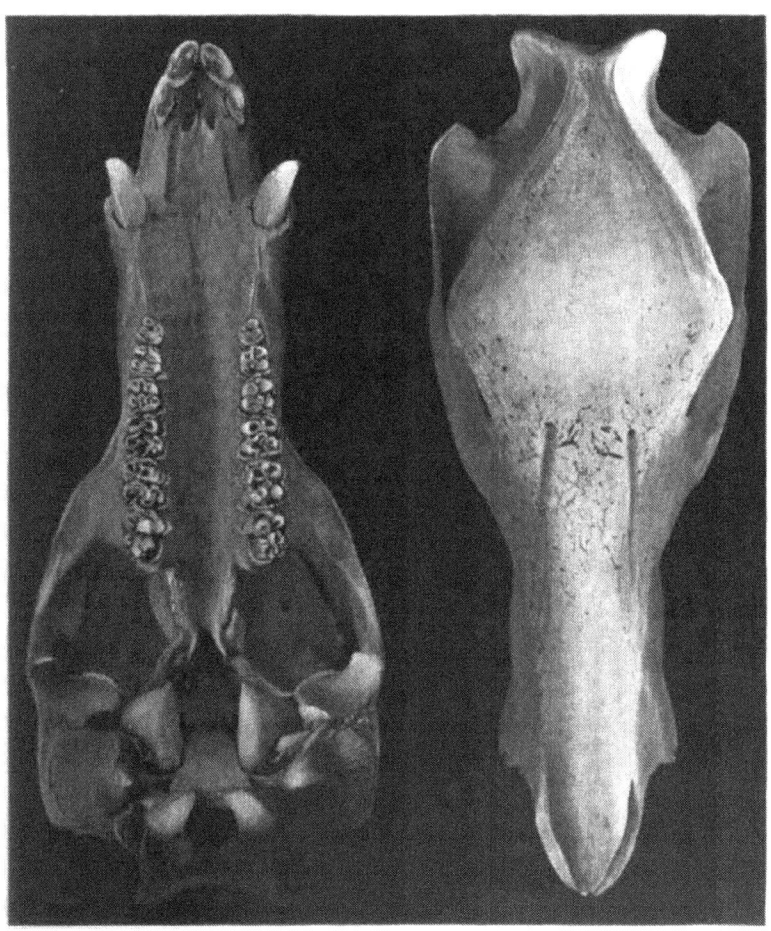

TAGASSU (OLIDOSUS) PECARI.
No. 14872 Am. Mus. Nat. Hist. Coll. ½ nat. size.

. *Genl. Char.* Size large, ears small; similar to *T. pecari*, but white face markings more extensive. Cranial characters distinctive. As compared with *T. pecari* the "parietal shield is narrower, elevated and bulging upward over posterior part of braincase; nasals more acute anteriorly; premaxillæ slightly longer; zygomata and posterior expansion of squamosals decidedly broader; palatal projection behind molars abruptly narrowed at post molar notch and continuing backward, with smooth parallel sides of equal breadth throughout; bullæ smaller and ending below in an elongated papilla pointing toward hamular process; basioccipital narrower between bullæ posteriorly." (Merr., l. c.)

Color. Above black, slightly grizzled with fulvous, most apparent on sides of neck and shoulders; muzzle, chin and lips yellowish white, extending on under jaw to beneath ears; under parts fulvous black; legs and feet blackish mixed with whitish near hoofs.

Measurements. Total length, 1180 (skin); hind foot, 229 (flesh), (Merr.). Skull: basal length, 242; Hensel, 231; occipito-nasal length, 270; zygomatic breadth, 112; breadth across squamosals posteriorly, 106; palatal length, 184; breadth of basi-occipital between bullæ, posteriorly, 20; length of upper molar series, 78.

The Pecora, or true Ruminants, is one of the best defined groups of the Mammalia. Its members are sometimes designated as the Solid-horned ruminants, in contradistinction to the Hollow-horned species of the Bovine group. The antlers, which are usually seen only on the male, are grown in a few months and then dropped, generally about the time the young are born. Some of these defensive structures are of great size, such as those carried by the Moose and Wapiti, and it seems almost incredible that their growth from a mere knob, that formed the base of the antlers of the previous season, to the sometimes immense perfected antlers when the velvet disappears, could have been accomplished in so brief a period. Nearly all lands of any extent, except Australia, possess representatives of this Family, and its members are probably familiar to more of the human race than those of most groups of Mammals. All sizes, from the lordly Moose to the diminutive Musk Deer, are found among them, and the shapes and styles of the horns are many and diverse. All climates are encountered by these animals, from the ice and snow-covered barrens of the Arctic regions to the sun-baked soil of tropical lands, but, wherever found, suitable modifications in structure and covering have been produced to fit them to resist and overcome any climatic influence hostile to their well being.

In the countries south of the Mexican and United States boundary only diminutive members of the CERVIDÆ are found, and but few species even of these. Most of them represent in miniature the White-tailed Deer of the United States, similar in color and style of antlers, though in certain species there is a tendency to a darkening of the coat and to the disappearance of the metatarsal gland and tuft, which, indeed, in some are entirely wanting. The darker color is merely characteristic of animals living in humid climates, but the absence of glands is not so easily explained.

Fam. II. Cervidæ. Deer.

Antlers solid, always present on the male, sometimes also on the female; first molar in upper and lower jaw brachyodont; lachrymal bone prevented from articulating with the nasals by an extensive antorbital vacuity; lachrymal duct with two orifices at or inside rim of orbit; upper canines often present, sometimes greatly elongated in the male; lateral hoofs nearly always present on all the feet; no gall bladder.

Subfam. I. Cervinæ.

28. Odontocœlus.

$$I.\frac{0-0}{4-4}; \ C.\frac{0-0}{0-0}; \ P.\frac{3-3}{3-3}; \ M.\frac{3-3}{3-3} = 32.$$

Odocoileus (*sic*) Rafin., Atlantic Jour., I, 1832, No. 3, p. 109, fig.
Type *Odocoileus! speleus* Rafin.=*Cervus americanus?* Erxleben.
Mazama H. Smith, Griff. Anim. King., v, 1827, p. 314. (nec Rafin.)
Dorcelaphus Gloger, Hand-u. Hilfsb. Naturg., 1841, p. 140.
Cariacus Less., Nouv. Tab. Règn. Anim., 1842, p. 173.
Oplacerus Haldeman, Proc. Acad. Scien. Phil., I, 1842, p. 188.
Reduncina Wagn., Schreb. Säugeth., IV, 1844, p. 373.
Macrotis Wagn., Schreb. Säugeth. Suppl., IV, 1844, p. 373.
Eucervus Gray, Ann. Mag. Nat. Hist., 3d Ser., XVIII, 1866, p. 338.
Otelaphus Fitzin., Sitzungsber, Math. Cl. K. Ak. Wiss. Wien., LXVIII, Abth. 1, 1873, p. 356.
Gymnotis Fitzin., Sitzungsber, Math. Cl. K. Ak. Wiss. Wien., LXXVIII, Abth. 1, 1879, p. 343.
Dama Zimm., Allen. Bull. Amer. Mus. Nat. Hist., 1902, pp. 18-20.

Size large; antlers on male only, large, with sub-basal snag, anterior prong of main fork more developed than the posterior one; metatarsal gland and tuft generally present; tail usually long, thickly haired beneath; face gland small; gland pit moderate; upper canines absent.

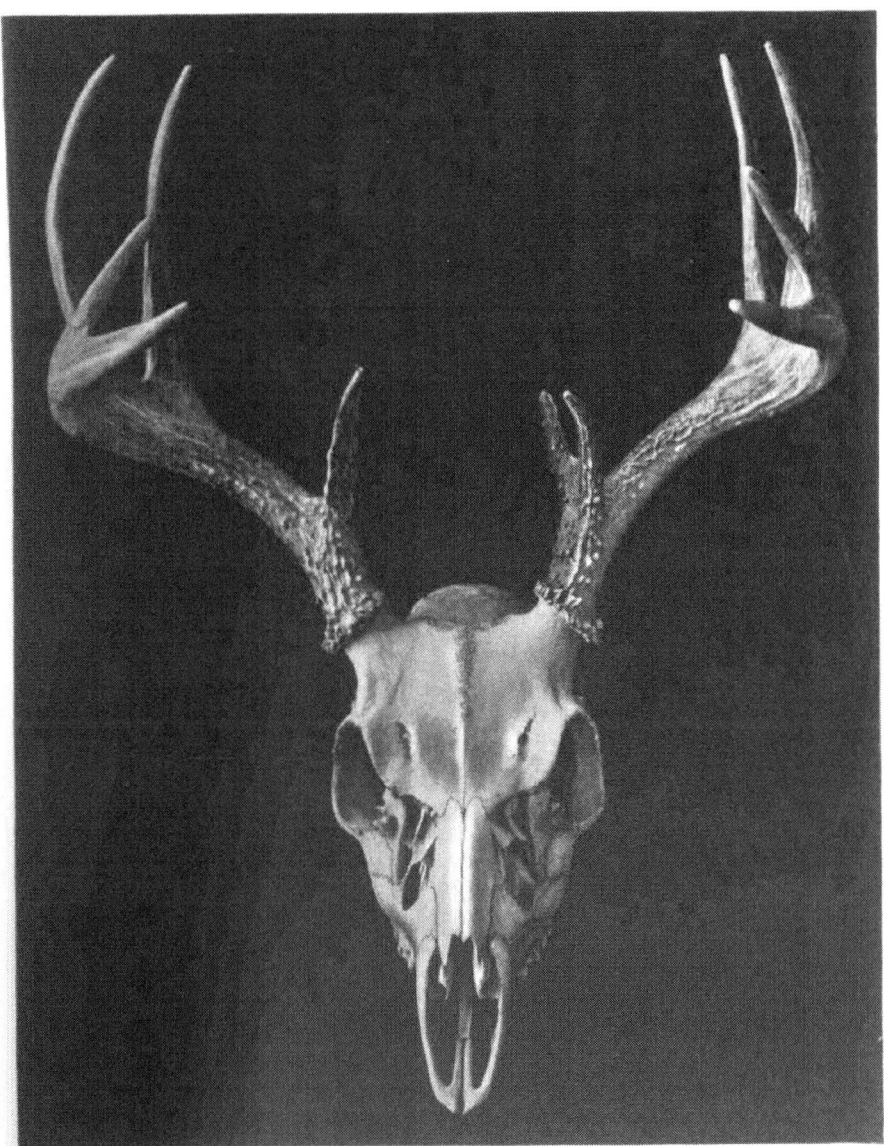

ODONTOCŒLUS A. TEXENSIS.
No. 7612 Field Columbian Mus Coll. ⅔ nat. size.

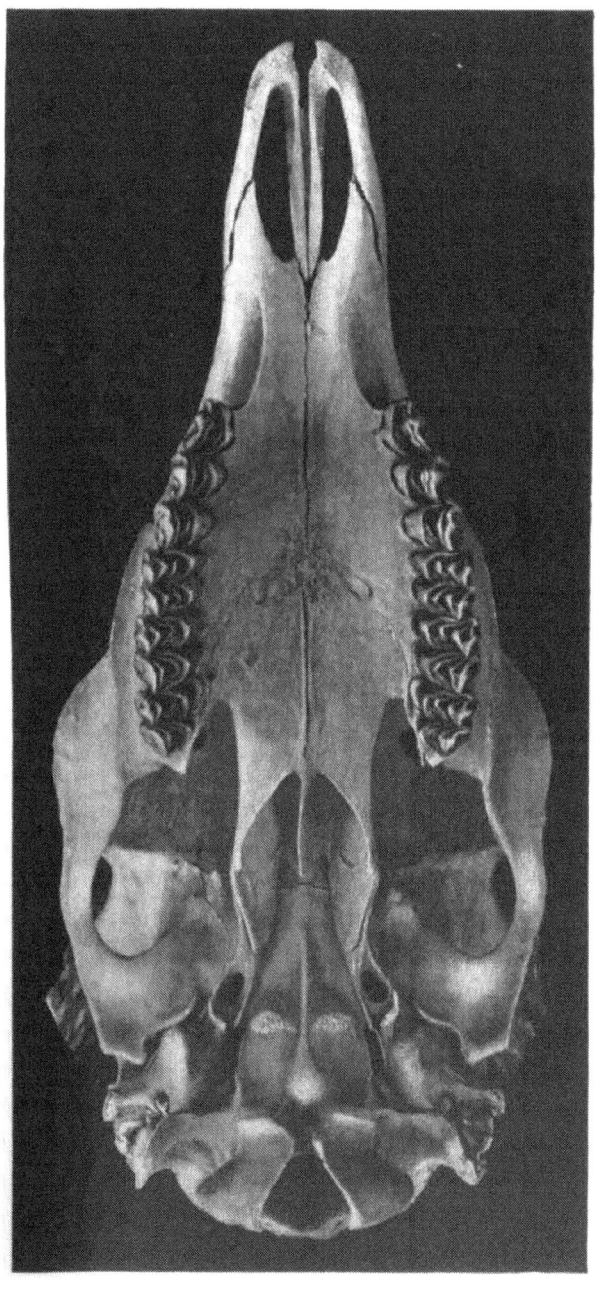

ODONTOCŒLUS A. TEXENSIS.
No. 7612 Field Columbian Mus. Coll. ¼ nat. size.

KEY TO THE SPECIES AND SUBSPECIES.

A. Size small.
  a. Horns similar in shape to those of *O. americanus.*   PAGE
    a.' Upper parts pale reddish; tail above black.*O. a. texensis* 70
    b.' Upper parts dull fawn; tail reddish brown...*O. a. couesi* 70
    c.' Upper parts gray brown; tail grizzled white
       and brown ....*O. battyi* 71
    d.' Upper parts speckled foxy red; tail foxy red
       .... *O. lichtensteini* 72
    e.' Upper parts brown, tipped with fawn;
       tail fawn, tip black ....*O. rothschildi* 72
  b. Horns slightly lyrate, beams straight.
    a.' Tarsal gland present.
      a." Upper parts bright chestnut; tail above
         tawny................................ *O. truii* 73
      b." Upper parts mixed black and buff; tail
         above cinnamon..................*O. costaricensis* 73
      c." Upper parts yellowish brown and gray;
         tail above dusky...................*O. nemoralis* 74
    b.' Tarsal gland absent.
      a." Upper parts dark chestnut brown; tail
         above brown........................ *O. toltecus* 74
  c. Horns sloping back on plane of face, tips
    curving inward and forward, with a spike
    from burr on inner side curving inward
    and forward...............................*O. thomasi* 75
  d. Horns small subcylindrical spikes ....*O. nelsoni* 75
B. Size large.
  a. Horns with a single branch from main tine..*O. cerrosensis* 76
  b. Horns with short sub-basal snag, and beam
    curving upward and forming a dichotomous
    fork and again dividing, the normal points
    being five on a side.
    a.'* Upper parts tawny; tail white with black
      tip .... *O. hemionus* 76
    b.'* Upper parts pale tawny; tail with dark
      median band above, tip black.......*O. h. californicus* 77
    c.'* Upper parts fulvous; tail whitish, tip
      black ....*O. h. eremicus* 77
    d.'* Upper parts drab gray; tail, basal half
      dark, middle part white, tip black.........*O. h. canus* 78

      ———
    * Summer pelage described.

c.† Horns simple spikes.   PAGE
   a.' Upper parts pale drab gray; tail whitish,
      tip black..........................*O. h. peninsulæ*   78
   b.'‡ Upper parts yellowish gray brown; tail
      bright rufous.........................*O. sinaloæ*   78

americanus texensis (*Dorcelaphus*), Mearns, Proc. Biol. Soc. Wash.,
    1898, p. 23.
   a. texensis (*Odocoileus!*), Elliot, Syn. N. Am. Mamm., 1901, p. 40.
TEXAN DEER. *Venado* in Mexico for all Deer.
   Type locality. Fort Clark, Kinney County, Texas.
   Geogr. Distr. Texas and northern Mexico.
   Genl. Char. Size small; tip and edges of ears black; horns small,
incurved; molar teeth large; color pale.
   Color.—Winter Pelage. Top of head black, sides light ash gray;
upper parts yellowish white and gray mixed; black line from crown
to root of tail; sides pale yellowish ash; chest fuliginous; rest of
under parts white; chin white, with black transverse cross-bar; jaws
light ash; throat white; legs reddish fawn mixed with gray and
black; tail black above, white below.
   Summer Pelage. Pale reddish.
   Measurements. Total length, 1585; tail vertebræ, 265; ear from
crown, 160; height at shoulder, 880.
   Antlers. Length of beam on outside curve, 440; widest expanse,
330; circumference of beam at base, 80.

americanus couesi (*Cervus*), Coues and Yarrow, in Wheel. Geog.
    and Geol. Surv. West 100th Merid., 1875, p. 72.
   americanus couesi (*Odocoileus!*), Elliot, Syn. N. Am. Mamm., 1901,
    p. 72.
   mexicanus Baird, N. Am. Mamm., 1875, p. 653. (Part.)
   virginianus var. Coues and Yarrow, in Wheel., Rep. Geog. and
    Geol. Surv. West 100th Merid., v, 1875, pp. 72–75.
   var. couesi Rothrock, Rep. Geog. and Geol. Surv. West 100th
    Merid., v, 1875, p. 72.
COUES' DEER.
   Type locality. Camp Crittenden, Pima County, Arizona.
   Geogr. Distr. Arizona and State of Sonora to City of Mexico.
   Genl. Char. Smaller than *O. americanus*, horns similar.
   Color.—Summer Pelage. Above pale dull fawn color, tinged with
ochraceous; dorsal area mouse gray; sides tawny or reddish brown;

---

† Animal not fully adult.
‡ Animal not fully adult.

FIG. XXIII. ODONTOCŒLUS A COUESI. COUES' DEER.

throat, under parts and inner side of limbs white; tail above reddish brown, fringed with white, beneath pure white.

Measurements. Height at withers, 812; at rump, 902; ears, 197. Skull: occipital condyles to apex of intermaxillæ, 210; width across orbits, 96; zygomatic width, 90; interparoccipital width, 45; length of nasals, 68; occipital condyles to anterior edge of intermaxillæ, 210.

50. battyi (*Odocoileus!*), Allen, Bull. Am. Mus. Nat. Hist., 1903, p. 591.

BATTY'S DEER.

Type locality. Rancho Santuario, State of Durango, Mexico.

Genl. Char. Similar to *O. a. couesi,* but skull with smaller ant-orbital vacuities, broader and less arched nasals; basisphenoid more cuneate, dentition heavier and antlers bent more sharply outward.

Color.—Winter Pelage. Upper parts gray brown, sometimes tinged with buff, top of head and dorsal line to tail darkest; flanks lighter; middle of throat white; sides of throat, cheeks, foreneck and chest pale grayish brown, sometimes with a buffy tinge; lower breast, axillæ and inside of fore legs, lower part of abdomen, inguinal region, inner side of thighs and hind leg, white; tarsal gland white around a deep orange center; a narrow white band above hoofs;

ears gray brown; black patch on side of nose and on lower lip; tail
at base above with the hairs dark brown, tipped with white; sometimes deep ochraceous with the base yellowish brown; edges and lower
surface white or mixed yellow and white; hind part of rump white;
ear externally gray brown, white internally.

Measurements. Male, total length, 1574; head and body, 1371;
tail vertebræ, 216; ear from crown, 190; from notch, 160. Skull:
total length, 248; occipito-nasal length, 201; Hensel, 220; zygomatic
breadth, 115; interorbital constriction, 61.5; mastoid breadth, 85.5;
greatest length of nasals, 77; greatest width of nasals, 31; length of
upper premolar series, 69.5; length of mandible, 192; height, at
condyle, 64; at coronoid, 96; length of lower premolar series, alveolar
border, 72. Antlers: length of main beam along external curvature,
353; from burr to top of fork of first point in straight line, 71; to
second, 179; from second to third, 103; between tips, 218; between
burrs, 56; greatest expanse, inside measurement, 340.

51. lichtensteini (*Cervus*), Allen, Bull. Amer. Mus. Nat. Hist., 1902,
 p. 20 (footnote).
 *mexicana Licht., Darstell. Thiere, 1827, p. 34, pl. XVIII. (nec
 Gmel. et Auct.)

MEXICAN DEER.

Type locality. Unknown.

Geogr. Distr. Southern Mexico, range unknown.

Genl. Char. Size small, resembling *O. americanus* in style of
horns and general appearance; tail short; metatarsal gland situated
in thick tuft of hair.

Color.—Winter Pelage. Grayish or ashy brown, under parts paler;
chin, throat, and inguinal regions white; chest reddish brown; tail
above at base like the back, tip and under part white.

Summer Pelage. Upper parts speckled foxy red; head and ears
dark grizzled gray, tawny behind and below ears; chin, lower jaw,
throat and under parts pure white; tail bright foxy red above,
beneath white.

Measurements. Height at shoulder, 2 feet 9 inches; antlers 11½
inches long to 13½ along the curve.

52. rothschildi (*Dama*), Thomas, Novitat. Zoöl., IX, 1902, p. 136.

ROTHSCHILD'S DEER.

Type locality. Coiba Island, off west coast of Panama.

Genl. Char. Size very small; horns short, two or three tined.
Skull small, delicate; antorbital vacuities large; no metatarsal gland.

---

*Cervus mexicanus Gmel. is indeterminable; see Allen, Bull. Am. Mus.
Nat. Hist., 1902, p. 20, footnote.

Color. Upper parts brown, hair tipped with fawn; dorsal line darker; whitish mark on each side of muzzle, and one above and below each eye; chin, throat, inner sides of the upper part of fore legs, inguinal region, and inner side of thighs white; rest of under parts rufous fawn, as are also the outer side of thighs and feet; tarsal glands reddish brown; tail above at base fawn, terminal portion black, beneath white.

Measurements. Head and body, 1120; tail, 100; hind foot, with hoofs, 290; ear from notch, 88 (skin). Skull: greatest length, 201; basal length, 180; greatest breadth, 86.5; nasals, 64×22.5; muzzle to orbit, 103; breadth of braincase, 58; muzzle to anterior premolar, 61; alveolar length of upper tooth row, 65; of lower tooth row, 26.

53. *truii (*Odocoileus!*), Merr., Proc. Biol. Soc. Wash., XII, 1898, p. 103 (note).

clavatus True, Proc. U. S. Nat. Mus., 1888, p. 417. (nec H. Smith).
TRUE'S DEER.

Type locality. Segovia River, Honduras.

Geogr. Distr. Honduras, Central America. Extent of range not known.

Genl. Char. Size medium; metatarsal gland present. Antlers simple spikes, directed backward in line of face.

Color.—Summer Pelage. Bright chestnut; dusky brown band from nose to forehead, which is darker than face; orbital ring whitish; white spot on each side of nostril and one on lower lip; head beneath and throat white; back bright chestnut; chest and flanks pale chestnut; neck pale grayish chestnut; abdomen, inguinal region and inside of fore legs and thighs white; tail above tawny, beneath white; hairs of tarsal gland white.

Antlers. Slightly lyrate in form; beams straight, slender, laterally compressed and pointed; basal two-thirds rugose.

Measurements. (Skin.) "Height at shoulder, 732, length of head, 246; tail to end of hairs, 239; length of antler, 88; juv." (True, l. c.) Skull: basal length, 200–220; length of upper tooth row, 66; lower tooth row, 73; length of three lower premolars, 28.

54. †costaricensis (*Odocoileus!*), Miller, Proc. Biol. Soc. Wash., XIV, 1901, p. 35.

---

* The animal with red pelage as described by Dr. True, will stand for the type *clavatus=truii.* The example mentioned by Dr. True, as in "Winter Pelage," proves on examination and on the statement to me of Mr. Townsend, who obtained it, to have been killed in July. It is a larger animal than *truii,* and very differently colored, and cannot be supposed to exhibit a pelage other than that of summer. It possibly represents a separate race.

† This may possibly be the same as *O. nemoralis,* the following species, examples of which from Costa Rica are stated to be in the British Museum. See Lydekker, Deer of all Lands, 1898, p. 265.

Costa Rica Deer.

Type locality. Talamanca, eastern side of Costa Rica, and the foot of the Cordilleras.

Geogr. Distr. Known only from Costa Rica.

Genl. Char. Larger than *O. truii*, color lighter.

Color. Above mixed black and buff, the hairs being black with a buff subterminal band; darkest on top of head, neck and fore part of back; sides lighter; throat whitish; under parts wood brown; inguinal region and line along belly to chest white; tail above cinnamon, tip dusky; beneath white; ears dark brownish gray outside, white inside; legs cinnamon.

Measurements. Total length, 1400; tail, 120; hind foot, 375; ear from crown, 110. Skull: greatest length, 250–264; basal length, 237–250; median palatal length, 155–165; width of palate between anterior molars, 38–46; interorbital constriction, 57–64; greatest width between lower rims of orbit, 101–112; zygomatic breadth, 94.6–108; mastoid breadth, 74–86; occipital depth, 57–58; length of mandible, 190–195; upper tooth row, 68; lower tooth row, 79–82; length of three lower premolars, 31–33.

55. nemoralis (*Cervus*), H. Smith, Griff. Anim. King., iv, 1827, p. 137.

Hamilton Smith's Deer.

Type locality. "Virginia"?

Geogr. Distr. Honduras to Panama, Central America.

Genl. Char. Similar to *O. truii*. Metatarsal gland very small. Antlers small; beams straight, with a small tine in front above the burr pointing upward; another tine at tip turned inward and forward, with a short posterior tine almost making a forked termination to the beam.

Color. Forehead and crown blackish; upper lip and patch on lower lip, black; sides of nostrils, lower lip, and chin, white; space around eye fawn; rest of upper parts and sides yellowish brown gray; under parts of buttocks white; limbs "ochery"; tail above dusky, beneath white.

Measurements. Height at withers, 98.5; antler, 90.5 in length; spread, 219.4.

56. toltecus (*Cervus*), Sauss., Rev. Mag. Zoöl., 2me Sér., xii, 1860, p. 247.

yucatanensis Hays, Ann. Lyc. Nat. Hist. N. Y., x, 1874, p. 218.

acapulcensis (*Cervus*), Caton, Antel. & Deer, Amer., 1877, p. 113.

Yucatan Deer.

Type locality. Near Orizaba, State of Vera Cruz, Mexico.

Geogr. Distr. State of Vera Cruz to Yucatan, southeastern Mexico.

Genl. Char. Size very small; tail long; antlers short, straight, semi-palmate; metatarsal gland wanting. Color same at all seasons.

Color. Upper parts dark chestnut brown; under parts white; face blackish; tail brown above, white beneath. Color of pelage does not change with the seasons.

Measurements. One-third smaller than *O. americanus.*

57. thomasi (*Odocoileus!*), Merr., Proc. Biol. Soc. Wash., XII, 1898, p. 102.

THOMAS' DEER.

Type locality. Huehuetan, State of Chiapas, Mexico.

Geogr. Distr. States of Oaxaca and Chiapas, southeastern Mexico. Limits of range unknown.

Genl. Char. Size large; metatarsal gland very small, midway between calcaneum and hoof. Skull and teeth similar to those of *O. truii* from Honduras.

Color. Winter Pelage. General color fulvous; forehead black, or black and fulvous; inside of thighs, middle of belly, hind part of fore legs, and inguinal region white; chin white, with the usual black patch on each side; tail above bright fulvous, beneath white.

Antlers slope backwards nearly on plane of face, with tips curving inward and forward. On inner side near burr of both beams is a spike 110 mm. in length, which curves backward and forward. Spread between tips of these spikes 50 mm. On the left beam 70 mm. from tip is a posterior prong 50 mm. long which projects backward and inward.

Measurements. Type. Total length, 1544; tail vertebræ, 153; hind foot, 425 (ex Merr., l. c.). Skull: basal length, 220–230; length of upper tooth row, 70; lower tooth row, 73; length of three lower premolars, 32.

58. nelsoni (*Odocoileus!*), Merr., Proc. Biol. Soc. Wash., XII, 1898, p. 103.

NELSON'S DEER.

Type locality. San Christobal, highlands of State of Chiapas, Mexico.

Geogr. Distr. State of Chiapas, southern Mexico.

Genl. Char. Size medium; top of head and dorsal band blackish.

Color. Above brownish gray, grizzled tips of hairs fulvous; black stripe from nose to forehead, which is also blackish; dorsal band from head to rump black; ears grizzled gray; chin white, crossed by black bar; inner and back side of fore legs, inner side of thighs, and inguinal regions white; sides of belly and legs pale fulvous; tail above fulvous, beneath white.

Measurements. Type. "Total length, 1250; tail vertebræ, 170; hind foot, 360" (Merr., l. c.). Animal probably not fully grown.

59. cerrosensis (*Odocoileus!*), Merr., Proc. Biol. Soc. Wash., xii, 1898, p. 101.

CERROS ISLAND DEER.

Type locality. Cerros or Cedros Island, off coast of Lower California, Mexico.

Geogr. Distr. Type locality only.

Genl. Char. Similar to *O. h. californicus*, but smaller.

Color. Above dark grizzled gray; blackish dorsal band from occiput to and over upper surface of tail; dusky spot on top of nose and one each side of nostrils; forehead dark; ears grizzled gray exteriorly, interiorly white; throat and neck dusky gray; middle of breast and fore legs blackish; sides of breast and belly like upper parts; abdomen, and inguinal region whitish; thighs, inner side of hind legs, and back of fore legs buffy; tail, dark band above, basal two-thirds whitish, remainder blackish.

Antlers. Small, bowed outward, tips incurved. A single branch projects backward and upward from upper third of main tine.

Measurements. Type. Total length, 1560; tail vertebræ, 180; hind foot, 380; ear from crown, 180 (ex Merr., l. c.).

60. hemionus (*Cervus*), Rafin., Amer. Month. Mag., i, 1817, p. 436.

macrotis Say, Long's Exped. Rocky Mts., ii, 1823, p. 88.

auritus Ward, Desc. États Unis, v, 1820, p. 540.

hemionus (*Odocoileus*), Elliot, Syn. N. Am. Mamm., 1901, p. 42.

MULE DEER.

Type locality. "Sioux River," probably on eastern border of South Dakota.

Geogr. Distr. Lower California north through Nevada to latitude of San Francisco, and west of the Missouri River from Fort George, south to Texas; including North and South Dakotas, Nebraska, Kansas, Colorado, and Wyoming, Montana, Idaho, Nevada, California, Oregon, and Washington.

Genl. Char. Size large, body heavy; ears very large, thickly haired; tail moderate, round, white, tipped with a black tuft, naked beneath basally. Metatarsal gland occupying upper half of outer side of canon bone; tarsal gland present.

Antlers with short sub-basal snag, the beam from this projecting outward and then upward, forking dichotomously, both prongs nearly equal, and then again dividing.

Color. Summer Pelage. Pale yellow, dull yellowish, or yellowish tawny; this is replaced in the early autumn by a bluish gray coat, growing lighter in color as the hairs lengthen during the winter. A dark brown patch on forehead between the eyes and extending below them on the face; remainder of face and throat white, as are also the abdomen, inner side of legs and buttocks; rest of under parts blackish brown; tail white, tip black; ear bordered with black anteriorly.

Measurements. Total length, male, 1983; tail vertebræ, 203.

*Antlers. Length along curve, 393–698; tip to tip, 189–369; widest inside, 369–483.

a.—californicus (Cariacus), Caton, Amer. Nat., 1876, p. 464.

h. californicus (Odocoileus!), Elliot, Syn. N. Am. Mamm., 1901, p. 43.
CALIFORNIA MULE DEER.

Type locality. St. Julian Ranch, Summit of Gaviota Pass, Coast Range, forty miles from Santa Barbara, California.

Geogr. Distr. In the Coast Range south of San Francisco and into Lower California, Mexico.

Genl. Char. Ears smaller than those of the type species; tail differs from that of *O. hemionus* by having a dark median stripe above; metatarsal gland very large.

Color. Similar to *O. hemionus,* the chief difference being the dark band on top of the tail embracing one-third of the circumference; under side of tail naked, tip black.

Measurements. Rather smaller in size than *O. hemionus.*

b.—eremicus (Dorcelaphus), Mearns, Proc. U. S. Nat. Mus., xx, 1897, p. 470.
DESERT MULE DEER.

Type locality. Sierra Seri, near the Gulf of California, State of Sonora, Mexico.

Geogr. Distr. Lower California and State of Sonora, Mexico. Limits of range unknown.

Genl. Char. Size large, color pale.

Color. Above pale drab gray; dark vertebral line from neck to tail, extending a short distance on latter; inguinal region, abdomen, and middle of tail all around white; chest sooty drab; hind part of legs pale cinnamon; upper side of tail at base dusky, middle portion all white; end black like true *O. hemionus.*

Antlers are stout, with beam of considerable length before forking; expanse between tips, 620–775.

Measurements. The type of this form was merely a flat skin

---

* Ward's horn measurements.

which is now used as a rug in a private house in Washington, D. C., and no measurements are available.

c.—canus (*Odocoileus!*), Merr., Proc. Wash. Acad. Nat. Scien., III, 1901, p. 560.
CHIHUAHUA MULE DEER.

Type locality. Sierra en Media, State of Chihuahua, Mexico.

Genl. Char. Similar to *O. hemionus*, but paler (winter pelage). Antlers similar but lighter and more slender.

Color. Above pale gray; top of head and face pale brown; chin white; breast black; beneath white; tail above dark on basal half, sometimes for the whole length, tip black.

Measurements. Total length, 1830; tail vertebræ, 230; hind foot, 500; height at withers, 955.

d.—peninsulæ (*Mazama*), Lydekk., Proc. Zoöl. Soc., 1897, p. 899–900.
LOWER CALIFORNIA DEER.

Type locality. La Paz, Lower California, Mexico.

Geogr. Distr. Cape region of Lower California.

Genl. Char. Size small, black dorsal line.

Color. Winter Pelage. Above iron gray; dorsal black band extending over the tail; flanks and legs chestnut; under parts blackish brown; tail white with black base and tip.

Antlers. Simple spikes, and basal snag.

Measurements. Smaller than *O. h. californicus*.

61. sinaloæ (*Odocoileus!*), Allen, Bull. Am. Mus. Nat. Hist., 1903, p. 613.
SINALOA WHITE-TAILED DEER.

Type locality. Escuinapa, State of Sinaloa, Mexico.

Color. Above yellowish gray brown; black band above nose; sides of nose, space behind nose band, and orbital ring gray; chin and throat buffy grayish white; axillary and inguinal regions, posterior surface of upper fore legs and inner side of thighs white; limbs buffy brown anteriorly; yellowish white on sides and hind parts below carpal and tarsal joints; tail above bright rufous, below white; ears whitish inside.

Measurements. Total length, 1435; tail vertebræ, 175; hind foot, 340; ear from notch, 117; from anterior base, 145. Skull: total length, 215; Hensel, 200; occipito-nasal length, 182; length of nasals, 56; zygomatic breadth, 91; width of frontals at anterior border of orbit, 54.5; width of constriction at base of horns, 69; mastoid breadth, 65; alveolar length of upper premolar-molar series, 70; young adult; antlers slender spikes, 45 and 88 mm. in length in two individuals.

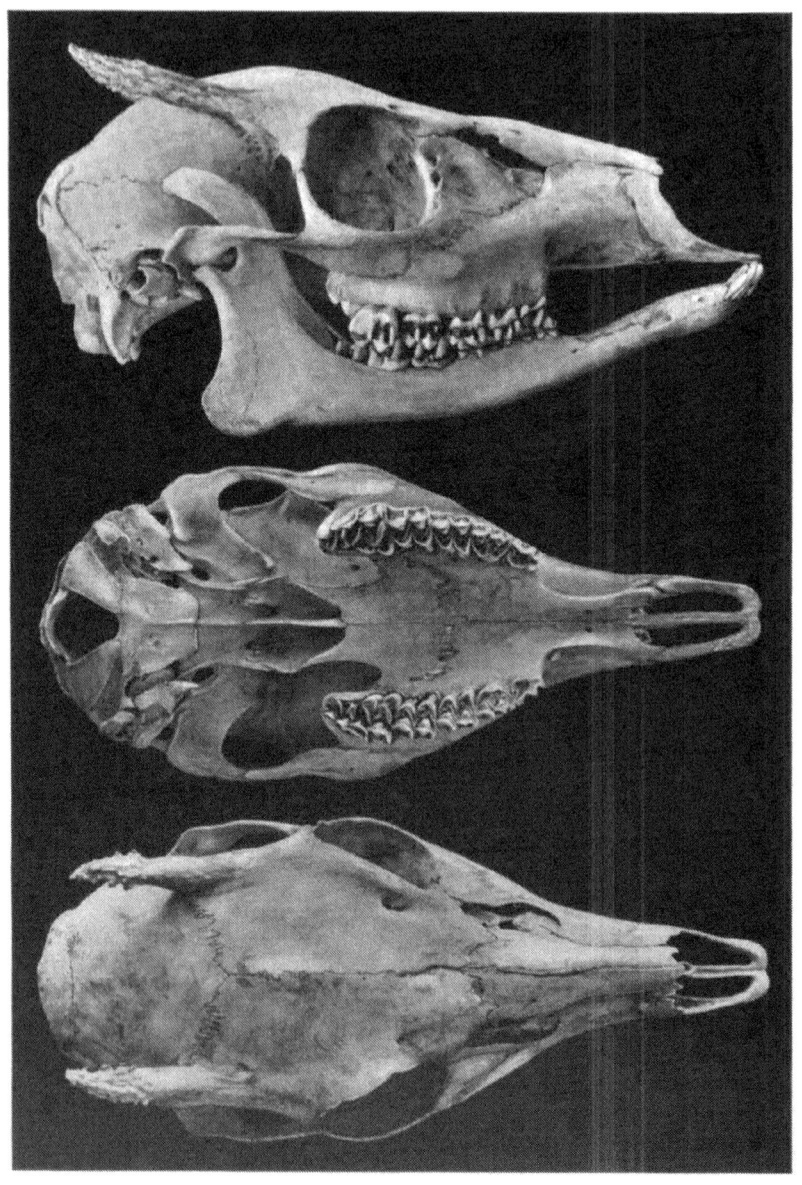

MAZAMA SARTORI.
No. 7631 Field Columbian Mus. Coll. ⅗ nat. size.

The Brockets are natives of Mexico, Central America, and South America. They are small in stature, and possess antlers in the form of spikes, without any branches. The metatarsal gland is wanting, and in certain instances (South American species) the tarsal gland and tuft also. They are peculiar little creatures, with the top of the head tufted, similarly to the Muntjac's, or to those of the diminutive Antelopes of the genus Madoqua, the Dik-Diks of Africa, with a rather heavy, ungraceful body and an arched back. The fawns are spotted with white, like those of the large species of deer, and canines are sometimes present in the males. Although fossil remains have been found in Brazil and Argentina, the Brockets are considered to be a modern, as well as a degenerate group of New World deer.

29. Mazama. Brockets.

$$\text{I.}\frac{0-0}{4-4}; \ \text{C.}\frac{0-0}{0-0} \text{ or } \frac{1-1}{0-0}; \ \text{P.}\frac{3-3}{3-3}; \ \text{M.}\frac{3-3}{3-3} = 32 \text{ or } 34.$$

Mazama Rafin., Amer. Month. Mag., 1, 1817, p. 363. Type *Cervus rufinus* Illiger.

Horns simple, unbranched, directed backward; ears and tail short, the former broad, rounded; upper canines occasionally present in male; metatarsal and sometimes the tarsal gland absent; hair on forehead forming a tuft; face gland small, exposed; gland pit deep, triangular; size small; fawns white spotted.

FIG. XXIV. MAZAMA SARTORI. CENTRAL AMERICAN BROCKET.

KEY TO THE SPECIES.

A. Antlers straight spikes, size small.   PAGE
  a. Upper parts brownish red; tail above brownish
     red..*M. sartori* 80
  b. Upper parts drab brown; tail dull fulvous.....*M. pandora* 80

62. sartori (*Cervus*), Sauss., Rev. Mag. Zoöl., 2me Sér., XII, 1860,
  p. 252.
SARTORI'S BROCKET. *Cabra del Monte* in Costa Rica.
  Type locality. Mirador, State of Vera Cruz, Mexico.
  Geogr. Distr. Southern Mexico and Central America.
  Genl. Char. Very similar to *Mazama tema* Rafin., from Ecuador,
S. A., but of smaller size, the height at withers being only 20½
inches, to 25½ inches in the other.
  Color. General color bright brownish red; throat, neck, and chest
fawn; abdomen white; lower part of face, outer side of hind legs and
front of fore legs shaded with bluish black; tail above like back,
below white; antlers whitish horn color.
  Measurements. Height at withers, 512.5. Skull, male: occipito-
nasal length, 144; basal length, 162; zygomatic width, 70; least interorbital width, 36; mastoid width, 32; length of nasals, 50; palatal
arch to incisive foramina, 66; length of upper tooth row, 42.5; width
of palate between last molars, 29; length of mandible, angle to
alveoli of incisors, 126.5; height at condyle, 40; at coronoid process,
62; length of lower tooth row, 51.

63. pandora (*Mazama*), Merr., Proc. Biol. Soc. Wash., XIV, 1901, p. 105.
TUNKAS BROCKET.
  Type locality. Tunkas, Yucatan, Mexico.
  Geogr. Distr. State of Campeche, and Yucatan, Mexico.
  Genl. Char. Antlers straight spikes, furrowed longitudinally.
Skull similar to that of *M. sartori*, but larger; foramen ovale broad and
opening downward; notch on each side of basioccipital in front of
occipital condyles.
  Color. Neck grayish; chin, lower lip, and front of upper lip,
inguinal region, inner sides of fore legs and thighs, and under side of
tail whitish; tail above, and anal region dull fulvous; forehead with
rusty red patches; fore legs, fore and hind feet dull fulvous; rest of
animal drab brown.
  Measurements. Total length, 1125; tail vertebræ, 140; hind foot,
273; height at shoulder, 572. Skull: basal length, 163; occipito-
nasal length, 157; zygomatic breadth, 82; interorbital constriction,
44; length of nasals, 59; length of upper molar series, 50; length of
antler, 113.

The Prong-horn Antelope, while allied to the Bovidæ, resembles the members of the Cervidæ by possessing horns with branches, and which are shed every year. The hair is very peculiar, being coarse and brittle, and breaks on the slightest pressure. The Prong-horn is an animal of the plains, and depends for its safety upon its keen eyesight and exceeding fleetness. Few animals can keep up with him as he bounds over the prairies, and his wary nature makes a near approach difficult of accomplishment. Yet his one great weakness, curiosity, often nullifies these advantages, and any strange object on his domains proves an irresistible attraction, and his desire to investigate it often costs him his life. Once numerous on our Western Plains, the Prong-buck has already vanished from many localities, and is now met with only in greatly reduced numbers in the comparatively few places it still frequents.

Fam. III. **Antilocapridæ. Prong-horn Antelope.**

Horns branched, deciduous; allied to the Bovidæ.

30. Antilocapra.

$$I.\frac{0-0}{4-4}; \ C.\frac{0-0}{0-0}; \ P.\frac{3-3}{3-3}; \ M.\frac{3-3}{3-3} = 32.$$

Antilocapra Ord, Jour. de Phys., LXXXVII, 1818, p. 149. Type *Antilope americana* Ord.

Dicranocerus H. Smith, Griff., Anim. King., 1827, p. 312.

FIG. XXV. ANTILOCAPRA A. MEXICANA. MEXICAN PRONG-HORN.

Horns compressed at base; flattened process in front, end conical, recurved; deciduous; lateral hoofs absent; hair stiff, coarse, brittle; nose hairy, save a narrow line in the center; tail very short; horns in the female rudimentary, or absent.

KEY TO THE SPECIES AND SUBSPECIES.

A. Horns flattened, recurved.   PAGE
    a. Color yellowish brown....................*A. americana* 82
    b. Color paler...........................*A. a. mexicana* 82

64. *americana (*Antilope*), Ord, Guth. Geog., 2d Am. ed., II, 1815, p. 292, descrip. p. 308.

americana (*Antilocapra*), Elliot, Syn. N. Am. Mamm., 1901, p. 43.
PRONG-HORN ANTELOPE.

Type locality. Plains east of the Missouri? Black Mountains?

Geogr. Distr. Valley of Saskatchewan, latitude 53°, south to Mexico, and from Missouri River on plains westward to Rocky Mountains and the Cascade Range in Oregon and Washington.

Genl. Char. Size of domestic sheep with much longer legs and neck; eyes large, gazelle like; no lachrymal gland; low mane on back of neck.

Color. Male. Upper parts and sides yellowish brown; band between eyes covering forehead, nose, and a spot below ear, liver brown; sides of head, spot behind ear, throat, front of neck extending in two triangles reaching the brown on each side; entire under parts and rump white; legs yellowish brown; horns, hoofs, and naked skin on nose black.

Measurements. Total length, 1245; tail, 178; height at withers, 780. Skull: occipito-nasal length, 240; breadth between outer edge of orbits, 136; width between orbits, 134; length of nasals, 96.5; palatal arch to incisive foramina, 136; length of upper tooth row, 68; width of palate between last molars, 56; length of mandible, 216; length of lower tooth row, 70.

a.—mexicana (*Antilocapra*), Merr., Proc. Biol. Soc. Wash., XIV, 1901, p. 31.
MEXICAN PRONG-HORN. *Berendo* in Mexico.

Type locality. Sierra en Media, State of Chihuahua, Mexico.

Geogr. Distr. Northern Mexico in States of Sonora, Chihuahua, and Tamaulipas. Lower California.

---

* This species may possibly cross the United States and Mexican boundary at some point in its range and go into Mexico, and is, therefore, included in this volume.

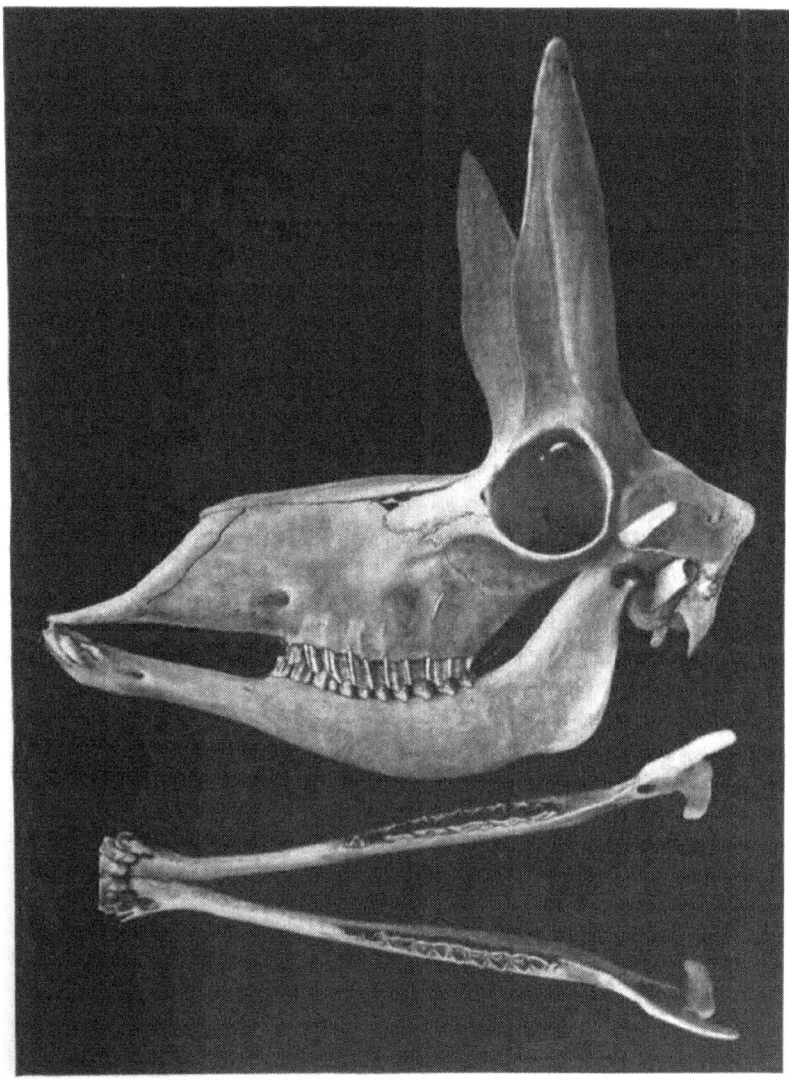

ANTILOCAPRA A. MEXICANA.
Old male from State of Chihuahua, Mexico. ⅔ nat. size.

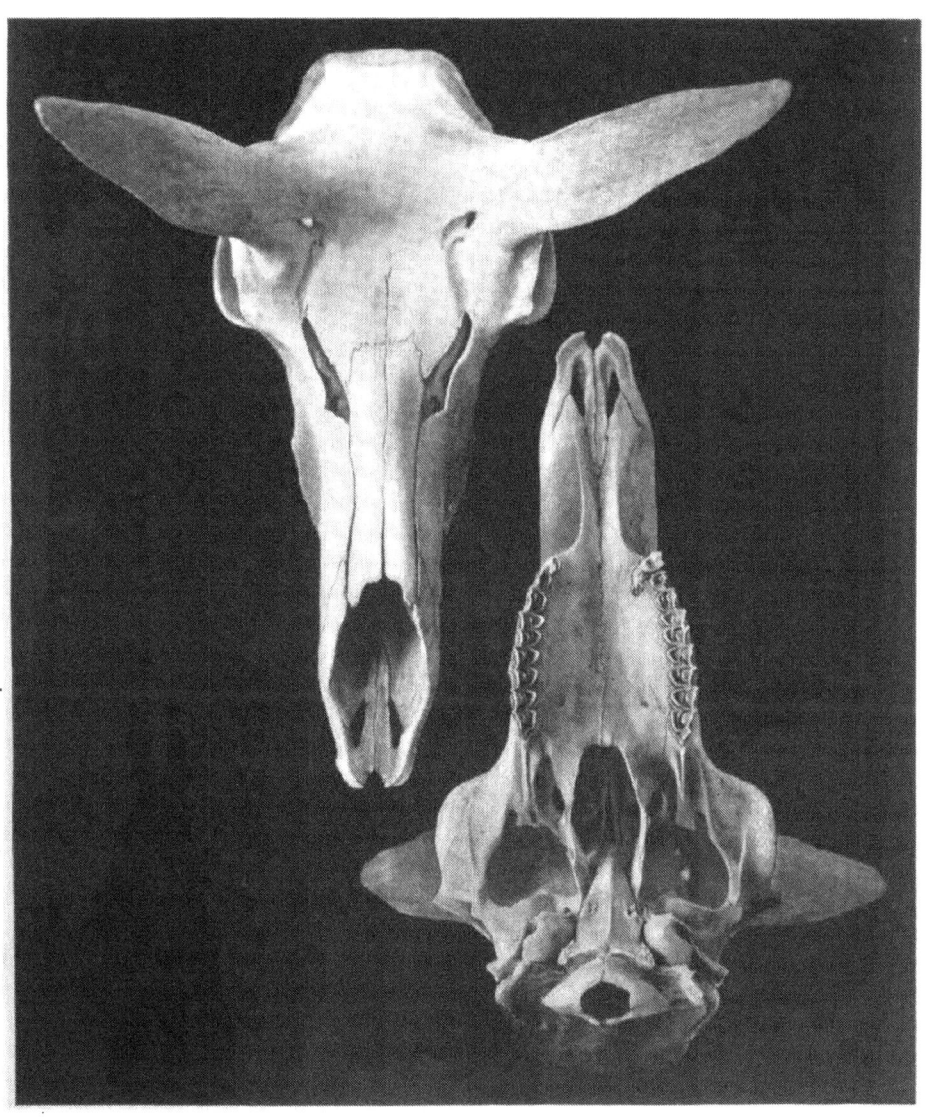

ANTILOCAPRA A. MEXICANA.
Old male from State of Chihuahua, Mexico. ⅞ nat. size.

Genl. Char. Colors pale. Skull similar to that of *A. americana*, orbits less protruding antero-inferiorly; premaxillæ and nose slender; bullæ thinner; lips of posterior nares longer.

Color. Similar to *A. americana*, but paler; median dark streak on neck, sometimes reaching shoulders; occiput whitish with median dark stripe.

Measurements. Total length, 1420; tail vertebræ, 145; hind foot, 410; height at shoulders, 830 (ex Merr., l. c.). Skull: occipito-nasal length, 216; breadth between outer edge of orbits, 125; width between orbits across frontals, 104; length of nasals laterally, 93; palatal arch to incisive foramina, 127; length of upper tooth row, 72; width of palate between last molars, 46.5; length of mandible, angle to alveoli of incisors, 204; height of condyle, 67; at coronoid process, 95; length of lower tooth row, 76. Skull of old male from State of Chihuahua, Mexico.

The BOVIDÆ or Hollow-horned Ruminants form an extensive family in the Old World, and are very generally distributed except in Australia. They are not represented in Central or South America, but certain forms are to be found in Mexico and northward to the Arctic Sea. One of the noblest members of the family, the American Bison, which at one time was found in millions on the plains of North America, is now practically extinct in the wild state. In this family are included the Antelopes, confined chiefly to Africa, in which continent a great number of species are still to be found. But some, which in herds like those of the Bison, once roamed the veldt in countless numbers, have disappeared before the hunter's rifle, and many species yet living will meet the same fate if government protection is not afforded them. In America, beside the Bison, now no longer to be considered as among the wild animals, there are the Musk Oxen and several varieties of Mountain Sheep. Of these last, two varieties of the Rocky Mountain species are found in the northern part of Mexico and Lower California, and are the only representatives of the *Bovidæ*, south of the United States boundary.

Fam. IV. **Bovidæ. Cattle, Sheep, Etc.**

31. **Ovis. Sheep.**

$$I.\frac{0-0}{4-4};\ C.\frac{0\cdot 0}{0-0};\ P.\frac{3-3}{3-3};\ M.\frac{3-3}{3-3} = 32.$$

Ovis Linn., Syst. Nat., 1, 1758, p. 70; and 1, 1766, p. 97. Type *Ovis aries* Linnæus.

Body stout; legs rather short; neck of moderate length; nose narrow, pointed, small naked space between nostrils, rest hairy; chin beardless; ears small, pointed, upright, hairy; tail short, pointed; lateral hoofs present; glands present between hoofs, and often on face below eyes; canon bones long and slender; skull broadest between eyes, then narrowing rapidly to nose; horns curving backward and then downward in a majestic sweep, tips everted, transverse ridges prominent.

KEY TO THE SUBSPECIES.   PAGE

A. Above whitey brown.....................*O. c. crcmnobates*   84
B. Above drab brown.......................*O. c. mexicanus*   86

FIG. XXVI. OVIS C. CREMNOBATES. OLD RAM.

cervina cremnobates (Ovis), Elliot, Pub. Field Columb. Mus., III, 1903, p. 239. Zoölogy.
LOWER CALIFORNIA MOUNTAIN SHEEP.

Type locality. Mattomi, San Pedro Martir Mountains, Lower California, Mexico.

Geogr. Distr. San Pedro Martir and probably the Laguna Mountains, Lower California, Mexico.

Genl. Char. Resembling *O. c. nelsoni* from the Grape Vine Mountains, boundary of Nevada and Lower California, but of a much lighter color, the head of a three-year-old ram being nearly white, with a very small caudal patch not divided from color of upper parts by any perceptible line; fore part of legs almost black,

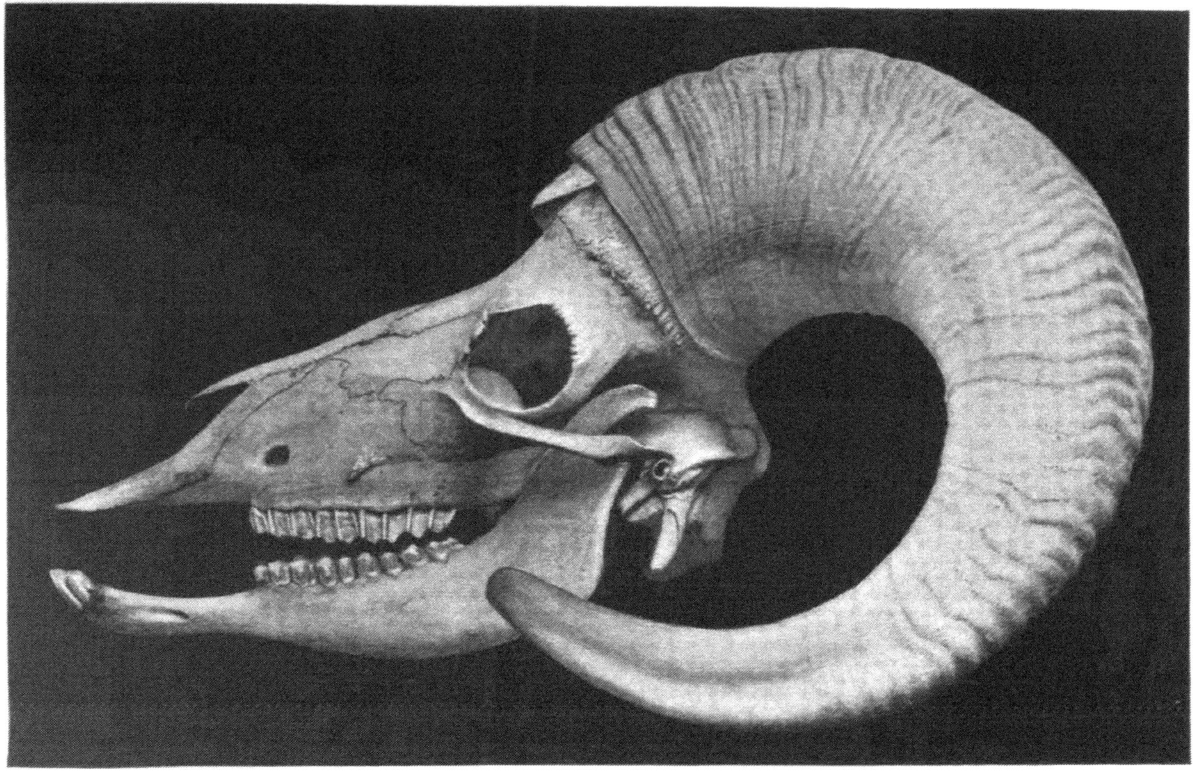

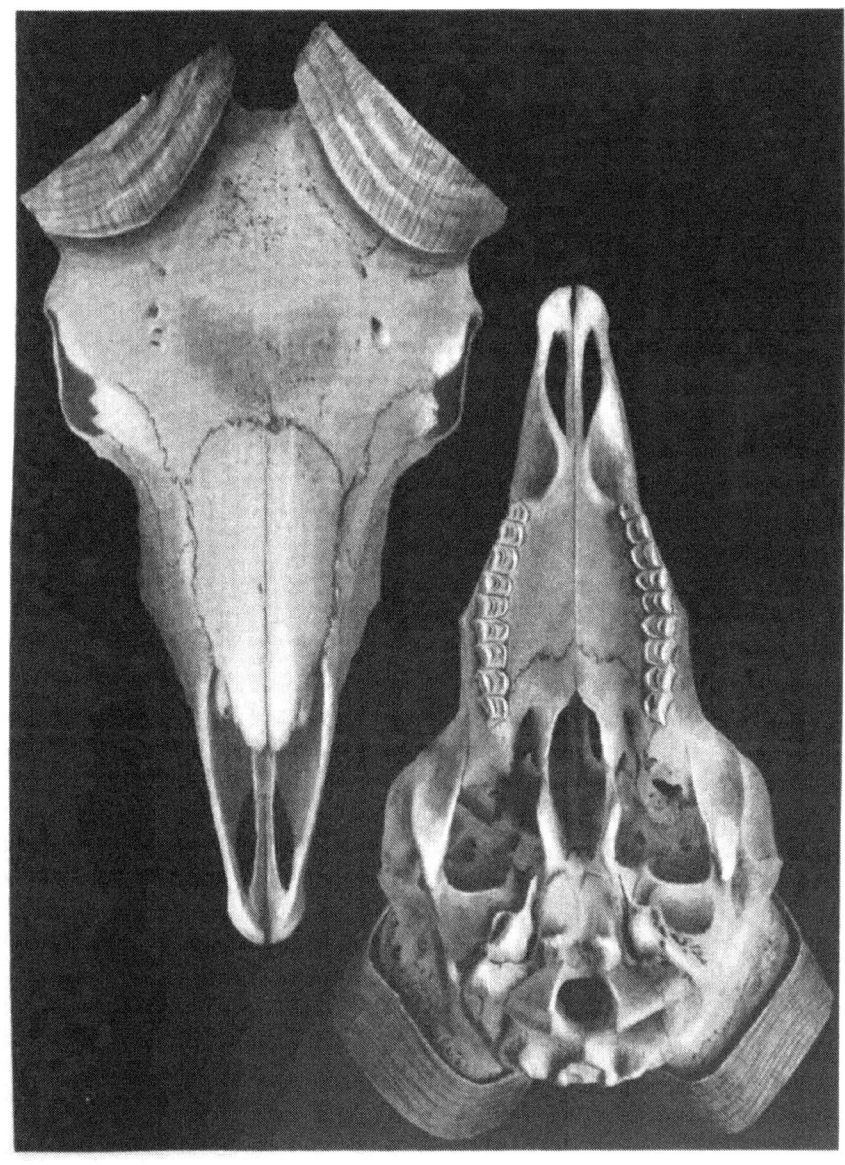

OVIS C. MEXICANA.
U. S. Nat. Mus. Coll.

FIG. XXVII. OVIS C. CREMNOBATES. YOUNG RAM AND OLD EWE.

similar to those of *O. stonii*; head very broad between orbits, from 20 to 25 mm. broader in old rams than the head of *O. c. nelsoni*; horns of adult rams very large and curving outward from the head; those of ewes with the points diverging widely apart.

Color. Upper parts and sides varying in individuals from drab gray or pale broccoli brown to hair and whitey brown; in some cases this sheep appears almost white; chest, line along ventral surface, and front of legs black or brownish black; head and neck hair brown, darker than back in some individuals, drab gray in the old ram; back part of legs and inside of hind legs, narrow line in center of ventral surface, caudal patch, nose around nostrils and inside of ears white; line across caudal patch from tail to darker color on rump (as in all Mountain Sheep), and the tail brownish black.

Measurements. Female. Total length, 1450; tail, 120; hind foot, 375; ear, 114. Skull: total length, 283; occipito-nasal length, 226; Hensel, 246; width between outer edge of orbits, 156; zygomatic width, 124; length of nasals, 109; palatal length, 148; length of upper tooth row, 84; length of half of mandible, 203; of lower tooth row, 82. Horns, total length along curve, 310; circumference at base, 144; spread at tip, 393. Head of old ram, total length, 330; width between orbits, inner edge, 180; circumference of horns at base, 395; length along outer curve, 850; spread at tips, 485.

cervina mexicanus (*Ovis*), (Merr.,) Proc. Biol. Soc. Wash., xiv, 1901, p. 30.

MEXICAN MOUNTAIN SHEEP. *Borrego Cimaron* in Mexico; *Tenatzali Taje* of Indians.

Type locality. Mountains about Lake Santa Maria, State of Chihuahua.

Geogr. Distr. Sierra Madre and Guadalupe Mountains of northern Mexico and southern New Mexico and Texas.

Genl. Char. Size large; color lighter than that of *O. cervina*, but of the same pattern; ears and tail long. Molars large; lips of posterior nares thin, everted.

Color. General color drab brown; no dorsal stripe; rump patch broad; throat, legs, and tail darker than back; chin, posterior and inner side of hind leg whitish.

Measurements. "Total length, 1530; tail vertebræ, 130; hind foot, 425; height at shoulder, 900." (Merr., l. c.)

The Tapirs are not a very extensive family, but have representatives in both Hemispheres. They are natives of tropical lands, and in the New World are not found north of Mexico. They dwell in the forests, generally near water, into which they often go for refuge, are nocturnal in their habits, and of a mild, inoffensive disposition. Their food consists of leaves, buds, and tender shoots of trees, and various vegetable substances. Tapirs of the Old and New Worlds, though living in regions widely separated, are nevertheless closely allied, but the Middle American species are distinguished by the more or less pronounced elongation of the ossification of the methesmoid, which in them extends beyond the nasal bones, but which in Old World forms does not go beyond these. Members of this family are not known to have existed previous to the Miocene epoch, and the animals of that and subsequent periods cannot be distinguished generically from those living at present, although they are specifically distinct. At one time doubtless the Tapirs had a wide distribution, extending from China through Europe, and in the United States as far north as South Carolina, thence westward to California. Tapirs have a massive body, with short, stout legs, and a long, prehensile upper lip, short ears, neck rather long, and a short tail. The front feet have four toes, but the outer one

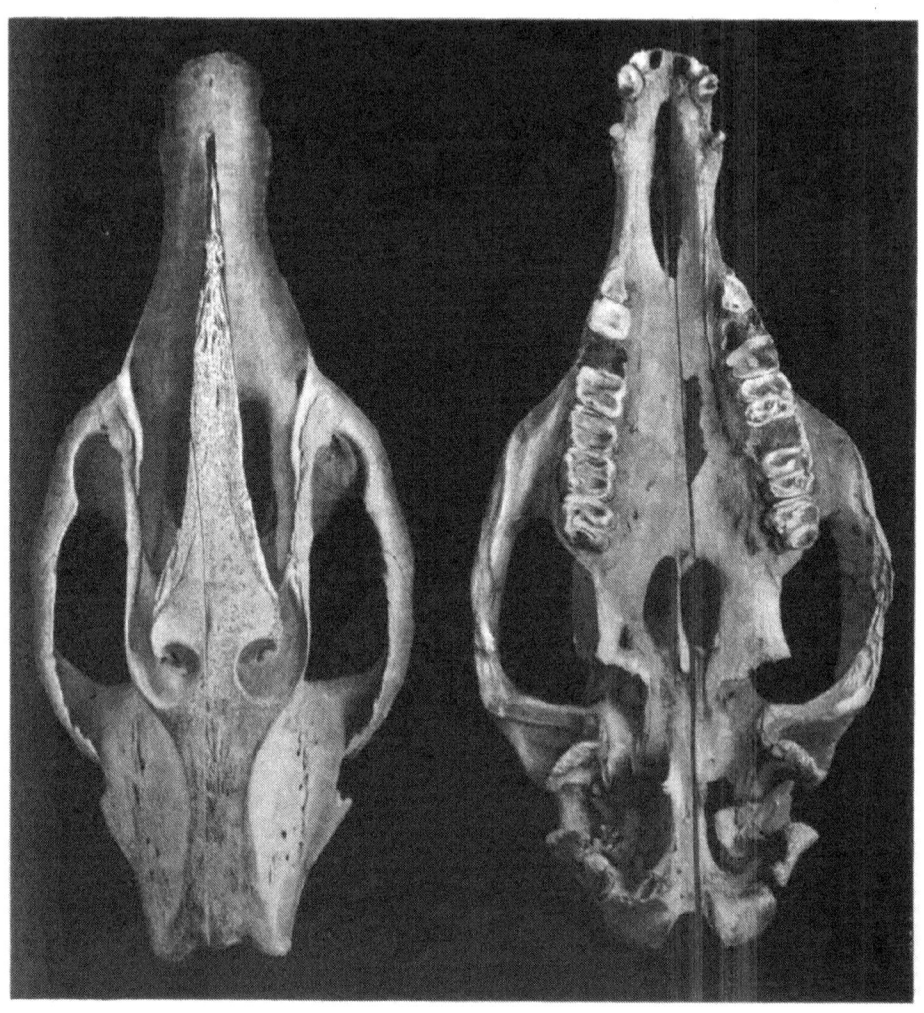

TAPIRELLA DOWI.
No. 6019 U. S. Nat. Mus. Coll.

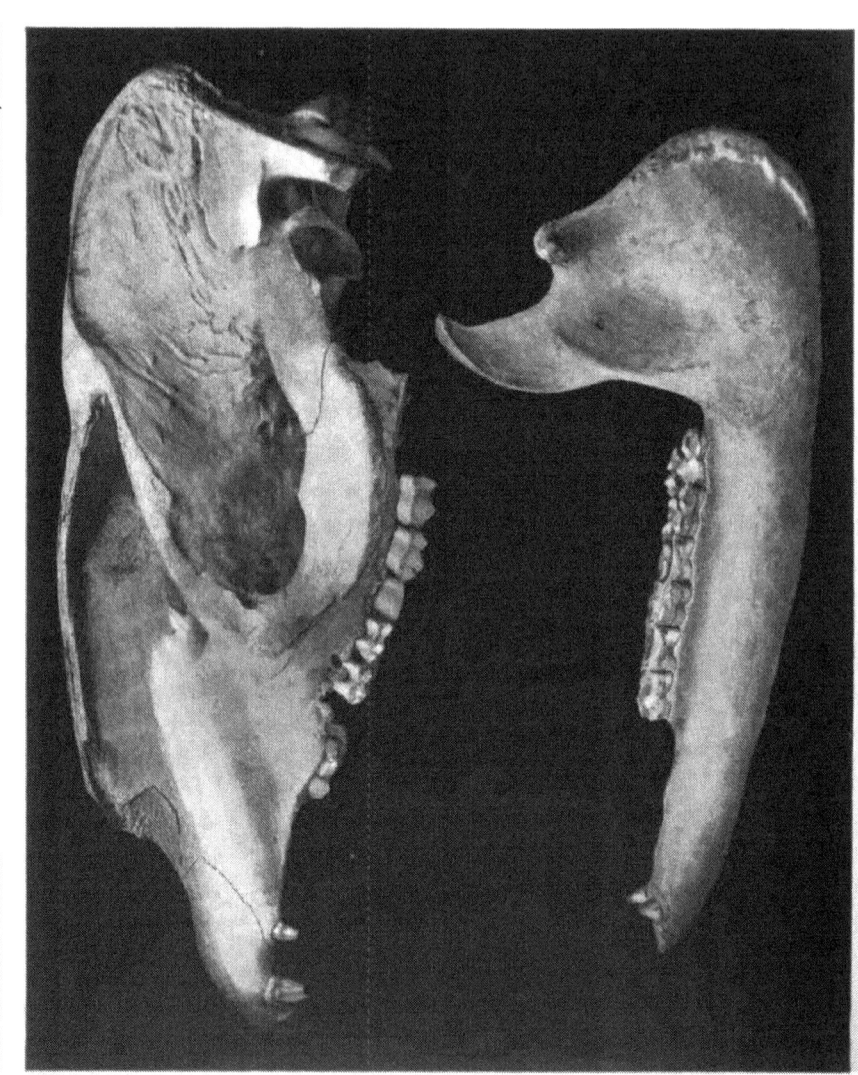

TAPIRELLA DOWI.

No. 6019 U. S. Nat. Mus. Coll.

does not render any support to the body. The young of the Tapirs are sometimes spotted or streaked with white. In the Andes there is one species that in its choice of locality differs widely from its relatives, as it makes its abode in elevated tracts of several thousand feet altitude; and, probably as a protection against the low temperature of these lofty heights, has the skin covered with hair.

Fam. V. **Tapiridæ. Tapirs.**

32. **Tapirella.**

$$I.\frac{3-3}{3-3}; \ C.\frac{1-1}{1-1}; \ P.\frac{4-4}{3-3}; \ M.\frac{3-3}{3-3} = 42.$$

Tapirella Palmer, Science, 1903, p. 873. (May.)

Elasmognathus Gill, Proc. Acad. Nat. Scien. Phil., 1865, p. 183. (nec Fieber, Hemiptera.) Type *Elasmognathus bairdi* Gill.

"Supra-maxillaries swollen above and in front of the infraorbital foramina, and thence extend upward and backward into a squamous portion, which embraces with its fellow a thick, bony nasal septum continuous with the vomer, and which is elevated to a line with the forehead, and has a widened upper edge, which still further enlarges behind and embraces the nasal bones. The grooves for the muscles of the proboscis are in front, straight, entirely confined to the frontals, and do not encroach upon the supra-maxillaries; while behind they describe a spiral curve around a pit between the nasals and frontals." (Gill l. c.)

KEY TO THE SPECIES.

A. Nose elongated, projecting beyond the mouth,
    flexible; body stout, heavy.
    a. Frontals not advancing between nasals.   PAGE
        Young spotted or streaked . . . . *T. bairdi*   87
    b. Frontals advancing and separating nasals.
        Young not spotted or streaked. . . . . *T. dowi*   88

65. bairdi (*Elasmognathus*), Gill, Proc. Acad. Nat. Scien. Phil., 1865, p. 183.

BAIRD'S TAPIR.

Type locality. Isthmus of Panama.

Geogr. Distr. Southern Mexico to Panama.

Genl. Char. "Nasals well developed, each ossified from a single center, separate through life, thick at their base, and articulated with one another for the greater part of their length."

Color. Dark reddish brown; throat and breast solid white; cheeks chestnut; ears margined with white. (Immature specimen.)

Measurements. Total length, 1050; height, 575.* (Immature.) Skull: occipito-nasal length about 455; Hensel, 409; zygomatic width 179; palatal length, 210; length of mandible, angle to tip of incisors, 375.

FIG. XXVIII. TAPIRELLA DOWI. DOW'S TAPIR.

66. dowi (*Elasmognathus*), Gill, Amer. Jour. Scien. Arts, L, 1870,
 p. 142.
DOW'S TAPIR.

Type locality. Gautemala.

Geogr. Distr. Gautemala and Nicaragua, Central America.

Genl. Char. In the young the basilar processes are "recurrent forward along the frontal bones, and as the animal advances in age the frontals grow forward, and force apart the nasals, which do not increase, and are fused with the frontals." Young without longitudinal whitish stripes.

Color. Face and nose rufous; cheeks pale brown; remainder of body and limbs blackish brown.

Measurements. About the same as those of *T. bairdi*.

---

* An adult should measure twice this size. One obtained by Mr. Heller in the State of Vera Cruz, Mexico, has the following measurements: Total length, 2020, tail vertebræ, 70; hind foot, 375; ear, 140.

Order VI. **Rodentia. Rodents.**

Coues and Allen, *Monographs of North American Rodentia*, U. S. Geol. Survey, 1877.

The Rodents constitute the largest Order of Mammals, and the numerous members possess a great diversity of form. They are readily distinguished among all mammals by their incisors, four in number, two above and two below, (except Hares and Rabbits, which have a supplementary upper hinder pair in adults,) curved hollow tubes filled with pulp, hardened at the surface, the portion beneath the gum curving and sometimes traversing the length of the jaw bone. The species are mostly small, the harvest mouse being the pigmy, from which genus the size increases until the comparatively great beaver is reached, and he is exceeded in bulk only by the capybara of South America. The majority, however, are small animals, and their habits are as diverse as their shapes, and we find among them not only terrestrial and aquatic creatures, but others which are provided with extensible membranes between the limbs and body, to enable them to traverse the air as if carried by parachutes. Rodents are cosmopolitan, the greatest number being found in South America, the fewest in Australia. They are mostly herbivorous, yet some, like the ordinary rat, are omnivorous. The incisors have a continuous growth and are worn away at the terminal portion by constant gnawing or by attrition. The molar teeth are usually rootless, and their crowns often present many varied, even intricate, patterns of enamel folds and loops. No canine exists in any rodent. Normally the species of this family generally have no premolars, although in some a small one is present, and among squirrels two on each side above, and one below are found, but the additional premolar is frequently deciduous. The diversity of form and habits is very great in the members of this order, and we have the tree-loving, graceful squirrel in countless colors, and its small imitator, the chipmunk; ground squirrels that live in burrows, and flying squirrels darting through the air; the innumerable field mice of many genera and species; rice and cotton rats, pouched rats that live under ground and tunnel long galleries like the moles; jumping mice with long hind legs and greatly lengthened tails that possibly may assist their owners in making the kangaroo-like leaps over the fields; aquatic rats whose home is in the water, and whose feet are formed more for swimming than walking— all these, and more, help to constitute the great order of the Gnawers.

The family first to be considered of this order is that of the Squirrels, and excepting the Australian region and the Island of Madagascar, these animals are found in nearly all the temperate and tropical regions of the world. They may be divided into two classes, the tree squirrels and the ground squirrels, with a kind of connecting link in the chipmunks of the genus TAMIAS, which, to a considerable extent, possess the habits of both. While the tree-squirrels are found in both temperate and tropical zones, the ground squirrels are dwellers of more northern climes, and some species are found even on the bleak shores of the Arctic Sea. In the tropics, however, the tree-squirrels attain their greatest diversity of coloration and highest development, and in the Oriental region they reach their greatest size and most brilliant hues. North America is perhaps the third on the list of those countries in which squirrels are found, being exceeded in number of species by the Indian and Ethiopian Regions. Europe and South America have comparatively few species of this family. Marmots, known usually as woodchucks or ground hogs, are the largest members of the *Sciuridæ*, but none are found within the limits of this work, and the little chipmunks are among the smallest. North America is probably the richest of all lands in ground squirrels, having a large number, varying greatly in size and coloration.

Fam. I. **Sciuridæ. Squirrels, Marmots, etc.**

Coues and Allen, *Monographs of North American Rodents*, U. S. Geol. Survey, 1877.

E. W. Nelson, *Review of the Squirrels of Mexico and Central America*, Proc. Wash. Acad. Scien., i, 1899, pp. 15–106.

Tail without scales, cylindrical, bushy, hairs long; distinct postorbital processes; infraorbital opening small; molars rooted, tubercular; first upper premolar small.

Subfam. I. **Sciurinæ.**

KEY TO THE GENERA AND SUBGENERA.

A. Upper incisors grooved.   PAGE
  a. Size medium. Premolars, $\frac{1-1}{1-1}$ . . . .*Synthetos(iurus*   91
B. Upper incisors not grooved.
  a. Size variable; tail flat, bushy, long. Skull
    short, broad; braincase more or less
    arched.   Premolars, $\frac{2-2}{1-1}$ or $\frac{1-1}{1-1}$. . . . .*Sciurus*   93
  b. Size small, total length under 450 mm.
    a.' Premolars, $\frac{1-1}{1-1}$ or $\frac{2-2}{1-1}$.
      a." Superior outline of skull greatly curved. .*Tamiasciurus*   132

The first genus of the *Sciuridæ* is remarkable for the slender lower incisors which project outwards, and the upper ones are grooved in the center. But one species is known.

33. *Synthetosciurus.

Syntheosciurus (*sic*) Bangs, Bull. Mus. Comp. Zoöl., 1902, Vol. xxxix, p. 25.

Premolars, $\frac{1-1}{1-1}$. Size small. Skull thin, papery; rostrum straight, audital bullæ small; incisors slender, the lower pair projecting outward, the upper with central groove.

67. brochus (*Syntheosciurus!*), Bangs, Bull. Mus. Comp. Zoöl., xxxix, 1902, p. 25.

PROJECTING-TEETH SQUIRREL.

Type locality. Boquete, Chiriqui, Panama. Altitude 7,000 feet.

Genl. Char. Size small; ears low, round, woolly; pelage long, soft, woolly; other characters as in genus.

---

*σύνθετια-σκίουρος=Synthetosciurus.

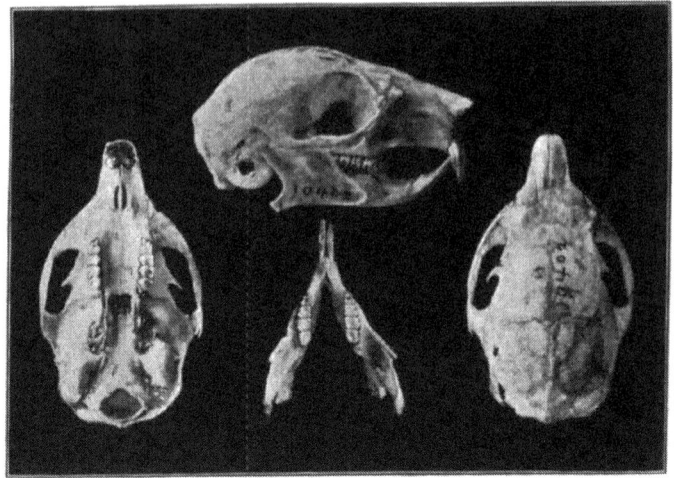

FIG. 12. SYNTHETOSCIURUS BROCHUS.
No. 10402 Mus. Comp. Zoöl. Coll. Nat. Size.

Color. Above mixed olivaceous bistre and dull tawny olive; under fur dark mouse gray; under parts orange rufous; tail above like back, less olivaceous beneath, fringed with pale rusty.

Measurements. Total length, 320; tail vertebræ, 150; hind foot, 46; ear, 17. Skull: basal length, 35.6; occipito-nasal length, 44; zygomatic width, 25.2; interorbital width, 12.6; palatal length, 20; to end of pterygoid, 27.4; length of nasals, 13; width of nasals, 5.8; length of upper molar series, 7.6; length of mandible, 27.

The next genus SCIURUS, with its subgenera, contains the tree squirrels whose lives are passed among the branches. It possesses the largest members of these animals in North America, and they are all remarkable for the long bushy tail, often exceeding the head and body in length, and which when elevated over the back, is both a beautiful ornament and a protecting shade. The genus is a very large one, and has representatives in many parts of the globe, and the species differ greatly in size and especially in coloration, in which there is almost endless variation; and as if it were not sufficient for distinct species to differ from each other, it was evidently deemed necessary that individuals of the same species should sometimes be totally unlike in the hues of their coats. It is this variation in color

among individuals that frequently makes it so difficult to correctly
determine a species, and any one who relies upon these numerous
hues to discriminate what species are before him, will probably, at a
later period, in the discovery of the blunders that have been made,
become a wiser and a sadder man. It will doubtless be a long time
before the exact status of our American squirrels is satisfactorily
ascertained. Melanism is of frequent occurrence among these
animals, and erythrism also; the latter perhaps less often; and
albinism is the rarest of all. Yet in spite of the endless variation in
colors, and the great difference frequently observed in the size of
species, as well as occasionally in their form, few would fail to recognize at once any of these sprightly creatures as not rightfully belonging to the family SCIURIDÆ.

The remaining genera contain those species familiarly known as
Gray Squirrels and their allies, although many of them have by no
means a gray pelage. In size, also, these graceful creatures are very
variable, and range from the little Bornean species *S. soricinus*, no
larger than a mouse, to the great Malayan long-tailed forms almost
as big as a cat. These last are placed in the genus *Ratufa*. As a rule,
squirrels have no especial nuptial dress, as birds have, but retain their
individual coloring throughout the year, the completed moult bring-
ing no change. But one exception to this is known, the *S. caniceps*
of India (northern Tenasserim), which assumes on the upper parts in
winter a bright orange hue, a dress strikingly different from the
ordinary gray or olive livery worn at other seasons of the year. Of
the countries embraced in this volume Mexico contains the greatest
number of these beautiful animals, astonishingly varied in the hues
and patterns of their coats, making accurate determination of their
specific relationship a matter at times of considerable difficulty, as
individuals of the same species, not infrequently, have a totally
different coloration.

34. Sciurus. Tree Squirrels.

Sciurus Linn., Sys. Nat., I, 1758, p. 63.

> *Guerlinguetus* Gray, Lond. Med. Repos., xv, 1821, p. 304.
>
> *Macroxus* F. Cuv., Dent's des Mamm., 1823, p. 162. *Id.* Dict.
> Class. Hist. Nat., x, 1826, p. 16. *Id.* Dict. Scien. Nat., LIX,
> 1829, p. 474.
>
> *Rheithrosciurus* (*sic*) Gray, Ann. Mag. Nat. Hist., 3d Ser., xx,
> 1867, p. 272.
>
> *Rhinosciurus* Gray, List Spec. Mamm. Brit. Mus., 1843, pp. xxv,
> 195, Ann. Mag. Nat. Hist., 3d Ser., xx, 1867, p. 286.
>
> *Neosciurus* Treuss., Le Nat., II, 1880, p. 292.

Parasciurus Treuss., Le Nat., II, 1880, p. 292.
Echinosciurus Treuss., Le Nat., II, 1880, p. 292.
Tamiasciurus Treuss., Le Nat., II, 1880, p. 292.
Microsciurus Allen, Bull. Am. Mus. Nat. Hist., 1895, p. 332.
Hesperosciurus, Nelson, Proc. Wash. Acad. Scien., I, 1899, p. 27.
Otosciurus, Nelson, Proc. Wash. Acad. Scien., I, 1899, p. 28.
Aræosciurus Nelson, Proc. Wash. Acad. Scien., I, 1899, p. 29.
Baiosciurus Nelson, Proc. Wash. Acad. Scien., I, 1899, p. 31.

Tail broad long, bushy, hairs mostly directed laterally; ears moderate, hairy, sometimes with long tufts at tip; no cheek pouches. Skull rather short, broad; postorbital processes directed downward and backward and well developed; one upper premolar, sometimes two; in the latter case the first is always very small; antorbital foramen slit-like, placed anteriorly to the zygomatic process of the maxillary.

KEY TO THE SPECIES AND SUBSPECIES.

*A. Size small, total length under 450 mm., but above 300 mm.   PAGE
   a. Premolars $\frac{1-1}{1-1}$; ears medium long, thinly haired.
      a.' Tail washed with bright ferrugineous....*S. æ. hoffmanni*   104
      b.' Tail washed with tawny............*S. æ. chiriquensis*   104
      c.' Tail washed with yellowish.............*S. richmondi*   105
   b. Premolars $\frac{2-2}{1-1}$;
      a.' Ears small or medium, not tufted.
         a." Ears short, rounded, thickly haired;
            total length under 300 mm.
            a.''' Tail washed with reddish.............*S. alfari*   99
            b.''' Tail washed with grayish white.......*S. browni*   100
            c.''' Tail washed with tawny olive....*S. boquetensis*   100
         b." Ears medium long, pointed, thinly
            haired; total length over 300 mm.
            a.''' Back grayish brown...............*S. negligens*   102
            b.''' Back reddish or yellowish brown.......*S. deppii*   101
            c.''' Back rusty reddish.................*S. d. vivax*   102
   c. Premolars $\frac{2-2}{1-1}$ or $\frac{1-1}{1-1}$.
      a.' Ears large, tufted.
         a." Back gray, belly white, lateral line
            black ....*S. d. mearnsi*   133

---

* The construction of an intelligible key for the endless variations displayed by the members of the genus *Sciurus* is an almost insuperable task, and may not be attempted unless specimens of all the species are available at one time. This not having been possible for the author to accomplish, the present key for the species, with some few additions, has been taken from E. W. Nelson's " Revision of the Squirrels of Mexico and Central America."

B. Size large, total length over 450 mm.
  a. Premolars $\frac{1-1}{1-1}$.
    a.′ Belly buffy or yellowish.
      a.″ Back gray. PAGE
        a.‴ Median area on back black; belly
          usually deep buff .... *S. oculatus*   107
        b.‴ Median area on back washed with
          blackish; belly buffy whitish........... *S. tolucæ*   107
      b.″ Back yellowish gray.
        a.‴ Belly rusty yellow; total length over
          500 mm........................... *S. apache*   110
        b.‴ Belly more dingy yellow; total
          length under 500 mm.............. *S. r. texensis*   110
    b.′ Belly white.
      a.″ Back golden buffy or yellowish, overlaid
        with grizzling of black and white ..... *S. nayaritensis*   108
      b.″ Back gray.
        a.‴ Back uniform gray or grayish brown;
          ears gray; total length under 500 mm ..... *S. alleni*   108
        b.‴ Back gray washed with yellowish on
          nape and back of shoulders; ears
          rusty.
          a.′ Total length under 500 mm....... *S. arizonensis*   109
          b.′ Total length over 500 mm........ *S. a. huachuca*   109
  b. Premolars $\frac{2-2}{1-1}$.
    a.′ Nape patch strongly marked, rump patch
    present or absent.
      a.″ Belly gray, rump patch well marked;
        back dark gray................... *S. a. frumentor*   117
      b.″ Belly white or buffy.
        a.‴ Rump patch poorly defined or absent.
          a.′ Back dull whitish; belly white or buffy. *S. socialis*   123
          b.′ Back pale gray; belly white or buffy ....
            .... *S. p. hernandezi*   118
        b.‴ Rump patch well defined.
          a.′ Middle of back gray; feet gray or
            blackish; belly white........... *S. p. colimensis*   119
          b.′ Middle of back whitish; feet whitish;
            belly white or buffy................. *S. s. cocos*   124
      c.″ Belly rufous.
        a.‴ Feet gray or blackish.
          a.′ Ribs, and sometimes shoulders, rufous.
            a.⁵ Back pale gray; top of head iron
              gray......................... *S. aureigaster*   115

a.⁶ Top of head paler than back.

A. Microsciurus, Allen.

"Size small; ear short, rounded; tail shorter than body, slender, rounded. Premolars, $\frac{2-2}{1-1}$. Skull short, broad, and deep at base; nasals narrow and shorter than interorbital breadth, upper end of premaxillæ very heavy; màlar broad and expanded vertically; post-palatal notch only a trifle posterior to last molar." (Nelson.)

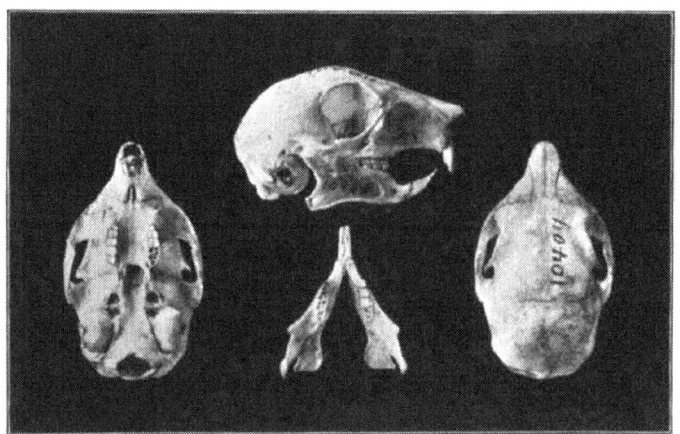

FIG. 13. SCIURUS (MICROSCIURUS) BROWNI.
No. 10404 Mus. Comp. Zoöl. Nat. Size.

KEY TO THE SPECIES OF THE SUBGENUS.

A. Size small; tail slender, round.   PAGE

    a. Upper parts dusky olivaceous, finely grizzled
       with yellowish rusty........................*S. alfari*   99

    b. Upper parts tawny olive and bistre............*S. browni*   100

    c. Upper parts olivaceous brown, shaded with
       yellowish...........................*S. boquetensis* 100

68. alfari (*Sciurus*), Allen, Bull. Amer. Mus. Nat. Hist., 1895, p. 333. ALFARO'S PIGMY SQUIRREL.

Type locality. Jimenez, Costa Rica, Central America.

Geogr. Distr. Northern, eastern, and southwestern Costa Rica.

Genl. Char. Size very small; ears short, rounded; pelage soft, dense; tail slender, much shorter than head and body.

Color. Upper parts and outer sides of arms, legs, hands, and feet dusky olivaceous, finely grizzled with yellowish rusty; top of head and narrow orbital ring rufous; chin, throat, breast, and inner side of arms pale ferrugineous shading into dark brown; middle of belly and inner side of thighs varying from fulvous gray to rufous; tail like back at base, remainder above and beneath grizzled dark reddish brown and black, tip black; hairs of under surface of tail reddish brown or chestnut, encircled with three bands.

Measurements. Total length, 290; tail vertebræ, 105; hind foot, 35; ear from crown, 9.   Skull: basal length, 29.5; palatal length, 14.2; interorbital width, 13; zygomatic width, 22; length of upper molar series, 6.

69. browni (*Sciurus*), Bangs, Bull. Mus. Comp. Zoöl., xxxix, 1902, p. 24.
BROWN'S SQUIRREL.
Type locality.   Bogava, Chiriqui, Panama.
Genl. Char.   Size small, pelage rather harsh, thin.
Color.   "Upper parts a fine mixture of tawny olive and bistre, produced by the dark brown bases and tawny olive tips of the hairs; nose, forehead, and orbital ring tawny; tail with the hairs dark reddish brown basally, then black and tipped with grayish white; a small, black pencil; under parts dull gray to grayish white, slightly washed with buffy or yellowish in some specimens (very slightly in the type on under side of neck and middle of belly); under sides of legs darker—more nearly like upper parts."
Measurements.   "Total length, 232–260; tail vertebræ, 110–120; hind foot, 36–38; ear, 13–14.   Skull: type, basal length, 29; occipito-nasal length, 36; zygomatic width, 21.2; interorbital width, 12.4; palatal length, to palatal notch, 13.4; to end of pterygoids, 20.2; length of nasals, 11; length of upper molar series, 5.8." (Bangs, l. c.)

70. boquetensis (*Sciurus*), Nelson, Proc. Biol. Soc. Wash., xvi, 1903, p. 121.
CHIRIQUI PIGMY SQUIRREL.
Type locality.   Boquete, Chiriqui, Panama, altitude 6,000 feet.
Genl. Char.   Pelage soft, thick, woolly; tail slender, flat.   Skull long and narrow; braincase arched.
Color.   Upper parts, sides of body, and upper surface of arms and legs olivaceous brown, shaded with yellowish; chin and throat dingy rusty; under side of neck and breast rusty rufous shading into dull grizzled brown; tail above and below dull tawny olive, washed and tipped with black and edged with pale yellowish; hands and feet washed with rusty reddish.
Measurements.   Total length, 257; tail, 116; hind foot, 37; (dried skin.)   Skull: palatal length, 15.5; interorbital breadth, 14; length of upper molar series, 7.

B. Baiosciurus.

Premolars, $\frac{2-2}{1-1}$.   Skull long and slender; braincase arched; rostrum broad, about equal to interorbital breadth; audital bullæ small.

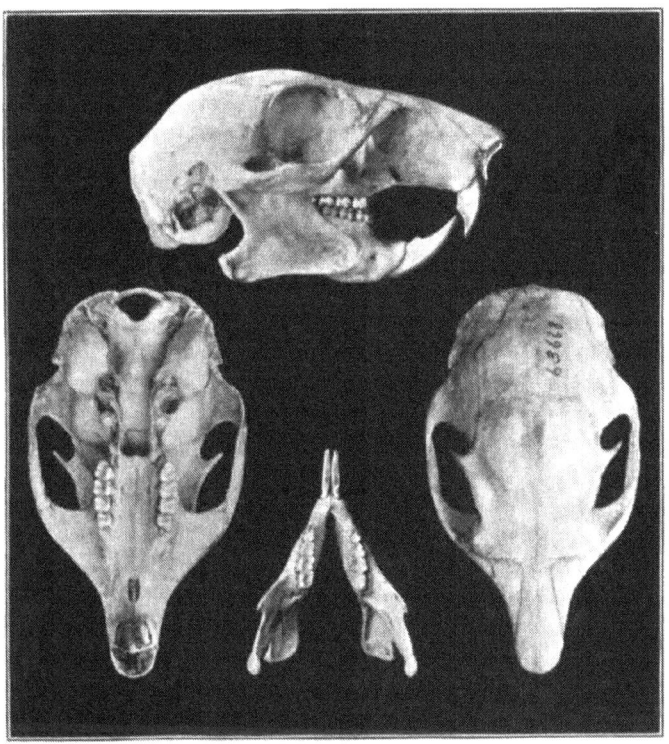

FIG. 14. SCIURUS (BAIOSCIURUS) DEPPII.
No. 63668 U. S. Nat. Mus. Coll. Nat. size.

KEY TO THE SPECIES AND SUBSPECIES OF THE SUBGENUS.

A. Size small.   PAGE
   a. Above grizzled rusty, or yellowish brown .... *S. deppii* 101
   b. Above rusty reddish .... *S. d. vivax* 102
   c. Above grizzled grayish brown, tinged with
      yellow or reddish .... *S. negligens* 102

71. deppii (*Sciurus*), Peters, Monatsb. K. Preuss. Akad. Wiss., Berl.,
    1863, p. 654.
    tephrogaster Gray, Ann. Mag. Nat. Hist., 3d Ser., xx, 1867, p. 431.
    tæniurus Gray, Ann. Mag. Nat. Hist., 3d Ser., xx, 1867, p. 431.
    griseogena Gray, Ann. Mag. Nat. Hist., 3d Ser., xx, 1867, p. 429.
DEPPE'S SQUIRREL.
    Type locality.   Papantla, State of Vera Cruz, Mexico.

Geogr. Distr. State of Vera Cruz from Papantla on east coast to Isthmus of Tehuantepec, Mexico, and into Guatemala, Central America. Altitude, 6,000-9,000 feet.

Color. Above grizzled rusty or yellowish brown; flanks and sides of head paler yellowish brown; ears like back, with whitish basal patches; outside of arms and hands dark gray; legs and feet similar to flanks in color, but darker: under parts grayish white to rusty fulvous; tail above black, washed with white, beneath grizzled reddish or yellowish brown, bordered with black and edged with white.

Measurements. Total length, 392; tail vertebræ, 188; hind foot, 54. Skull: occipito-nasal length, 56; Hensel, 47; zygomatic width, 33; across postorbital processes, 27; length of nasals, 15; palatal length, 27; length of upper tooth row, 11; length of lower tooth row, 9.

a.—vivax (Sciurus), Nelson, Proc. Biol. Soc. Wash., XIV, 1901, p. 131.
APAZOTE SQUIRREL.

Type locality. Apazote, State of Campeche, Mexico.

Geogr. Distr. Lowland forest in the States of eastern Tabasco, southern Campeche, and southern and eastern Yucatan, Mexico.

Genl. Char. Similar to *S. deppii*, but paler; rostrum heavy; nasals broader; audital bullæ smaller and more nearly round.

Color. Above rusty reddish; under parts white or grayish white; shoulders, arms, and hands gray; feet like back but washed with gray; tail above heavily washed with white.

Measurements. Total length, 373; tail vertebræ, 168; hind foot, 52.

72. negligens *(Sciurus)*, Nelson, Proc. Biol. Soc. Wash., XII, 1898, p. 147.

arizonensis Alston, Biol. Cent. Amer., Mamm., I, 1880, p. 125.

deppei Allen, Bull. Am. Mus. Nat. Hist., 1891, p. 222. (Part.)
LITTLE GRAY SQUIRREL.

Type locality. Alta Mira, State of Tamaulipas, Mexico.

Geogr. Distr. Southern Tamaulipas, through eastern San Luis Potosi into northern Vera Cruz, Mexico.

Genl. Char. Similar to *S. deppii*, but paler.

Color. Upper parts grizzled grayish brown, tinged with yellow or reddish; sides of neck, nape, and head yellowish brown; ears ferrugineous with basal white patches; shoulders outside of arms and hands gray; outside of legs and feet grizzled grayish brown; under parts white, varying to fulvous; chin and throat white; tail above black, washed with white, beneath grizzled grayish, or yellowish brown, bordered with black and edged with white.

Measurements. Total length, 384; tail vertebræ, 189; hind foot, 54. Skull: average of four; basal length, 41.7; palatal length, 21.4; interorbital width, 15.3; zygomatic width, 28.4; length of upper molar series, 9.1.

O. Guerlinguetus.

Size small; ears long; tail shorter than body, flat, bushy. Skull broad, braincase not highly arched, expanded at parietal region; bullæ small; rostrum broad, deep; nasals long, widest anteriorly; post-palatal notch behind last molar. Premolars, $\frac{1-1}{1-1}$.

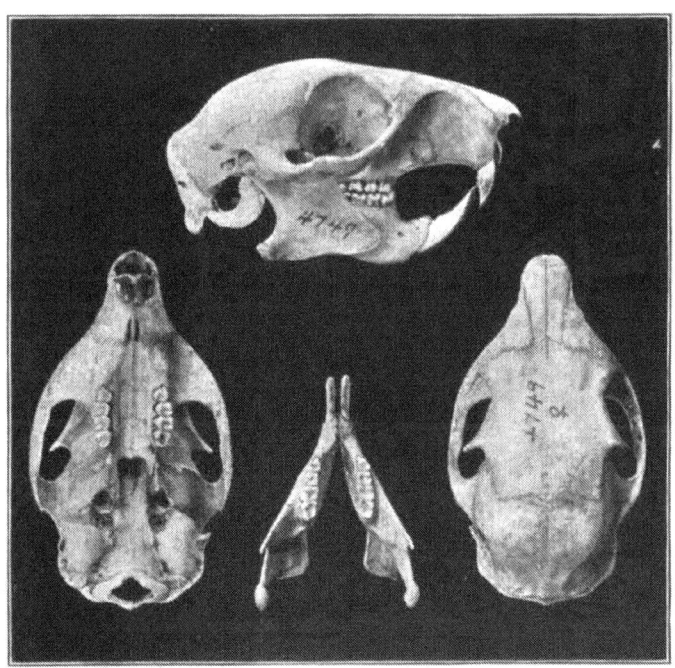

FIG. 15. SCIURUS (GUERLINGUETUS) Æ. HOFFMANNI.
No. 4749 Field Columbian Mus. Coll. Nat. size.

KEY TO THE SPECIES AND SUBSPECIES OF THE SUBGENUS.

A. Size small; tail flat, bushy.   PAGE

œstuans hoffmanni (*Sciurus*), Peters, Monatsb. K. Preuss. Akad.
Wiss. Berl., 1863, p. 654.

xanthotus Gray, Ann. Mag. Nat. Hist., 3d Ser., xx, 1867, p. 429.

griseogena Gray, Ann. Mag. Nat. Hist., 3d Ser., xx, 1867, p. 430.
(Part. Costa Rica.)

rufoniger Allen, Mon. N. Am. Rodent., 1877, p. 757–763. (Part.
Costa Rica.)

griseogenys Alston, Proc. Zoöl. Soc., 1878, p. 667. (Part. Costa
Rica, Veragua, Panama.)

HOFFMAN'S SQUIRREL.

Type locality. Costa Rica, Central America.

Geogr. Distr. Costa Rica south to upper Cauca River, Colombia.

Genl. Char. Similar to *S. œstuans* in size, but darker; tail broad, flat.

Color. Upper parts grizzled rusty brown, sometimes blackish on
top of head and median line of back; orbital ring dark buff; chin and
throat yellowish buff; under parts rusty buff to deep ferrugineous,
outside of arms and legs like back, inner side like under parts; hands
and feet similar to back, but more inclined to yellowish; tail above
black, washed with ferrugineous, beneath grizzled black and yellowish
brown, and broadly edged with ferrugineous; ears thinly haired;
darker than head, basal patch small, dull fulvous.

Measurements. Total average length, 426.6; tail vertebræ, 187;
hind foot, 54.3. Skull: average of five; basal length, 43.2; palatal
length, 23.3; interorbital width, 17; zygomatic width. 31.3; length of
upper molar series, 9.

œstuans chiriquensis (*Sciurus*), Bangs, Bull. Mus. Comp. Zoöl.,
XXXIX, 1902, p. 22.

CHIRIQUI SQUIRREL.

Type locality. Divala, Chiriqui, Panama.

Genl. Char. Very similar to *S. œ. hoffmanni;* under parts paler.

Color. "Upper parts finely mixed blackish brown and tawny,
the tawny color predominating on sides, the dark brown color along
middle of back; orbital ring, back of ear, and a small spot just behind
ear clear tawny; under parts tawny, becoming yellower, about raw
sienna, on under side of neck and head, and often the breast similarly colored; tail much the same as back, but with the tawny annulations wider; deeply fringed along sides with clear tawny, under side
darker than upper."

Measurements. "Type. Total length, 400; tail vertebræ, 190;
hind foot, 52; ear, 20." Skull: basal length, 46.2; occipito-nasal

length, 54; zygomatic width, 31.4; length of nasals, 16.4; palatal length, 23.2. (Bangs, l. c.)

73. richmondi (*Sciurus*), Nelson, Proc. Biol. Soc. Wash., XII, 1898, p. 146. RICHMOND'S SQUIRREL.

Type locality. Escondido River, fifty miles above Bluefields, Nicaragua.

Geogr. Distr. Tropical lowland forests along the Escondido River, Nicaragua.

Genl. Char. Similar to *S. æ. hoffmani*, but more ochraceous, under parts and tail washed with yellow.

Color. Upper parts and base of tail dark ochraceous brown, darkest on crown and median part of back; sides of head yellowish brown; orbital ring buffy; outer side of arms, hands, and sides of neck, ochraceous; outer side of thighs dark ochraceous brown, feet similar but more ochraceous; under parts buffy yellow to dingy ferrugineous; tail above black, washed with yellowish, beneath grizzled yellowish brown narrowly bordered with black and edged with dull yellow; ears dark ochraceous brown, with small yellow basal patch, but the latter not always present.

Measurements. Total length, 368; tail vertebræ, 164; hind foot, 53.5. Skull: average of five; basal length, 42.1; palatal length, 22.1; interorbital breadth, 16.3; zygomatic breadth, 30.4; length of upper molar series, 8.3.

variabilis morulus (*Sciurus*), Bangs, N. Eng. Zoöl. Club, II, 1900, p. 43.
variabilis True, Proc. U. S. Nat. Mus., VII, 1884, p. 596.
LION HILL SQUIRREL.

Type locality. Loma del Leon, Panama.

Genl. Char. Skull similar to that of *S. variabilis* from Colombia, but wider and heavier.

Color. Upper parts mixed yellow ferrugineous and blackish brown; dorsal region darker, blackish at base of tail; upper surface of legs like back of arms, ferrugineous; chin, lips, and cheeks tawny olive; under parts bright ferrugineous; tail above blackish at base and tip, remainder bright ferrugineous, beneath tawny olive and blackish, outer margin ferrugineous, tip black.

Measurements. Male. Total length, 435–490; tail vertebræ, 200–235; hind foot, 55; ear, 20–25. Skull: basal length, 46.2; occipito-nasal length, 55.2; zygomatic width, 34; mastoid width, 23.6; interorbital width, 17.8; width behind postorbital processes, 20.2; length of nasals, 17; length of palate to palatal notch, 25; upper tooth row, 9.4; lower tooth row, 10; mandible, 32.2.

D. Aræosciurus.

Premolars, $\frac{1-1}{1-1}$. Skull broad, depressed between orbits; superior outline curved greatly at occipital region; orbital region very broad; postorbital process curving downward, and pointed; nasals long, extending posteriorly to end of premaxillæ.

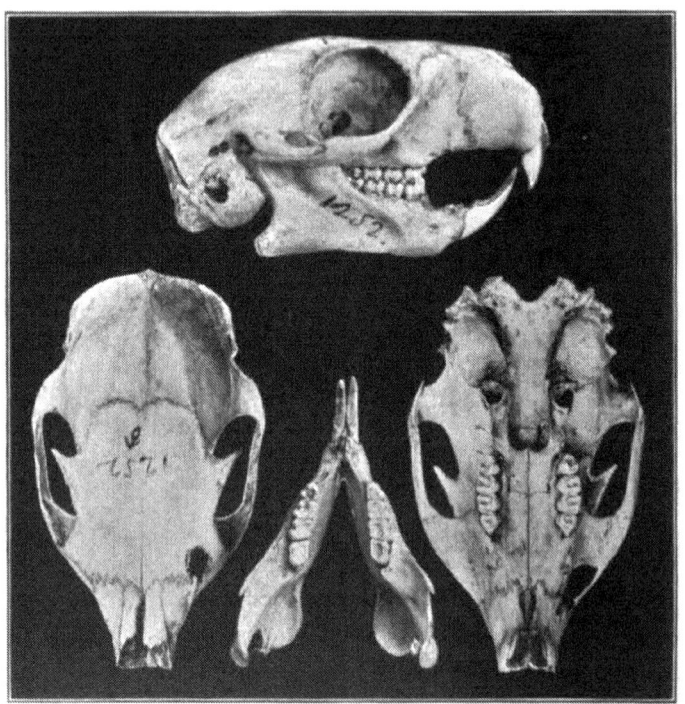

FIG. 16. SCIURUS (ARÆOSCIURUS) NAYARITENSIS.
No. 4741 Field Columbian Mus. Coll. Nat. size.

KEY TO THE SPECIES AND SUBSPECIES OF THE SUBGENUS.

A. Size large.   Premolars, $\frac{1-1}{1-1}$.   PAGE
    a. Back with longitudinal black band...........*S. oculatus*   107
    b. Back without longitudinal black band.
        a.' Under parts white.
            a." Above gray, washed with blackish; tail
                beneath yellowish gray or brown ....*S. o. tolucæ*   107
            b." Above grizzled yellowish brown; tail
                beneath grizzled yellowish gray ....*S. alleni*   108

74. oculatus (*Sciurus*), Peters, Monatsb. K. Preuss. Akad. Wiss. Berl., 1863, p. 653.

capistratus Licht., Abh. K. Akad. Wiss., 1827, p. 116. (nec Bosc.)

carolinensis Sauss., Rev. Mag. Zoöl., 2me Sér., XIII, 1861, p. 4. (nec Gmel.)

hypopyrrhus Allen, Bull. U. S. Geol. Surv. Terr., IV, 1878, p. 881. (Part.)

niger melanonotus Thomas, Proc. Zoöl. Soc., 1870, p. 73, pl. VI.

BLACK-BACKED SQUIRREL.

Type locality. Las Vigas? State of Vera Cruz, Mexico.

Geogr. Distr. Northwestward through States of Vera Cruz, Hidalgo, Queretaro to southeastern San Luis Potosi, Mexico. Altitude, 6,000-12,000 feet.

Genl. Char. Size large; skull broad and heavy.

Color. Upper parts dark gray, with a black band from middle of crown to base of tail; orbital ring whitish; cheeks and sides of neck grizzled gray with a buff tinge, not always, however, present; ears dull gray with white or buff basal patches; hands and feet grizzled gray or buffy; outside of arms and thighs gray, sometimes rusty brownish on the thighs; under parts white with buffy tinge to ochraceous buff; tail above black washed with white, beneath grizzled gray, tinged with yellowish, bordered with black and edged with white.

Measurements. Total length, 544; tail vertebræ, 254; hind foot, 69. Skull: average of three; basilar length, 54.6; palatal length, 28.5; interorbital width, 20.3; zygomatic width, 36.3; length of upper molar series, 11.

a.—tolucæ (*Sciurus*), Nelson, Proc. Biol. Soc. Wash., XII, 1898, p. 148.

TOLUCA SQUIRREL.

Type locality. North slope of the Volcano of Toluca, State of Mexico, Mexico.

Geogr. Distr. Sierra Madre from Toluca Volcano, State of Mexico, to border of State of Michoacan; southern and western parts of State of Queretaro; central and eastern parts of State of Guanajuato, and southern part of State of San Luis Potosi, Mexico.

Genl. Char. Similar to *S. oculatus*, but paler. Skull: nasals narrow.

Color. Top of head and back gray, washed with blackish; flanks and outside of arms and legs grizzled gray, tinged with yellowish; orbital ring grayish white; sides of head and ears gray, tinged with buff; white basal patches on ears; under parts whitish; hands and feet grayish white, tinged with buff; tail black above, washed with white, beneath yellowish gray or brown, bordered with black and edged with white.

Measurements. Total length, 520; tail vertebræ, 260; hind foot, 66. Skull: average of two; basilar length, 54; palatal length, 29; interorbital width, 20.5; zygomatic width, 36.7; length of upper molar series, 11.

75. alleni (*Sciurus*), Nelson, Proc. Biol. Soc. Wash., XII, 1898, p. 147.

carolinensis? Baird, N. Am. Mamm., 1857, p. 263.
carolinensis Allen, Mon. N. Am. Roden., 1877, p. 706.
carolinensis Alston, Proc. Zoöl. Soc., 1878, p. 658.
arizonensis Allen, Bull. Am. Mus. Nat. Hist., N. Y., 1891, p. 222. (Part.)

ALLEN'S SQUIRREL.

Type locality. Monterey, State of Nuevo Leon, Mexico.

Geogr. Distr. From Monterey, State of Nuevo Leon, into State of Tamaulipas, Mexico. Altitude, 2,000–8,500 feet.

Genl. Char. Similar to *S. carolinensis*, back uniform in color.

Color. Above grizzled yellowish brown, grayest on flanks; arms, hands, and feet whitish gray; thighs like flanks; orbital ring whitish; under parts white; faint grayish lateral line; tail like back all around the base, above black washed with white, beneath grizzled yellowish gray, bordered with black and edged with white; ears brownish gray.

Measurements. Total length, 471; tail vertebræ, 217; hind foot, 60. Skull: average of five; basilar length, 50.4; palatal length, 26.3; interorbital width, 18.4; zygomatic width, 33.7; length of upper molar series, 10.3.

76. nayaritensis (*Sciurus*), Allen, Bull. Am. Mus. Nat. Hist., N. Y., 1890, p. VII, footnote, and p. 185.

alstoni Allen, Bull. Am. Mus. Nat. Hist., N. Y., 1889, p. 167. (nec Anderson.)

NAYARIT SQUIRREL.

Type locality. Sierra de Valparaiso, State of Zacatecas, Mexico.

Geogr. Distr. States of Jalisco and Zacatecas, Mexico.

Genl. Char. Similar to *S. griseiflavus*; ears high and broad.

Color. Above gray, as are also the sides; outer surface of limbs pale gray; rest of pelage white; orbital ring grayish white; ears grayish; tail above black and white mixed, fringed on sides with white, beneath cinnamon rufous.

Measurements. Total length (skin), 260–304; tail vertebræ, 254–287; ear: height, 23–25; width at base, 21.5–24. Skull: average of five; basilar length, 55.5; palatal length, 28.4; interorbital width, 37.1; length of upper molar series, 11.7.

77. arizonensis (*Sciurus*), Coues, Am. Nat., 1867, p. 357. Elliot, Syn. N. Am. Mamm., 1901, p. 59.

colliæi Allen, Mon. N. Am. Rod., 1877, p. 738. (nec Rich.)
ARIZONA GRAY SQUIRREL.

Type locality. Fort Whipple, Yavapai County, Arizona.

Geogr. Distr. State of Nuevo Leon, Mexico (San Pedro Mines, Allen), to Arizona. Texas?

Genl. Char. Smaller than *S. carolinensis*; tail as long as head and body; soles naked to heel.

Color. Above mixed gray, black, white, and tawny, the latter predominating; sides and limbs outside grizzled gray and white; beneath and inside of limbs pure white; tail above basally gray and white, remainder black mixed with white and fringed broadly with white, beneath tawny in the center, bordered with black and fringed with white.

Measurements. Total length, 457.4; tail vertebræ, 241.3; hind foot, 58.4; height of ear, 20.3. Skull: occipito-nasal length, 63; Hensel, 48; zygomatic width, 35; interorbital width, 20; palatal length, 21; length of upper molar series, 11.

a.—huachuca (*Sciurus*), Allen, Bull. Am. Mus. Nat. Hist., 1894, p. 349. Elliot, Syn. N. Am. Mamm., 1901, p. 60.
HUACHUCA SQUIRREL.

Type locality. Huachuca Mountains, southern Arizona.

Geogr. Distr. Huachuca Mountains, Arizona, into State of Sonora, Mexico.

Genl. Char. Similar to *S. arizonensis*, but nearly uniform gray above, only a trace of fulvous dorsal stripe.

Color. Upper parts grizzled gray; dorsal stripe nearly obsolete; sides lighter gray; nape patch pale fulvous; under parts of body and limbs pure white; tail above black sprinkled with white and fringed broadly with white, beneath pale chestnut, bordered with black and broadly fringed with white.

Measurements. Total length, 540; tail vertebræ, 265; hind foot, 70; ear, 34. Skull: average of four; basal length, 53.5; palatal length,

27.5; interorbital width, 19.9; zygomatic width, 36.4; length of upper molar series, 11.4.

78. apache (*Sciurus*), Allen, Bull. Am. Mus. Nat. Hist., 1893, p. 29.
    Elliot, Syn. N. Am. Mamm., 1901, p. 58.
    griseiflavus Thomas, Proc. Zoöl. Soc., 1882, p. 372.   (nec Gray.)
    niger ludovicianus Thomas, Proc. Zoöl. Soc., 1890, p. 73 (footnote).
APACHE SQUIRREL.
    Type locality.   Mountains of northwestern Chihuahua, Mexico.
    Geogr. Distr.   Sierra Madre in States of western Durango, northwestern Chihuahua, eastern Sonora, and northeastern Sinaloa, Mexico; also in Chiricahua Mountains, southern Arizona.
    Genl. Char.   Similar in size and color to *S. r. texensis Bach.*, but darker and with ferrugineous legs, and other distribution of hues.
    Color.   Upper parts iron gray, usually washed with yellow; crown and back blackish; sides of head mixed gray, black and fulvous; orbital ring buffy white or fulvous; flanks washed with pale yellowish; outside of legs suffused with rusty; outside of arms and under parts varying from buffy yellow to orange yellow; tail at base like back, above black washed with yellow and fringed with white, beneath orange or rusty rufous with a black border and fringed with pale yellowish; ears gray with a buff tinge; hands buffy or orange yellow; feet darker.
    Measurements.   Total length, 565; tail vertebræ, 279; hind foot, 79.   Skull: average of five; basilar length, 56.4; palatal length, 28.7; interorbital width, 21.5; zygomatic width, 37.5; length of upper molar series, 11.9.

E. Parasciurus.

Premolars, $\frac{1-1}{1-1}$.   Skull: braincase narrow at occiput, not inflated over parietal region, long, narrow; rostrum long, broad; nasals broad; molar series large, heavy.

rufiventer texensis (*Sciurus*), (Bach.), Proc. Zoöl. Soc., 1838, p. 86.
    ludovicianus limitis Baird. Proc. Acad. Nat. Scien. Phil., 1855,
        p. 331.   Elliot, Syn. N. Am. Mamm., 1901, p. 53.
TEXAS FOX SQUIRREL.
    Type locality.   Devil's River, Valverde County, Texas.
    Geogr. Distr.   Texas and adjacent parts of the States of Nuevo Leon and Coahuila, Mexico.
    Genl. Char.   Smaller than *S. carolinensis*; hairs short and close pressed; upper molars four; feet small; colors pale.
    Color.   Upper parts mixed cinnamon and black; sides of head, limbs, and under parts, light cinnamon brown; feet above tinged with

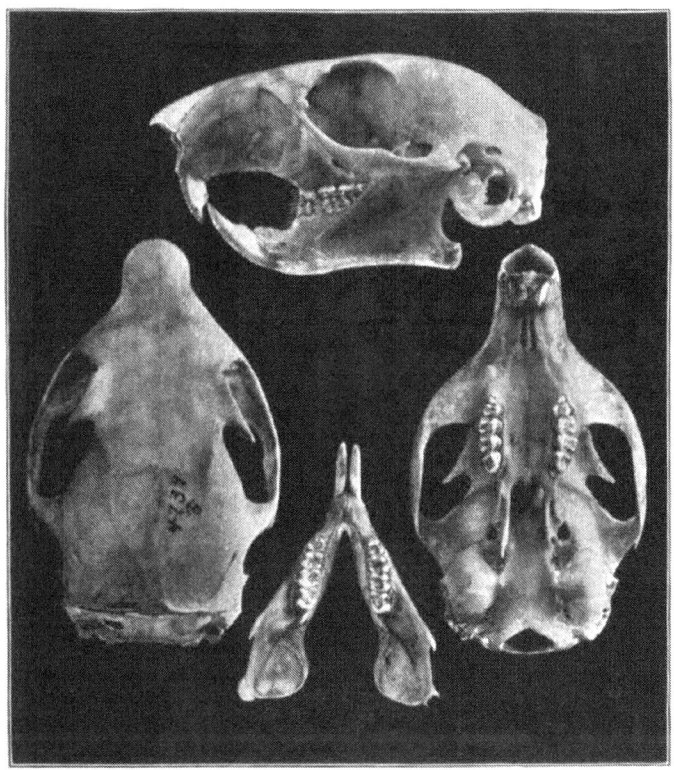

FIG. 17. SCIURUS (PARASCIURUS) RUFIVENTER TEXENSIS.
No. 4739 Field Columbian Mus. Coll. Nat. size.

rusty; tail above similar to back but more yellowish, beneath uniform cinnamon, darker than belly. The color of ventral surface varies from pure white to deep orange among individuals.

Measurements. Total length, 532; tail to end of hairs, 280; hind foot, 64.5. Skull: average of two; basal length, 50; palatal length, 25.5; interorbital width, 18; zygomatic width, 33.5; length of upper molar series, 10.5.

F. Otosciurus, Nelson.

Premolars, $\frac{2-2}{1-1}$. Skull short and very broad, with curved superior outline; rostrum rather short, compressed; nasals narrow posteriorly and extending beyond premaxillæ; molars heavy.

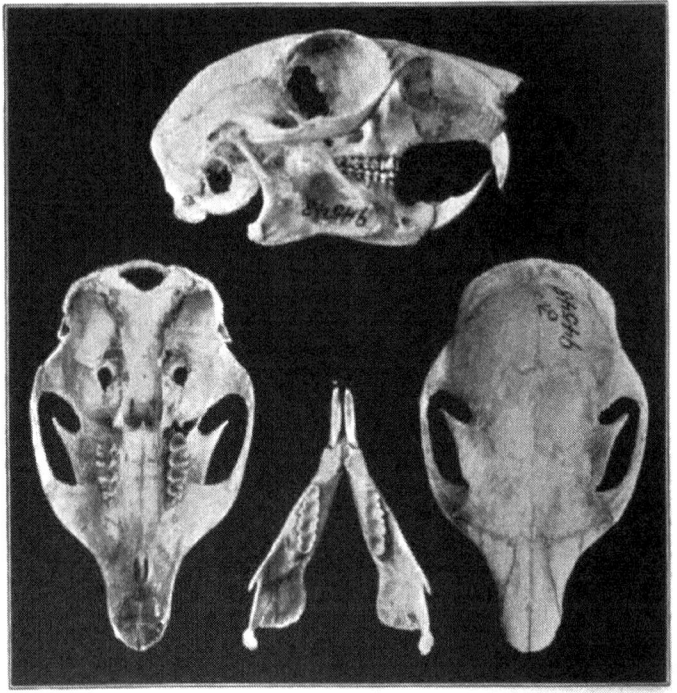

FIG. 18. SCIURUS (OTOSCIURUS) DURANGI.
No. 94548 U. S. Nat. Mus. Coll.   Nat. size.

79. durangi (*Sciurus*), Thomas, Ann. Mag. Nat. Hist., 6th Ser., XI,
    1893, p. 49.
aberti Thos., Proc. Zoöl. Soc., 1882, p. 372. (nec Woodhouse.)
DURANGO SQUIRREL.

Type locality.   Ciudad Ranch (100 miles west of Durango City),
State of Durango, Mexico.

Geogr. Distr.   Sierra Madre, west part of States of Durango and
Chihuahua, Mexico.

Genl. Char.   Similar to *S. aberti*, but less red on back; tail
beneath uniform grizzled gray; dorsal stripe not reaching base of tail.

Color.   Above gray with chestnut dorsal stripe from shoulders to
rump; flanks and outside of hind legs grizzled gray; outside of fore
legs paler gray; lateral black line ill defined; orbital ring brownish
white; ear gray, tufts black; under parts white, base of tail like back,
remainder above black washed with white, beneath grizzled gray,

with black border and subapical bar and white edging; hands white or whitish; feet grizzled gray.

Measurements. Total length, 500; tail vertebræ, 247; hind foot, 70. Skull: average of five; basal length, 50.1; palatal length, 27.3; interorbital width, 19.8; zygomatic width, 35.1; length of upper molar series, 11.2.

G. Echinosciurus.

Premolars, $\frac{2-2}{1-1}$. Skull: broad, depressed between orbits; superior outline curved, sharpest decline posteriorly; occipital region widest; interorbital constriction very slight; rostrum broad and short; outer side of nasals reaching ends of premaxillæ; the nasals shorter than width between orbits; basioccipital and basisphenoid broad, widely separating audital bullæ.

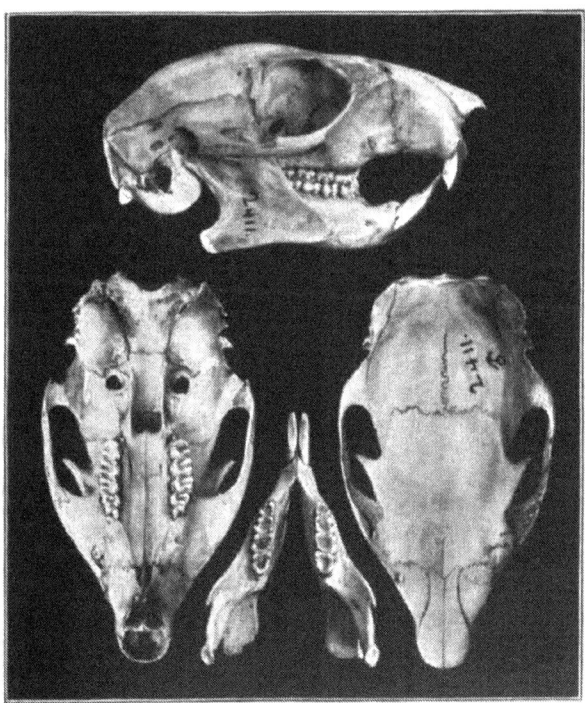

FIG. 19. SCIURUS (ECHINOSCIURUS) AUREIGASTER.
No. 4744 Field Columbian Mus. Coll. Nat. size.

KEY TO THE SPECIES AND SUBSPECIES OF THE SUBGENUS.

A. Size large.   Premolars, $\frac{2-2}{1-1}$.

v. Above pale iron gray; rump and middle of
    back washed with black . . . . *S. goldmani* 130

FIG. XXIX. SCIURUS AUREIGASTER. GOLDEN-BELLIED SQUIRREL.

80. aureigaster (*Sciurus*), F. Cuv., Hist. Nat. Mamm., VI, 1829,
    Livr. LIX, pl. Text.
niger Erxl., Syst. Regn. Anim., 1777, p. 417.
variegatus Desm., Nouv. Dict. d'Hist. Nat., x, 1817, pp. 103–104.
    (nec Erxl., 1777.)
rufiventer Licht., Abh. K. Ak. Wiss. Berl., 1827, p. 116.
leucogaster! F. Cuv., Suppl. d'Hist. Nat. Buffon, I, 1831, pp.
    300–301.
mustelinus Aud. & Bach., Proc. Acad. Nat. Scien. Phil., 1841,
    p. 100.
ferruginiventris ! Aud. & Bach., Proc. Acad. Nat. Scien. Phil., 1841,
    p. 101.
aurogaster ! Aud. & Bach., Quad. N. Am., III, 1851, p. 344.
hypoxanthus I. Geoff., Voy. de la Vénus, Zoöl., 1855, p. 158.
aureogaster! Gray, Ann. Mag. Nat. Hist., 3d Ser., 1867, p. 423.
hypopyrrhus Allen, Bull. Am. Mus. Nat. Hist., N. Y., III, 1891,
    p. 222. (nec Wagl.)
leucops Allen, Bull. Am. Mus. Nat. Hist., N. Y., IX, 1897, p. 198.
GOLDEN-BELLIED SQUIRREL.   *Ardilla* in Mexico, common name for
    all squirrels.
Type locality.   Eastern Mexico.
Geogr. Distr.   Eastern Mexico from southern part of State of
Tamaulipas to northern side of Isthmus of Tehuantepec through

States of Vera Cruz, eastern San Luis Potosi, Queretaro, Puebla, and northeastern Hidalgo and northern Oaxaca, up to 4,000 and 8,000 feet.

Genl. Char. Size large; coloring varied, under parts ferrugineous to dark rufous; tail long.

Color. Above varying from iron gray to whitish gray; nape yellowish brown or rusty rufous; orbital ring deep buff; between eye and ear yellowish brown; chin and cheeks grizzled gray; arms and hands iron gray; feet varying from blackish to iron gray; under parts and sometimes a band over shoulders and outer side of arms, bright ferrugineous; ears gray to rusty red; tail at base like back, remainder above black, washed with white, beneath ferrugineous, with a black border edged with white.

Measurements. Total length, 539; tail vertebræ, 265; hind foot, 67. Skull: average of five; basal length, 49.8; palatal length, 26.5; interorbital width, 19.1; zygomatic width, 34.5; length of upper molar series, 11.3.

a.—hypopyrrhus (*Sciurus*), Wagler, Isis, 1831, p. 510.
   morio Gray, Ann. Mag. Nat. Hist., 3d Ser., xx, 1867, p. 424.
   maurus Gray, Ann. Mag. Nat. Hist., 3d Ser., xx, 1867, p. 425.
   variegatus Sumichrast, La Naturaleza, vii, 1887, p. 360.
   aureogaster! Allen, Bull. Am. Mus. Nat. Hist., iii, 1890, p. 181.
     (nec Cuv.)

Fire-bellied Squirrel.

Type locality. Mexico. State of Vera Cruz ?

Geogr. Distr. Southern part State of Vera Cruz, and States of Tabasco, Oaxaca, and Chiapas, Mexico.

Genl. Char. Darker than *S. aureigaster.*

Color. Above grizzled with black, rusty or grayish white; fore part of crown and nose iron gray; nape grizzled rufous or brownish; orbital ring brownish buff; chin and cheeks grayish; under parts deep ferrugineous, this color covering arms and forming a band on shoulder; outside of thighs grizzled like back; hands and feet black or grizzled with gray; tail at base like back, rest black washed with white, beneath ferrugineous with a black border and white edge; sometimes for ferrugineous black is substituted; ears gray or reddish brown, sometimes with black border, and a basal patch in winter, grayish white.

Measurements. Total length, about 522; tail vertebræ, 266; hind foot, 67. Skull: average of five; basal length, 52.6; palatal length, 27.1; interorbital breadth, 18.4; zygomatic breadth, 34.8; length of upper molar series, 11.4.

b.—*frumentor* (*Sciurus*), Nelson, Proc. Biol. Soc. Wash., XII, 1898,
    p. 154.
PEROTE SQUIRREL.
    Type locality. Las Vigas, State of Vera Cruz, Mexico.
    Geogr. Distr. "East and north base of the Cape de Perote, and
eastern slope of the Cordillera near Las Vigas, State of Vera Cruz,
Mexico, in pine and oak forests, at 6,000–8,000 feet."
    Genl. Char. Nuchal and rump patches yellowish or rufous brown,
and under parts gray, sometimes washed with rufous.
    Color. Upper parts (except nape and rump), nose, forehead,
sides, and outside of arms and legs, grizzled iron gray; patches
on nape and rump, and sides of head pale brown or reddish,
tinged with black, palest on sides of head; chin and throat grayish;
under parts gray, sometimes tinged with rust red; tail above like
back at base, remainder black washed with white, beneath yellowish
or dark ferrugineous, with a black border and white edge; ears
varying in color, sometimes grayish, again similar to the nuchal
patch, occasionally having a black border; hands and feet iron gray
to black.
    Measurements. Total length, "average of five adults, 504.6; tail
vertebræ, 249.2; hind feet, 68.8." (Nelson.) A topotype: total
length, 502; tail, 250; hind foot, 69. Skull: average of four; basal
length, 52.5; palatal length, 26.1; interorbital width, 19.7; zygomatic width, 34.4; length of upper molar series, 11.2.

81. poliopus (*Sciurus*), Fitz., Sitzung. K. Akad. Wiss. Wien, I, 1867,
    p. 478.
    albipes Wagn., Abhandl. Math.–Phys. Cl. K. Bayer Akad. Wiss.
        München, II, 1837, pp. 501–506. (nec Kerr.)
    varius Wagn., Suppl. Schreb. Säugeth., III, 1843, p. 168.
    rufipes Fitz., Sitzung. K. Akad. Wiss. Wein, I, 1867, p. 478.
    leucops Gray, Ann. Mag. Nat. Hist., 3d Ser., XX, 1867, p. 427.
    variegatus Alston, Proc. Zoöl. Soc., 1878, p. 660.
    wagneri Allen, Bull. Am. Mus. Nat. Hist., N. Y., 1898, p. 453.
OAXACA SQUIRREL.
    Type locality. Cerro San Felipe, State of Oaxaca, Mexico.
    Geogr. Distr. Mountains about Valley of Oaxaca, except those
on the west, 7,500–11,000 feet altitude.
    Genl. Char. Size large.
    Color. Back and outside of arms and legs gray, mixed with
yellowish or brownish hairs; nose and forehead iron gray; patch on
nape and rump pale brown tinged with black; sides of head gray;
orbital ring, chin, and throat grayish white; under parts ferrugineous

red; tail at base all around like back; above black washed with white, beneath mixed yellowish, or reddish brown bordered with black and edged with white; hands and feet white.

Measurements. Total length, 523; tail vertebræ, 263; hind foot, 71. Skull: average of five; basal length, 52.2; palatal length, 27.3; interorbital width, 18.6; zygomatic width, 34.5; length of upper molar series, 10.7.

a.—hernandezi (*Sciurus*), Nelson, Proc. Wash. Acad. Scien., I, 1899, p. 48.
    albipes quercinus Nels., Proc. Biol. Soc. Wash., XII, 1898, p. 150. (nec Erxl.)
    wagneri quercinus Allen, Bull. Am. Mus. Nat. Hist., 1898, p. 453.
    albipes hernandezi Nels., Science, N. Ser., VIII, 1898, p. 783.
OAK WOODS SQUIRREL.

Type locality. Mountains 15 miles west of the city of Oaxaca, State of Oaxaca, Mexico.

Geogr. Distr. Mountains west of Oaxaca Valley into southern part of State of Puebla and southeastern portion of State of Guerrero, Mexico. Altitude, 8,000–9,000 feet.

Genl. Char. Similar to *S. poliopus*, but paler.

Color. Above and outside of arms and legs pale gray and yellowish; nose and forehead grizzled gray tinged with black, faint yellowish and black patch on nape; ears gray, with white basal patch; orbital ring whitish; chin and throat white; under parts white, sometimes tinged with buff; hands and feet white; tail above and below at base like back, remainder of upper part black washed with white, beneath yellowish gray or rusty grizzled, with an indistinct black border and broad white edging.

Measurements. Total length, 540; tail vertebræ, 273; hind foot, 68. Skull: average of five; basal length, 51.5; palatal length, 26.5; interorbital width, 19.5; zygomatic width, 34.6; length of molar series, 11.

b.—nemoralis (*Sciurus*), Nels., Proc. Wash. Acad. Scien., I, 1899, p. 50.
    albipes nemoralis Nels., Proc. Biol. Soc. Wash., XII, 1898, p. 151.
    wagneri nemoralis Allen, Bull. Am. Mus. Nat. Hist., 1898, p. 454.
MICHOACAN SQUIRREL.

Type locality. Patzcuaro, State of Michoacan, Mexico.

Geogr. Distr. Volcano of Toluca, State of Mexico, to Nahuatzin, State of Michoacan. Altitude, 7,000–12,000 feet.

Genl. Char. Similar to *S. p. hernandezi*, but darker above Possessing a melanistic phase in certain localities.

Color. Back and outside of arms and legs iron gray, occasionally with a rusty tinge; nose and forehead iron gray tinged with black; nape patch grizzled yellowish brown to chestnut brown; rump patch paler; ears grizzled gray, basal patch white; orbital ring and sides of head whitish; feet and hands white; under parts white, sometimes yellow; tail above black washed with white, beneath varying from grizzled gray to yellowish brown, indistinct black border and white edge.

Measurements. Total length, 530; tail vertebræ, 265; hind foot, 70. Skull: average of three; basal length, 51; palatal length, 26.8; interorbital width, 18.6; zygomatic width, 34.3; length of upper molar series, 11.

c.—cervicalis (Sciurus), Allen, Bull. Am. Mus. Nat. Hist., 1890,
     p. 183.
leucops Allen, Mon. N. Am. Rod., 1877, p. 753. (nec Gray.)
variegatus Alston, Biol. Cent. Am., Mammalia, 1, 1880, p. 127.
aureogaster leucops Allen, Bull. Am. Mus. Nat. Hist., 1889, p. 166.
wagneri cervicalis Allen, Bull. Am. Mus. Nat. Hist., 1898, p. 454.
poliopus cervicalis Nels., Proc. Wash. Acad. Nat. Scien., 1, 1899,
     p. 51.

COLIMA MOUNTAIN SQUIRREL.

Type locality. Hacienda San Marcos, Tonila, State of Jalisco, at east base of Sierra Nevada de Colima, Mexico.

Geogr. Distr. From the Sierra Nevada de Colima along high mountains north to Ameca, State of Jalisco, and east into western part of State of Michoacan, Mexico. Altitude, 6,000–12,000 feet, and down occasionally to 4,000.

Genl. Char. Dark upper parts, and iron gray under surface of tail.

Color. Back, outside of arms and legs, iron gray, sometimes with yellowish hairs intermixed; nose and forehead showing black and grayish white; nape and rump patches fulvous or rusty brown grizzled with black; orbital ring whitish; under parts white, in some specimens grizzled with black; hands and feet gray, paler than arms and legs; tail like back at base, rest above black washed with white, beneath iron gray with a black border and white edge.

Measurements. Total length, 515; tail vertebræ, 248; hind foot, 68. Skull: average of five; basal length, 51.2; palatal length, 27.6; interorbital width, 19.8; zygomatic width, 34.9; length of upper molar series, 11.7.

d.—colimensis (Sciurus), Nelson, Proc. Wash. Acad. Scien., 1, 1899,
     p. 52.
leucops Allen, Mon. N. Am. Roden., 1877, p. 753. (Part.)

albipes colimensis Nels., Proc. Biol. Soc. Wash., XII, 1898, p. 152.
wagneri colimensis Allen, Bull. Am. Mus. Nat. Hist., 1898, p. 454.
COLIMA SQUIRREL.

Type locality. Hacienda Magdalena, State of Colima, Mexico.

Geogr. Distr. Coast region, State of Colima, Mexico.

Color. Back grizzled gray; nose and forehead grizzled iron gray; nape and patch on rump yellowish brown to rusty, tinged with black; sides of head grizzled gray; orbital ring buff; outside of arms, legs, and hands iron gray; feet blackish, mixed with gray; ears yellowish brown or rusty, with a whitish patch at base; under parts white; tail at base same as back, remainder black washed with white, beneath grizzled iron gray, bordered with black and edged with white.

Measurements. Total length, 500; tail vertebræ, 258; hind foot, 68. Skull: average of five; basal length, 50.2; palatal length, 26.7; interorbital width, 19.1; zygomatic width, 38.8; length of molar series, 11.2.

e.—effugius (*Sciurus*), Nelson, Proc. Wash. Acad. Scien., I, 1899, p. 54.
albipes effugius Nels., Proc. Biol. Soc. Wash., XII, 1898, p. 152.
wagneri effugius Allen, Bull. Am. Mus. Nat. Hist., N. Y., 1898,
 p. 454.
GUERRERO SQUIRREL.

Type locality. Mountains west of Chilpancingo, State of Guerrero, Mexico.

Geogr. Distr. Forests of Cordillera, State of Guerrero, Mexico. Altitude, 7,500–9,500 feet.

Genl. Char. Similar to *S. poliopus*, but under parts darker.

Color. Above grizzled iron gray, mixed with rusty; nape and indistinct patch on rump rusty rufous, the latter sometimes mixed with black and white; sides of head, base of ears, and around eyes pale rufous; orbital ring whitish; ears reddish brown, with white patches at bases; chin and throat white; outside of arms and legs gray; hands and feet white, sometimes washed with rufous; under parts dark rufous; tail at base same as rump, remainder above black washed with white, beneath dark rufous with black border and white edge.

Measurements. Total length, 498; tail vertebræ, 249; hind foot, 68. Skull: average of four; basal length, 50.9; palatal length, 26.9; interorbital width, 19.1; zygomatic width, 34.5; length of upper molar series, 11.

82. nelsoni (*Sciurus*), Merr., Proc. Biol. Soc. Wash., VIII, 1893,
 p. 144.

NELSON'S SQUIRREL.

Type locality. Huitzilac, State of Morelos, Mexico.

Geogr. Distr. Mountains south and west of the Valley of Mexico, and south of Valley of Toluca in the States of Mexico and Morelos.

Genl. Char. Size large, colors dark.

Color. Above blackish, grizzled with yellowish brown; top of head, ears, outside of arms and legs, hands and feet, black; sides of head grayish black, or yellowish brown; under parts blackish brown tinged with yellowish or rusty; tail all around at base like back, rest of upper part black washed with grayish white, beneath, yellowish gray and black with white edge; feet black.

Measurements. Total length, 520; tail vertebræ, 256; hind foot, 70. Skull: average of five; basal length, 50.7; palatal length, 26.7; interorbital width, 19.4; zygomatic width, 34.9; length of upper molar series, 11.9.

a.—hirtus (*Sciurus*), Nelson, Proc. Biol. Soc. Wash., XII, 1898, p. 153.

MOUNTAIN SQUIRREL.

Type locality. Tochimilco, State of Puebla, Mexico.

Geogr. Distr. Mounts Popocatepetl and Iztaccihuatl, Mexico. Altitudes 8,000–12,000 feet, in States of Mexico, Puebla, and Morelos.

Genl. Char. Similar to *S. nelsoni*; gray area on back; tail broad, bushy.

Color. Top of head and upper parts except middle of back, grayish brown; nose and middle of back bluish gray grizzled; ears, like crown, with white patch at the bases; chin and cheek gray; under parts ferrugineous; tail black above, washed with white, beneath grizzled rusty brown, bordered with black and edged with white; hands and feet gray.

Measurements. Total length, 525; tail vertebræ, 260; hind foot, 68. Skull: average of five; basal length, 50.2; palatal length, 26.5; interorbital width, 19.4; zygomatic width, 35.1; length of upper molar series, 11.8.

83. colliæi (*Sciurus*), Rich., Voy. Blossom, Zoöl., 1839, p. 8.

hypopyrrhus Alston, Proc. Zoöl. Soc., 1878, p. 662. (Part.)

COLLIE'S SQUIRREL.

Type locality. San Blas, Territorio de Tepic, Mexico.

Geogr. Distr. From northern border of Territorio de Tepic to Bay of Banderas, State of Jalicso, Mexico, below 2,500 feet.

Genl. Char. Size large; pelage thin; tail long.

Color. Above yellowish gray, tinged with black; nose iron gray, crown darker than back; outside of arms and legs dark gray; hands and feet whitish; orbital ring whitish; ears yellowish brown, with whitish basal patches; flanks and shoulders on sides gray; under parts white; tail at base all around like back, remaining portion above black, washed with white, beneath grizzled black and yellowish gray, or black, gray, and yellowish brown, bordered narrowly with black and edged with white.

Measurements. Total length, 508; tail vertebræ, 265; hind foot, 65. Skull: average of four; basal length, 49; palatal length, 25.6; interorbital width, 18.4; zygomatic width, 32.6; length of upper molar series, 11.

a.—nuchalis (Sciurus), Nelson, Proc. Wash. Acad. Scien., 1, 1899, p. 59.

MANZANILLO SQUIRREL.

Type locality. Manzanillo, State of Colima, Mexico.

Geogr. Distr. From State of Michoacan to Bay of Banderas, State of Jalisco, Mexico, and inland to mountains near San Sebastian in the same State, below 3,000 feet.

Genl. Char. Like *S. collicæi,* but ears rusty rufous; nape and shoulders buffy yellow; rump washed with black.

Color. Above yellowish gray, lower part of back and rump washed with black; nape mixed with black; crown like nape; ears rusty rufous, with rusty white basal patches; cheeks and sides of nose yellowish brown; orbital ring yellowish; flanks gray, paler than back; outside of arms, hands, and feet iron gray; shoulders washed with gray; under parts white; base of tail like back, remainder above black washed with white; beneath grizzled reddish yellow, bordered with black and edged with white.

Measurements. Total length, 525; tail vertebræ, 272; hind foot, 71. Skull: average of five; basal length, 52.3; palatal length, 27.6; interorbital width, 20.6; zygomatic width, 35.9; length of upper molar series, 11.4.

84. sinaloensis *(Sciurus),* Nelson, Proc. Wash. Acad. Scien., 1, 1899, p. 60.

SINALOA SQUIRREL.

Type locality. Mazatlan, State of Sinaloa, Mexico.

Geogr. Distr. Southern and central parts of the State of Sinaloa, Mexico, below 2,500 feet.

Genl. Char. Similar to *S. collicæi,* but paler.

Color. Crown and rest of upper parts pale reddish yellow mixed with black; nose and forehead grizzled yellowish gray; sides of neck, shoulders, flanks, arms, legs, hands, and feet grayish white; orbital ring and side of nose gray; cheeks yellowish brown; ears yellowish with basal brownish yellow patches; under parts white; tail above grizzled brownish yellow, washed with white, beneath similar in color, bordered with black and edged with white.

Measurements. Total length, 524; tail vertebræ, 255; hind foot, 62. Skull: basal length, 53; palatal length, 26.3; interorbital width, 20; zygomatic width, 34; length of upper molar series, 11.

85. truii (*Sciurus*), Nelson, Proc. Wash. Acad. Scien., I, 1899, p. 61. TRUE'S SQUIRREL.

Type locality. Camoa, Rio Mayo, State of Sonora, Mexico.

Geogr. Distr. Southwestern part of the State of Sonora and northern portion of the State of Sinaloa, Mexico.

Genl. Char. Similar to *S. colliæi*, but back dark yellowish and skull differently proportioned.

Color. Upper parts grizzled dark yellowish; sides of head yellowish gray; ears rusty; basal patches same color; outside of arms, legs, hands and feet dark gray; under parts white; tail like back at base, remainder above mixed black and dark yellowish tinged with white; beneath similar, with an ill-defined black border and white edge.

Measurements Total length, 485; tail vertebræ, 254; hind foot, 66. Skull: average of four; basal length, 47.7; palatal length, 24.3; interorbital width, 18.3; zygomatic width, 33.2; length of upper molar series, 11.

86. socialis (*Sciurus*), Wagn., Abh. Math. Phys. Cl. K. Bayer. Akad. Wiss., München, II, 1837, pp. 501–7, pl. v.

aureigaster Allen, Mon. N. Am. Roden., 1877, p. 750. (Part.)

leucops Allen, Mon. N. Am. Roden., 1877, p. 753. (Part.)

variegatus Alston, Proc. Zoöl. Soc., 1878, p. 660. (Part.)

TEHUANTEPEC SQUIRREL.

Type locality. Near Tehauntepec City, State of Oaxaca, Mexico.

Geogr. Distr. Puerto Angel, State of Oaxaca, south to Tonala and Tuxtla, State of Chiapas, Mexico, below 3,000 feet.

Genl. Char. Color variable; *nape patch, when present, rusty rufous; under parts from white to ferrugineous.

Color. Above mixed white and black tinged with yellowish or rufous; nape blackish rusty rufous; rump patch sometimes exhibiting rusty rufous; cheeks, space around eyes, and bases of ears yellowish brown; orbital ring dull fulvous; under parts white; tail like rump all around the base, remainder above black washed with white, beneath from fulvous gray to orange, bordered by black and edged with white; hands and feet like back.

---

* Specimens from near Tonala are usually without nape patch.

Measurements. Total length, 540; tail vertebræ, 280; hind foot, 65. Skull: average of five; basal length, 51.8; palatal length, 26.9; interorbital width, 18.1; zygomatic width, 33.9; length of upper molar series, 11.3.

a.—cocos (*Sciurus*), Nelson, Proc. Biol. Soc. Wash., XII, 1898, p. 155.
  aureogaster! I. Geoff., Voy. Vénus, Zoöl., Atlas, 1846, pl. 10. (nec Cuv.)
ACAPULCO SQUIRREL.
  Type locality. Acapulco, State of Guerrero, Mexico.
  Geogr. Distr. From Acapulco, State of Guerrero to Jamiltepec, State of Oaxaca, Mexico, below 1,500 feet.
  Genl. Char. Nape and rump patches distinct; under parts variable, white to ferrugineous.
  Color. Back, outside of arms, legs, hands, and feet grayish or creamy white; nape and rump blackish rufous or chestnut; sides of head and base of ears and around eyes yellowish; under parts white to dark ferrugineous, varying among individuals; ears rufous; tail above black washed with white, sometimes tinged with rufous, beneath reddish buff to dark ferrugineous, bordered with black and edged with white.
  Measurements. Total length, 520; tail vertebræ, 261; hind foot. 68. Skull: average of five; basal length, 50.8; palatal length, 26.5; interorbital width, 19.3; zygomatic width, 33.7; length of upper molar series, 10.6.

87. griseiflavus (*Macroxus*), Gray, Ann. Mag. Nat. Hist., 3d Ser., XX, 1867, p. 427.
  ludovicianus Tomes, Proc. Zoöl. Soc., 1861, p. 281. (nec. Custis.)
  aureigaster Allen, Mon. N. Am. Roden., 1877, p. 750.   (Part.)
  affinis Alston, Proc. Zoöl. Soc., 1878, p. 660.
GUATEMALA SQUIRREL.
  Type locality. Guatemala.
  Geogr. Distr. Southeastern part of State of Chiapas, Mexico, into Guatemala, 7,000–10,000 feet altitude.
  Genl. Char. Size large; color variable, usually grizzled yellowish brown and gray.
  Color. Upper parts grizzled yellowish brown, occasionally washed with white, this most conspicuous in winter specimens; outside of arms, legs, and flanks like back; hands and feet grizzled yellow or yellowish brown; sides of head and ears gray, the latter with grayish basal patches; chin and throat grayish fulvous; under parts rufous

or gray washed with fulvous; base of tail like back, rest above black washed with white, beneath grizzled yellowish or brownish, bordered with black and edged with white.

Measurements. Total length, 547; tail vertebræ, 270; hind foot, 68. Skull: average of four; basal length, 51.8; palatal length, 27.2; interorbital width, 19; zygomatic width, 34.4; length of upper molar series, 11.4.

a.—chiapensis (*Sciurus*), Nelson, Proc. Wash. Acad. Scien., I, 1899, p. 69.

CHIAPAS SQUIRREL.

Type locality. San Cristobal, State of Chiapas, Mexico.

Geogr. Distr. Mountains of central portion of State of Chiapas, Mexico. Altitude, 7,500–9,500 feet.

Genl. Char. Similar to *S. griseiflavus*, but grayer above, and feet whitish.

Color. Upper parts and outside of arms and legs grizzled gray; ears grayish white, with white basal patches; orbital ring pale fulvous; sides of head fulvous brown; sides of nose, lower part of cheeks, chin and throat, whitish gray; hands and feet grayish white; under parts rufous; base of tail all around like back, above black, washed with white, beneath yellowish brown or rufous, bordered with black and edged with white.

Measurements. Total length, 506; tail vertebræ, 253; hind foot, 70. Skull: average of five; basal length, 50.9; palatal length, 26.6; interorbital width, 19.1; zygomatic width, 34.3; length of upper molar series, 11.4.

88. yucatanensis (*Sciurus*), Allen, Bull. Am. Mus. Nat. Hist., 1897, p. 5. *carolinensis yucatanensis* Allen, Mon. N. Am. Roden., 1877, p. 705. *carolinensis* Alston, Proc. Zoöl. Soc., 1878, p. 658.

YUCATAN SQUIRREL.

Type locality. Merida, Yucatan, Mexico.

Geogr. Distr. Peninsula of Yucatan, Mexico.

Genl. Char. Pelage coarse, bristly; ear tufts sometimes present.

Color. Upper parts and flanks grizzled gray and black, sometimes tinged with yellowish brown; sides of head pale gray; orbital ring whitish; outside of arms and legs, with feet and hands, like back; ears dark gray; some specimens have whitish basal patches; tufts yellowish white, not always present; under parts whitish or grayish white; base of tail all around like back, remainder above black, washed with white, beneath grizzled gray, with a narrow black border and white edge.

Measurements. Total length, 460; tail vertebræ, 230; hind foot, 55. Skull: basal length, 45; palatal length, 23; interorbital width, 16.2; zygomatic width, 19; length of upper molar series, 9.5.

a.—baliolus (*Sciurus*), Nelson, Proc. Biol. Soc. Wash., XIV, 1901, p. 131.

SWARTHY SQUIRREL.

Type locality. Apazote, State of Campeche, Mexico.

Geogr. Distr. Southern part of State of Campeche, and eastern part of State of Tabasco, Mexico.

Genl. Char. Similar to *S. yucatanensis*, but darker.

Color. Above blackish gray suffused with buff; beneath iron gray; legs and hands grizzled with buff or gray; feet black; tail black, washed with gray.

Measurements. Total length, 464; tail vertebræ, 238; hind foot, 59.

89. thomasi (*Sciurus*), Nelson, Proc. Wash. Acad. Scien., I, 1899, p. 71.

> *boothiæ* (nec Gray), Allen, Mon. N. Am. Rodent., 1877, pp. 741–746. (Part. Costa Rica.)
> *hypopyrrhus* Alston, Proc. Zoöl. Soc., 1878, pp. 662–664. (Part. Costa Rica specimens.)

THOMAS' SQUIRREL.

Type locality. Talamanca, Costa Rica.

Geogr. Distr. "Humid tropical forests of eastern Costa Rica."

Genl. Char. Hair on back coarse, stiff, glossy; tail long, narrow.

Color. Crown and upper parts of body, sides of neck, outer side of arms and legs, and base of tail black, the yellowish under color appearing; chin and sides of head grayish brown; under parts ferrugineous, irregularly varied with white; tail above black, washed with white, beneath grizzled black and yellowish brown, bordered with black and edged with white; feet black or grizzled ferrugineous; ears blackish, with basal ferrugineous patch, and black tufts.

Measurements. Average total length, 517.5; tail, 246.5; hind foot, 62.2. Skull: basal length, 50; palatal length, 26.2; interorbital width, 22; zygomatic width, 35.5; length of upper molar series, 11.5.

90. adolphei (*Macroxus*), Less., Nouv. Tabl. Règn. Anim. Mamm., 1842, p. 112.

> *boothiæ* Allen, Mon. N. Am. Rodent., 1877, p. 741. (Part. Nicaragua.)
> *hypopyrrhus* Alston, Biol. Centr. Amer. Mamm., I, 1880, p. 128. (Part.)

NICARAGUA SQUIRREL.

Type locality. Realejo, Nicaragua.

Geogr. Distr. Lowlands on west coast of Nicaragua.

Genl. Char. Hairs of back coarse, stiff, shiny; under fur short.

Color. Upper parts and upper portion of thighs, dark brown, washed with grayish; flanks paler; crown iron gray; cheeks, sides of neck, and nape grayish brown; under parts, arms, part of shoulders, and inside of thighs dark reddish chestnut; chin, throat, axillar region, and between hind legs white; tail above black washed with white, beneath rusty brown bordered with black and edged with white; hands and feet grizzled chestnut; ears iron gray, with white patches at base.

Measurements. Total length, 440; tail vertebræ, 199; hind foot, 65.

a.—dorsalis (*Sciurus*), Gray, Proc. Zoöl. Soc., 1848, p. 138.

rigidus Peters, Monatsb. K. Preuss. Akad. Wiss. Berlin, 1863, p. 652.

intermedius Gray, Ann. Mag. Nat. Hist., 3d Ser., xx, 1867, p. 421.

nicoyana Gray, Ann. Mag. Nat. Hist., 3d Ser., xx, 1867, p. 423.

boothiæ Allen, Mon. N. Am. Rodent., 1877, p. 741. (Part. Costa Rica specimens.)

hypopyrrhus Allen, Mon. N. Am. Rodent., 1877, p. 746. (Part. West Costa Rica specimens.)

BANDED-BACK SQUIRREL.

Type locality. Liberia, Costa Rica nec Caracas, Venezuela.

Geogr. Distr. Western Costa Rica, Alajuela to Liberia, Peninsula of Nicoya, and possibly adjacent parts of Nicaragua.

Genl. Char. Tail long, narrow; colors brighter than those of *S. adolphei.* Hairs coarse, stiff.

Color. Dorsal band from nape to tail blackish to grizzled yellowish brown; sides grayish; crown and sides of head paler than back; under parts, arms, legs, hands, and feet varying from white or yellowish to rufous; chin, cheeks, and sides of neck grizzled gray to grayish brown; tail above black, washed with white, beneath dark rufous to rusty orange, or grayish white bordered by black and edged with white; ears grizzled gray or brown, with black border, and basal white patches. This is a very variable species and has several color phases, individuals presenting quite a different appearance from each other.

Measurements. Total length, 510; tail vertebræ, 248.5; hind foot, 62.2. Skull: average of two; basal length, 51.7; palatal length, 27.7; interorbital width, 21.8; zygomatic width, 35.2; length of upper molar series. 11.7.

91. melania (*Sciurus*), Gray, Ann. Mag. Nat. Hist., 3d Ser., xx,
    1867, p. 425.
GRAY'S BLACK SQUIRREL.
    Type locality. Point Burica, Costa Rica.
    Geogr. Distr. Costa Rica, Chiriqui, Panama. (Bangs.) 2,000
feet altitude.
    Color. Back dark chocolate; rest of pelage and tail black.
    Measurements. Total length, 500; tail vertebræ, 260; hind foot,
63; ear, 30.

92. boothiæ (*Sciurus*), Gray, List. Spec. Mamm. Brit. Mus., 1843,
    p. 139.
    richardsoni Gray, Ann. Mag. Nat. Hist., 1st Ser., x, 1842, p. 264.
    (nec Bachman.)
    fuscovariegatus Schinz., Syn. Mamm., II, 1845, p. 15.
    hypopyrrhus Alston, Proc. Zoöl. Soc., 1878, p. 662. (Part. Hon-
    duras specimens.)
HONDURAS SQUIRREL.
    Type locality. Honduras. San Pedro Sula?
    Geogr. Distr. Humid coast forests of northern Honduras and
border of Guatemala.
    Genl. Char. Pelage thin; hair coarse; tail nearly as long as head
and body, narrow.
    Color. Above grizzled grayish brown, occasionally tinged with
reddish; chin and sides of head grayish brown; outside of arms and
legs, hands and feet dark grayish brown, sometimes nearly black;
tail at base like back, remainder above black, washed with white,
beneath dark grizzled brown, tinged with yellowish and bordered
with black and edged with white; ears with black border.
    Measurements. Total length, 524; tail vertebræ, 255; hind foot,
63.

a.—belti (*Sciurus*), Nelson, Proc. Wash. Acad. Scien., I, 1899, p. 78.
ESCONDIDO RIVER SQUIRREL.
    Type locality. Escondido River, fifty miles above Bluefields,
Nicaragua.
    Geogr. Distr. East coast region of Nicaragua, north to Segovia
River, Honduras.
    Genl. Char. Similar to *S. boothiæ*, but under parts rusty rufous,
and rusty yellowish hands and feet.
    Color. Above, including nose and base of tail, grizzled grayish
brown, washed with black; chin and sides of head yellowish brown
to brownish gray; under parts rusty rufous; outer side of arms and
legs suffused with rusty yellow; hands and feet rusty yellowish and

black; tail above black, washed with white, beneath grizzled yellowish brown or rusty, bordered with black and edged narrowly with white; ears and basal patches, rusty; border of ears black.

Measurements. Total length, 506; tail vertebræ, 258; hind foot, 60. Skull: average of four; basal length, 48.8; palatal length, 25.5; interorbital width, 19.6; zygomatic width, 33.7; length of upper molar series, 11.4.

93. variegatoides (*Sciurus*), Ogilby, Proc. Zoöl. Soc., 1839, p. 117.
  pyladei Less., Nouv. Tabl. Règn. Anim., Mamm., 1842, p. 112.
  colliæi Gray, Ann. Mag. Nat. Hist., 3d Ser., xx, 1867, p. 746. (Part. var. 1.)
  hypopyrrhus Allen, Mon. N. Am. Rodent., 1877, pp. 746-750. (Part. Salvador specimens.)
VARIEGATED SQUIRREL.

Type locality. Salvador, west coast Central America.

Geogr. Distr. Tropical forests of Salvador, West Coast of Central America.

Genl. Char. Tail slender; base of tail and upper parts yellowish gray.

Color. Upper parts, outer side of arms and legs above, and base of tail grizzled yellowish gray; under parts buffy ochraceous; lower part of arms and legs, and hands and feet dark ochraceous buff; chin and cheeks grayish buff; tail above black, washed with white, beneath grizzled buffy gray, bordered by black and edged with white, ears pale rusty; basal patches buff, bordered faintly with black.

Measurements. Total length, 545; tail vertebræ, 280; hind foot, 66 Skull: basal length, 49; palatal length, 26; interorbital breadth, 20; zygomatic width, 34; length of upper molar series, 11.

94. managuensis (*Sciurus*), Nelson, Proc. Wash. Acad. Scien., 1, 1899, p. 81.
  boothiæ managuensis Nelson, Proc. Biol. Soc. Wash., xii, 1898, p. 150.
RIO MANAGUA SQUIRREL.

Type locality. Managua River, Guatemala.

Geogr. Distr. Northern Guatemala, along Managua River.

Genl. Char. Pelage dense; hair coarse; tail long, narrow.

Color. Above with base of tail grizzled yellowish brown; crown and median dorsal region washed with black; sides and outer sides of arms and legs suffused with yellowish; chin and sides of head yellowish brown; under parts dingy yellow to reddish buff; tail above black, washed with white, beneath grizzled grayish or yellowish brown, bordered with black and edged with white; hands and feet grizzled buffy yellow; ears edged with black, basal patches buffy yellow; tufts rusty.

Measurements. Total length, 537; tail vertebræ, 270; hind foot, 62.7. Skull: average of two; basal length, 50.2; palatal length, 27; interorbital width, 19; zygomatic width, 33.9; length of upper molar series, 11.5.

95. goldmani (*Sciurus*), Nelson, Proc. Biol. Soc. Wash., XII, 1898, p. 149.

GOLDMAN'S SQUIRREL.

Type locality. Huehuetan, State of Chiapas, Mexico.

Geogr. Distr. Southeastern part of State of Chiapas, Mexico, into Guatemala.

Genl. Char. Ear tufts present; pelage thin; hairs coarse.

Color. Upper parts pale iron gray; rump and middle of back washed with black; top of head iron gray, tinged with yellow and washed with black; shoulders yellowish gray; outside of arms dark gray; hands paler; outside of thighs and flanks pale yellowish iron gray; feet grizzled iron gray; ears gray, with rufous tufts and white basal patches; orbital ring brownish buff; under parts white; tail above black, washed with white, beneath iron gray to grizzled yellowish, narrowly bordered with black on sides and subterminally, and edged with white.

Measurements. Total length, 570; tail vertebræ, 305; hind foot, 68. Skull: average of five; basal length, 51.2; palatal length, 26.9; interorbital width, 19.8; zygomatic width, 33.9; length of upper molar series, 11.3.

H. Hesperosciurus.

Premolars, $\frac{2-2}{1-1}$. Skull long, broad, especially across parieta region, where it is depressed; zygomatic process of squamosal horizontal, arch obliquely ascending; rostrum rather broad; nasals long, terminating equally with posterior ends of premaxillæ; molars large, heavy.

96. griseus (*Sciurus*), Ord, Guth., Geog., 1815, p. 292. Elliot, Syn. N. Am. Mamm., 1901, p. 55.

fossor Peale, U. S. Expl. Exped., Mamm. & Birds, 1848, p. 55.

hermanni LeConte, Proc. Acad. Nat. Scien. Phil., 1852, p. 149.

leporinus Hensh., Ann. Rep. Engin., 1876, p. 25.

anthonyi Mearns, Proc. U. S. Nat. Mus., XX, 1898, p. 501.

CALIFORNIA GRAY SQUIRREL.

Type locality. The Dalles, Columbia River, Wasco County, Oregon.

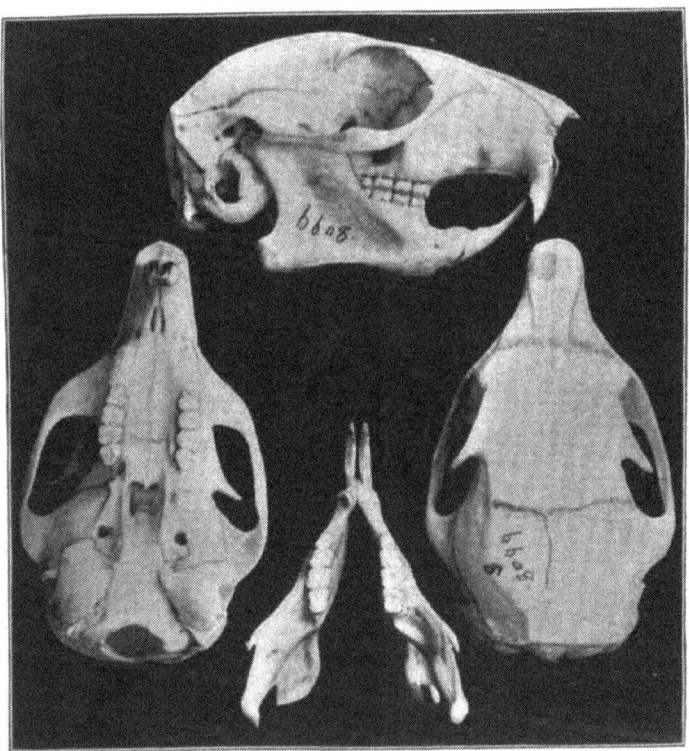

FIG. 20. SCIURUS (HESPEROSCIURUS) GRISEUS.
No. 6608 Field Columbian Mus. Coll. Nat. size

Geogr. Distr. Oregon and northern California, into Lower California, Mexico.

Genl. Char. Size large; tail vertebræ as long as body and head; upper premolars and molars, five; skull slender, elongate.

Color. Above and outside of arms and legs grizzled bluish gray and black; under parts pure white; tuft at base of ear chestnut; tail above mixed gray, white, and black, with a white border, beneath grizzled grayish white, with sometimes an ill-defined border and white edge; hands grayish white to grizzled gray; feet gray to blackish gray.

Measurements. Total length, 565; tail vertebræ, 257; hind foot, 77; ear, 30½. Skull: average of five; basal length, 56.3; palatal length, 32.1; interorbital width, 20.8; zygomatic width, 38.7; length of upper molar series, 11.7.

The Chickarees, or Red Squirrels, comprising the last sub-genus of *Sciurus*, are boreal in their range and are found as far north as the limit of trees. So far as known, only one subspecies penetrates into Mexico, and this is confined to the San Pedro Martir Mountain range of Lower California, where probably by a lofty altitude it counteracts the effects of a more southern latitude. The tail of the Red Squirrel is smaller than that of the gray squirrel group and its allies, and the ears are often tufted or penciled in winter, and the species represent in America the common squirrel of Europe. Like many species of Tamias, at certain seasons the Chickarees have a black line along the side dividing the color of the upper and under parts.

I. Tamiasciurus, Chickarees.

Premolars, $\frac{1-1}{1-1}$ or $\frac{2-2}{1-1}$. Skull short, broad, depressed between orbits; superior outline greatly curved, highest point between post-orbital process of the frontal; rostrum short, broad; nasals broad, not

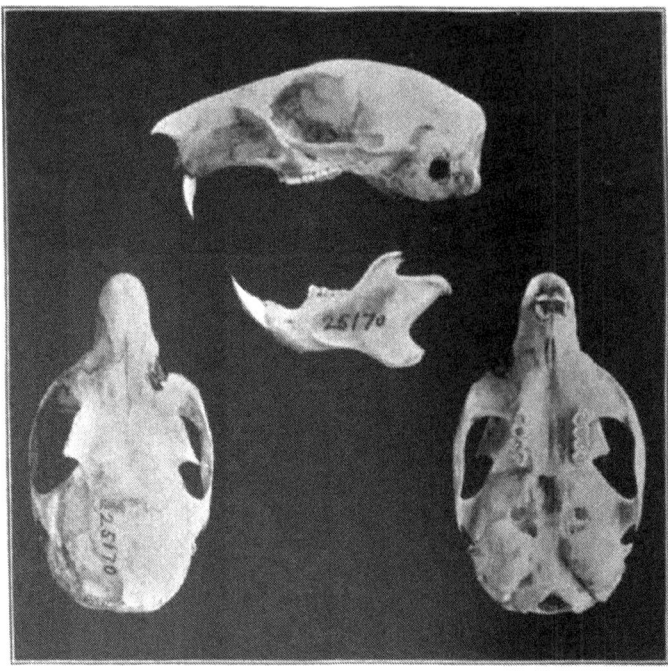

FIG. 21. SCIURUS (TAMIASCIURUS) D. MEARNSI.
No. 25170 U. S. Nat. Mus. Coll. Nat. size.

equaling interorbital width; squamosal process of zygoma projecting outward, curving gradually downward; molar series rather heavy.

douglasi mearnsi (*Sciurus*), Towns., Proc. Biol. Soc. Wash., XI, 1897, p. 146.

MEARNS' SQUIRREL.

Type locality. San Pedro Martir Mountains, Lower California, Mexico. Altitude, 7,000 feet.

Geogr. Distr. Forest of San Pedro Martir Mountains, Lower California, Mexico. Altitude about 7,000 feet.

Genl. Char. Similar to *S. d. albolimbatus*, but grayer, with pale colored hands and feet.

Color. Upper parts pale gray, tinged with yellowish; sides of head grizzled gray; orbital ring whitish; ears gray; broad lateral line black; under parts whitish; basal half of tail above pale gray, suffused with rusty and edged with white, remaining portion black, washed with white, beneath grizzled gray, bordered with black and edged with white; hands and feet pale buff; ears gray; tufts black.

Measurements. Total length, 308–346; tail vertebræ, 111–130; hind foot, 51–55. Skull: basal length, 41; palatal length, 25; inter-orbital constriction, 14.5; zygomatic width, 28; length of upper molar series, 8.

The sprightly and gayly colored little Chipmunks, north of the Mexican boundary, are among the commonest of American Rodents. They frequent the forests or rocky places, often bare of trees, and make their nests in holes either in the ground or in stumps of trees, or in the interstices of rocks. Brushheaps are much frequented by these lively creatures as affording a facile means of escape, and also for observing the movements of anything that has excited their fears. While dwellers of the ground, they readily climb trees, but rarely leap from branch to branch, as do the tree squirrels. They feed on seeds, nuts, and grain, and diligently provide an ample store against the coming of winter, carrying it to the various hiding places in their cheek pouches. They are pretty animals, usually possessing bright colored stripes, and the many forms into which the genus has been divided bear a general resemblance to each other, so close indeed that in not a few instances it is not an easy matter for even the expert to distinguish them. Comparatively few species are found south of the Mexican and United States boundary line, and of these some are more plentifully represented in the more northern land.

35. Tamias. Chipmunks.

$$I.\frac{1-1}{1-1};\ P.\frac{2-2}{1-1}\ or\ \frac{1-1}{1-1};\ M.\frac{3-3}{3-3} = 20\ or\ 22.$$

J. A. Allen. *A review of some of the North American Ground Squirrels.* Bull. Am. Mus. Nat. Hist., 1890, p. 45.

Tamias Illig., Prodr. Syst. Mamm. et Av., 1811, p. 83. Type *Sciurus striatus* Linnæus.

Tenotis Rafin., Am. Month. Mag., I, 1817, p. 362.

Eutamias Trouess., Le Nat., II, 1880, p. 86.

Tail short, not bushy, narrow; cheek pouches large; ears without tufts. Skull narrow anteriorly; superior outline convex; postorbital processes slender, directed backward and downward; antorbital foramen oval; zygomata expanded and depressed anteriorly; upper premolars either two or one.

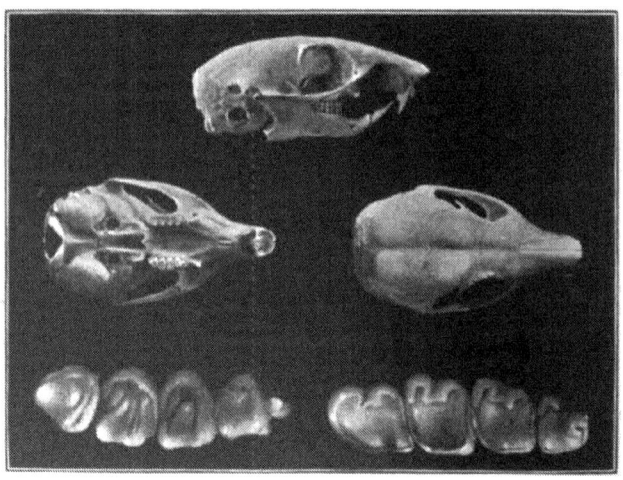

FIG. 22. TAMIAS OBSCURUS.
Field Columbian Mus. Coll. Nat. size.

UPPER TOOTH ROW.   LOWER TOOTH ROW.
Enlarged 8 times.   Enlarged 8 times.

A. Eutamias.

Skull with superior outline highest over parietals; orbital foramen a narrow, slit-like opening; rostrum compressed; nasals short, slightly narrowed posteriorly; pterygoid fossa long and wide; audital bullæ large.

KEY TO THE SPECIES AND SUBSPECIES.

97. dorsalis (*Tamias*), Baird, Proc. Acad. Nat. Scien. Phil., 1855,
    p. 332. Elliot, Syn. N. Am. Mamm., 1901, p. 68.

    quadrivittatus pallidus Coues & Yarr., in Wheeler's Rep. Geogr.
      & Geol. Expl. and Surv. West of 100 Merid., v, 1876, p. 118.

GILA CHIPMUNK.

  Type locality. Fort Webster, Grant County, New Mexico.

  Geogr. Distr. Northern Utah to northern Mexico; west to the
Sierra Nevada; east to the Mimbres in New Mexico.

  Genl. Char. Single dorsal stripe alone conspicuous.

  Color. Upper parts hoary mixed with rusty and brown; flanks
and hips dull rusty; dorsal stripe dark brown; a faint stripe of grayish
on flanks; under parts dull grayish white; tail above mixed black,
chestnut, and white, hairs chestnut at base; face with usual *Tamias*
stripes.

  Measurements. Total length, 237; tail vertebræ, 105; hind foot,
22; ear, 20. Skull: occipito-nasal length, 37; Hensel, 29; zygomatic
width, 20; interorbital width, 9; palatal length, 16; length of upper
molar series, 5.

FIG. XXX. TAMIAS OBSCURUS. DUSKY CHIPMUNK.

98. obscurus (*Tamias*), Allen, Bull. Am. Mus. Nat. Hist., 1890,
p. 70.

LOWER CALIFORNIA CHIPMUNK.

Type locality. San Pedro Martir Mountains, Lower California,
Mexico.

Geogr. Distr. San Pedro Martir and Hanson Laguna Mountains,
Lower California, Mexico.

Genl. Char. Size small; pelage soft, color dark.

Color. Post-breeding Pelage. Above dark brownish gray; five
indistinct dorsal stripes dull chestnut, posterior half of median one
blackish; intermediate light stripes whitish gray; central portion of
flanks yellowish brown; shoulders, arms, thighs, and legs like upper
parts; top of head blackish gray; ears similar, with large white
patches at base; sides of face whitish, with three chestnut stripes
from nose to ears; under parts whitish; tail above black or dark
brown washed with black, beneath rufous edged with black and with
white hairs intermingled; hands like arms; feet buffy or brownish
gray.

Winter Pelage. Dull; light stripes on dorsal region very faint,
almost obsolete; dark stripes also faintly perceptible; upper parts
iron gray tinged with brown, with the light gray stripes, and darker
brown ones faintly indicated; middle of sides fulvous; top of head
like back; two white stripes on sides of face, one from tip of nose
over and beyond eye, the other from beneath eye to base of ear;
shoulders and sides like back; upper parts of arms and hands, legs
and feet paler brownish or brownish gray; under parts white, plumbeous of under fur showing through; tail above black, beneath in
center dark rufous, bordered with black and edged with yellowish
white; anal region very dark rufous.

Measurements. Total length, 230; tail vertebræ, 80–103; hind
foot, 32–34. Skull: occipito-nasal length, 37; Hensel, 16; zygomatic
width, 20; interorbital width, 8; palatal length, 15.5; length of upper
molar series, 5.5.

99. bulleri (*Tamias*), Allen, Bull. Amer. Mus. Nat. Hist., 1889,
p. 173.

MEXICAN CHIPMUNK.

Type locality. Sierra de Valparaiso, State of Zacatecas, Mexico.

Geogr. Distr. State of Zacatecas, Mexico.

Genl. Char. Similar to *T. merriami*. Ears tricolor; black patch
between eye and ear.

Color. Upper parts and sides pale gray tinged with buff; five
dark dorsal stripes, three distinct, the three median ones seal brown

mixed with yellowish chestnut; outermost ones yellowish chestnut mixed with blackish; median light stripes gray, outer ones grayish white; facial stripes white and black, bordered with rusty; ears on center at base black, edge margined with rusty and tipped with white; large white patch at base of ear; under parts and upper surface of hands and feet grayish white; tail above black mixed with buff, and hairs white-tipped, beneath ochraceous, bordered with black and fringed with white.

Measurements. Total length, 250; tail vertebræ, 98; hind foot, 33. Skull: average total length, 36.6; greatest breadth, 23.

100. durangæ (*Eutamias*), Allen, Bull. Am. Mus. Nat. Hist., 1903, p. 595.
DURANGO CHIPMUNK.

Type locality. Arroyo de Bucy, State of Durango, Mexico.

Genl. Char. Similar to *T. bulleri* but larger and paler.

Color. "Similar to *E.* (*T.*) *bulleri* from southwestern Zacatecas, but larger and paler, with the white markings on the head broader, and the white post-auricular patch larger; rump, basal portion of the tail, and flanks faintly suffused with a very pale tinge of buff instead of being gray as in *bulleri*; the dark dorsal stripes are similar in extent and in color, but the intervening light stripes are suffused with pale cinnamon instead of being nearly clear white as in *bulleri*, and the rufous of the flanks is much paler. (Allen, l. c.)

Measurements. Total length, 238; tail vertebræ, 98; hind foot, 32; ear from notch, 19. Skull: total length, 39; zygomatic width, 21.

101. merriami (*Tamias*), Allen, Bull. Am. Mus. Nat. Hist., 1890, p. 84. Elliot, Syn. N. Am. Mamm., 1901, p. 71.
asiaticus merriami Allen, Bull. Am. Mus. Nat. Hist., 1889, p. 176.
MERRIAM'S CHIPMUNK.

Type locality. San Bernardino Mountains, California.

Geogr. Distr. From Tulare and Monterey Counties, California, south in the mountains into Lower California, Mexico.

Genl. Char. Size large; color pale; facial and dorsal marking not strongly contrasted; ears high, pointed.

Color. Above dull yellowish gray mixed in places with fulvous; dorsal stripes pale yellowish to fulvous brown, the light ones pale gray; sides and rump pale olivaceous; tail above blackish, fringed with gray, beneath reddish chestnut, bordered with black and fringed with whitish.

Post-breeding Pelage. Flanks golden rufous; dark dorsal stripes yellowish rufous; light ones silver gray; tail beneath orange chestnut, above orange yellow.

Measurements. Average total length, 285; tail vertebræ, 117, hind foot, 34. Skull: occipito-nasal length, 41; Hensel, 31; zygomatic width, 21; interorbital width, 9; palatal length, 16; length of upper molar series, 5.5.

102. quadrivittatus (*Sciurus*), Say, Long's Exped. Rocky Mts., II.
      1823, p. 45.
      quadrivittatus (*Tamias*), Elliot, Syn. N. Am. Mamm., 1901, p. 75.
COLORADO CHIPMUNK.

Type locality. Arkansas River, Colorado, "near where it breaks through the foothills," Park County (?)

Geogr. Distr. State of Durango, Mexico, north to southern boundary of Colorado, northward through Wyoming to and including the Yellowstone National Park.

Genl. Char. Rather small; general color gray.

Color. Breeding Pelage. Above gray, sides washed with pale yellowish brown; beneath grayish white; dark dorsal stripes black and rufous; light ones ashy; outer white.

Post-breeding Pelage. Above rufous; thighs plumbeous gray; dark dorsal stripes black and rufous; outer light stripes whitish; flanks yellowish rufous; under parts grayish white; dark facial stripes rusty brown; light ones grayish white; tail above black and buff, beneath buffy ochraceous bordered and fringed with black.

Measurements. Average total length, 223; tail vertebræ, 82; hind foot, 31. Skull: occipito-nasal length, 35; Hensel, 14; zygomatic width, 18; interorbital width, 7; palatal length, 15; length of upper molar series, 4.

The genus CITELLUS, containing the Spermophiles, is represented in North America by a considerable number of species and varieties, which exhibit the extremes of form from that of a rather small, stout, short-tailed animal, to a large, more slenderly and gracefully shaped creature with a long, bushy, squirrel-like tail. In many places they are known as "gophers," and like the real gopher, they are great diggers, and live in burrows, into which they scurry at the least alarm. They are gregarious and sociable, living in communities, and in certain districts of our country are veritable pests when making their abodes in cultivated ground. Very active and industrious, they lay up great stores of food against the winter, and in spite of their troublesome propensities, are pleasing objects in a landscape, as they flit over the ground waving their bushy tails, or sit upright at the mouths of their burrows, vigilant watchmen of danger's approach. The different species are striped, spotted, or otherwise marked in various colors, and some are exceedingly pretty creatures.

36. Citellus.

$$I.\frac{1-1}{1-1};\ P.\frac{2-2}{1-1};\ M.\frac{3-3}{3-3} = 22.$$

Citellus Oken, Lehrb. der Zoöl., ii, 1816, p. 842. Type *Mus citellus* Linnæus.

Spermophilus F. Cuv., Mem. Mus., vi, 1825, p. 293.

Spermatophilus Wagl., Syst. Av. Amphib., 1830, p. 22.

Citillus Lichten., Darst. neuer oder wenig bekannt. Säugeth., Heft 5, 1827–34, pl. xxxi, fig. 2.

Colobotis Brandt, Bull. Classe Phys-math. Acad. Imp. Scien. St. Petersb., ii, 1844, p. 365.

Otospermophilus Brandt, Bull. Classe Phys-math. Acad. Imp. Scien. St. Petersb., ii, 1844, p. 379.

Otocolobus Brandt, Bull. Classe Phys-math. Acad. Imp. Scien. St. Petersb., ii, 1844, p. 382.

Ictidomys Allen, Mon. N. Am. Rod., 1877, p. 821.

Spermophilopsis Blasius, Tag. Deut. Nat. Vers., 1884, p. 324.

Ammospermophilus Merr., Proc. Biol. Soc. Wash., vii, 1892, p. 27.

Xerospermophilus Merr., Proc. Biol. Soc. Wash., vii, 1892, p. 27.

Calospermophilus Merr., Proc. Biol. Soc. Wash., ii, 1897, p. 189 (note).

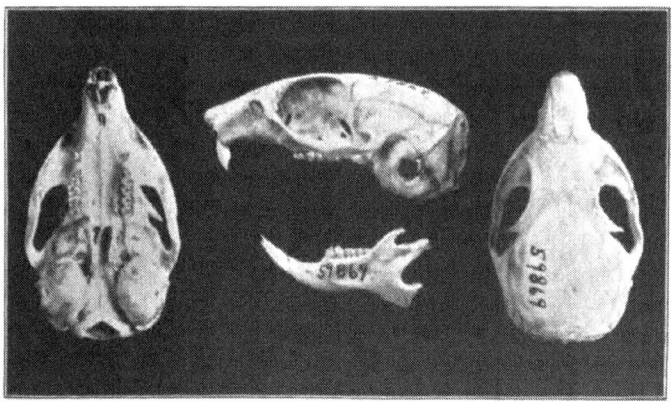

FIG. 23. CITELLUS (AMMOSPERMOPHILUS) H. SAXICOLA.
No. 59869 U. S. Nat. Mus. Coll. Nat. size.

Body rather slender; tail either long, moderate, or short, varying greatly in its length among the different species; ears large to rudimentary, not tufted; cheek pouches large; skull variable, short and broad like *Sciurus*, or long and narrow; postorbital processes strong, with a downward direction; antorbital foramen circular, with a well-developed tubercle on outer lower border.

KEY TO THE SUBGENERA.

A. Coronoid process of mandible broad and heavy.   PAGE
   a. Nasals equal in length to the premaxillæ
     . . . .*Ammospermophilus* 141
   b. Nasals not so long as the premaxillæ. . .*Xerospermophilus* 143
B. Coronoid process of mandible long and slender.
   a. Nasals longer than the premaxillæ.
     a.′ Tail short, flat. . . . .*Callospermophilus* 147
     b.′ Tail long and bushy. . . . .*Otospermophilus* 148

KEY TO THE SPECIES AND SUBSPECIES.

A. Size small; tail short, narrow.
   a. Upper parts not spotted.
     a.′ Tail beneath white on central portion.   PAGE
       a.″ Above grizzled grayish brown and vinaceous. . . . . *C. harrisi* 141
       b.″ Above pale yellowish brown and white. *C. h. saxicola* 142
       c.″ Above grizzled gray. . . . . *C. leucurus* 142
       d.″ Above iron gray. . . . .*C. l. peninsulæ* 143
       e.″ Above grizzled gray and vinaceous . . . . *C. interpres* 143
     b.′ Tail not white beneath.
       a.″ Above grizzled gray and yellowish brown. . . . . *C. tereticaudus* 144
       b.″ Above fawn color. . . . . *C. t. sonoriensis* 144
       c.″ Above grizzled yellowish brown and black. . . . .*C. perotensis* 145
   b. Upper parts spotted.
     a.′ Above reddish brown. . . . .*C. spilosoma* 145
     b.′ Above russet brown, white spots bordered with dusky. . . . .*C. s. microspilotus* 145
     c.′ Above dark reddish or yellowish brown. . . .*C. mexicanus* 146
     d.′ Above pale reddish brown. . . . .*C. m. parvidens* 146
B. Size large, tail long, bushy.
   a. Tail not over one-fourth total length . . . .*C. madrensis* 147
   b. Tail less than half total length.
     a.′ Tail not annulated.

A. Ammospermophilus, Merr.

Size small. Skull broad; interorbital constriction slight; nasals short, broad, reaching posterior end of premaxillæ; first premolar less than half the second in size; tail, one-third the length of the head and body.

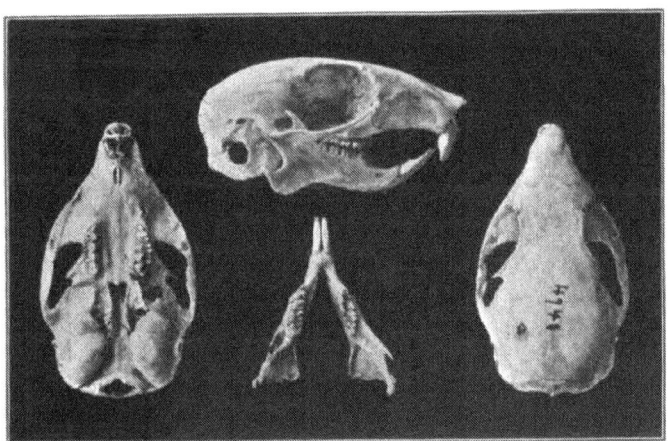

FIG. 24. CITELLUS (AMMOSPERMOPHILUS) L. PENINSULÆ.
No. 4948 Am. Mus. Nat. Hist. Coll. Nat. size.

103. harrisi (*Spermophilus*), Aud. & Bachm., Quadr. N. Am., III, 1854, p. 267, pl. 144, Fig. 1. Elliot, Syn. N. Am. Mamm., 1901, p. 85.

HARRIS'S SPERMOPHILE.

Type locality. Not determined.

Geogr. Distr. Northeastern part of State of Sonora, Mexico, to southern Utah and Nevada, into California.

Genl. Char. Ears small, tail short, black dorsal stripe wanting; angle of mandibular ramus much developed.

Color. Above grizzled grayish brown, tinged with vinaceous posteriorly; narrow white stripe on sides; flanks tinged with pale chestnut; orbital ring, and lower parts of body yellowish white; tail flat, above black and white, edged with white, beneath white bordered with black and edged with white. Some specimens exhibit little or no difference between the upper and lower sides of the tail.

Measurements. Total length, 260; tail vertebræ, 88; hind foot, 41. Skull: occipito-nasal length, 39.5; Hensel, 32; zygomatic width, 23; interorbital width, 10; palatal length, 19; length of upper molar series, 8.

a.—saxicola (Spermophilus), Mearns, Proc. U. S. Nat. Mus., 1896, p. 444. ˙ Elliot, Syn. N. Am. Mamm., 1901, p. 86.

ROCK SPERMOPHILE.

Type locality. Tinajas Atlas, Gila Mountains, Yuma County, Arizona.

Geogr. Distr. State of Sonora, Mexico, north to Gila Mountains, southwestern Arizona.

Genl. Char. Colors pallid; tail rather long.

Color. Like *C. harrisi,* but all hues much paler, and tail longer. Above grayish brown, all the hairs tipped with yellowish white giving a grizzled appearance; a narrow whitish stripe on side from shoulder to rump; flanks speckled with brown and rufous; shoulders and limbs pale chestnut, hair tipped with whitish; under parts white; tail black mixed with white and a narrow edging of the same; hands and feet white tinged with buff.

Measurements. Total length, 245; tail vertebræ, 93; hind foot, 40. (ex Type.) Skull: occipito-nasal length, 40; Hensel, 35; zygomatic width, 23.5; interorbital width, 10; palatal length, 20; length of upper molar series, 8.

104. leucurus *(Tamias),* Merr., N. Am. Faun., No. 2, 1889, p. 20.

leucurus (Spermophilus), Elliot, Syn. N. Am. Mamm., 1901, p. 86.

WHITE-TAILED SPERMOPHILE.

Type locality. San Gorgonio Pass, Riverside County, California.

Geogr. Distr. Lower California, Mexico, to California, Utah, Arizona, and New Mexico.

Genl. Char. Smaller than *C. harrisi,* tail shorter, below white.

Color. Above grizzled gray, vinaceous on head and rump; outside of legs salmon; white stripe on side of back; under parts white; tail above iron gray, with indistinct white border, beneath white, bordered with black.

Measurements. Total length, 209; tail vertebræ, 69; hind foot, 38. Skull: occipito-nasal length, 40.5; Hensel, 18; zygomatic width, 23.5; interorbital width, 9.5; palatal length, 18; length of upper molar series, 6.

a.—peninsulæ (*Tamias*), Allen, Bull. Am. Mus. Nat. Hist., 1893, p. 197.

LOWER CALIFORNIA SPERMOPHILE.

Type locality. San Telmo, Lower California, Mexico.

Geogr. Distr. Lower California, Mexico, range unknown.

Genl. Char. Color darker than *C. harrisi*; tail very short.

Color. Upper part of back gray; top of head and lower back dusky or reddish brown; outside of arms and thighs reddish cinnamon; lateral stripe and under parts white; tail above iron gray, beneath white with subapical black band.

Measurements. Total length, 213; tail vertebræ, 50; hind foot, 35. Skull: occipito-nasal length, 41.5; Hensel, 34; zygomatic width, 24; interorbital width, 10; length of nasals, 10; palatal length, 19.5; length of upper molar series, 7; length of mandible, 19; length of lower molar series, 6.

105. interpres (*Tamias*), Merr., N. Am. Faun., No. 4, 1890, p. 21.

interpres (*Spermophilus*) Elliot, Syn. N. Am. Mamm., 1901, p. 86.

TRADER SPERMOPHILE.

Type locality. El Paso, El Paso County, Texas.

Geogr. Distr. State of Chihuahua, Mexico, north to Utah.

Genl. Char. Similar to *C. leucurus*; tail long; lateral hairs with two black bands.

Color. Winter Pelage. Above grizzled gray tinged with vinaceous; shoulder, hips, outer surface of legs ochraceous buff; white stripe on eyelids and on each side of back; under parts white; tail above grizzled gray tinged with fulvous, edge whitish, beneath white.

Measurements. Total length, 226; tail vertebræ, 80; hind foot, 37.

B. Xerospermophilus.

Size small. Skull short, broad; interorbital constriction considerable; nasals not reaching posterior ends of premaxillæ. First molar one-third the size of second. Tail one-third the length of body and head.

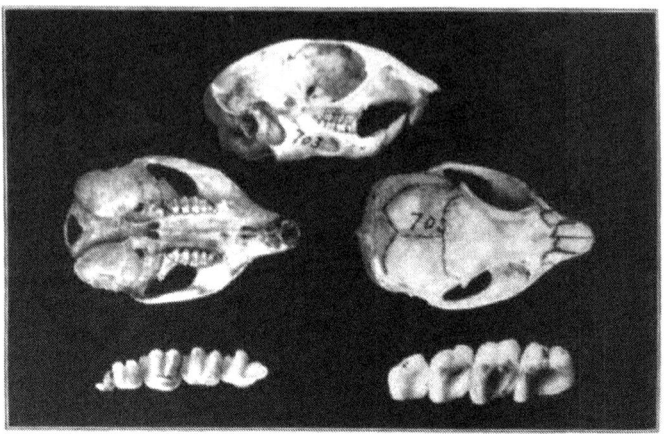

FIG. 25. CITELLUS (XEROSPERMOPHILUS) S. MICROSPILOTUS.
No. 703 Field Columbian Mus. Coll.   Nat. size.

UPPER TOOTH ROW.   LOWER TOOTH ROW.
Enlarged 4 times.   Enlarged 4 times.

106. tereticaudus (*Spermophilus*), Baird, N. Am. Mamm., 1857, p.
315.   Elliot, Syn. N. Am. Mamm., 1901, p. 98.
ROUND-TAILED SPERMOPHILE.

Type locality.   Old Fort Yuma, San Diego County, California.

Geogr. Distr.   Central California to Lower California, Arizona,
and State of Sonora, Mexico.

Genl. Char.   Tail as long as four-fifths head and body; no spots
on body; feet broad, soles hairy.

Color.   Above grizzled gray and yellowish brown; beneath brownish white; tail like back, brown at end, tip yellowish.

Measurements.   Total length, 248; tail vertebræ, 112; hind foot,
35.   Skull: occipito-nasal length, 37; Hensel, 29; zygomatic width,
23; interorbital width, 9; palatal length, 17; length of upper molar
series, 7.

a.—sonoriensis (*Spermophilus*), Ward, Amer. Nat., xxv, 1891, p.
158.
SONORAN SPERMOPHILE.

Type locality.   Hermosillo, State of Sonora, Mexico.

Geogr. Distr.   State of Sonora, range unknown.

Genl. Char.   Similar to *C. cryptospilotus:* body without spots.

Color.   Upper parts fawn color, hairs ringed with black, straw
yellow, and walnut brown, and tipped with cream buff; sides paler

fawn; under parts and inner sides of legs and hind feet white, the last washed with rufous; tail above at base like back, remainder fawn color bordered with black and fringed with whitish rufous, beneath pale rufous.

Measurements. Total length, 220; tail vertebræ, 65; hind foot, 33. Skull: occipito-nasal length, 37.5; Hensel, 30; zygomatic width, 24; interorbital width, 9; length of nasals, 6; palatal length, 18; length of upper tooth row, alveolar border, 7.

107. perotensis (*Spermophilus*), Merr., Proc. Biol. Soc. Wash., VIII, 1893, p. 131.

PEROTE SPERMOPHILE.

Type locality. Perote, State of Vera Cruz, Mexico.

Geogr. Distr. State of Vera Cruz, Mexico; extreme eastern border of tableland. Altitude 8,000–10,000 feet.

Genl. Char. Similar to *C. elegans.* Skull large, heavy; molars with broad crowns.

Color. Above grizzled yellowish brown irregularly lined with black posteriorly; under parts, hands, and feet buffy; tail above mixed black and yellowish brown, beneath ochraceous buff with a subapical black band.

Measurements. Total length, 253; tail vertebræ, 68; hind foot, 38.

108. spilosoma (*Spermophilus*), Bennett, Proc. Zoöl. Soc., 1833, p. 40. Elliot, Syn. N. Am. Mamm., 1901, p. 96.

  mexicanus, Aud. & Bachm., Quad. N. Am., III, 1853, p. 42, pl. CIX.

SPOTTED SPERMOPHILE.

Type locality. "That part of California that adjoins Mexico." Western Texas?

Geogr. Distr. Janos, State of Sonora, Mexico, north into Texas and New Mexico. Altitude, 5,600 feet.

Genl. Char. Size small; ears very short; tail about half the length of body.

Color. Above reddish brown spotted with white (bordered with black posteriorly in the young); under parts yellowish white; tail above like back, with subterminal black bar, tip yellowish brown, beneath brownish yellow, bordered with black and fringed with yellowish.

Measurements. Total length, 255; tail vertebræ, 75; hind foot, 32.

a.—microspilotus (*Spermophilus*), Merr., N. Am. Faun., No. 4, 1890, p. 38. Elliot, Syn. N. Am. Mamm., 1901, p. 96.

SMALL-SPOTTED SPERMOPHILE.

Type locality. Oracle, Pinal County, Arizona.

Geogr. Distr. State of Sonora, Mexico, north to Arizona.

Genl. Char. Size medium; dorsal spots large, separate, distinct.

Color. Above russet brown; dorsal spots white bordered posteriorly with dusky; under parts whitish; tail above, basal half like back, remainder mixed buff and black, bordered with buff, beneath pale ochraceous buff, with indistinct submarginal black band.

Measurements. Total length, 220; tail vertebræ, 74; hind foot, 30. Skull: occipito-nasal length, 36; Hensel, 27; zygomatic width, 21.5; interorbital width, 7; palatal length, 16; length of upper molar series, 7.

109. mexicanus (*Sciurus*), Erxl., Syst. Règn. Anim., 1, 1777, p. 428.

mexicanus (*Citillus*), Licht., Darst. Neu. Säugeth., 1830, pl. 31.
    Fig. 2. (ex Toluca.)

mexicanus (*Spermophilus*), Elliot, Syn. N. Am. Mamm., 1901, p. 98.

MEXICAN SPERMOPHILE. *Urion, Huron,* in Mexico.

Type locality. Toluca, Mexico.

Geogr. Distr. South central Mexico.

Genl. Char. Tail about half as long as body; ears short; size medium.

Color. Above dark yellowish to reddish brown, with nine or eleven lines of white spots; beneath yellowish white; head above mixed white, black and yellowish; orbital ring and lower side of cheek white; tail above black and yellowish white, bordered with black and edged with yellowish white, beneath brownish white, bordered with black, and fringed with brownish white.

Measurements. Total length, 305; tail vertebræ, 100; hind foot, 42.

*a.—parvidens (*Spermophilus*), Mearns, Proc. U. S. Nat. Mus., XVIII, 1896, p. 443. Elliot, Syn. N. Am. Mamm., 1901, p. 99.

SMALL-TOOTHED SPERMOPHILE.

Type locality. Fort Clark, Kinney County, Texas.

Geogr. Distr. Texas into Mexico?

Genl. Char. Similar to *C. mexicanus,* but smaller; colors paler; teeth weak.

Color. Pattern like *C. mexicanus,* colors paler, under parts white; hairs of tail with two black rings instead of three, and tips grayish instead of yellowish; back yellowish broccoli brown.

Measurements. Total length, 325; tail vertebræ, 130; hind foot, 44. Skull: occipito-nasal length, 41; Hensel, 19; zygomatic width,

*This form doubtless passes the Mexican boundary, but up to the present time it has not been recorded from that country.

24; interorbital width, 9; palatal length, 19; length of upper molar series, 7.

O. * Callospermophilus.

Skull broad; interorbital space broad; rostrum broad, nasals longer than premaxillæ; first premolar smaller than second. Tail less than half the length of head and body.

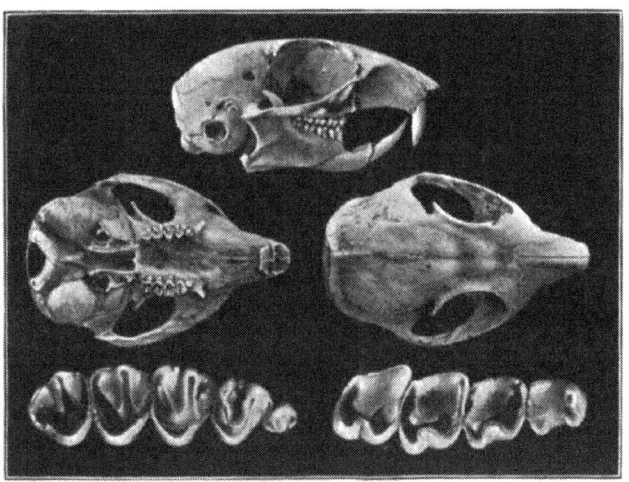

FIG. 26. CITELLUS (CALLOSPERMOPHILUS) MADRENSIS.
U. S. Nat. Mus. Coll. Nat. size.

UPPER TOOTH ROW.   LOWER TOOTH ROW
Enlarged 5 times.   Enlarged 5 times.

110. madrensis (*Callospermophilus*), Merr., Proc. Wash. Acad. Scien., III, 1901, p. 563.

SIERRA MADRE SPERMOPHILE.

Type locality. Sierra Madre, near Guadalupe y Calvo, State of Chihuahua, Mexico.

Genl. Char. "Similar to *C. lateralis*, but tail about half as long; rostrum less swollen."

Color. Fall Pelage. Above grizzled grayish; rump dull grayish fulvous; light lateral stripe grizzled buffy gray, black stripes faint, nearly obsolete; under parts whitish; tail above grizzled grayish fulvous and black, beneath yellowish, the plumbeous base of hairs showing through; hands and feet whitish.

*Some writers accord this subgenus, generic rank. It does not seem to possess sufficient claims for this distinction.

Measurements. Total length, 233; tail vertebræ, 58; hind foot, 39. Skull: total length, 41; Hensel, 33; zygomatic width, 25; interorbital width, 15; palatal length, 18; width of braincase above zygomata, 20; length of upper tooth row, 8; length of mandible, angle to alveolus of incisor, 19; length of lower tooth row, 7.5.

D. Otospermophilus, Brandt.

"Skull broad; molars small; edge of outer wall of antorbital foramina not thickened; coronoid processes of lower jaw long and slender; ears very large; tail long, full, bushy."

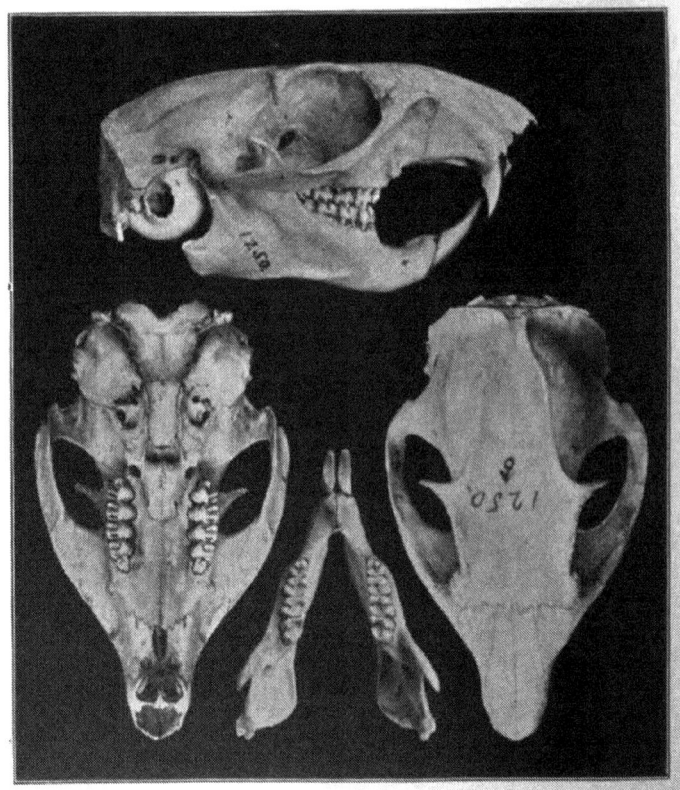

FIG. 27. CITELLUS (OTOSPERMOPHILUS) VARIEGATUS.
No. (1250) 4944 Field Columbian Mus. Coll. Nat. size.

111. variegatus (*Sciurus*), Erxl., Syst. Règn. Anim., I, 1777, p. 421
Nelson, Science, 1898, p. 898.

macrourus (Spermophilus), Bennett, Proc. Zoöl. Soc., 1833, p. 41.
buccatus Licht., Abh. K. Akad. Wiss. Berl., 1827, p. 115.
LONG-TAILED SPERMOPHILE.
   Type locality. "California adjoining Mexico." State of Jalisco?
Geogr. Distr. State of Jalisco, Mexico. (J. A. Allen.)
Genl. Char. Size large, color dark.

FIG. XXXI. CITELLUS (OTOSPERMOPHILUS) VARIEGATUS.
LONG-TAILED SPERMOPHILE.

   Color. Above mixed black and brownish gray; crown black; lips
and chin ferrugineous; under parts ferrugineous varied with black;
tail black washed with white, with an irregular black border edged
with whitish, and subapical black band.
   Measurements. Total length, 470–485; tail vertebræ, 178; to
end of hairs, 216; hind foot, 56–63. Skull: occipito-nasal length, 65;
Hensel, 53; zygomatic width, 39.5; interorbital width, 16.5; palatal
length, 31; length of upper molar series, 13.5.

a.—grammurus (Sciurus), Say, Long's Exped. Rocky Mts., II,
      1823, p. 72.
   couchi Baird, N. Am. Mamm., 1855, p. 311, pl. 81.
   grammurus (Spermophilus) Elliot, Syn. N. Am. Mamm., 1901, p. 88.
BUSHY-TAILED SPERMOPHILE.
   Type locality. Purgatory Creek, Colorado, lat. 37° 32'; long.
103° 30'.
   Geogr. Distr. State of Sonora, Mexico, north to Parks of Central
Colorado.
   Genl. Char. Tail long, full, bushy; ears large; body large, stout;
soles of feet smooth.
   Color. Crown speckled black and white, or brown, black and
white; above gray, mottled with brownish white and black, washed
posteriorly with brown; sides of neck and flanks purer gray; inside
of limbs brownish white or buff; under parts pale yellowish white;

hands and feet gray or buff; tail grayish white and black above, beneath with sometimes central area buff but usually like upper part.

Measurements. Total length, 530; tail vertebræ, 200; hind foot, 60. Skull: occipito-nasal length, 57; Hensel 45; zygomatic width, 36; interorbital width, 10; length of upper molar series, 6.

b.—fisheri (Spermophilus), Merr., Proc. Biol. Soc. Wash., VIII, 1893, p. 133. Elliot, Syn. N. Am. Mamm., 1901, p. 88.
FISHER'S SPERMOPHILE.

Type locality. Kern/Valley, twenty-five miles above Kernville, Tulare County, California.

Geogr. Distr. Lower California, Mexico, central and southern California to western border of Nevada.

Genl. Char. Like *C. v. beecheyi*, but paler; white shoulder stripes longer.

Color. Similar to *C. v. beecheyi*, but sides of neck and shoulder stripes silvery white; body spotted on sides with whitish bordered with dusky; lower part of face whitish; under parts and feet buffy.

Measurements. Total length, 415; tail vertebræ, 175; hind foot, 58. Skull: occipito-nasal length, 53; Hensel, 43; zygomatic width, 32.5; interorbital width, 8.5; palatal length, 25; length of upper molar series, 11.

c.—atricapillus (Spermophilus), Bryant, Proc. Calif. Acad. Scien., 2d Ser., II, 1889, p. 26.
BLACK-HEADED SPERMOPHILE.

Type locality. Comondu, Lower California, Mexico.

Geogr. Distr. Lower California. Northern Mexico (?)

Genl. Char. Crown black, scapular region blackish.

Color. Similar to *C. v. grammurus*, but crown black (varying in extent); orbital ring white; neck, scapulars, and interscapulars black, mixed with white and buff; rest of upper parts mixed buff and black; sides grayish or buffy white; under parts whitish; hands and feet buff; tail black above, the hairs tipped with buff at base and whitish on remaining portions, and edged with white, beneath black washed with white.

Measurements. Total length, 535; tail, 235; hind foot, 50; (skin.)

d.—rupestris (Citellus), Allen, Bull. Am. Mus. Nat. Hist., 1903, p. 595.
ROCK SPERMOPHILE.

Type locality. Rio Sestin, State of Durango, Mexico.

Genl. Char. Similar to *C. v. grammurus*, but larger, and crown and nape black.

Color. Front, top, and sides of head black; nose and eyes grayish brown; white patch above and below eye; upper parts mixed blackish brown and whitish, darkest on anterior half of dorsal region and suffused with yellowish brown; sides paler and grayer; throat, upper breast, and axial region ochraceous buff; rest of under parts yellowish buff; fore feet yellowish gray; hind feet more strongly yellow; ear black; tail above grizzled black and white, the hairs ringed with black and whitish and tipped with white, beneath pale yellowish white, margined on each side with three black bands, the outer one the broadest.

Measurements. Total length, 520; head and body, 279; tail vertebræ, 241; hind foot, 64; ear, 25. Skull: total length, 66; zygomatic width, 40; length of nasals, 23; length of upper tooth row, 13.

112. annulatus (*Spermophilus*), Aud. & Bachm., Jour. Acad. Nat. Scien. Phil., VIII, 1842, p. 319.
RING-TAILED SPERMOPHILE.

Type locality. None given; probably western Mexico.

Geogr. Distr. Plains of State of Colima, and Territorio de Tepic, western Mexico; extent of range unknown.

Genl. Char. Body squirrel-like; tail long, rather bushy, ringed; ears broad and rather high, rounded; claws short, curved; pelage coarse, stiff.

Color. Top of head black, speckled with deep buff; entire upper parts and sides mixed black and pale yellow; sides of head, neck, outer surface of arms and hands, legs, and feet reddish brown, nearly chestnut; inner side of thighs, and under parts straw yellow; basal portion of hairs on abdomen black; tail at base like back, rest of upper part alternately banded with black and pale yellow, beneath reddish cinnamon, with a narrow interrupted black border edged with yellow.

Measurements. Total length, 405; tail vertebræ, 200; hind foot, 57.

a.—goldmani (*Spermophilus*), Merr., Proc. Biol. Soc. Wash., XV, 1902, p. 69.
GOLDMAN'S SPERMOPHILE.

Type locality. Santiago, Territorio de Tepic, Mexico.

Genl. Char. "Similar to (*C.*) *annulatus*, but hind foot smaller (averaging 52.5 instead of 56.5); whitish eyelids clearer and more distinct; ferrugineous of face, neck, thighs, and tail less extensive and usually less intense."

Measurements. "Type. Total length, 430; tail vertebræ, 216; hind foot, 52." (Merr., l. c.)

113. adocetus (*Citellus*), Merr., Proc. Biol. Soc. Wash., 1903, p. 79.
PLAIN-TAILED SPERMOPHILE.

Type locality. La Salada, 40 miles south of Uruapan, State of Michoacan, Mexico.

Genl. Char. Near *C. annulatus,* but smaller; tail without rings; pelage harsh; ears short, tail long. Skull has broader jugal, and broad frontal; long postorbital processes decurved.

Color. Upper parts grizzled grayish and black, top of head darker; superciliary stripe buffy, sometimes washed with pale fulvous; buffy band under eye; under parts yellowish buff; occasionally fulvous on throat and chin; fore legs, hands, and feet dull pale fulvous; sides of neck washed with fulvous; tail grizzled black and buffy, terminal half bordered with black and edged with buffy fulvous, median line of distal half beneath pale fulvous. At certain seasons the upper parts of body are dull ochraceous brown.

Measurements. Total length, 350; tail vertebræ, 156; hind foot. 48. Skull: basal length, 41; palatal length, 24; postpalatal length, 17; zygomatic breadth, 26; interorbital breadth, 13; length of tooth row on alveolus, 8.25; on crowns, 7.5.

The "Prairie-dogs," as their name implies, are dwellers of the plains, where they congregate in such large numbers that their countless burrows are known as "towns." The presence of any one approaching one of these is immediately announced by the barking of the "dogs," which, sitting bolt upright at the mouths of their burrows, by shrill staccato cries express their disapproval of the intrusion. Not very brave are the "dogs," for when a near approach is threatened, each one disappears into the nearest hole, and does not come out again until satisfied that all danger is past. In form this Marmot is rather chunky, with short tail and coarse short hair, the tips of which have been worn away by constant rubbing against the soil in their mining operations. The dentition is powerful and the fore paws are formed for digging. The galleries in their "towns" ramify in all directions and cover a vast extent of ground, and it would be a useless effort to try and dig out one of these animals. They are very animated, incessantly in motion, and when barking the tail is jerked upward with a spasmodic action as if the creature were moved by springs. Owls and rattlesnakes are fellow-boarders with the "dogs" in these towns, by no means dwelling in amity, as supposed by some, for the snakes and owls destroy the various young, those of the "dogs" being the chief sufferers, and doubtless they would be only too happy to be rid of their unwelcome neighbors had they the power to cause their removal.   These animals seem to be independent of water, possibly the dew that often falls heavily in the districts in which they live satisfying their moderate needs.   Prairie-dogs are not easily caught, and when captured, are difficult to tame, the wild, free life of the plains, causing by comparison, that of a captive to be unsupportable.   They live on seeds and grasses, and their cheek pouches are small.

37. Cynomys.

$$I.\frac{1-1}{1-1};\ P.\frac{2-2}{1-1};\ M.\frac{3-3}{3-3} = 22.$$

Cynomys Rafin., Amer. Month. Mag., II, 1817, p. 45.   Type *Arctomys ludovicianus* Ord.

Anisonyx Rafin., Am. Month. Mag., II, 1817, p. 45.   (nec Latreille, 1807, Coleopt.)

Lipura Illig., Prodr. Syst. Mamm., et. Av., 1811, p. 95.

Monax Warden, Statist. Polit. Hist. Acc. U. S., I, 1819, pp. 225-228.

Cheek pouches shallow; ears rudimentary; tail very short, flat; feet with claws on all five toes; pollex large, nail well developed; pelage short, bristly; dentition very heavy: molars large with three transverse grooves on their crowns; first and second premolars nearly equal in size; outline of molar series curved, divergent anteriorly, approximating posteriorly; postorbital processes strong, well developed, decurved; antorbital foramina large, subtriangular, the tubercle at end large and visible when viewed from above, and projecting beyond the superior outline of skull; palate greatly contracted posteriorly; occipital and saggital crests present.

KEY TO THE SPECIES.

A. Size large. Check pouches present, shallow; palate greatly contracted posteriorly; postorbital processes long, pointed.

  a. Tail short, flat; pelage bristly.

    a.' Under parts white; tail with subterminal   PAGE
bar of broccoli brown..................*C. arizonensis* 154

    b.' Under parts yellowish white; tail with
apical third black....................*C. ludovicianus* 155

    c.' Under parts pale fulvous; tail with apical
half mixed black and white ....*C. gunnisoni* 156

    d.' Under parts buffy; tail with apical half
black............................ ..*C. mexicanus* 156

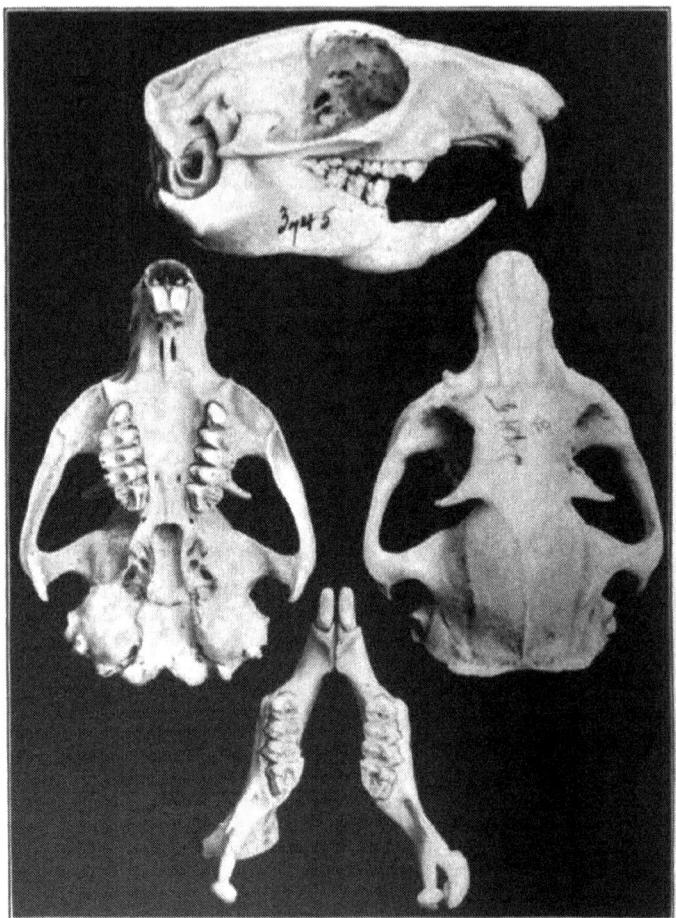

FIG. 28. CYNOMYS LUDOVICIANUS.
No. 3745 Field Columbian Mus. Coll. Nat. size.

114. arizonensis (*Cynomys*), Mearns, Bull. Am. Mus. Nat. Hist., 1890, p. 305. Elliot, Syn. N. Am. Mamm., 1901, p. 104.
ARIZONA PRAIRIE-DOG.

Type locality. Point of Mountain, near Wilcox, Cochise County, southern Arizona.

Geogr. Distr. State of Chihuahua, Mexico, into Arizona.

Genl. Char. Size large; tail nearly twice as long as that of *C. gunnisoni.*

Color. Summer Pelage. Above vinaceous cinnamon; below whitish; tail with a narrow subterminal bar of broccoli brown.

Winter Pelage. Above pale sandy buff; below white tinged with buff.

Measurements. Total length, 376; tail vertebræ, 84; hind foot, 61. Skull: basilar length, 54.3; total length, 66; interorbital width, 14; length of nasals, 25; zygomatic width, 43.5; length of upper molar series, 16.3.

FIG. XXXII. CYNOMYS LUDOVICIANUS. PRAIRIE-DOG.

115. ludovicianus (*Arctomys*), Ord, Guth., Geog., II, 1815, 2d Am. ed., p. 292.

socialis Rafin., Am. Month. Mag., II, 1817, p. 45.

grisea Rafin., Am. Month. Mag., II, 1817, p. 45.

missouriensis Warden, Descrip. États Unis, v, 1820, p. 627.

latrans Harlan, Faun. Am., 1825, p. 306.

ludovicianus (*Cynomys*) Elliot, Syn. N. Am. Mamm., 1901, p. 102.

PRAIRIE-DOG, *Perrito, Perrito del Campo,* in Mexico.

Type locality. Plains of the Upper Missouri.

Geogr. Distr. State of Chihuahua, Mexico, north into western Texas and Kansas to 49th parallel.

Genl. Char. Body stout; tail short; ears very small; claws long.

Color. Summer Pelage. Above reddish brown, varied with gray and black hairs; beneath yellowish white; tail like the back with the apical third black.

Winter Pelage. Above pale vinaceous buff, grizzled and mixed with black; below pale buff.

Measurements. Total length, 393; tail vertebræ, 88; hind foot, 57. Skull: occipito-nasal length, 65; Hensel, 55; zygomatic width, 47; interorbital constriction, 15; palatal length, 33; length of nasals, 23; length of upper molar series, 17.

116. gunnisoni (*Cynomys*), Baird, Proc. Acad. Nat. Scien. Phil., 1855, p. 334. Elliot, Syn. N. Am. Mamm., 1901, p. 103.
GUNNISON'S PRAIRIE-DOG.

Type locality. Cochetopa Pass, Rocky Mountains, Saguache County, Colorado.

Geogr. Distr. San Diego, State of Chihuahua, Mexico, into New Mexico and Arizona to Colorado.

Genl. Char. Smaller than *C. ludovicianus*; tail one-sixth length of body.

Color. Summer Pelage. Above tawny fulvous, mixed with black; under parts pale fulvous; tail like back on basal half, rest mixed black and white bordered and tipped with white.

Winter Pelage. Above pale buff, mixed with black; under parts pale yellow to fulvous.

Measurements. Average total length, 360; tail vertebræ, 69; hind foot, 60. Skull: occipito-nasal length, 57.5; Hensel, 28; zygomatic width, 41; interorbital constriction, 11.5; palatal length, 28.5; length of nasals, 20; length of upper molar series, 14.

117. mexicanus (*Cynomys*), Merr., Proc. Biol. Soc. Wash., VII, 1892, p. 157.
MEXICAN PRAIRIE-DOG.

Type locality. La Ventura, State of Coahuila, Mexico.

Geogr. Distr. Eastern and northern Mexico.

Genl. Char. Size large. Similar to *C. ludovicianus*, tail longer and blacker; nasals as long as distance from anterior edge of foramen magnum to posterior edge of palate.

Color. Above grizzled buffy fulvous mixed with long black hairs; under parts, hands, and feet buffy; tail, basal half above like back, bordered with black, remainder black, beneath basal half buffy, remainder black and buff grizzled.

Measurements. Total length, 419; tail vertebræ, 107; hind foot, 63.

The Flying Squirrels are so called, not because they are capable of any true flight, but on account of a fold of skin attached to the front and hind limbs and body, which when stretched by the extension of the arms and legs enables the animal to sail in a descending line for a considerable distance as if carried by a parachute. They are beautiful creatures with velvety fur and large, expressive eyes, nocturnal in their habits, and live in nests or holes in trees. Their aerial flights occur usually about dusk, and at this time in the localities they frequent, several may be seen gliding from lofty branches to the base of a distant tree, up the trunk of which they hasten until the top is nearly reached, when the voyager is ready for another trip through space.

Subfam. II. **Pteromyinæ.**

38. Sciuropterus.

$I.\frac{1-1}{1-1}$; $P.\frac{2-2}{1-1}$; $M.\frac{3-3}{3-3} = 22$.

Sciuropterus F. Cuv., Dent's du Mamm., 1825, p. 255. *Id.* Ann. du Mus., x, 1825, p. 126, pl. x. Type *Sciurus volans* Linnæus.

Limbs connected by a furred membrane extending outwardly from the sides, and supported by a process from the olecranon. Tail depressed, flattened, thick; ears large; pelage of velvet softness.

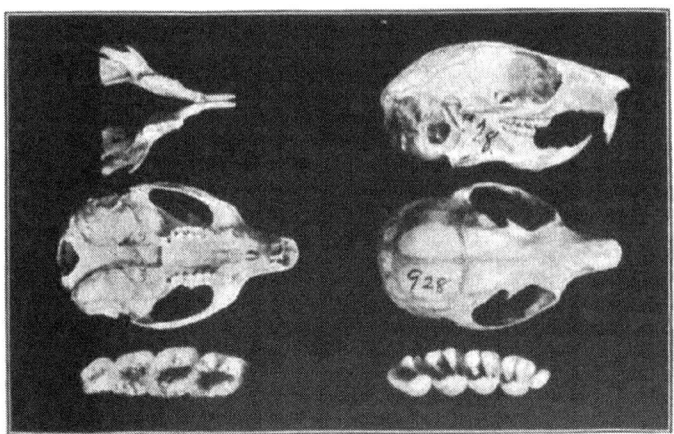

FIG. 29. SCIUROPTERUS VOLANS.
No. 928 Field Columbian Mus. Coll. Nat. size.

UPPER TOOTH ROW.   LOWER TOOTH ROW.
Enlarged 4 times.   Enlarged 4 times.

118. volans (*Mus.*) Linn., Syst. Nat., 1, 1758, p. 63; 1, 1776, p. 85.
  (nec *Sciurus*, p. 88.)
  volucella Pall., Nov. Spec. Glires, 1788, p. 351.
  volans (*Sciuropterus*), Elliot, Syn. N. Am. Mamm., 1901, p. 109.
FLYING SQUIRREL.

Type locality. Virginia.

Geogr. Distr. From northern border of Mexico into Guatemala, Central America. In United States to northern New York and southern New Hampshire; not in Florida.

Genl. Char. Size medium; winter and summer pelage alike in color; hairs of under parts white to base.

FIG. XXXIII. SCIUROPTERUS VOLANS. FLYING SQUIRREL.

Color. Upper parts drab shaded with russet, tail slightly darker; hands above grayish white, feet drab; black orbital ring; under parts pure white, washed in some specimens with buff.

Measurements. Total length, 234.5; tail vertebræ, 99.6; hind foot, 31.4. Skull: occipito-nasal length, 34; Hensel, 27; zygomatic width, 20; interorbital width, 7; palatal length, 9; length of upper molar series, 6.

Largest of North American rodents, attaining a weight of fifty pounds or more, the Beaver, which at one time extended its range over nearly all forest-covered land in the northern Hemisphere, has now become extinct in the majority of localities. Its skin and scent bags were too valuable commercially to preserve it from man's rapacity. It is probably one of the best-known rodents in the land, and most persons have some knowledge of the beaver's house and dam, or have seen the trunks of trees that have been cut down by the wonderful adze-like incisors. Clothed in a dense furry coat impervious to water, and provided with paddle-like hind feet and a broad rudder-like tail, the beaver is at home in the lake or river, where most of its life is passed. It shuns the vicinity of man, and exists only in the virgin wilderness.

Fam. II. **Castoridæ. Beavers.**

Skull massive, no postorbital processes, superior outline nearly straight; molars single-rooted, with re-entering of enamel folds, and decreasing in size posteriorly; the molar series is not parallel but converges anteriorly, and the palate is arched, contracted anteriorly. Incisors large, powerful, the lower much longer than the upper, with chisel-like edges, and a deep orange red color exteriorly. Lower jaw massive; angle of mandible rounded.

39. Castor. Beaver.

$$I.\frac{1-1}{1-1}; \ P.\frac{1-1}{1-1}; \ M.\frac{3-3}{3-3} = 20.$$

Castor Linn., Syst. Nat., 1, 1758, p. 58; 1, 1766, p. 78. Type *Castor fiber* Linnæus.

Feet four-toed, hind feet large, webbed; upper molars subequal, with one inner and two outer enamel folds; tail broad, flat, scaly; molars with dentinal pulp persisting until quite late in life.

canadensis frondator (*Castor*), Mearns, Proc. U. S. Nat. Mus., 1897, p. 502. Elliot, Syn. N. Am. Mamm., 1901, p. 116. SONORAN BEAVER.

Type locality. San Pedro River, State of Sonora, Mexico, near monument No. 98, Mexican boundary.

Geogr. Distr. From State of Sonora, Mexico, to Wyoming and Montana.

Genl. Char. Size large; scaly portion of tail less than twice as long as wide.

Color. Upper parts russet, chocolate at root of tail; under parts grayish cinnamon to ferrugineous beneath tail; sides wood brown varied with tawny olive; feet burnt sienna.

Measurements. Total length, 1070; tail vertebræ, 360; scaly portion of tail 290×125; hind foot, 185. Skull: occipito-nasal length, 134; Hensel, 122; zygomatic width, 97; interorbital constriction, 22,

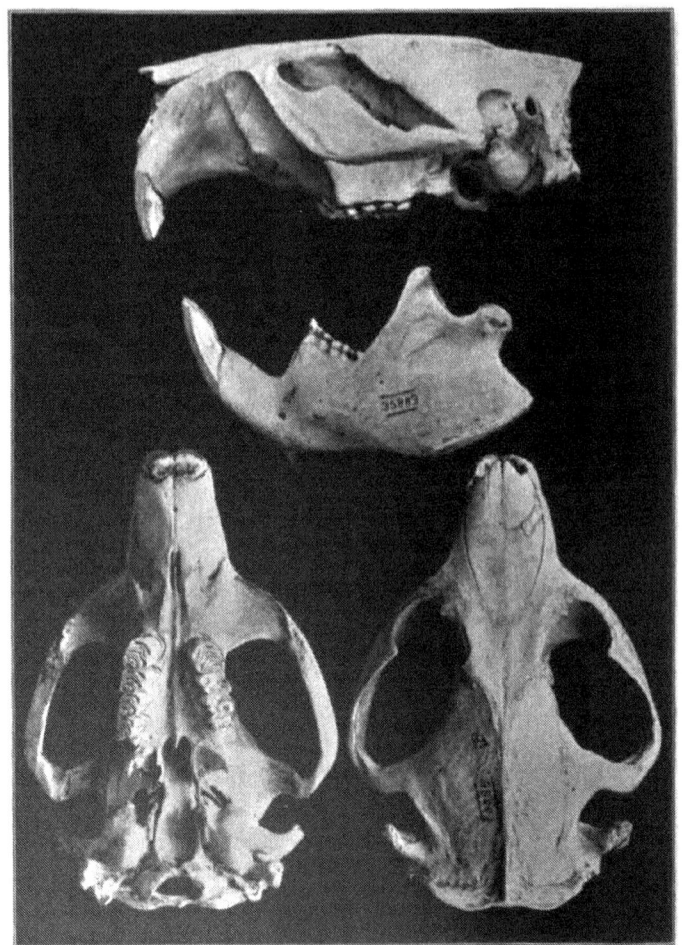

FIG. 30. CASTOR C. FRONDATOR.
No. 35883 U. S. Nat. Mus. Coll. Type.

palatal length, 78; length of nasals, 46; length of upper molar series, 32; length of mandible, angle to alveolus of incisors, 96; height at condyle, 22; at coronoid process, 52.

FIG. XXXIV. CASTOR C. FRONDATOR. SONORAN BEAVER.

The MURIDÆ is the largest family of the rodents and is cosmo-politan, some members, such as the Brown Rat being met with throughout the world, carried from place to place in ships. It includes a vast number of genera, embracing both terrestrial and aquatic animals of varied structure and habits, while the size of the numerous species ranges from that of the pigmy Harvest Mouse (genus *Rhithro-dontomys*) to that of the Musk-rat. Every land possesses its own peculiar species, and North America has a large number indigenous to it. They are of a great variety and are dwellers of the woods, cultivated fields, prairie lands, swamps, lakes, and rivers, each seeking, after its kind, localities best suited to its mode of life.

The subfamily *Murinæ* is typified by the Rat and Mouse of our houses, and these have their representatives in other subfamilies of many varied forms and structures. Some are possessed of cheek pouches. The tubercular teeth have their crowns worn by constant use to a flat surface and they then exhibit various tracery patterns, and the consequent angles and loops shown, more readily indicate the relationship of their owner to other forms. Members of the *Muridæ* have a certain family resemblance to each other, in the more or less lengthened tail, generally naked and scaly, bright eyes, and a modest coloration suitable for concealing them from their foes.

Fam. III. **Muridæ. Rats, Mice, Voles, etc.**

Premolars none; molars with or without roots, tuberculate, or with enamel folds; lower incisors compressed; frontals greatly contracted anteriorly.

Subfam. I. **Murinæ.**

Molars rooted, tuberculate; root of under incisor creating a swelling on outer side of mandible between processes of the condyle and coronoid; descending process of mandible below the plane of the molars. Palate nearly flat.

The genus Mus, has more members than any other of the Mammalia and its representatives are found throughout the world, except in Madagascar, and possibly other islands whose faunæ are unknown. The habits of the various species are similar, although some are arboreal and others even aquatic.

40. Mus. Mice, Rats.

$$I.\frac{1-1}{1-1};\ M.\frac{3-3}{3-3} = 16.$$

Mus Linn., Syst. Nat., I, 1758, p. 59; II, 1766, p. 79.   Type *Mus rattus* Linnæus.

Ears large, prominent; tail long, scaly; nose acute; molars with transverse series of tubercles, three in a series, longitudinal.

KEY TO THE SPECIES AND SUBSPECIES.

A. Size small; tail moderately long.   PAGE
  a. Pelage above grayish brown and black,
    beneath ashy plumbeous ....*M. musculus*   162
  b. Pelage above black, beneath buffy gray.....*M. m. jalapæ*   163
B. Size large, tail very long, naked, scaly.
  a. Pelage black on upper parts..................*M. rattus*   163
  b. Pelage brownish on upper parts.
    a.' Under parts ashy white................*M. norvegicus*   164
    b.' Under parts yellowish white..........*M. alexandrinus*   164

119. musculus (*Mus*), Linn., Syst. Nat., I, 1758, p. 62; II, 1766, p. 83.   Elliot, Syn. N. Am. Mamm., 1901, p. 118.

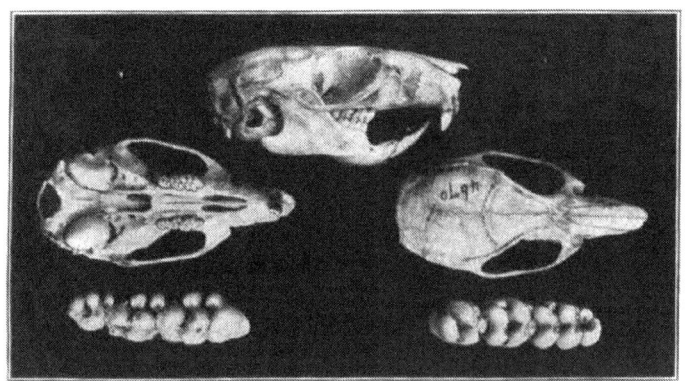

FIG. 31. MUS RATTUS.
No. 4670 Field Columbian Mus. Coll. Nat. size.
UPPER TOOTH ROW.   LOWER TOOTH ROW.
Enlarged 4 times.   Enlarged 4 times.

HOUSE MOUSE.

Type locality. Sweden.

Geogr. Distr. Cosmopolitan.

Genl. Char. Tail longer than body; soles naked; size small.

Color. Above grayish brown, lined with blackish; beneath ashy plumbeous, tinged with reddish; tail dusky; feet ashy brown.

Measurements. Total length, 170; tail vertebræ, 83; hind foot, 18; ear, 13.5. Skull: occipito-nasal length, 20; Hensel, 15; zygomatic width, 10; interorbital constriction, 3.5; length of nasals, 6; palatal length, 8; length of upper tooth row, 3.

a.—jalapœ (*Mus*), Allen & Chapman, Bull. Am. Mus. Nat. Hist., 1897, p. 198.

JALAPA HOUSE MOUSE.

Type locality. Jalapa, State of Vera Cruz, Mexico.

Geogr. Distr. State of Vera Cruz, Mexico, range unknown.

Genl. Char. General hue dark, dorsal band broad.

Color. Above black; sides mouse color; under parts buffy gray; tail black; feet and ears brown.

Measurements. Total length, 164; tail vertebræ, 82; hind foot, 18; ear, 14.

120. rattus (*Mus*), Linn., Syst. Nat., 1, 1758, p. 61; 1, 1766, p. 83. Elliot, Syn. N. Am. Mamm., 1901, p. 117.

FIG. XXXV. Mus RATTUS. BLACK RAT.

BLACK RAT.
Type locality.   Sweden.
Geogr. Distr.   Cosmopolitan.
Genl. Char.   Tail little longer than head and body.
Color.   Above sooty black; beneath plumbeous; feet brown.
Measurements.   Total length, 368; tail vertebræ, 190.   Skull: occipito-nasal length, 37; Hensel, 30; zygomatic width, 17.5; interorbital constriction, 5.5; length of nasals, 12; palatal length, 16.5; length of upper tooth row, 6.

121. norvegicus (*Mus*), Erxl., Syst. Règ. Anim., 1777, p. 381.
Elliot, Syn. N. Am. Mamm., Suppl., 1901, p. 428.
decumanus Pall., Glir., 1778, p. 91.   Elliot, Syn. N. Am. Mamm., 1901, p. 117.
NORWAY RAT.
Type locality.   Norway.
Geogr. Distr.   Cosmopolitan.   Introduced into North America.
Genl. Char.   Tail little shorter than head and body, sparsely haired; annuli about 200.
Color.   Above rusty grayish brown; sides grayer; beneath ashy white; tail above dusky, beneath paler.
Measurements.   Total length, 310; tail vertebræ, 146; hind foot, 38.   Skull: occipito-nasal length, 44; Hensel, 21; zygomatic width, 22; interorbital constriction, 6; length of nasals, 14; palatal length, 20.5; length of upper tooth row, 6.

122. alexandrinus (*Mus*), I. Geoff., Descr. Egypt, II, 1818, p. 733, Atlas, pl. v, fig. 1.   Elliot, Syn. N. Am. Mamm., 1901, p. 118.
tectorum Savi, Nov. Giorn., 1825.

Brown Rat.

Type locality. Alexandria, Egypt.

Geogr. Distr. Cosmopolitan.

Genl. Char. Smaller than *M. norvegicus*; tail considerably longer than head and body; annuli about 240.

Color. Above yellowish brown tinged with reddish; flanks grayish; under parts and upper surface of feet yellowish white; tail dusky.

Measurements. Total length, 356; tail vertebræ, 198; hind foot, 35. Skull: occipito-nasal length, 43; Hensel, 34; zygomatic width, 19; interorbital constriction, 6.5; length of nasals, 15; palatal length, 20; length of upper tooth row, 6.5.

The next genus contains the Mole Mice, little creatures with a soft, velvety pelage and rather short tails. They have usually a pale coloration, with white or whitish belly, hands, and feet.

41. Onychomys. Mole Mice.

Onychomys Baird, N. Am. Mamm., 1857, p. 458.   Type *Hypudæus
    leucogaster*, Max.

Hypudæus. Max., Reise, N. Am. 1841, p. 99. (nec Auct.)

"Form arvicoline; tail less than half the body in length; claws very large, fossorial, the anterior longest; soles with only four tubercles, the two posterior of the other groups wanting; posterior two-thirds of soles densely furred. Skull without orbital crest; the upper margin of orbit sharp."

KEY TO THE SPECIES AND SUBSPECIES.

A. Tail slightly over one-fourth the total length   PAGE
    of head and body.
    a. Upper parts tawny cinnamon and black...*O. melanophrys*   166
    b. Upper parts pale cinnamon and little black.*O. m. pallescens*   166
    c. Upper parts grayish vinaceous buff...........*O. ramona*   167
B. Tail one-third or over the total length of head
    and body.
    a. Upper parts yellowish brown...............*O. torridus*   167
    b. Upper parts drab gray..................*O. t. arenicola*   168
    c. Upper parts cinereous................*O. t. perpallidus*   168
    d. Upper parts ashy vinaceous...............*O. macrotis*   169

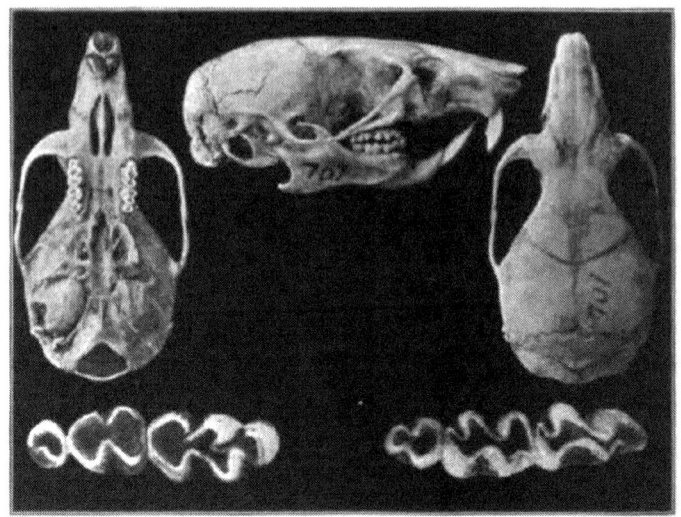

FIG. 32. ONYCHOMYS TORRIDUS.
No. 701 Field Columbian Mus. Coll. Twice nat. size.

UPPER TOOTH ROW.   LOWER TOOTH ROW.
Enlarged 10 times.   Enlarged 10 times.

123. melanophrys (*Onychomys*), Merr., N. Am. Faun., No. 2, 1889, p. 2, and No. 3, 1890, p. 61. Elliot, Syn. N. Am. Mamm., 1901, p. 120.

BLACK-BROWED MOLE MOUSE.

Type locality. Kanab, Kane County, Utah.

Geogr. Distr. State of Sonora, Mexico, north to Utah.

Genl. Char. Size of *O. leucogaster*; ear a little smaller; hind foot densely furred to base of toes.

Color. Above rich tawny cinnamon, well mixed with black-tipped hairs on the back, and brightest on the sides; a distinct black ring around the eye, broadest above. "This ring is considerably broader and more conspicuous than the very narrow ring of *leucogaster*."

Measurements. Total length, 154; tail, 41; hind foot, 21; ear from crown, 10. Skull: Hensel, 22.3; zygomatic breadth, 15.4; interorbital constriction, 5.2; length of nasals, 10.7; length of upper molar series, 4.8; length of mandible, 15.7. (Merr., l. c.)

a.—pallescens (*Onychomys*), Merr., N. Am. Faun., No. 3, 1890, p. 61. Elliot, Syn. N. Am. Mamm., 1901, p. 121.

PALE MOLE MOUSE.

Type locality. Moki Pueblos, Apache County, Arizona.

Geogr. Distr. State of Chihuahua, Mexico, into Arizona.

Genl. Char. Size large, exceeded only by *O. longipes* of Concho County, Texas; ears, feet, and tail much as in *O. melanophrys*; fur full, long, and soft; orbital ring absent or inconspicuous; lanuginous tuft at base of ear well developed.

Color. Above pale tawny cinnamon, palest anteriorly, and brightest on the flanks and rump, not noticeably mixed with black-tipped hairs; below pure white to the roots of hairs.

Measurements. Total length, 168; head and body, 125; tail, 45; hind foot, 22; ear from crown, 12; from anterior root, 16.5. (Merr., l. c.) Skull: occipito-nasal length, 20; Hensel, 23; zygomatic width, 15; interorbital constriction, 5; length of nasals, 11; palatal length, 11; length of upper molar series, 4.

124. ramona (*Onychomys*), Rhoads, Am. Nat., XXVII, 1893, p. 833. Elliot, Syn. N. Am. Mamm., 1901, p. 121.

RAMONA MOLE MOUSE.

Type locality. San Bernardino Valley, California.

Geogr. Distr. Lower California, Mexico, and southern California.

Genl. Char. Larger than *O. torridus*, with larger ears.

Color. Similar to *O. longicaudus*; above grayish vinaceous buff, dorsal part darker; beneath white; tail bicolor, dark above, lighter below.

Measurements. Total length, 147; tail vertebræ, 48; hind foot, 18; ear, 12. Skull: Hensel, 19; zygomatic breadth, 32.6; parietal breadth, 11.5; interorbital constriction, 5; length of nasals, 10; from foramen magnum to incisive foramina, 13.4; length of upper molar series, alveolar border, 3.8; length of mandible, 15; height of coronoid process, 6.8.

FIG. XXXVI. ONYCHOMYS TORRIDUS. TROPICAL MOLE MOUSE.

125. torridus (*Onychomys*), Coues, Proc. Acad. Nat. Scien. Phil., 1874, p. 183. Elliot, Syn. N. Am. Mamm., 1901, p. 122.

TROPICAL MOLE MOUSE.

Type locality. Camp Grant, Graham County, Arizona.

Geogr. Distr. States of Chihuahua and Sonora, Mexico, north to Upper Missouri.

Genl. Char. Similar to *O. leucogaster*; tail longer; ears larger; colors more yellowish.

Color. General color more yellowish than *O. leucogaster*; no dorsal stripe; feet and under parts tawny white; tail above dusky, beneath white.

Measurements. Total length, 70 tail vertebræ, 51; hind foot, 20; ear, 19. Skull: Hensel, 18.5; zygomatic width, 12.5; interorbital constriction, 4.2; length of nasals, 9.6; length of upper molar series, 3.5.

a.—arenicola (*Onychomys*), Mearns, Proc. U. S. Nat. Mus., XIX, 1896, p. 139. Elliot, Syn. N. Am. Mamm., 1901, p. 122.

SAND-LOVING MOLE MOUSE.

Type locality. Rio Grande near El Paso, El Paso County, Texas.

Geogr. Distr. State of Chihuahua, Mexico, into Texas.

Genl. Char. Similar to *O. torridus*, but slightly smaller, with relatively smaller ears and a very much paler coloration.

Color. Above drab gray, inclining to fawn color on sides; dorsal area with very little admixture of black-edged or black-tipped hairs; a conspicuous tuft of white hairs at anterior base of ears; dark spot on anterior band of ear, drab (not black); whiskers more white than black; under parts, feet, and end of tail white; basal two-thirds of upper side of tail drab, some of the hairs with hoary tips.

Measurements. Total length, 137; tail vertebræ, 53 (to end of pencil, 57); hind foot, 21. Skull: 25.5×13.5. (Mearns, l. c.)

b.—perpallidus (*Onychomys*), Mearns, Proc. U. S. Nat. Mus., XIX, 1896, p. 140. Elliot, Syn. N. Am. Mamm., 1901, p. 122.

DRAB GRAY MOLE MOUSE.

Type locality. Colorado River at Monument No. 204, Mexican boundary line, Yuma County, Arizona.

Geogr. Distr. Lower California and State of Sonora, Mexico, to Gila City across Yuma and Colorado Deserts to Coast Range of Mountains.

Genl. Char. Larger than *O. torridus*, with relatively larger ears, longer tail, and a much paler coloration.

Color. Above drab gray, becoming more cinereous anteriorly; sides and rump barely tinged with fawn color; dusky line on basal three-fourths of tail nearly obsolete, much obscured by whitish hairs; ears less densely clothed than in the other forms of *O. torridus*, and without a well-defined dusky spot; whiskers mostly white or colorless; under parts, feet, and end of tail white.

Measurements. Total length, 157; tail vertebræ, 57; ear from crown, 16; hind foot, 22. Skull, 26 × 13.7. (Mearns, l. c.)

126. macrotis (*Onychomys*), Elliot, Pub. Field Columb. Mus., III, 1903, p. 155. Zoölogy.

LARGE-EARED MOLE MOUSE.

Type locality. Head of San Antonio River, Lower California, Mexico.

Geogr. Distr. San Quentin to San Antonio River, Trinidad Valley and plain of El Alamo, to about 5,000 feet elevation in the San Pedro Martir Mountains, Lower California, Mexico.

Genl. Char. Size medium, color pale; ears and hind feet long; tail rather long, without white tip; no black spot on ear.

Color. Above ashy vinaceous finely lined on dorsal surface and top of head with blackish brown, causing these parts to be darker than the rest of upper surface; sides of body ashy vinaceous; orbital ring black; lips, cheeks, under parts, hands, and feet pure white; tail dusky above, white beneath; ears pale grayish brown, outer edge blackish; tuft of white hairs at base of ears.

Measurements. Total length, 155; tail vertebræ, 55; hind foot, 21; ear, 20.5; Skull: occipito-nasal length, 26; Hensel, 20; zygomatic breadth, 13.5; interorbital constriction, 4.5; length of nasals, 7.5; palatal arch to alveoli of incisors, 10; greatest width of braincase, 11.

The next genus, PEROMYSCUS, contains numerous species, and is well represented in North America, and for its most characteristic member may be selected the White-footed or Deer Mouse, with its various races. This little animal with white feet, large ears, subdued coloring, bright eyes like shining black beads, and velvet fur is most attractive. It has small cheek pouches into which it stores the seeds and grain while on its foraging expeditions for providing a food supply against the approaching winter; and it is not particular as to its choice of abode, often taking up its residence in man's habitation if situated sufficiently convenient to its beloved fields and woods. Among the large number of species comprising this genus, there are, as may be supposed, those of all sizes and varieties of coloration. The most striking perhaps among them all is the Golden Mouse, *P. nuttalli*, of the Central and Southern United States, with its golden cinnamon coat bordered with white beneath. The majority, however, are modestly dressed in the hue for which their familiar family name has provided an appellation—mouse color, varied with shades of black, russet, and numerous tints of yellows and browns, with white harmoniously applied and blended. Usually the under parts are white, as are the hands and feet also; but again these latter are often plumbeous in different shades. They are the gleaners of our fields and woods, often graceful of shape and always agile of foot, the "small deer" of our land.

No careful revision of *Peromyscus* has as yet been made, and until that is done, the status of many of the forms now deemed distinct and the arrangement of the species cannot be satisfactorily determined.

42. Peromyscus. Field Mice, Deer Mice.

$$I.\frac{1-1}{1-1}; \ M.\frac{3-3}{3-3} = 16.$$

Peromyscus Gloger, Handb. und Hilfsb. Naturg., 1841, p. 95. Type *Peromyscus arboreus Gloger=Mus sylvaticus noveboracensis Fischer.*

Calomys Aud., Quad. N. Amer., II, 1851, p. 303. (nec Waterh., P. Z. S., 1837, p. 21.)

Vesperimus Coues, Proc. Acad. Nat. Scien. Phil., 1874, p. 178.

Baiomys True, Proc. U. S. Nat. Mus., 1894, XVI, p. 758.

Trinodontomys Rhoads, Proc. Acad. Nat. Scien. Phil., 1894, p. 257.

Haplomylomys Osgood, Proc. Biol. Soc. Wash., XVII, 1904, p. 53.

Size moderate, eyes rather prominent; face rather long, nose pointed; ears thin, rather rounded, in some species very large. Feet small, digits slender, palms naked; hind feet long, soles with six tubercles; tail terete, tapering, slender, hairy, sometimes longer than head and body, and occasionally tufted; pelage soft, frequently glossy. Skull thin, papery; braincase broad, rather flat, superior outline curving both ways from highest point just behind orbits; zygomata slender, threadlike, dipping midway to level of the palate, zygomatic arch composed mainly of processes of the maxillary and squamosal. Orbital foramina just above the level of the alveolus; interorbital constriction considerable, but wider than rostrum; nasals and intermaxillæ project beyond the incisors; auditory bullæ small, thin, and obliquely situated; lower jaw straight; coronoid very short; molar series short, narrow, the teeth decreasing in size from front to rear; upper molars with three roots each, the lower with two, and the unworn teeth have a double series of conical tubercles, which gradually are reduced by abrasion, and the pattern varies constantly.

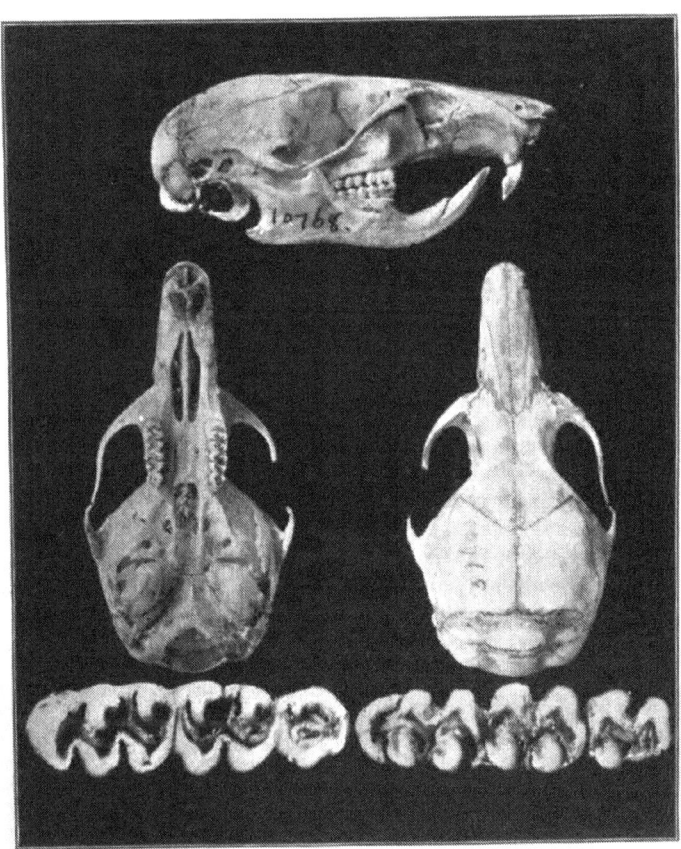

FIG. 33. PEROMYSCUS FURVUS.
No. 10768 Am. Mus. Nat. Hist. Coll.　Twice nat. size.

UPPER TOOTH ROW.	LOWER TOOTH ROW.
Enlarged 10 times.	Enlarged 10 times.

KEY TO THE SPECIES AND SUBSPECIES.

A.　Size small, not over 190 mm.
　　a.　Tail equal to or less than half the length of
　　　　the body and head.
　　　　a.' Above drab gray.　　　　　　　　　　　　PAGE
　　　　　a." Under parts white, ears large........*P. tiburonensis*　175
　　　　　b." Under parts buffy, ears small.
　　　　　　a."' Smaller ....*P. musculus*　175

127. tiburonensis (*Peromyscus*), Mearns, Proc. U. S. Nat. Mus.,
      1897, p. 720.
ISLAND OF TIBURON MOUSE.
  Type locality. Tiburon Island, Gulf of California, Mexico.
  Geogr. Distr. Known only from type locality.
  Genl. Char. Size very small; ears nearly naked.
  Color. Upper parts drab gray and black; sides ochraceous cin-
namon; under parts white; tail blackish brown; hands and feet
white; ears dark brown.
  Measurements. Total length, 175; tail vertebræ, 85; hind foot,
19; ear, 16. Skull: total length, 24; greatest width, 12. (Mearns,
l. c.)

*127a. allex (*Peromyscus*), Osgood, Proc. Biol. Soc. Wash., XVII,
      1904, p. 76.
DWARF MOUSE.
  Type locality. Colima, State of Colima, Mexico.
  Genl. Char. Like *P. musculus*, but smaller; nasals short; bullæ
very small.
  Color. Exactly like *P. musculus.*
  Measurements. Total length, 113; tail vertebræ, 47; hind foot,
14; ear from notch, 9.7. Skull: greatest length, 18.4; Hensel, 14.6;
zygomatic width, 9.6; interorbital constriction, 3.2; length of nasals,
6.3; interparietal, 5.5 × 1.3; palate, 3; length of upper molar series, 3.

128. musculus (*Sitomys*), Merr., Proc. Biol. Soc. Wash., VIII, 1892,
      p. 170.
COLIMA CITY MOUSE.
  Type locality. Near Colima City, State of Colima, Mexico.
  Geogr. Distr. States of Colima and Jalisco, Mexico.
  Genl. Char. Similar to *P. taylori*, but lighter and larger.

---

*Descriptions of numerous forms of *Peromyscus* by Mr. Osgood (l. c.) were
published too late to be included here in their regular numerical order.

Color. Above drab gray and black, tinged with tawny; under parts buffy; tail above like back, beneath paler.

Measurements. Total length, 123; tail vertebræ, 48; hind foot, 17; ear, 5.5 (dried skin).

a.—brunneus (*Peromyscus*), Allen & Chapman, Bull. Amer. Mus. Nat. Hist., 1897, p. 203.

JALAPA BROWN MOUSE.

Type locality. Jalapa, State of Vera Cruz, Mexico.

Geogr. Distr. State of Vera Cruz, Mexico; range unknown.

Genl. Char. Smaller than *P. musculus*, and darker.

Color. Upper parts drab gray thickly flecked with buff, tinged with rufous, and varied with black; sides similar but paler; under parts yellowish; ears dark brown; tail above dark brown, beneath paler; hands and feet whitish.

Measurements. Total length, 121; tail vertebræ, 48; hind foot, 15; ear, 13. Skull: total length, 15; basal length, 12.5; zygomatic breadth, 11; interorbital constriction, 4; length of nasals, 7.5; palatal length, 7; length of upper tooth row, 3.2. (ex Type.)

b.—nigrescens (*Peromyscus*), Osgood, Proc. Biol. Soc. Wash., XVII, 1904, p. 76.

SOOTY MOUSE.

Type locality. Valley of Comitan, State of Chiapas, Mexico.

Genl. Char. Similar to *P. musculus*, but darker.

Color. Upper parts mixed Vandyke brown and sooty blackish; under parts cream buff; tail dusky above, paler beneath.

Measurements. Total length, 113–120; tail vertebræ, 40–45; hind foot, 14.5–16. Skull: greatest length, 20.1; Hensel, 15.2; zygomatic width, 10.5; interorbital constriction, 3.4; length of nasals, 8; interparietal, 6.4; palate, 2.8; length of upper molar series, 3.2.

129. paulus (*Peromyscus*), Allen, Bull. Am. Mus. Nat. Hist., 1903, p. 598. TINY MOUSE.

Type locality. Rio Sestin, State of Durango, Mexico.

Genl. Char. Smaller than *P. musculus*, color lighter and grayer.

Color. Upper parts gray brown suffused with pinkish buff; under parts grayish white, base of hairs plumbeous, tips whitish; belly sometimes tinged with buff.

Measurements. Total length, 108; head and body, 62; tail vertebræ, 44; hind foot, 14; ear from notch, 13; from crown, 11.

130. thurberi.

americanus thurberi (*Sitomys*), Allen, Bull. Am. Mus. Nat. Hist., 1893, p. 185.

texanus! medius Mearns, Proc. U. S. Nat. Mus., 1895, p. 446.
texensis medius Elliot, Syn. N. Am. Mamm., 1901, p. 130.
THURBER'S FIELD MOUSE.

Type locality. San Pedro Martir Mountains, Lower California, Mexico. Altitude, 7,000 feet.

Geogr. Distr. Region of San Pedro Martir Mountains, Lower California, Mexico, north into southern California to Colorado desert at least.

Genl. Char. General color pale grayish fulvous lined with black.

Color. Upper parts grayish fulvous, darkest on dorsal line where it is blackish; sides fulvous; lower sides and under parts and feet white; ears dusky, at base a tuft colored like head; tail blackish above, white below.

Measurements. Total length, 177; tail vertebræ, 78; hind foot, 21; ear, 18. Skull: occipito-nasal length, 25.1; Hensel, 99.2; zygomatic width, 13; interorbital constriction, 4; length of nasals, 9.2; width of braincase above roots of zygomata, 11.5; palatal length, 10; length of upper molar series, 4; length of lower molar series, 4.

a.—mesomelas (*Peromyscus*), Osgood, Proc. Biol. Soc. Wash., XVII, 1904, p. 57.

Type locality. Orizaba, State of Vera Cruz, Mexico.

Genl. Char. Similar to *P. thurberi*, darker; tail shorter; hind foot larger; pectoral spot present.

Color. Upper parts Prout's brown; under parts creamy white; pectoral spot fawn; hands and feet white; ankles dusky brownish; tail bicolor; ears dusky, edges whitish.

Measurements. Total length, 169; tail vertebræ, 76; hind foot, 23. Skull: greatest length, 26.5; Hensel, 20.2; zygomatic width, 13.6; interorbital constriction, 4; interparietal, 8.6 × 2.3; length of nasals, 10.4; palatal length, 3.8; palatal foramina, 5.2 × 2; postpalatal length, 9.1; length of upper molar series, 3.7.

b.—castaneus (*Peromyscus*), Osgood, Proc. Biol. Soc. Wash., XVII, 1904, p. 58.

YOHALTUN MOUSE.

Type locality. Yohaltun, State of Campeche, Mexico.

Genl. Char. Similar to *P. mesomelas*, but smaller; no pectoral spot.

Color. Upper parts between Prout's brown and burnt umber; under parts white; no lateral line; hands and feet white; ankles dusky.

Measurements. "Total length, average of ten adults, 163 (156–169); tail vertebræ, 73 (68–79); hind foot, 21.5 (20–22). Skull of type: greatest length, 25.3; Hensel, 19; zygomatic breadth, 13; inter-

orbital constriction, 4; interparietal, 8 × 2.3; length of nasals, 9.3;
palatal length, 4; palatine slits, 4.2 × 19; diastema, 6.2; postpalatal
length, 9.5; upper molar series, 3.5."

131. labecula (*Peromyscus*), Elliot, Pub. Field Columb. Mus., III,
     1902, p. 143. Zoölogy.
WHITE-SPOT DEER MOUSE.
   Type locality. Ocotlan, State of Jalisco, Mexico.
   Genl. Char. Similar to *P. thurberi* in color, but cranial characters
very different; tail short; skull with braincase more nearly square-
shaped than that of *P. thurberi*, outer edge of frontals more curved;
interorbital constriction greater; nasals shorter and broader anteriorly;
palatine foramina shorter and narrower; pterygoids shorter; maxillary branch of zygoma broader and heavier; pelage soft, thick.
   Color. Above dark grayish fulvous; top of nose grayish buff;
conspicuous white spot at posterior base of ears; shoulders mixed
buff and black; lateral line from upper lip to thigh tawny ochraceous;
lips and under parts white; thighs like sides; arms, hands, and feet
white; tail hairy, above black, sides and beneath white.
   Measurements. Total length, 144; tail, 57; hind foot, 18. (Skin.)
Skull: greatest length, 25; Hensel, 20; zygomatic width, 13; mastoid
width, 11; length of nasals, 10; greatest width of rostrum, 4; palatal
length, 4; length of upper tooth row, 4.

132. cecilii (*Peromyscus*), Thomas, Ann. Mag. Nat. Hist., 7th Ser.,
     XI, 1903, p. 486.
CECIL'S MOUSE.
   Type locality. Santa Barbara Camp, southern slope of Mt. Orizaba, State of Vera Cruz, Mexico. Altitude, 12,500 feet.
   Genl. Char. Size small; ears medium; tail heavily haired.
   Color. Upper parts dark grayish fulvous, dorsal area heavily
lined with black; sides dull fulvous brown; under parts dull gray;
tail black above, white on sides and beneath; hands and feet dull
whitish, ankles dusky; ears black with white edges.
   Measurements. Total length, 169; tail vertebræ, 75; hind foot,
20; ear, 18. Skull: greatest length, 26.5; basilar length, 20; nasals,
11×3; interorbital constriction, 3.9; palatal length, 10.9; palatal
foramina, 5.8×2; length of upper molar series, 3.7.

133 melanotis (*Peromyscus*), Allen & Chapman, Bull. Am. Mus.
     Nat. Hist., 1897, p. 203.
BLACK-EARED FIELD MOUSE.
   Type locality. Las Vigas, State of Vera Cruz, Mexico.
   Geogr. Distr. State of Vera Cruz, Mexico.

Genl. Char. Similar to *P. rufinus*; colors paler, ears larger.

Color. Above yellowish brown, with a tinge of reddish, darkest on dorsal line; sides brighter; under parts and feet white; tail black above, rest white; ears black, edged with white.

Measurements. Total length, 160–164; tail vertebræ, 66–71; hind foot, 20; ear, 20. Skull: total length, 27; basal length, 22; zygomatic breadth, 14; interorbital constriction, 4; length of nasals, 11.5; palatal length, 11; length of upper tooth row, 4. (Type.)

a.—zamelas (*Peromyscus*), Osgood, Proc. Biol. Soc. Wash., xvii, 1904, p. 59.
JET MOUSE.

Type locality. Colonia Garcia, State of Chihuahua, Mexico. Altitude, 6,700 feet.

Genl. Char. Similar to *P. melanotis*, but more sooty; size small.

Color. Broad black stripe from top of head to base of tail; sides dark cinnamon rufous suffused with sooty; orbital ring and base of whiskers black; sides of face sooty; patch below eye and lateral line rufous; under parts white, plumbeous under fur showing; tail above black, beneath white; hands and feet white.

Measurements. Total length, 160; tail vertebræ, 63; hind foot, 20. Skull: greatest length, 25.9; Hensel, 19.3; zygomatic width, 13; interorbital constriction, 4; interparietal, 8×2.2; length of nasals, 11; palatal length, 3.8; palatine foramina, 5.3×2.1; postpalatal length, 8.5; length of upper molar series, 3.3.

134. spicilegus (*Peromyscus*), Allen, Bull. Amer. Mus. Nat. Hist., 1897, p. 50.
JALISCO SMALL-EARED MOUSE.

Type locality. Mineral San Sebastian, Mascota, State of Jalisco, Mexico.

Geogr. Distr. State of Jalisco, Mexico; range unknown.

Genl. Char. Similar to *P. floridanus*; ears smaller, tail longer.

Color. Above yellowish brown and dusky, darkest on dorsal line; sides yellowish; lateral line ochraceous; under parts white; outer surface of arms to wrist fulvous; tarsus dusky, hands and feet white; ears dusky; tail above brown, beneath white.

Measurements. Total length, 188; tail vertebræ, 92; hind foot, 20. Skull: total length, 28; basal length, 22; interorbital constriction, 5; length of nasals, 10. (ex Type.)

a.—evides (*Peromyscus*), Osgood, Proc. Biol. Soc. Wash., xvii, 1904, p. 64.
JUQUILA MOUSE.

Type locality. Juquila, State of Oaxaca, Mexico.

Genl. Char. Similar to *P. spicilegus;* dusky area on hind foot from ankle to base of toes. Skull heavier; teeth larger.

Color. Upper parts tawny mixed with dusky; orbital ring and base of whiskers black; pectoral spot tawny; under parts creamy white; tail above blackish, below white; forearm sooty to wrist; hands white; ankle and proximal half of foot above sooty.

Measurements. Average of five adults: Total length, 211; tail vertebræ, 106; hind foot, 25. Skull: greatest length, 29; Hensel. 22; zygomatic width, 14.4; interorbital constriction, 4.6; interparietal 9.3×3.2; length of nasals, 11.5; palate, 4.7; postpalatal length, 9.1: length of upper molar series, 5.

b.—simulus (*Peromyscus*), Osgood, Proc. Biol. Soc. Wash., XVII,   1904, p. 64.
MIMIC MOUSE.

Type locality. San Blas, Territorio de Tepic, Mexico.
Genl. Char. Similar to *P. spicilegus.* Skull smaller; nasals shorter.
Color. Upper parts cinnamon rufous; dorsal area darker; under parts white; pectoral spot rufous; tail bicolor; hands and feet white; ankles dusky.

Measurements. Average of three adults: Total length, 208; tail vertebræ, 111; hind foot, 23. Skull: greatest length, 26.3; Hensel. 20.3; zygomatic width, 14; interorbital constriction, 4.1; length of nasals, 9.4; palate, 3.6; postpalatal length, 9.4; length of upper molar series, 3.8.

135. affinis (*Hesperomys*), Allen, Proc. U. S. Nat. Mus., 1891, p. 195.
ALLIED FIELD MOUSE.

Type locality. Barrio, State of Oaxaca, Mexico.
Geogr. Distr. Southeastern Mexico.
Genl. Char. Similar to *P. melanophrys*, but darker.
Color. Above mixed fulvous brown and black, dorsal area darkest; flanks fulvous; under parts white tinged with yellow; ears brownish with white edges; hands and feet white; tail above brown, beneath paler.

Measurements. Total length, 157–171; tail vertebræ, 76–83; hind foot, 19.8; ear, 4.5. Skull: total length, 26.4; basal length, 25.9; length of nasals, 11.4; length of mandible, 16; height at angle, 6.4. (ex Type.)

a.—musculoides (*Peromyscus*), Merr., Proc. Biol. Soc. Wash., XII,   1898, p. 124.
ALLIED HOUSE MOUSE.

Type locality. Cuicatlan, State of Oaxaca, Mexico.

Geogr. Distr. State of Oaxaca, Mexico; range unknown.

Genl. Char. Size small; similar to *Mus musculus*.

Color. Above drab gray, sides brownish; sides of nose, lips, under parts, hands, and feet white; tail brownish above, white beneath.

Measurements. Total length, 187; tail vertebræ, 88; hind foot, 2.5.

136. cozumelæ (*Peromyscus*), Merr., Proc. Biol. Soc. Wash., xiv, 1901, p. 103.

ISLAND OF COZUMEL MOUSE.

Type locality. Island of Cozumel, Yucatan, Mexico.

Genl. Char. Size medium; ears large; similar to *P. affinis*.

Color. Above from grayish to fulvous brown; beneath white; tail brownish dusky above, whitish beneath; wrists and ankles dusky; hands and feet whitish.

Measurements. Total length, 180; tail vertebræ, 80; hind foot, 23.

137. geronimensis (*Peromyscus*), Allen, Bull. Amer. Mus. Nat. Hist., 1898, p. 156.

SAN GERONIMO ISLAND FIELD MOUSE.

Type locality. San Geronimo Island, Lower California, Mexico.

Geogr. Distr. Lower California, Mexico; Island form.

Genl. Char. Similar to *P. texensis*; size larger.

Color. Above fulvous brown and black, darker dorsal band perceptible; sides pale fulvous; under parts white; tail above dusky, sides beneath whitish.

Measurements. Total length, 175; tail vertebræ, 83; hind foot, 20. Skull: total length, 25; basal length, 21; mastoid breadth, 11; interorbital constriction, 3.8; length of nasals, 8. (ex Type.)

138. dubius (*Peromyscus*), Allen, Bull. Amer. Mus. Nat. Hist., 1898, p. 157.

TODOS SANTOS ISLAND MOUSE.

Type locality. Todos Santos Island, Lower California, Mexico.

Geogr. Distr. Lower California, Mexico.

Genl. Char. Size rather large; color dark.

Color. Above grayish fulvous brown, and black; sides fulvous; lateral line brownish fulvous; under parts white; hands and feet whitish; tail, basal third blackish brown, sides and beneath whitish.

Measurements. Total length, 175; tail vertebræ, 82; hind foot, 18. Skull: total length, 26; basal length, 21.5; zygomatic breadth, 12; mastoid breadth, 11; interorbital constriction, 3.5; length of nasals, 9.5. (ex Type.)

leucopus sonoriensis (*Hesperomys*), Le Conte, Proc. Acad. Nat. Scien. Phil., 1853, p. 413.

americanus sonoriensis (*Peromyscus*), Elliot, Syn. N. Am. Mamm.,
    1901, p. 125.
SONORA WHITE-FOOTED FIELD MOUSE.

Type locality. Santa Cruz, State of Sonora, Mexico, boundary
line Mexico and United States.

FIG. XXXVII. PEROMYSCUS L. SONORIENSIS.
SONORA WHITE-FOOTED FIELD MOUSE.

Geogr. Distr. From Zapotlan, Sierra Nevada de Colima, State of
Jalisco, north to British Columbia west of Rocky Mountains.

Genl. Char. Medium size; tail short; color pale.

Color. Above dark cinereous, mixed with brownish gray; under
parts whitish; feet brownish white; tail above dark brown, beneath
paler.

Measurements. Total length, 166; tail vertebræ, 69; hind foot,
21.5; ear, 20. Skull: occipito-nasal length, 24; Hensel, 18; zygomatic width, 11.5; interorbital constriction, 4; palatal length, 10;
length of nasals, 10; length of upper tooth row, 3.

leucopus blandus.

sonoriensis blandus (*Peromyscus*), Osgood, Proc. Biol. Soc. Wash.,
    XVII, 1904, p. 56.
GENTLE MOUSE.

Type locality. Escalon, State of Chihuahua, Mexico.

Genl. Char. Similar to *P. l. sonoriensis*, but smaller; tail shorter.

Color. Upper parts vinaceous buff sprinkled with dusky; lateral
line vinaceous buff; under parts creamy white; hands and feet white;
ankles with traces of dusky or buffy; ears whitish.

Measurements. Total length, 145; tail vertebræ, 61; hind foot,
21. Skull: greatest length, 25.4; Hensel, 19.7; zygomatic width,
12.5; interorbital constriction, 4; interparietal, 8×1.9; length of nasals,

9.8×2.5; palatal length, 3.5; palatine foramina, 5.7×2; postpalatal length, 9.1; length of upper molar series, 3.8.

leucopus fulvus.
> *sonoriensis fulvus (Peromyscus)*, Osgood, Proc. Biol. Soc. Wash., XVII, 1904, p. 57.

FULVOUS MOUSE.

Type locality. Oaxaca City, State of Oaxaca, Mexico.

Genl. Char. Similar to *P. l. sonoriensis*, but darker. "Skull with anterior part of zygoma heavier, more deeply notched by infraorbital foramen."

Color. Upper parts russet; middle of back mars and Prout's brown; under parts creamy white; ear tufts buffy or pale cream color; hands, arms, and feet white; outer side of ankles brownish; tail above brown, beneath white.

Measurements. Total length, 167; tail vertebræ, 68; hind foot, 22. Skull: greatest length, 25; Hensel, 19.5; zygomatic width, 12.8; interorbital constriction, 4; interparietal, 8×2.1; length of nasals, 10; length of palate, 3.7; palatal foramina, 5.6×2; postpalatal length, 8.7; length of upper molar series, 3.8.

139. exiguus *(Peromyscus)*, Allen, Bull. Amer. Mus. Nat. Hist., 1898, p. 157.

SAN MARTIN ISLAND MOUSE.

Type locality. San Martin Island, Lower California, Mexico.

Geogr. Distr. Lower California, Mexico.

Genl. Char. Similar to *P. t. clementis*, smaller, tail shorter.

Color. Head and back dark rufous brown, darkest on upper back, but no dorsal stripe; flanks pale rufous; upper lip, chin, and under parts pure white; orbital ring black; ears like back; tail above dark brown, beneath white; feet white.

Measurements. Total length, 158; tail vertebræ, 69; hind foot, 20. Skull: total length, 26.5; basal length, 22; mastoid breadth, 11.5; interorbital constriction, 3.5; length of nasals, 10. (ex Type.)

140. cherrii *(Hesperomys)*, Allen, Bull. Amer. Mus. Nat. Hist., 1891, p. 211.

CHERRIE'S COSTA RICAN MOUSE.

Type locality. La Carpintera, Costa Rica.

Geogr. Distr. Costa Rica.

Genl. Char. Similar to *P. nuttalli*, but whiter beneath; ears larger, tail much longer.

Color. Above dark cinnamon rufous, lined with black on the head and dorsal region; flanks rusty cinnamon; upper lip on a line

from nose, under parts of arms, legs, and body pure white; orbital
ring and line between eyes and nose black; limbs dusky brown; tail
dusky brown, naked; hands and feet dusky brown; fingers and toes
whitish; ears naked, dusky.

Measurements. Total length, 182.4–187.10; tail, 108–114.3; hind
foot, 18.3; ear, 12.7. Skull: basal length, 20.3; total length, 24.6;
interorbital constriction, 4.3; length of nasals, 9.4; length of upper
molar series, 4.6; length of mandible, 13.5.

141. aztecus (*Hesperomys*), Sauss., Rev. Zoöl., 2me Sér., 1860, p. 105.
AZTEC MOUSE.

Type locality. "Mexico."

Geogr. Distr. Cape St. Lucas, Lower California, Sierra de Juan-
acutlan, State of Jalisco, Mexico. Range unknown.

Genl. Char. Coloring rich and dark, tail long.

Color. Upper parts and sides rusty red or rusty orange, extending
over the arms to the wrists and on the hind leg onto the metatarsus;
dorsal area brownish black; top of head not so dark as back; sides of
lips, under parts, hands, and feet white; tail above blackish, beneath
paler.

Measurements. Total length, 190; tail vertebræ, 107; hind
foot, 21.

142. beatæ (*Peromyscus*), Thomas, Ann. Mag. Nat. Hist., 7th Ser.,
        XI, 1903, p. 485.
XOMETLA MOUSE.

Type locality. Xometla Camp, Mt. Orizaba, State of Vera Cruz,
Mexico. Altitude 8,500 feet.

Genl. Char. Size small; ears large; tail longer than head and
body, well haired. Skull: supraorbital edges square; interparietal large.

Color. Upper parts brownish fulvous, grayer on head and fore-
quarters and lined with black on dorsal surface; sides dull fulvous;
under surface dull gray, with a buff pectoral spot; hands and feet dull
white, ankles dusky; tail blackish above, dull white below; ears
blackish with faint white edges.

Measurements. Total length, 215; tail, 118; hind foot, 21; ear, 20.
Skull: greatest length, 28.2; basilar length, 21.6; nasals, 11.5×3.3;
interorbital constriction, 4.2; interparietal, 3.9×9.4; palatal length,
11.5; palatal foramen, 6.8×2.3; length of upper molar series, 4.5.

143. fraterculus (*Vesperimus*), Miller, Amer. Nat., 1892, p. 261.

fraterculus (*Peromyscus*), Elliot, Syn. N. Am. Mamm., 1901, p. 136.
RELATED WHITE-FOOTED MOUSE.

Type locality. Dulzura, San Diego, California.

Geogr. Distr. Southern and Lower California, Mexico.

Genl. Char. Medium; tail longer than head and body, thinly haired; soles naked; ears large.

Color. Above yellowish wood brown mixed with black, darkest on median line; flanks fulvous with distinct lateral stripe; feet white; under parts yellowish white with fulvous pectoral spot; tail above brownish, paler beneath.

Measurements. Total length, 185; tail vertebræ, 113; hind foot, 20; ear, 18. Skull: occipito-nasal length, 25; Hensel, 18.5; zygomatic breadth, 12; interorbital constriction, 4; length of nasals, 9; palatal length, 9; width of braincase, 11.5.

144. propinquus.

eremicus propinquus (*Peromyscus*), Allen, Bull. Amer. Mus. Nat. Hist., 1898, p. 154.

ALLIED DESERT MOUSE.

Type locality. San Pablo Point, San Pablo Bay, Lower California, Mexico.

Geogr. Distr. Lower California, Mexico.

Genl. Char. Similar to *P. fraterculus,* but blacker.

Color. Above mixed blackish brown and grayish fulvous; lateral line deep fulvous; under parts white; hands and feet white; tail brownish above, paler beneath.

Measurements. Total length, 190; tail vertebræ, 100; hind foot, 18.5.

145. cedrosensis (*Peromyscus*), Allen, Bull. Amer. Mus. Nat. Hist., 1898, p. 154.

CERROS ISLAND MOUSE.

Type locality. Cerros or Cedros Island, Lower California, Mexico.

Geogr. Distr. Lower California, Mexico.

Genl. Char. Similar to *P. fraterculus,* but darker.

Color. Upper parts brown, tinged with fulvous and blackish; lateral line broad, ochraceous; under parts white with pectoral fulvous line, sometimes reaching the abdomen; hands and feet white; ears large, pale brown; tail pale brown above, lighter beneath.

Measurements. Total length, 194; tail vertebræ, 107; hind foot, 20. Skull: total length, 26; basal length, 21; zygomatic width, 10; mastoid width, 10.5; interorbital constriction, 3.5; length of nasals, 9.6. (ex Type.)

146. anthonyi (*Hesperomys*), Merr., Proc. Biol. Soc. Wash., IV, 1887, p. 5.

ANTHONY'S FIELD MOUSE.

Type locality. Camp Apache, Grant County, New Mexico.

Geogr. Distr. States of Sonora and Chihuahua, Mexico, north into New Mexico.

Genl. Char. Small; tail longer than head and body; ears large; soles naked.

Color. Above ash gray, lined with black; sides buffy fulvous; under parts white; tail above dark brown, beneath whitish.

Measurements. Total length, 144; tail vertebræ, 80; hind foot. 18.5; ear, 12. Skull: basilar length, 20.3; Hensel, 18; zygomatic width, 12.8; interorbital constriction, 3.9; length of nasals, 7.8; length of upper molar series, 3.8; length of mandible, 12.9.

146a. goldmanl (*Peromyscus*), Osgood, Proc. Biol. Soc. Wash., XVII, 1904, p. 75.
GOLDMAN'S MOUSE.

Type locality. Alamos, State of Sonora, Mexico.

Genl. Char. Similar to *P. anthonyi;* tail long, hairy; heel slightly hairy.

Color. Upper parts and sides ochraceous buff mixed with black; under parts creamy white; pectoral spot buff.

Measurements. Total length, 217; tail vertebræ, 117; hind foot, 24. Skull: greatest length, 27.3; Hensel, 21.1; zygomatic width, 14.2; interorbital constriction, 4; interparietal, 8.6×3.2; length of nasals. 9.6; palate, 4.2; postpalatal length, 10; length of upper molar series, 4.

147. texensis (*Hesperomys*), Woodh., Proc. Acad. Nat. Scien. Phil., 1853, p. 242.
texensis (*Peromyscus*), Elliot, Syn. N. Am. Mamm., 1901, p. 130.
TEXAN FIELD MOUSE.

Type locality. Rio Grande near El Paso, El Paso County, Texas.

Geogr. Distr. Southern Texas, into State of Chihuahua, Mexico.

Genl. Char. Small; tail equal to head and body; ears small.

Color. Above cinereous mixed with pale brown; lower sides, feet and under parts white; tail above brown, beneath white.

Measurements. Total length, 106.6; tail vertebræ, 53.3; hind foot, 44; ear, 10. Skull: occipito-nasal length, 23; Hensel, 17.5; zygomatic width, 12; interorbital constriction, 3.5; length of nasals, 7.5; palatal length, 9.5.

a.—arizonæ.

americanus arizonæ (*Sitomys*), Allen, Bull. Am. Mus. Nat. Hist., 1894, p. 321.
americanus arizonæ (*Peromyscus*), Elliot, Syn. N. Am. Mamm., 1901, p. 125.
ARIZONA FIELD MOUSE.

Type locality. Fairbank, Cochise County, Arizona.

Geogr. Distr. State of Chihuahua, north to White, Chiricahua and Graham Mountains of Arizona.

Genl. Char. Similar to *P. gambeli*, but smaller; longer ears and shorter tail.

Color. Above dark plumbeous slate; below whitish; tail bicolor.

Measurements. Total length, 158; tail vertebræ, 67; hind foot, 24. Skull: occipito-nasal length, 24; Hensel, 10; zygomatic width, 12; interorbital constriction, 4; length of nasals, 10; palatal length, 10; length of upper molar series, 3.

b.—flaccidus (*Peromyscus*), Allen, Bull. Am. Mus. Nat. Hist., 1903, p. 599.

FEEBLE MOUSE.

Type locality. Rio Sestin, State of Durango, Mexico. Altitude, 7,500 feet.

Genl. Char. Similar to *P. t. arizonæ*, but paler; tail longer; hind foot shorter.

Color. Upper parts dark fawn brown, blackish on median area; lateral line fulvous; head grayer than body; under parts pure white, under fur plumbeous; fore legs white to shoulder; body color extending on hind leg to tarsal joint; tail above dark brown on upper third, rest grayish white, beneath paler; ear gray brown margined narrowly with white.

Measurements. Total length, 177; tail vertebræ, 79; hind foot, 20; ear from crown, 14; from notch, 18.

c.—clementis (*Peromyscus*), Mearns, Proc. U. S. Nat. Mus., 1896, p. 446. Elliot, Syn. N. Am. Mamm., 1901, p. 130.

SAN CLEMENTE ISLAND MOUSE.

Type locality. San Clemente Island, California.

Geogr. Distr. Coronodos Islands, Mexico, and San Clemente Island, California.

Genl. Char. Much blacker than *P. thurberi*, and of a more reddish coloration, save on the head.

Color. Top of head drab gray; upper parts drab, tinged with burnt umber; ears black; feet and under surface white; tail bicolor.

Measurements. Total length, 177; tail vertebræ, 77; hind foot, 21; ear, 17.

d.—coolidgii (*Peromyscus*), Thomas, Ann. Mag. Nat. Hist., 7th Ser., 1, 1898, p. 45.

COOLIDGE'S FIELD MOUSE.

Type locality. Santa Anita, Cape Region, Lower California, Mexico.

Geogr. Distr. Cape Region, Lower California.

Genl. Char. Like *P. l. nebrascensis*, tail slightly haired.

Color. Upper parts dull buffy, darker on dorsal line, lighter on sides; tail brown above, white below. (O. Thomas in litt., ex Type.)

Measurements. Total length, 167; tail, 76; hind foot, 22; ear, 20. Skull: basilar length, 19.5; greatest breadth, 13; length of nasals, 9.5; length of upper molar series, 3.7.

e.—deserticola.

> *leucopus deserticola* (*Hesperomys*), Mearns, Bull. Am. Mus. Nat. Hist., II, 1890, p. 285, Desc. p. 287.
>
> *americanus deserticola* (*Peromyscus*), Elliot, Syn. N. Am. Mamm., 1901, p. 125.

DESERT WHITE-FOOTED MOUSE.

Type locality. Mojave Desert, San Bernardino County, California.

Geogr. Distr. Lower California and State of Sonora, Mexico, and desert regions of California and Arizona.

Genl. Char. Ears medium; tail long; pelage short.

Color. Above pale cinereous drab, slightly darker on median line, light fulvous on sides and rump; tail narrowly striped above with dark brown.

Measurements. Total length, 168; tail, 78.7; hind foot, 20.8. Skull: occipito-nasal length, 25.5; Hensel, 20; zygomatic width, 12.5; interorbital constriction, 4; length of nasals, 5.5; palatal length, 10.

148. tornillo (*Peromyscus*), Mearns, Proc. U. S. Nat. Mus., 1896, p. 445. Elliot, Syn. N. Am. Mamm., 1901, p. 126.

RIO GRANDE WHITE-FOOTED MOUSE.

Type locality. Rio Grande, six miles above El Paso, El Paso County, Texas.

Geogr. Distr. State of Chihuahua, north into Texas and New Mexico.

Genl. Char. Similar to *P. t. arizonæ*, but paler, ears smaller, body stouter. (Mearns, l. c.)

Color. Above light broccoli brown; feet and under parts pure white; tail bicolor.

Measurements. Total length, 192; tail vertebræ, 90; hind foot, 23; ear, 12.

149. gymnotis (*Peromyscus*), Thomas, Ann. Mag. Nat. Hist., 6th Ser., XIV, 1894, p. 365.

NAKED-EARED MOUSE.

Type locality. Guatemala.

Geogr. Distr. Guatemala, Central America.

Genl. Char. Size medium; ears long, naked; tail shorter than head and body.

Color. Upper parts bistre brown, beneath slaty buff; tail dark brown above and beneath; hands and feet silvery white.

Measurements. Total length, 191; tail, 92; hind foot, 22; ear, 17. Skull: greatest length, 30.5; basal length, 25; Hensel, 23.1; greatest breadth, 15.4; length of nasals, 12×3.3; interorbital constriction, 4.6; palatal length, 12.4; length of upper molar series, 4.1; length of mandible, condyle to tip of incisor, 18; height of coronoid process, 7.3.

rowleyi pinalis (*Sitomys*), Miller, Bull. Am. Mus. Nat. Hist., 1893, p. 331.

rowleyi pinalis (*Peromyscus*), Elliot, Syn. N. Am. Mamm., 1901, p. 135.

ROWLEY'S PINE MOUSE.

Type locality. Granite Gap, Grant County, New Mexico.

Geogr. Distr. State of Sonora, Mexico, north to New Mexico and Arizona.

Genl. Char. Smaller than *P. rowleyi*, more yellowish in color.

Color. Above olive buff, darker on dorsal region, and grayer on head and face; hairs sepia-tipped; indistinct orbital ring; feet and under parts pure white; tail brown above, white beneath.

Measurements. Total length, 196; tail, 104; hind foot, 23; ear, 20. (ex Type.) Skull: occipito-nasal length, 26; Hensel, 19; zygomatic width, 13; interorbital constriction, 5; palatal length, 10.5; length of nasals, 9.5; length of upper molar series, 4.

150. martirensis (*Sitomys*), Allen, Bull. Amer. Mus. Nat. Hist., 1893, p. 187.

SAN PEDRO MARTIR MOUNTAINS MOUSE.

Type locality. San Pedro Martir Mountains, Lower California, Mexico. Altitude, 7,000 feet.

Geogr. Distr. San Pedro Martir and Hanson Laguna Mountains, Lower California, Mexico. Altitude above 5,000 feet.

Genl. Char. Similar to *P. truii*; tail longer.

Color. Above pale yellowish brown, varied with blackish; sides tawny; under parts white; sometimes a tawny pectoral spot; orbital ring blackish; ears naked, dusky; hands and feet white; tail above blackish, beneath grayish white.

Measurements. Total length, 195; tail vertebræ, 102; hind foot, 22. Skull: total length, 28; basilar length, 23.4; Hensel, 21; zygomatic width, 13.7; interorbital constriction, 5; length of nasals, 10; palatal length, 11; width of braincase, 13; length of upper molar series, 4; length of mandible, 9.5; length of lower molar series, 4.

151. banderanus (*Peromyscus*), Allen, Bull. Am. Mus. Nat. Hist., 1897, p. 51.

BANDERAS FIELD MOUSE.

Type locality.   Terro Tepic, Valle de Banderas, State of Jalisco, Mexico.

Geogr. Distr.   State of Jalisco, Mexico; range unknown.

Genl. Char.   Above pale yellowish brown, dorsal region with black-tipped hairs; lateral line fulvous; outside of arms, under parts, hands, and feet white; outer side of legs grayish; ears brownish; tail above brown, beneath paler.

Measurements.   Total length, 226; tail vertebræ, 112; hind foot, 24. Skull: total length, 31; basal length, 25; interorbital constriction, 5; length of nasals, 11.5. (ex Type.)

a.—vicinior (*Peromyscus*), Osgood, Proc. Biol. Soc. Wash., xvii, 1904, p. 68.

LA SALADA DEER MOUSE.

Type locality.   La Salada, State of Michoacan, Mexico.

Genl. Char.   Similar to *P. banderanus,* but darker; skull narrower; anterior palatine foramina nearly elliptical.

Color.   "Slightly darker and more vinaceous than *P. banderanus* in worn or summer pelage; decidedly darker in winter pelage, with a definite dusky median dorsal area; markings about eyes, whiskers and ankles sooty black instead of brown; upper side of tail sooty instead of brownish.

Measurements.   Total length, 216; tail vertebræ, 107; hind foot, 27. Skull: greatest length, 31–32; Hensel, 23.3–24.1; zygomatic width, 14–14.3; interorbital constriction, 4.8–5; length of nasals, 11.8–12.4; interparietal, 3.7×10.2–4.5×10.2; length of upper molar series, 4.4–4.6.

b.—angelensis (*Peromyscus*), Osgood, Proc. Biol. Soc. Wash., xvii, 1904, p. 69.

PUERTO ANGEL MOUSE.

Type locality.   Puerto Angel, State of Oaxaca, Mexico.

Genl. Char.   Similar to *P. banderanus,* but larger; supra-orbital bead nearly obsolete.

Color.   Like *P. banderanus,* possibly slightly darker.

Measurements.   Total length, 235; tail vertebræ, 123; hind foot, 26.5. Skull: greatest length, 31.3–33.4; Hensel, 23.4–24.9; zygomatic width, 15–15.4; interorbital constriction, 5–5.2; length of nasals, 11.7–12.8; length of upper molar series, 4.6.

152. stephensi (*Peromyscus*), Mearns, Proc. U. S. Nat. Mus., 1897, p. 721. Elliot, Syn. N. Am. Mamm., 1901, p. 136.

STEPHEN'S FIELD MOUSE.

Type locality. Cañon at east base of Coast Range Mountains, in San Diego County, California, near Mexican boundary.

Geogr. Distr. Southern California into Lower California, Mexico.

Genl. Char. Similar to *P. eremicus*, but smaller, tail longer, and colors paler.

Color. Above grayish cream buff; sides and rump pale ochraceous buff; feet and under parts white; tail above dusky, below white.

Measurements. Total length, 193; tail vertebræ, 108; hind foot, 19; ear, 18.5. Skull: occipito-nasal length, 24; Hensel, 18; zygomatic width, 12; interorbital constriction, 3.5; palatal length, 9; length of upper molar series, 3.

153. eremicus (*Hesperomys*), Baird, Mamm. N. Am., 1857, p. 479.

eremicus (*Peromyscus*), Elliot, Syn. N. Am. Mamm., 1901, p. 136.

DESERT MOUSE.

Type locality. Old Fort Yuma, San Diego County, California.

Geogr. Distr. Lower California and State of Sonora, Mexico, into New Mexico, Arizona, and California.

Genl. Char. Ears very large; tail longer than head and body; palms and soles naked.

Color. Above pale yellowish gray, mixed with black; pale fulvous band on cheeks and sides; tail obscurely bicolor, above little darker than dorsal region; feet whitish.

Measurements. Total length, 190; tail vertebræ, 96.5; hind foot, 22; ear, 19. Skull: occipito-nasal length, 26.5; Hensel, 19; zygomatic width, 13; interorbital constriction, 4; palatal length, 10; length of nasals, 10; length of upper molar series, 3.

a.—arenarius (*Peromyscus*), Mearns, Proc. U. S. Nat. Mus., 1896, p. 138. Elliot, Syn. N. Am. Mamm., 1901, p. 136.

SAND-LOVING MOUSE.

Type locality. Rio Grande, near El Paso, El Paso County, Texas.

Geogr. Distr. State of Chihuahua, Mexico, into southern Texas.

Genl. Char. Similar to *P. eremicus*, without dark dorsal line.

Color. Above and sides pale ochraceous drab mixed with black; tail above dusky drab and hoary, below pure white; head grayish; orbital ring dusky.

Measurements. Total length, 198; tail vertebræ, 106; hind foot, 21.5; ear, 15.

b.—phæurus (*Peromyscus*), Osgood, Proc. Biol. Soc. Wash., XVII, 1904, p. 75.

BLACK-TAILED MOUSE.

Type locality. Hacienda La Parada, State of San Luis Potosi, Mexico.

Geogr. Distr. Middle portion of tableland in States of San Luis Potosi, Zacatecas and Nuevo Leon.

Genl. Char. Similar to *P. eremicus*, but darker; tail unicolor.

Color. Shades of buff deeper than in *P. eremicus* and upper parts more heavily mixed with black; under parts white; tail blackish brown above and below; hands and feet white; ankles dusky.

Measurements. Total length, 176–195; tail vertebræ, 92–103; hind foot, 21.

attwateri pectoralis (*Peromyscus*), Osgood, Proc. Biol. Soc. Wash., XVII, 1904, p. 59.

JALPAN MOUSE.

Type locality. Jalpan, State of Queretaro, Mexico.

Genl. Char. Similar to *P. attwateri;* pectoral spot prominent.

Color. Upper parts pale ochraceous buff thickly sprinkled with dusky; sides of head behind eyes grayish; a narrow black orbital ring; pectoral spot buffy ochraceous; under parts white; tail above dusky, beneath white; hands and feet white.

Measurements. Total length, 200; tail vertebræ, 114; hind foot. 22. Skull: greatest length, 27; Hensel, 19.7; zygomatic width, 13.7; interorbital constriction, 4.2; length of nasals, 9.9; palatal length, 3.7; palatine foramina, 4.9×2; postpalatal length, 9.4; length of upper molar series, 3.8.

attwateri eremicoides (*Peromyscus*), Osgood, Proc. Biol. Soc. Wash., XVII, 1904, p. 60.

ALLIED MOUSE.

Type locality. Mapimi, State of Durango, Mexico.

Genl. Char. Similar to *P. eremicus*, but molar enamel pattern having accessory cusps.

Color. Upper parts mixed pinkish buff and dusky; lateral line pinkish buff; facial region grayish; under parts pale creamy white; tail above dusky, beneath white; hands, feet, and ankles white.

Measurements. Total length, 180–195; tail vertebræ, 102–111; hind foot, 20–21. Skull: greatest length, 24; Hensel, 18; zygomatic width, 12; interorbital constriction, 3.9; interparietal, 8.3×3; length of nasals, 8.5; palate, 3.5; palatine foramina, 4.5×1.5; postpalatal length, 8.5; length of upper molar series, 3.5.

154. metallicola Elliot, Pub. Field Columb. Mus., III, 1903, p. 245. Zoölogy.

Type locality. Providentia Mines, northwestern Sonora, Mexico.

Genl. Char. Similar to *P. eremicus*, but tail hairy and with a pencil; sides deep orange buff instead of pale fulvous.

Color. Upper parts mixed black and orange buff; forehead and nose gray and buff mixed; sides of face, shoulders, sides, and rump about base of tail deep orange buff; orbital ring black; lips and entire under parts, hands, and feet pure white; tail above dusky, sides and beneath white; ears brown.

Measurements. Total length, 190.5; tail vertebræ, 101.6; hind foot, 25. Skull: occipito-nasal length, 26; Hensel, 20; zygomatic width, 13; interorbital constriction, 4.5; width of braincase, 12; length of nasals, 10; palatal length, 10.5; length of upper tooth row, 4; length of mandible, angle to alveolus of incisor, 10; length of lower tooth row, 4.

155. difficilis (*Vesperimus*), Allen, Bull. Amer. Mus. Nat. Hist., 1891, p. 298.

TROUBLESOME MOUSE.

Type locality. Sierra de Valparaiso, State of Zacatecas, Mexico.

Geogr. Distr. State of Zacatecas, Mexico; range unknown.

Genl. Char. Allied to *P. megalotis*; ears smaller; hind feet larger; color darker.

Color. Above dusky brown tinged with pale fulvous; nape and shoulders ashy; sides buffy cinnamon; orbital ring blackish; under parts white; tail above blackish, beneath whitish; hands and feet whitish; ears large, naked, with gray edges.

Measurements. Total length, 201; tail vertebræ, 103; hind foot, 25.4. Skull: total length, 29.7; basal length, 24.1; zygomatic width, 14.7; length of mandible, 18.3. (ex Type.)

155a. bullatus (*Peromyscus*), Osgood, Proc. Biol. Soc. Wash., XVII, 1904, p. 63.

PEROTE MOUSE.

Type locality. Perote, State of Vera Cruz, Mexico.

Genl. Char. Similar to *P. truii;* audital bullæ greatly inflated; ears very large; tail shorter than head and body.

Color. Upper parts and sides tawny ochraceous; middle of back dusky; top of head and nose broccoli brown; sides of head grayish; orbital ring dusky; under parts creamy white; tail bicolor; hands and feet white.

Measurements. Total length, 200; tail vertebræ, 93; hind foot, 23. Skull: greatest length, 28.9; Hensel, 22; zygomatic width, 14.5; interorbital constriction, 4.5; interparietal, 10×3; length of nasals,

10.4; palate, 4.2; postpalatal length, 10; length of upper molar series, 4.3; greatest diameter of audital bullæ, 6.5.

156. sagax (*Peromyscus*), Elliot, Pub. Field Columb. Mus., III, 1903, p. 142. Zoölogy.

LA PALMA FIELD MOUSE.

Type locality. La Palma, State of Michoacan, Mexico.

Genl. Char. Similar in color to *P. difficilis*, Allen, but skull very different; ears large; braincase broad, nearly square; interorbital constriction considerable. Compared with that of *P. difficilis*, the skull is shorter and narrower, with shorter nasals, but of about equal width; bullæ smaller and closer together, and molars much smaller.

Color. Top of head and dorsal region mixed grayish black and buff, the former predominating; sides grayish brown and buff, with an indistinct buff lateral line; orbital ring and spot behind nose black, with a buff spot between this and the eye; sides of head and shoulders buffy gray; under parts white; upper side of arms buffy gray; thighs like sides; hands and feet white; tail black above, white beneath; ears large, brown at base, blackish at tip, with narrow white edges.

Measurements. Total length, 192; tail vertebræ, 107; hind foot, 22. Skull: total length, 26; Hensel, 20; zygomatic width, 13; mastoid width, 12; length of nasals, 10; greatest width of rostrum, 4; palatal length, 4; interorbital constriction, 4.3; length of upper molar series, 4.

157. yucatanicus (*Peromyscus*), Allen & Chapman, Bull. Amer. Mus. Nat. Hist., 1897, p. 8.

YUCATAN MOUSE.

Type locality. Chichen Itza, Yucatan, Mexico.

Geogr. Distr. Yucatan, Mexico.

Genl. Char. Tail naked; size medium.

Color. Above fulvous mixed with black; lateral line bright fulvous; under parts white; outer surface of limbs like back; hands and feet white; tail naked, unicolor, brown; ears brownish.

Measurements. Total length, 210; tail vertebræ, 100; hind foot, 22; ear, 20. Skull: total length, 31; basal length, 25; interorbital constriction, 5; length of nasals, 12. (ex Type.)

a.—badius (*Peromyscus*), Osgood, Proc. Biol. Soc. Wash., XVII, 1904, p. 70.

APAZOTE MOUSE.

Type locality. Apazote, State of Campeche, Mexico.

Genl. Char. Similar to *P. yucatanicus*, but darker.

Color. Dorsal area and sides strongly mixed with black; lateral line cinnamon rufous; orbital ring black; under parts suffused with yellow; tail above blackish brown, beneath yellowish white blotched with dusky; hands and feet white.

Measurements. Total length, 193.4; tail vertebræ, 96.7; hind foot, 23.5. Skull: greatest length, 28.2; Hensel, 20.7; zygomatic width, 14.1; interorbital constriction, 4.7; interparietal, 9.2×3.1; length of nasals, 10.4; palate, 4.2; postpalatal length, 9.9; length of upper molar series, 4.1.

158. cineritius (*Peromyscus*), Allen, Bull. Am. Mus. Nat. Hist., 1898, p. 155.
ASHY GRAY FIELD MOUSE.

Type locality. San Roque Island, Lower California, Mexico.

Geogr. Distr. Known only from type locality.

Genl. Char. Size large, colors pale.

Color. Above ash gray tinged with fulvous, with black hairs intermixed; under parts white; hands and feet white; tail above brown, sides and beneath whitish.

Measurements. Total length, 191; tail vertebræ, 83; hind foot, 21. Skull: total length, 27.5; basal length, 23; mastoid breadth, 11; interorbital width, 3.3; length of nasals, 9.5. (ex Type.)

159. levipes (*Peromyscus*), Merr., Proc. Biol. Soc. Wash., XII, 1898, p. 123.
MOUNT MALINCHE MOUSE.

Type locality. Mount Malinche, State of Tlaxcala, Mexico. Altitude, 8,400 feet.

Geogr. Distr. State of Tlaxcala, Mexico; range unknown.

Genl. Char. Size small; ears large; tail longer than head and body. Similar to *P. gratus*, but rostrum heavier, nasals longer; bullæ smaller.

Color. Above grayish brown; sides and cheeks buffy fulvous; under parts, hands, and feet white; salmon spot on breast; tail above dusky, beneath whitish.

Measurements. Total length, 200; tail vertebræ, 102; hind foot, 23.5.

160. boylii (*Hesperomys*), Baird, Proc. Acad. Nat. Scien. Phil., 1855, p. 335.
gilberti, Allen, Bull. Am. Mus. Nat. Hist., 1893, p. 188.
boylii (*Peromyscus*), Elliot, Syn. N. Am. Mamm., 1901, p. 132.
BOYLE'S MOUSE.

Type locality. Middle fork of American River, Eldorado County, California.

Geogr. Distr. Lower California, Mexico, north to California.

Genl. Char. Body stout; ears very large; tail longer than the body.

Color. Above mixed glossy brown and pale yellowish brown; hands, feet, and lower parts white; white on sides bordered by a line of reddish buff, which grades into the color of the upper parts; tail above dusky, beneath white.

Measurements. Total length, 195; tail vertebræ, 101; hind foot, 23; ear, 22. Skull: occipito-nasal length, 29; Hensel, 22; zygomatic width, 14; interorbital constriction, 4; length of nasals, 11; palatal length, 11; length of upper molar series, 4.

*a.—penicillatus (Peromyscus), Mearns, Proc. U. S. Nat. Mus., 1896, p. 139. Elliot, Syn. N. Am. Mamm., 1901, p. 133.
FRANKLIN MOUNTAINS MOUSE.

Type locality. Franklin Mountains, near El Paso, El Paso County, Texas.

Geogr. Distr. Texas, probably into State of Chihuahua, Mexico.

Genl. Char. Nasal bones of skull truncate posteriorly, and ending in front of the posterior ends of premaxillæ.

Color. Above drab gray tinged with ochraceous buff on rump and sides; hands, feet, and under parts pure white; tail dusky drab above, white below.

Measurements. Total length, 202; tail vertebræ, 115; hind foot, 22; ear, 14.

160a. polius *(Peromyscus)*, Osgood, Proc. Biol. Soc. Wash., XVII, 1904, p. 61.
GRIZZLED MOUSE.

Type locality. Colonia Garcia, State of Chihuahua, Mexico.

Genl. Char. Similar to *P. boylii rowleyi*, but larger and grayer.

Color. Upper parts grayish broccoli brown; lateral line pinkish buff; orbital ring dusky; under parts white; tail bicolor; hands, feet, and ankles white.

Measurements. Total length, 210–234; tail vertebræ, 111–120; hind foot, 25–26. Skull: greatest length, 29.9; Hensel, 22.9; zygomatic width, 14.8; interorbital constriction, 4.5; interparietal, 10.5×2.8; length of nasals, 11.6; palate, 4.4; postpalatal length, 10; length of upper molar series, 4.7.

161. gratus *(Peromyscus)*, Merr., Proc. Biol. Soc. Wash., X, 1898, p. 123.
TLALPAM FIELD MOUSE.

Type locality. Tlalpam, Valley of Mexico.

---

* Dr. E. A. Mearns informs me that this mouse has been taken within a mile of the Mexican border.

Geogr. Distr. State of Mexico, range unknown.

Genl. Char. Size small; ears large; similar to *P. truii.*

Color. Above pale buffy fulvous and black; side buffy fulvous; under parts white as are also the hands and feet; tail dusky, the sides slightly paler.

Measurements. Total length, 209; tail vertebræ, 114; hind foot, 23.

a.—gentilis (*Peromyscus*), Osgood, Proc. Biol. Soc. Wash., XVII, 1904, p. 61.

FOREIGN MOUSE.

Type locality. Lagos, State of Jalisco, Mexico.

Genl. Char. Similar to *P. gratus*, but paler; molars slightly smaller.

Color. Upper parts pale ochraceous buff and dusky; sides of head ochraceous buff; under parts white; tail above blackish, beneath white; hands and feet white; ankles dusky.

Measurements. Total length, 194–210; tail vertebræ, 103–120; hind foot, 23–24.5. Skull: greatest length, 27.2; zygomatic width, 14; length of nasals, 9.5; interorbital constriction, 4.4; length of upper molar series, 4.

162. pavidus (*Peromyscus*), Elliot, Pub. Field Columb. Mus., III, 1903, p. 142. Zoölogy.

SHY FIELD MOUSE.

Type locality. Patzcuaro, State of Michoacan, Mexico.

Genl. Char. Tail long, ears large, colors dark. Skull with very broad braincase, mastoid width being apparently the greatest; rostrum long and rather slender, its greatest width being less than the least interorbital width, the palatine foramina broad, their posterior ends just reaching the anterior line of the first premolar. Skulls of all the specimens are badly broken. Somewhat similar to *P. gratus* in color, but larger in all its dimensions.

Color. Upper parts mixed buff and brownish black, palest on hind neck; orbital ring black; lateral line from lip to rump ochraceous buff, sides buff; under parts white, tinged with buff on chest; arms buff; legs and ankles dusky; hands and feet white; tail above blackish or brownish black, beneath white; ears brown.

Measurements. Total length, 238; tail vertebræ, 130; hind foot, 23.

163. madrensis (*Peromyscus*), Merr., Proc. Biol. Soc. Wash., XII, 1898, p. 16.

MARIA MADRE ISLAND MOUSE.

Type locality. Maria Madre Island, Tres Marias Islands, State of Jalisco, Mexico.

Geogr. Distr.   Tres Marias Islands, State of Jalisco, Mexico.

Genl. Char.   Size large, tail long; similar to *P. spicilegus*, but larger.   No superciliary ridges on skull.

Color.   Above and on sides pale ochraceous buff, dorsal region darkest; under parts white with salmon spot on breast; tail above and tip dark, beneath whitish.

Measurements.   Total length, 222; tail vertebræ, 119; hind foot, 26.

164. eva (*Peromyscus*), Thomas, Ann. Mag. Nat. Hist., 7th Ser., 1, 1898, p. 44.

EVE'S MOUSE.

Type locality.   San José del Cabo, Lower California, Mexico.

Geogr. Distr.   Cape region of Lower California, Mexico.

Genl. Char.   Size large; tail long.

Color.   Above sandy rufous, lined with black on dorsal region; sides rufous; under parts white, base of hairs slate; hands and feet silvery white; ankles dusky; tail above brown, paler beneath; ear pale brown.

Measurements.   Total length, 196–216; tail vertebræ, 108–128; hind foot, 21; ear, 17.   Skull: "basal length, 20.3; basilar length, 18.8; greatest breadth, 12.8; length of nasals, 9.6; interorbital constriction, 4; interparietal, 3.7 x 9; width of braincase at squamosals, 11.7; palate, length from henselion, 10; diastema, 6.6; palatal foramina, 4.8 × 2.1; upper molar series, 3.6."

165. nudipes (*Hesperomys*), Allen, Bull. Amer. Mus. Nat. Hist., 1891, p. 213.

LA CARPINTERA FIELD MOUSE.

Type locality.   La Carpintera, Costa Rica.

Genl. Char.   Ears, tail, and feet naked, the first very large.   Skull has rostrum very narrow, lengthened.

Color.   Above dark brown, hairs tipped with rufous; sides deeper rufous; nose grayish; dusky spot at base of whiskers, and another in front of eye; under parts grayish white, breast crossed by a band of pale chestnut; hands and feet naked, flesh color; tail naked, grayish brown, paler at tip.

Measurements.   Total length, 230.9; tail, 139.7; hind foot, 28.5; ear from crown, 20.3.   Skull: total length, 36.9; basal length, 31.2; greatest breadth, 16.5; length of nasals, 14.5; length of mandible, 22.4; height at condyle, 7.9; length of upper molar series, 5.1.

166. merriami (*Peromyscus*), Mearns, Proc. U. S. Nat. Mus., 1896, p. 138.

SONOYTA DESERT MOUSE.

Type locality.   Sonoyta, on Sonoyta River, State of Sonora, Mexico.

Geogr. Distr. State of Sonora, Mexico; extent of range unknown.

Genl. Char. Similar in color to *P. eremicus*; soles bare; ears naked; tail long and scantily haired.

Color. Above pale yellowish gray mixed with black; cheeks, outer side of arms, and sides pale fulvous or cinnamon; under parts white with buff spot on chest; hands and feet white; tail above little darker than back, beneath paler.

Measurements. Total length, 217; tail vertebræ, 113; hind foot, 23.

167. hemionotis (*Peromyscus*), Elliot, Pub. Field Columb. Mus., III,
    1903, p. 157. Zoölogy.
MULE-EARED MOUSE.

Type locality. Rosarito Divide, San Pedro Martir Mountains, Lower California, Mexico.

Genl. Char. Similar in color to *P. stephensi*; ears enormous; tail very long.

Color. Nose and tufts in front of ears dark gray; top of head and upper parts cream buff finely lined with black; bright ochraceous lateral line from lips to root of tail; under parts pure white, base of hairs plumbeous showing through; hands and feet white; tail above brownish black, beneath white; ears dark brown.

Measurements. Total length, 222; tail vertebræ, 126; hind foot, 22.5; ear, 25.5. Skull: occipito-nasal length, 27; Hensel, 20.3; zygomatic width, 13; interorbital constriction, 4.5; breadth of braincase, 13; palatal arch to alveoli of incisors, 10.5; length of nasals, 10; greatest breadth of rostrum, 4; length of upper tooth row, 4; length of mandible, angle to alveolus of incisors, 11.

168. gaurus (*Peromyscus*), Elliot, Pub. Field Columb. Mus., III,
    1903, p. 157. Zoölogy.
FRISKY MOUSE.

Type locality. San Antonio, San Pedro Martir Mountains, Lower California, Mexico.

Genl. Char. Size large, tail very long, colors pale. Skull with broad braincase; nasals long; interorbital space broad.

Color. Above ochraceous buff, finely lined with black; deepest and brightest on lower back and rump, becoming grayish cream buff; lined with black on upper back and top of head and nose, cheeks, sides, thighs, and upper parts of arms, legs, and entire under parts pure white, the plumbeous under fur showing through; hands and feet white; tail above dark wood brown, beneath cream color; ears blackish brown.

Measurements. Total length, 222; tail vertebræ, 122; hind foot, 22; ear, 20. Skull: occipito-nasal length, 27; Hensel, 21; zygomatic

width, 14; interorbital constriction, 4; length of nasals, 9; palatal
length, 11; postpalatal length, 9; greatest breadth of rostrum, 4;
length of mandible, 11; height at coronoid process, 7.

169. homochroia (*Peromyscus*), Elliot, Pub. Field Columb. Mus.,
      III, 1903, p. 158.   Zoölogy.
SAN QUENTIN MOUSE.
   Type locality.   San Quentin, Lower California, Mexico.
   Genl. Char.   Similar in color to *P. gaurus*, but smaller; tail shorter,
and black above instead of wood brown; ears comparatively longer.
Skull is shorter, braincase narrower, nasals shorter and more pointed
interorbital constriction greater, pterygoids and palate narrower.
   Color.   Above similar to *P. gaurus*, beneath white with buff
patch on chest between arms; sides ochraceous; hands and feet
white; tail above black with apical third black all round, beneath
whitish; in some specimens the tail is black above and beneath; ears
brownish black.
   Measurements.   Total length, 185; tail vertebræ, 100; hind foot,
20; ears, 21.5.   Skull: occipito-nasal length, 24.2; Hensel, 18; zygomatic width, 8; interorbital constriction, 3.8; width of braincase, 6.6;
length of nasals, 6.5; palatal length, 9; postpalatal length, 8.2; greatest
breadth of rostrum, 3; length of mandible, 10.

170. oresterus (*Peromyscus*), Elliot, Pub. Field Columb. Mus., III,
      1903, p. 159.   Zoölogy.
MOUNTAIN MOUSE.
   Type locality.   Vallecitos, San Pedro Martir Mountains, Lower
California, Mexico, 9,000 feet elevation.
   Genl. Char.   Color pale; tail medium; skull with braincase and
nasals broad; rostrum heavy.
   Color.   Upper parts pinkish buff, lined with black on top of head
and dorsal region; line from nose to below the eye; shoulders and
upper part of sides of body pinkish buff; end of nose, lips, cheeks,
lower parts of sides of body, and entire under parts pure white; base
of fur plumbeous; tail above dusky, sides and beneath white; hands
and feet white; ears brownish black with tufts of creamy buff hairs
on basal half.
   Measurements.   Total length, 201; tail vertebræ, 96; hind foot,
21; ear, 20.5.   Skull: occipito-nasal length, 26; Hensel, 20; zygomatic width, 12; interorbital constriction, 4; length of nasals, 9;
breadth of rostrum, 4; palatal length, 10; postpalatal length, 7; width
of braincase, 11.

171. leucurus (*Peromyscus*), Thomas, Ann. Mag. Nat. Hist., 6th
      Ser., 1894, p. 364.

WHITE-TAILED MOUSE.

Type locality. Tehuantepec, Mexico.

Geogr. Distr. Southern Mexico.

Genl. Char. Size about that of *P. californicus*; ears large; tail long, distal half white.

Color. Above cinereous gray lined with black; under parts white; hands and feet white; tail, proximal half brown, distal half white; ears blackish brown.

Measurements. Total length, 260; tail, 142; hind foot, 25.5; ear from notch, 17.5. Skull: greatest length, 30.3; greatest breadth, 15; length of nasals, 11.2; breadth of nasals, 3.4; interorbital constriction, 5.1; interparietal breadth, 10; interparietal length, 3.9; palatal length, 11.9; length of upper molar series, 4.5; length of mandible, condyle to tip of incisor, 18.2; height of coronoid process, 7.4.

a.—gadovi (*Peromyscus*), Thomas, Ann. Mag. Nat. Hist., 7th Ser., XI, 1903, p. 484.

GADOW'S MOUSE.

Type locality. San Carlos, Yautepec, State of Oaxaca, Mexico. Altitude, 2,250 feet.

Genl. Char. Similar to *P. leucurus*, color darker, ears larger.

Color. Upper parts pale brown tinged with buff; sides like back; under parts soiled grayish; chin white; hands and feet white; tail black above and at tip, beneath white.

Measurements. Total length, 265; tail, 150; hind foot, 27.6; ear, 25. Skull: greatest length, 31.5; basilar length, 24; zygomatic breadth, 4.2; breadth of braincase, 14; interparietal, 4.2×11.5; palatal length, 12.5; palatal foramina, 6.6×2.6; length of upper molar series, 4.7.

172. melanophrys (*Hesperomys*), Coues, Proc. Acad. Nat. Scien. Phil., 1874, p. 181.

BLACK-BROWED MOUSE.

Type locality. Santa Efigenia, Tehuantepec, State of Oaxaca, Mexico.

Geogr. Distr. Southern Mexico, Tehuacan and Tehuantepec, States of Puebla and Oaxaca.

Genl. Char. Size large; tail longer than head and body; ears large. Skull: palate ending opposite last molars.

Color. Above gray, tinged with fulvous, darker on dorsal line; sides fulvous; head gray; orbital ring black; under parts pure white; hairs at base plumbeous; hands and feet white; tail above like the back, beneath gray.

Measurements. Total length, 245; tail vertebræ, 135; hind foot,

26; ear above notch, 19.4. (ex Type.) Skull: total length. 30; width, 17.

a.—zamorœ (*Peromyscus*), Osgood, Proc. Biol. Soc. Wash., XVII, 1904, p. 65.
ZAMORA MOUSE.

Type locality. Zamora, State of Michoacan, Mexico.

Genl. Char. Similar to *P. melanophrys;* tawny band across pectoral region.

Color. Like *P. melanophrys,* but darker; tawny pectoral band: tail above black.

Measurements. Total length, 260; tail vertebræ. 141; hind foot. 29. Skull: adults, greatest length, 31.3–32; Hensel, 25–25.9; zygomatic width, 16.5–16.9; length of nasals, 12–12; length of upper molar series, 4.7–4.8.

b.—consobrinus (*Peromyscus*), Osgood, Proc. Biol. Soc. Wash., XVII, 1904, p. 66.
KINDRED MOUSE.

Type locality. Berriozabal, State of Zacatecas, Mexico.

Genl. Char. Similar to *P. melanophrys,* but tail shorter. Skull with larger audital bullæ; shorter and broader nasals; brain case more bulging.

Color. Upper parts and sides tawny ochraceous thickly lined with black; lateral line tawny; orbital ring black; sides of face grayish; pectoral spot tawny; under parts creamy white; tail white beneath: hands and feet creamy white; ankles dusky.

Measurements. Total length, 250; tail vertebræ, 131; hind foot, 26.5. Skull: greatest length, 30.8; Hensel, 25.3; zygomatic width, 16.3; interorbital constriction, 4.9; length of nasals, 11.1; palate. 4.4; length of upper molar series, 4.7.

172a. xenurus (*Peromyscus*), Osgood, Proc. Biol. Soc. Wash., XVII, 1904, p. 67.
ODD-TAILED MOUSE.

Type locality. Durango, State of Durango, Mexico.

Genl. Char. Tail black, with ventral white line; large pectoral spot.

Color. Upper parts grayish fawn mixed with black; rump fawn: cheeks mixed fawn and gray; large pectoral spot bright fawn; under parts white; tail black all around, except a narrow white stripe beneath; hind feet dusky brown to base of toes, latter creamy white.

Measurements. Total length, 246–248; tail vertebræ, 140–142; hind foot, 28. Skull: greatest length, 30; Hensel, 23.8; zygomatic width, 115.5; interorbital constriction, 4.9; length of nasals, 10.2; length of upper molar series, 4.9.

172b. zelotes (*Peromyscus*), Osgood, Proc. Biol. Soc. Wash., XVII, 1904, p. 67.

ENERGETIC MOUSE.

Type locality. Querendaro, State of Michoacan, Mexico.

Genl. Char. Similar to *P. melanophrys;* tail longer than head and body.

Color. "Similar in general to *P. melanophrys*, but facial region more suffused with tawny and the gray very much reduced; somewhat similar to *P. levipes*, but paler throughout; no trace of a pectoral spot."

Measurements. Total length, 218; tail vertebræ, 115; hind foot, 23. Skull: greatest length, 28.3; Hensel, 21.6; zygomatic width, 14.2; interorbital constriction, 4.9; length of nasals, 10.6; length of upper molar series, 4.6.

173. insignis (*Peromyscus*), Rhoads, Proc. Acad. Nat. Scien. Phil., 1895, p. 33. Elliot, Syn. N. Am. Mamm., 1901, p. 138.

SACHEM OR CHIEF MOUSE.

Type locality. Dulzura, San Diego County, California.

Geogr. Distr. Southern California, and Lower California, Mexico.

Genl. Char. Size large; tail very long, exceeding head and body.

Color. Above light brownish gray, mingled with black; darkest on back, brownest on rump and thighs; flanks ochraceous; hands, feet, and under surface grayish white; tail above sooty, below grayish white.

Measurements. Total length, 233; tail vertebræ, 132; hind foot, 26; ear, 23. Skull: occipito-nasal length, 29; Hensel, 11; zygomatic width, 9; interorbital constriction, 4; length of nasals, 11; palatal length, 11; length of upper molar series, 4.

174. californicus (*Mus*), Gambel, Proc. Acad. Nat. Scien. Phil., 1848, p. 78.

californicus (*Peromyscus*), Elliot, Syn. N. Am. Mamm., 1901, p. 137

CALIFORNIA MOUSE.

Type locality. Monterey, Monterey County, California.

Geogr. Distr. Lower California, Mexico, north along the coast of California.

Genl. Char. Size large; tail long; ears large, sparsely haired.

Color. Above dark gray mixed with light brown; sides fulvous; under parts grayish buff; tail blackish brown above, whitish below.

Measurements. Total length, 265; tail vertebræ, 142; hind foot,

28; ear, 26. Skull: occipito-nasal length, 29; Hensel, 22; zygomatic width, 15; interorbital constriction, 4; width of braincase, 13.2; length of nasals, 10; palatal length, 12.

175. hylocetes (*Peromyscus*), Merr., Proc. Biol. Soc. Wash., XII, 1898, p. 124.
DRYAD MOUSE.

Type locality. Patzcuaro, State of Michoacan, Mexico. Altitude 8,000 feet.

Geogr. Distr. State of Michoacan, Mexico; range unknown.

Genl. Char. Tail shorter than head and body; size large.

Color. Above buffy gray, darkest on dorsal region; sides buffy ochraceous; under parts, lips, and hands whitish; basal part of hind foot, wrists, and ankles dusky; tail above dusky, below white.

Measurements. Total length, 238; tail vertebræ, 114; hind foot, 25.

176. lepturus (*Peromyscus*), Merr., Proc. Biol. Soc. Wash., 1898, p. 118.
MOUNT ZEMPOALTEPEC FIELD MOUSE.

Type locality. Mount Zempoaltepec, State of Oaxaca, Mexico. Altitude, 8,200 feet.

Geogr. Distr. State of Oaxaca, Mexico, range unknown.

Genl. Char. Size large; tail long as head and body, nearly hairless; molars large.

Color. Above brownish, dorsal area dusky, almost black; cheeks and sides brownish fulvous; orbital ring dusky; under parts white; tail above dusky, paler beneath; hands whitish; wrists, ankles, and hind feet dusky.

Measurements. Total length, 238; tail vertebræ, 114; hind foot, 28. (ex Type.)

176a. lophurus (*Peromyscus*), Osgood, Proc. Biol. Soc. Wash., XVII, 1904, p. 72.
CRESTED-TAILED MOUSE.

Type locality. Todos Santos, Guatemala.

Genl. Char. Similar to *P. lepturus*, but smaller and paler; tail long, hairy, penicillate.

Color. Upper parts between wood brown and fawn; middle of back dusky; lateral line pale ochraceous buff; under parts white; orbital ring dusky; tail sepia brown, unicolor; hands white; forearm dusky; feet dusky brownish to base of toes, the latter white.

Measurements. Average of four adults. Total length, 208; tail vertebræ, 105; hind foot, 24.5. Skull: greatest length, 27.5; Hensel, 20.8; zygomatic width, 14.7; interorbital constriction, 4.3; interparietal, 10×4.5; length of nasals, 10; palate, 4; postpalatal length, 9.6; length of upper molar series, 4.7.

176b. simulatus (*Peromyscus*), Osgood, Proc. Biol. Soc. Wash. XVII, 1904, p. 72.

IMITATOR MOUSE.

Type locality. Jico, State of Vera Cruz, Mexico.

Genl. Char. Miniature of *P. lophurus*; tail long, hairy, crested; audital bullæ relatively large.

Color. Like *P. lophurus;* tail brown, narrow line of white on under surface.

Measurements. Total length, 169; tail vertebræ, 87; hind foot, 21; ear from notch, 14.3. Skull: greatest length, 24.4; Hensel, 18; zygomatic width, 12.5; interorbital constriction, 4.3; interparietal 8.2 × 3; length of nasals, 9; palate, 3.5; postpalatal length, 8; length of upper molar series, 8.9.

177. felipensis (*Peromyscus*), Merr., Proc. Biol. Soc. Wash., XII, 1898, p. 122.

SAN FELIPE FIELD MOUSE.

Type locality. Cerro San Felipe, State of Oaxaca, Mexico. Altitude, 10,200 feet.

Geogr. Distr. State of Oaxaca, Mexico; range unknown.

Genl. Char. Similar to *P. difficilis*, but larger and darker.

Color. Above dusky gray, blackish on dorsal region; lips, sides of nose, under parts, hands, and feet white; flanks grayish brown; spot on breast salmon; orbital ring black; tail above brown, beneath whitish.

Measurements. Total length, 238; tail vertebræ, 125; hind foot, 27.5.

177a. amplus (*Peromyscus*), Osgood, Proc. Biol. Soc. Wash., XVII, 1904, p. 62.

CHUBBY MOUSE.

Type locality. Coixtlahuaca, State of Oaxaca, Mexico.

Genl. Char. Similar to *P. felepensis*, but paler.

Color. Upper parts clay color, being a mixture of ochraceous buff and dusky; lateral line ochraceous buff; forehead and orbital region grayish; under parts creamy white; pectoral spot ochraceous buff; tail above dusky brownish, beneath white; hands and feet white; ankles dusky.

Measurements. Average of ten adults: total length, 248 (235–260); tail vertebræ, 136 (128–145); hind foot, 27 (26–28). Skull of type: zygomatic width, 10.4; interorbital constriction, 4.5; interparietal, 10.4 × 3.7; length of nasals, 11.3; palate, 4.6; postpalatal length, 10.2; upper molar series, 4.8.

178. tehuantepecus (*Peromyscus*), Merr., Proc. Biol. Soc. Wash., XII, 1898, p. 122.

TEHUANTEPEC FIELD MOUSE.

Type locality. Tehuantepec, State of Oaxaca, Mexico.

Geogr. Distr. State of Oaxaca, Mexico; range undetermined.

Genl. Char. Size medium; similar to *P. mexicanus*, but paler.

Color. Above pale fulvous lined with black, darkest on back; cheeks and sides buffy fulvous; under parts whitish, with a salmon pectoral spot, and sometimes tinged with buff; tail dusky, sides orange chrome; hands and feet whitish.

Measurements. Total length, 243; tail vertebræ, 124; hind foot, 26.

179. oaxacensis (*Peromyscus*), Merr., Proc. Biol. Soc. Wash., XII, 1898, p. 122.

OAXACA FIELD MOUSE.

Type locality. Cerro San Felipe, State of Oaxaca, Mexico. Altitude, 10,000 feet.

Geogr. Distr. State of Oaxaca, Mexico; limits of range unknown.

Genl. Char. Size medium; tail long. Skull similar to that of *P. mexicanus*, but bullæ larger, and first lower molar with a "supplementary enamel loop on outer side in front of posterior cusp."

Color. Upper parts fulvous, darkest on dorsal region; orbital ring dusky; cheeks and sides cinnamon rufous; under parts and hands white; feet white, their basal portion dusky; tail above dusky, beneath white.

Measurements. Total length, 242; tail vertebræ, 122; hind foot, 27.

180. comptus (*Peromyscus*), Merr., Proc. Biol. Soc. Wash., XII, 1898, p. 120.

CHILPANCINGO MOUSE.

Type locality. Mountains near Chilpancingo, State of Guerero, Mexico.

Geogr. Distr. State of Guerero, Mexico; range unknown.

Genl. Char. Size large; similar to *P. auritus*, but more fulvous; tail long.

Color. Upper parts golden fulvous lined with black, dorsal region darkest; orbital ring blackish; under parts white, some specimens tinged with yellowish; wrists and ankles dusky; tail dusky above, whitish beneath; hands and feet whitish.

Measurements. Total length, 285; tail vertebræ, 150; hind foot, 31.

181. mexicanus (*Hesperomys*), Sauss., Rev. Mag. Zoöl., 1860, 2me
Sér., p. 103, pl. IX, figs. i. ia.

MEXICAN MOUSE.

Type locality. State of Vera Cruz, Mexico.

Genl. Char. Size large; ears large, longer than wide; tail long.

Color. Above dark brownish mouse gray; silvery shade on back;
flanks yellowish gray; cheeks rusty gray; lips and chin yellowish
gray; under parts grayish white, with a yellow wash on breast;
hands white; hind feet brown; toes white; tail black above, white
beneath.

Measurements. Total length, 180; tail vertebræ, 78; hind foot, 26.

a.—teapensis (*Peromyscus*), Osgood, Proc. Biol. Soc. Wash., XVII,
1904, p. 69.

TEAPA MOUSE.

Type locality. Teapa, State of Tabasco, Mexico.

Genl. Char. Similar to *P. m. totontepecus*, but sides brighter.

Color. Dorsal region blackish; sides chestnut; orbital ring black;
under parts creamy white; pectoral spot often present; tail black,
with a few yellowish white spots beneath; hands white; feet white,
with a dusky area from ankles nearly to base of toes.

Measurements. Total length, 234–254; tail vertebræ, 121–136;
hind foot, 27–28.5. Skull: greatest length, 33; Hensel, 24.6; zygomatic width, 16.2; length of nasals, 12.7; interorbital constriction,
5.4; palate, 4.7; postpalatal length, 11.9; length of upper molar series,
4.5.

b.—orizabæ (*Peromyscus*), Merr., Proc. Biol. Soc. Wash., XII, 1898,
p. 121.

ORIZABA FIELD MOUSE.

Type locality. Orizaba, State of Vera Cruz, Mexico. Altitude,
4,200 feet.

Geogr. Distr. State of Vera Cruz, Mexico; range unknown.

Genl. Char. Similar to *P. mexicanus*, but larger; tail and hind
feet longer.

Color. Top of head and upper parts dark seal brown; sides and
cheeks chestnut fulvous; orbital ring and nose dusky; lips, under
parts, wrists, and hands white; salmon tinge on breast; hind feet
basally dusky, rest whitish; tail above dusky, beneath yellowish white.

Measurements. Total length, 257; tail vertebræ, 139; hind
foot, 29.5.

c.—saxatilis (*Peromyscus*), Merr., Proc. Biol. Soc. Wash., XII, 1898,
p. 121.

Rock Mouse.

Type locality.   Jacaltenango, Huehuetenango, Guatemala.

Geogr. Distr.   Guatemala; range unknown.

Genl. Char.   Similar to *P. mexicanus*; paler.

Color.   Above grayish fulvous, lined with black; dorsal area dark, almost black; sides fulvous; lips, wrists, hands, feet, and under parts white; ankles dusky; orbital ring blackish; cheeks pale fulvous; tail above dusky, beneath whitish.

Measurements.   Total length, 245.5; tail vertebræ, 127.5; hind foot, 27.5.

d.—totontepecus (*Peromyscus*), Merr., Proc. Biol. Soc. Wash., xii, 1898, p. 120.

Totontepec Field Mouse.

Type locality.   Totontepec, State of Oaxaca, Mexico.   Altitude, 6,500 feet.

Geogr. Distr.   State of Oaxaca, Mexico; range unknown.

Genl. Char.   Similar to *P. mexicanus*, but larger and darker.

Color.   Above dusky brown; black on dorsal region; sides and cheeks fulvous brown; under parts whitish, with sometimes a salmon patch on breast; tail above dusky, yellowish white below; hands whitish; ankles and basal portions of hind feet dusky.

Measurements.   Total length, 261; tail vertebræ, 136; hind foot, 28.

181a. allophylus (*Peromyscus*), Osgood, Proc. Biol. Soc. Wash., xvii, 1904, p. 71.

Alien Mouse.

Type locality.   Huehuetan, State of Chiapas, Mexico.

Genl. Char.   Tail shorter than head and body; tail scaly as in *Oryzomys;* proximal soles of hind feet finely haired.

Color.   Dorsal area blackish brown; sides mummy brown; under parts yellowish white; slate of under fur showing; orbital ring and antorbital spot black; tail unicolor, dusky blackish; hands and feet whitish; ankles dusky.

Measurements.   Total length, 202; tail vertebræ, 95; hind foot, 25.   Skull: greatest length, 29.8; Hensel, 22.5; zygomatic width, 14.5; interorbital constriction, 5; length of nasals, 11; palate, 4; post-palatal length, 10.5; length of upper molar series, 4.

182. furvus (*Peromyscus*), Allen & Chapman, Bull. Amer. Mus. Nat. Hist., 1897, p. 201.

Jalapa Field Mouse.

Type locality.   Jalapa, State of Vera Cruz, Mexico.

Geogr. Distr. State of Vera Cruz, Mexico.

Genl. Char. Skull large, strong; .rostrum broad, inflated anteriorly, bell-shaped; nasals pointed posteriorly and extending beyond the intermaxillæ; palate with slightly upturned posterior border; anterior palatine foramina very broad.

Color. Above dark brown, washed with grayish, blackish on dorsal line; inclined to reddish on sides; under parts and hind feet grayish white; fore feet white; rufous patch on breast; ears and tail dark brown, naked; tip of tail sometimes whitish.

Measurements. Total length, 248–282; tail vertebræ, 123–145; hind foot, 26–29; ear, 20–23. Skull: total length, 35; basal length, 29; zygomatic width, 16.7; interorbital constriction, 5; length of nasals, 8; length of upper tooth row, 5. (ex Type.)

183. zarhynchus (*Peromyscus*), Merr., Proc. Biol. Soc. Wash., XII, 1898, p. 117.

LONG-NOSED MOUSE.

Type locality. Tumbala, State of Chiapas, Mexico.

Geogr. Distr. State of Chiapas, Mexico; limit of range unknown.

Genl. Char. Size large; ears large; tail long, naked; hind feet long. Skull: rostrum elongated, bullæ small; interparietal narrow.

Color. Above dusky with a chestnut tinge; flanks seal brown or chestnut; under parts buff, with a chestnut tinge on breast, extending in some specimens over the belly; tail above dusky, beneath yellowish white; hands and feet whitish.

Measurements. Total length, 324; tail vertebræ, 176; hind foot, 35.

a.—cristobalensis (*Peromyscus*), Merr., Proc. Biol. Soc. Wash., XII, 1898, p. 117.

SAN CRISTOBAL FIELD MOUSE.

Type locality. San Cristobal, State of Chiapas, Mexico.

Geogr. Distr. State of Chiapas, Mexico; range unknown.

Genl. Char. Similar to *P. zarhynchus*, but lighter and more rufous.

Color. Above dusky brown and fulvous mixed, darkest on dorsal line; cheeks and sides fulvous; under parts whitish, with chest tinged with chestnut; tail dusky above, yellowish white beneath; hands and feet whitish.

Measurements. Total length, 322; tail vertebræ, 170; hind foot, 34.

184. auritus (*Peromyscus*), Merr., Proc. Biol. Soc. Wash., XII, 1898, p. 119.

LARGE-EARED FIELD MOUSE.

Type locality. Mountains west of Oaxaca City, State of Oaxaca, Mexico.

Geogr. Distr. State of Oaxaca, Mexico; limits of range unknown.

Genl. Char. Similar to *P. megalops*; ears and bullæ larger.

Color. Above grayish brown, back blackish; cheeks and sides washed with fulvous; under parts whitish; hands and feet whitish; wrists and ankles dusky; tail above dusky, beneath whitish.

Measurements. Total length, 288; tail vertebræ, 148; hind foot, 30.

185. megalops (*Peromyscus*), Merr., Proc. Biol. Soc. Wash., xII, 1898, p. 119.

OZOLOTEPEC FIELD MOUSE.

Type locality. Mountains near Ozolotepec, State of Oaxaca, Mexico.

Geogr. Distr. State of Oaxaca, Mexico; range unknown.

Genl. Char. Size large; ears short; tail long.

Color. Above mixed fulvous and black, darkest between ears and on dorsal region; cheeks and sides of body salmon fulvous; under parts whitish, breast tinged with salmon fulvous; hands white; wrists, ankles, and hind feet dusky.

Measurements. Total length, 282; tail vertebræ, 150; hind foot, 31.

185a. melanocarpus (*Peromyscus*), Osgood, Proc. Biol. Soc. Wash., xVII, 1904, p. 73.

BLACK-FOOTED MOUSE.

Type locality. Mount Zempoaltepec, State of Oaxaca, Mexico. Altitude, 8,000 feet.

Genl. Char. Similar to *P. megalops*, but smaller and darker; hands blackish to digits.

Color. Upper parts blackish and mummy brown, darkest on middle of back; under parts blackish slate, washed with creamy white; pectoral region cinnamon rufous; black line from nostril through eye; tail blackish, slightly paler on under side; hands and feet dusky brownish to base of toes.

Measurements. Total length, 241; tail vertebræ, 125; hind foot, 27. Skull: greatest length, 31.6; Hensel, 24.3; zygomatic width, 15.2; length of nasals, 12; interorbital constriction, 5.4; length of upper molar series, 5.

185b. altilaneus (*Peromyscus*), Osgood, Proc. Biol. Soc. Wash., xVII, 1904, p. 74.

FLUFFY MOUSE.

Type locality. Todos Santos, Guatemala. Altitude, 10,000 feet.

Genl. Char. Similar to *P. melanocarpus*, but smaller and with a shorter and less hairy tail; hands white.

Color. Like *P. melanocarpus*; tail blotched with yellowish white on under side; hands and part of forearm white; hind foot with V-shaped dusky mark from ankle half-way to base of toes; rest of foot white; pectoral spot strongly developed.

Measurements. Total length, 228; tail vertebræ, 115; hind foot, 28. Skull: greatest length, 31; Hensel, 24; zygomatic width, 14.6; interorbital constriction, 5; length of nasals, 11.5; palate, 4.8; post-palatal length, 11.2; length of upper molar series, 4.6.

186. guatemalensis (*Peromyscus*), Merr., Proc. Biol. Soc. Wash., XII, 1898, p. 118.

TODOS SANTOS MOUSE.

Type locality. Todos Santos, Guatemala. Altitude, 10,000 feet.

Geogr. Distr. State of Chiapas, Mexico, into Guatemala; range unknown.

Genl. Char. Size large; tail long, hairs scanty; fur long, soft.

Color. Above mixed dusky and grayish, blackish on dorsal region; indistinct line from nose to ear; flanks brownish fulvous, as are also the cheeks; salmon fulvous patch on breast; under parts and hands white; hind feet dusky, then white; wrists and ankles blackish; tail above dusky, beneath whitish.

Measurements. Total length, 273; tail vertebræ, 141; hind foot, 31. (ex Type.) Skull: zygomatic width, 16.5; interorbital constriction, 5; length of nasals, 14.5; palatal length, 14; length of upper tooth row, 4; length of mandible, angle to alveolus of incisor, 15; height at coronoid process, 6.2; length of lower tooth row, 5.

187. cacabatus (*Peromyscus*), Bangs, Bull. Mus. Comp. Zoöl., XXXIX, 1902, p. 29.

SOOTY MOUSE.

Type locality. Boquete, Chiriqui, Panama. Altitude, 5,000 feet.

Genl. Char. Allied to *P. guatemalensis*, but paler and with a shorter tail.

Color. Dorsal region sooty; sides brownish grading into dull orange buff on lower sides; sides of nose buffy white; top of nose and orbital ring black; under parts varying from dull grayish white to pinkish buff; pectoral collar orange buff; feet and hands white; tail above dusky, beneath pale yellowish gray; ear dusky.

Measurements. Total length, 252-270; tail vertebræ, 120-135; hind foot, 25-27; ear, 20-21. Skull: basal length, 28.8; occipito-nasal length, 32.4; zygomatic width, 15.6; mastoid width, 13.6; length of nasals, 13; width of nasals, 3.4; length of palatal slits. 6.2; width of palatal slits, 3.2; length of palate to palatal notch, 12.8; to end of pterygoid, 19; length of upper molar series, 5; length of single half mandible, 17.4.   (Bangs, l. c.)

188. mecisturus (*Peromyscus*), Merr., Proc. Biol. Soc. Wash., XII. 1898, p. 124.

GREAT-TAILED MOUSE.

Type locality. Chalchicomula, State of Puebla, Mexico. Altitude, 8,400 feet.

Geogr. Distr. State of Puebla, Mexico.

Genl. Char. Size large; tail very long; ears large.

Color. Above anteriorly gray lined with black, and suffused with fulvous; rump pale fulvous; orbital ring dusky; chin whitish; under parts whitish buff; breast buffy; wrists dusky; hands white; sides of ankles, and the toes white; tail above dusky, paler beneath.

Measurements. Total length, 249; tail vertebræ, 155; hind foot, 24.

43. Megadontomys.

$$I.\frac{1-1}{1-1}; \ M.\frac{3-3}{3-3} = 16.$$

Megadontomys Merr., Proc. Biol. Soc. Wash., XII, 1898, p. 115. Type *Megadontomys thomasi* Merriam.

Size large; ears and tail long, scantily haired; pelage long, soft, and very dense. Skull similar in general to that of *Peromyscus*. but very large and massive; rostrum and nasals much produced, the latter expanded anteriorly and projecting far beyond incisors. Molars very large and heavy, with short tubercles which wear off when the animal is still young, leaving flat crowns; first and second lower molars with a supplementary narrow enamel loop on each side; third lower molar with three salient and two reëntrant angles on each side.   Plantar tubercles, 7; Mammæ, 6; pectoral $\frac{1-1}{1-1}$, inguinal $\frac{2-2}{2-2}$. (Merriam, l. c.)

KEY TO THE SPECIES.

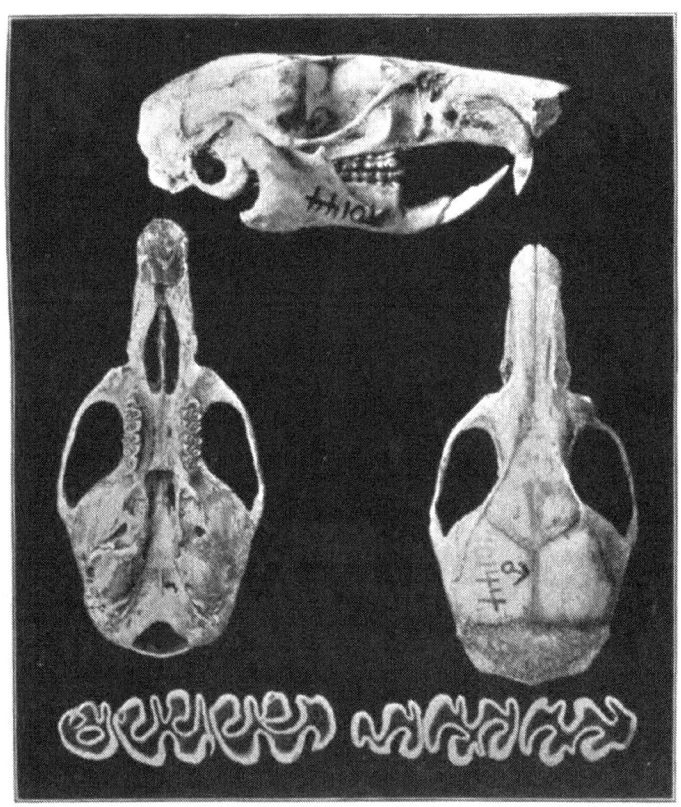

FIG. 34. MEGADONTOMYS THOMASI.
No. 70144 U. S. Nat. Mus. Coll. Twice nat. size.

| UPPER TOOTH ROW. | LOWER TOOTH ROW. |
| Enlarged 7 times. | Enlarged 7 times. |

189. thomasi (*Peromyscus*), Merr., Proc. Biol. Soc. Wash., XII, 1898, p. 116.

THOMAS' GUERRERO FIELD MOUSE.

Type locality. Mountains near Chilpancingo, State of Guerrero, Mexico. Altitude, 9,700 feet.

Geogr. Distr. State of Guerrero, Mexico; range unknown.

Genl. Char. Size large; tail very long; ears large, both nearly naked.

Color. Above fulvous, mixed with black on back; beneath white; breast sometimes tinged with yellowish buff; orbital ring blackish; hands and feet white; tail above dusky, almost black, beneath paler.

Measurements. Total length, 350; tail vertebræ, 188; hind foot, 34. Skull: occipito-nasal length, 65; Hensel, 54; zygomatic width, 32; interorbital constriction, 9; length of nasals, 26; palatal length, 33.5; length of upper tooth row, 11.

190. nelsoni (*Peromyscus*), Merr., Proc. Biol. Soc. Wash., XII, 1898, p. 116.

NELSON'S FIELD MOUSE.

Type locality. Jico, State of Vera Cruz, Mexico. Altitude, 6,000 feet.

Geogr. Distr. State of Vera Cruz, Mexico; range unknown.

Genl. Char. Similar to *M. thomasi*, but darker.

Color. Above and sides grayish brown, darkest on dorsal region; under parts white; tail dusky; hands white; hind feet dusky.

Measurements. Total length, 302; tail vertebræ, 172; hind foot, 35.

191. flavidus (*Megadontomys*), Bangs, Bull. Mus. Comp. Zoöl., XXXIX, 1902, p. 27.

BOQUETE MOUSE.

Type locality. Boquete, Chiriqui, Panama. Altitude, 4.000 feet.

Genl. Char. Braincase rounded and elevated; palatal slits very wide; audital bullæ small; ears small.

Color. Above brownish cinnamon, inclined to rusty on the rump; sides orange buff; black patch at base of whiskers; under parts white; pectoral collar buffy; hands and feet whitish; tail above dusky, beneath grayish; ears dusky inside, silvery outside.

Measurements. Total length, 320–375; tail vertebræ, 155–205; hind foot, 31–33; ear, 20–24. Skull: basal length, 35.4; occipito-nasal length, 40.2; zygomatic width, 19.6; mastoid width, 15; length of nasals, 17.8; width of nasals, 4.8; length of palatal slits, 7.4; width of palatal slits, 3.4; length of palate to palatal notch, 17; to end of pterygoid, 24.4; length of upper molar series, 5.6; length of single half mandible, 21.8. (Bangs, l. c.)

44. Nyctomys. Vesper Rats.

$$I.\frac{1-1}{1-1}; \ M.\frac{3-3}{3-3} = 16.$$

Nyctomys Sauss., Rev. Mag. Zoöl., 2me Sér., 1860, p. 106. Type *Hesperomys sumichrasti* Saussure.

Myoxomys Tomes, Proc. Zoöl. Soc., 1861, p. 284, pl. XXXI.

Muzzle short; ears not hidden in fur, which is short and fine; feet short, broad; tail as long as body, hairy. Skull: rostrum very slender, short braincase and zygomata expanded; orbital space broad and a well-developed supraorbital crest; antorbital foramen large, opening forwards; palate only reaching to the forward margin of molar series; palatal slits short; mammæ four.

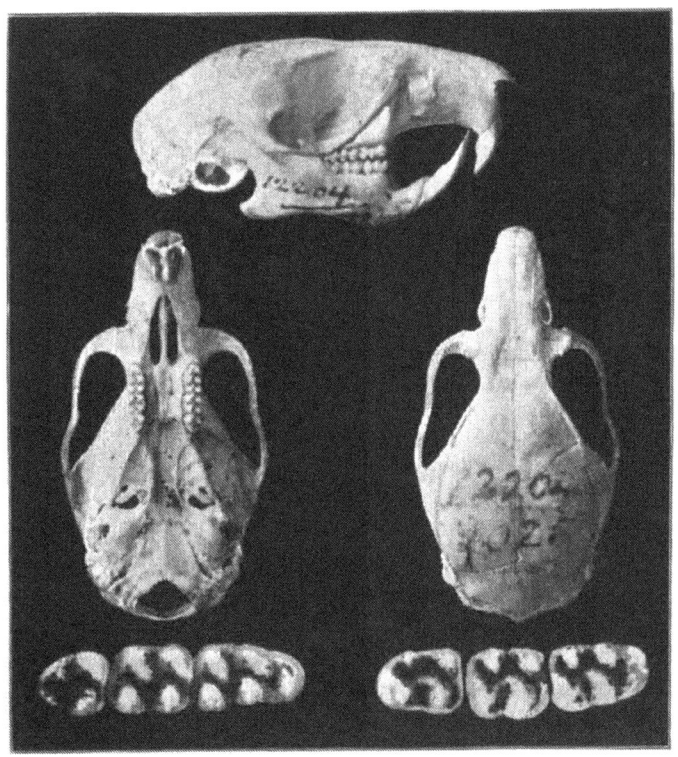

FIG. 35. NYCTOMYS SUMICHRASTI.
No. 12204 U. S. Nat. Mus. Coll. Twice nat. size.

UPPER TOOTH ROW.   LOWER TOOTH ROW
Enlarged 8 times.   Enlarged 8 times.

KEY TO THE SPECIES.

192. sumichrasti (*Hesperomys*), Sauss., Rev. Mag. Zoöl., 2me Sér..
    1860, p. 107, pl. 9, fig. 2.
SUMICHRAST'S VESPER RAT.
Type locality. "Habite le versant oriental de la Cordilière,".
State of Vera Cruz, ? Mexico.
Geogr. Distr. Eastern Mexico, State of Vera Cruz to Isthmus of
Tehuantepec.
Genl. Char. Ears long, higher than wide; nose pointed; hind
feet short.
Color. Upper parts isabella or pale orange, clearer on flanks;
under parts, chin, and lower jaw on side of face pure white; tail
brownish red; feet pale yellowish brown.
Measurements. Total length, 230; tail, 106; hind foot, 23. Skull:
occipito-nasal length, 58; Hensel, 49; zygomatic width, 30.5; interorbital constriction, 10; length of nasals, 16; palatal length, 25;
length of upper molar series, 10.

193. decolorus (*Sitomys*), True, Proc. U. S. Nat. Mus., XVI, 1894, p. 689.
FADED VESPER RAT.
Type locality. Rio de las Piedras, Honduras.
Geogr. Distr. Tehuantepec, Mexico, into Honduras, Central
America.
Genl. Char. Similar to *R. sumichrasti*, but paler, tail shorter;
ear naked, prominent; soles naked; tail covered with long hairs.
Color. Above brownish isabelline, middle of back washed with
gray; under parts and cheeks white; orbital ring dark brown; tail
chocolate brown; hands like back, feet dusky, toes whitish; ears
chocolate brown.
Measurements. Total length, 193; tail vertebræ, 85; hind foot,
23; ear, 14.

194. nitellinus (*Nyctomys*), Bangs, Bull. Mus. Comp. Zoöl., XXXIX.
    1902, p. 30.
BOQUETE VESPER RAT.
Type locality. Boquete, Chiriqui, Panama. Altitude, 4,000
feet.
Genl. Char. Similar to *R. decolorus*, but larger and darker.
Color. Above yellowish cinnamon, dorsal region darker; lower
sides shaded with orange buff; orbital ring, and space between eyes
and nose black; under parts white; tail blackish, unicolor; hands
white; feet, toes, and sides of tarsus white; middle of tarsus dark
brown; ears dusky.

Measurements. Total length, 250–260; tail vertebræ, 120–125; hind foot, 25; ear, 17. Skull: basal length, 28; occipito-nasal length, 32.4; zygomatic width, 18; mastoid width, 13; interorbital width, 11; length of nasals, 10.6; width of nasals, 3.2; length of palatal slits, 4.6; width of palatal slits, 2.2; length of palate to palatal notch, 12.4; length of upper molar series, 4.8; length of single half mandible, 18. (Bangs, l. c.)

45. Tylomys.

$$\text{I.}\frac{1-1}{1-1}; \text{ M.}\frac{3-3}{3-3} = 16.$$

Tylomys Peters, Monatsb. K. Preuss. Akad. Wiss. Berlin, 1866, p. 404. Type *Hesperomys nudicaudus* Peters.

"Ears rather large, naked; tail and soles of feet naked. Skull with no raised supraorbital ridges, but with a broad horizontal edge over orbits; antorbital foramen not visible from above."

KEY TO THE SPECIES.

195. nudicaudus (*Hesperomys*), Peters, Monatsb. K. Preuss. Akad. Wiss. Berlin, 1866, p. 404, pl. 1, figs. 1–4.

NAKED-TAILED RAT.

Type locality. Guatemala.

Geogr. Distr. Southeastern Mexico (State of Chiapas), into Guatemala.

Color. Upper parts ferrugineous, base of hairs plumbeous, under parts white; tail naked, basal half black, remainder yellow; hands and feet dusky.

Measurements. Total length, 403.75; tail, 207.5. Skull: occipito-nasal length, 46; Hensel, 38; zygomatic width, 24; interorbital constriction, 8.6; length of nasals, 14.8; palatal length, 19.3; length of mandible, angle to alveolus of incisor, 35; length of upper tooth row, 8.6.

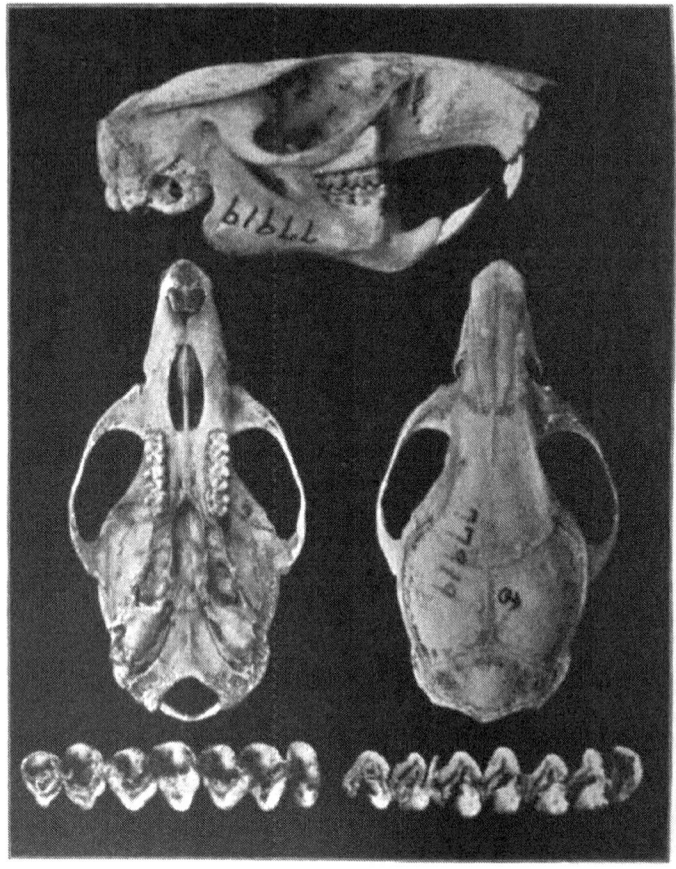

FIG. 36. TYLOMYS NUDICAUDUS.
No. 77919 U. S. Nat. Mus. Coll. Enlarged ½.

UPPER TOOTH ROW.   LOWER TOOTH ROW.
Enlarged 6 times.   Enlarged 6 times.

196. tumbalensis (*Tylomys*), Merr., Proc. Wash. Acad. Scien., III,
1901, p. 560.

TUMBALA RAT.

Type locality. Tumbala, State of Chiapas, Mexico.

Geogr. Distr. State of Chiapas, Mexico.

Genl. Char. Size large; ears large, naked; tail long. Skull similar to that of *T. nudicaudus*, but more slender and weaker; molar series large.

Color. Above dark gray, blackish on posterior half; orbital ring dusky; chin, breast, and patch between hind legs white; throat and belly plumbeous tinged with fulvous; hands and feet dark brown; tail blackish above on basal half, yellow on remainder.

Measurements. Total length, 448; tail vertebræ, 234; hind foot, 46.

197. bullaris (*Tylomys*), Merr., Proc. Wash. Acad. Scien., III, 1901, p. 561.

TUXTLA RAT.

Type locality. Tuxtla, State of Chiapas, Mexico.

Genl. Char. Unique specimen too young for reliable characters.

Color. "Similar to the young of *T. nudicaudus*, but grayer; under parts white; upper lip and patch on side of nose whitish; hands brown; hind feet dark brown; toes white.

Measurements. Total length juv., 324; tail, 158; hind foot, 37.5.

198. watsoni (*Tylomys*), Thomas, Ann. Mag. Nat. Hist., 7th Ser., IV, 1897, p. 278.

WATSON'S RAT.

Type locality. Bogava, Chiriqui, Panama. Altitude, 8,000 feet.

Genl. Char. Similar to *T. nudicaudus*, but more rufous; size medium; fur glossy; tail shorter than head and body.

Color. Above rufous fawn, lined with blackish; cheeks and sides paler and more gray; space between eyes and ears black; belly and inner side of hind limbs whitish buff; rest of under parts and inner side of fore limbs white; tail dark on basal, white on distal half; toes white.

Measurements. Total length, 493; tail, 243; hind foot, 38. Skull: greatest length, 54; Hensel, 42.5; greatest breadth, 26.5; length of nasals, 18; interorbital constriction, 10.5; palatal length from henselion, 22.2; length of upper molar series, 8.4.

199. panamensis (*Neomys*), Gray, Ann. Mag. Nat. Hist., 4th Ser., XII, 1873, p. 417.

PANAMA RAT.

Type locality. Panama.

Geogr. Distr. Panama, Southern Central America.

Color. Upper parts mouse color, lined with black; sides paler; throat and under parts and inner side of arms and legs white; tail black, tip white; hands and feet brownish; claws covered with white hairs.

Measurements. Total length, 412.5; tail, 200.

46. Ototylomys.

$$I.\frac{1-1}{1-1};\ M.\frac{3-3}{3-3}=16.$$

Ototylomys Merr., Proc. Wash. Acad. Scien., III, 1901, p. 561.
Type *Ototylomys phyllotis* Merriam.

Rat-like; ears large, thin, and naked; tail long and naked. Skull:
superciliary ridge present and reaching to occiput; bullæ with axes
parallel to that of skull; width and depth equal, and no anterior
prolongation; maxillary root of zygoma notched above; anterior opening of antorbital vacuity vertical; incisive foramina equal in width at

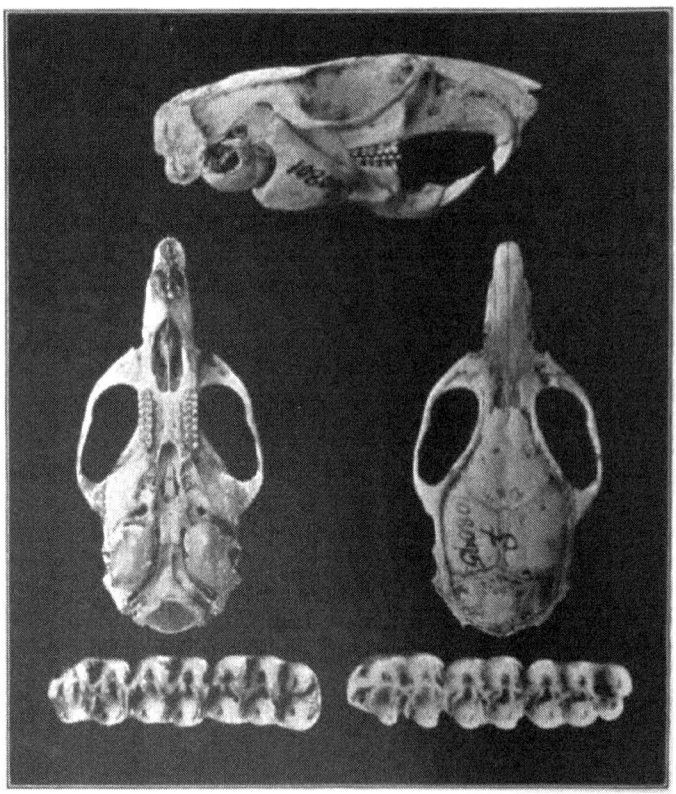

FIG. 37. OTOTYLOMYS PHYLLOTIS.
No. 108095 U. S. Nat. Mus. Coll. Enlarged ½.

UPPER TOOTH ROW.　　　　　　　LOWER TOOTH ROW
Enlarged 7 times.　　　　　　　　Enlarged 7 times.

both ends; mandible with angle excavated posteriorly, a backward projecting point; infracondyloid notch long and deep; coronoid process minute; postcoronoid notch flat, nearly horizontal.   (ex Merr., l. c.)

KEY TO THE SPECIES.

A. Ears very large; tail long as head and body,
   naked, scaly.
   a. Above grayish or fulvous brown.   PAGE
     a.' Size large; back uniform in color..........*O. phyllotis* 221
     b.' Size smaller; back varying in color........*O. p. phæus* 221

200. phyllotis (*Ototylomys*), Merr., Proc. Wash. Acad. Scien., III,
   1901, p. 562.
TUNKAS RAT.
   Type locality. Tunkas, Yucatan, Mexico.
   Genl. Char. Similar to *Tylomys* in appearance; characters those of the genus.
   Color. Above grayish or fulvous brown, sometimes mixed with black-tipped hairs; under parts and inner sides of legs white; cheeks and about eyes fulvous; tail above brownish dusky, beneath yellowish; hands and feet whitish; wrists and ankles dark; ears flesh color at base, rest black.
   Measurements. Total length, 303; tail, 148; hind foot, 28. (ex Type.)   Skull: occipito-nasal length, 39; Hensel, 31.8; zygomatic width, 24.3; interorbital constriction, 5.6; length of nasals, 14; palatal length, 15.6; length of upper molar series, 5.3; length of mandible, angle to alveolus of incisor, 18.6.

a.—phæus (*Ototylomys*), Merr., Proc. Wash. Acad. Scien., III, 1901,
   p. 563.
APAZOTE RAT.
   Type locality. Apazote, near Yohaltun, State of Campeche, Mexico.
   Genl. Char. Similar to *O. phyllotis*, but smaller; tail shorter; colors darker.
   Color. Above dark grayish brown, mixed with black, hinder part of dorsal region dusky; anterior part of back and sides grayish brown, tinged with fulvous; arms white; legs grayish brown; hands and feet white; tail dark brown above, yellow beneath; ears flesh color at base, rest black.
   Measurements. Total length, 266; tail, 136; hind foot, 26.5. (ex Type.)

The Genus HOLOCHILUS introduced here by Miller and Rehn, Syst. Res. N. Am. Mamm., 1902, p. 89, to follow TYLOMYS, has no species north of the Isthmus of Panama, and *H. pilorides*, Pallas, is a native of the island of Ceylon.

SIGMODON contains the well-known Cotton Rats of the Southern States and Mexico. They have a coarse, grizzled coat, harsh in feeling, and a bicolor tail. There is a very close general resemblance between the longest known species, *S. hispidus*, and the several races that have been separated from it, and they are not always easy to be distinguished, especially if the locality of a specimen is unknown. They are rather short, thickset animals, about half the size of a fully grown house rat, and are often found in large colonies in the localities they frequent.

47. Sigmodon.   Cotton Rats.

$$I.\frac{1-1}{1-1}; \; M.\frac{3-3}{3-3} = 16.$$

V. Bailey. *Synopsis of North America Species of Sigmodon*, Proc. Biol. Soc. Wash., xv, 1902, pp. 101–116.

Sigmodon Say & Ord, Journ. Acad. Nat. Scien. Phil., 1825, p. 352, pl. xxii, figs. 5–8.

Skull short and wide, length less than twice the zygomatic width; rostrum short, swollen; superior outline of skull arched; pointed process of lamellar plate of maxillary nearly dividing the lower part of the antorbital foramen from the upper; an azygos median process on palatal arch; prominent bead on the supraorbital border extending obliquely backwards to occiput; audital bullæ small; upper molars three-rooted; front lower molar four-rooted; second and third lower molars three-rooted; sometimes minute accessory fangs are present; upper molars with two reëntrant folds; the front one has two similar interior folds, the others only one each; front lower molar has two exterior and three interior reëntrant folds; the last two lower molars have generally but one reëntrant lobe on each exterior and interior side; pelage coarse, bristly; form stout; tail generally shorter than the body; ears large; front feet small; hind feet very long, soles naked.

KEY TO THE SPECIES AND SUBSPECIES.

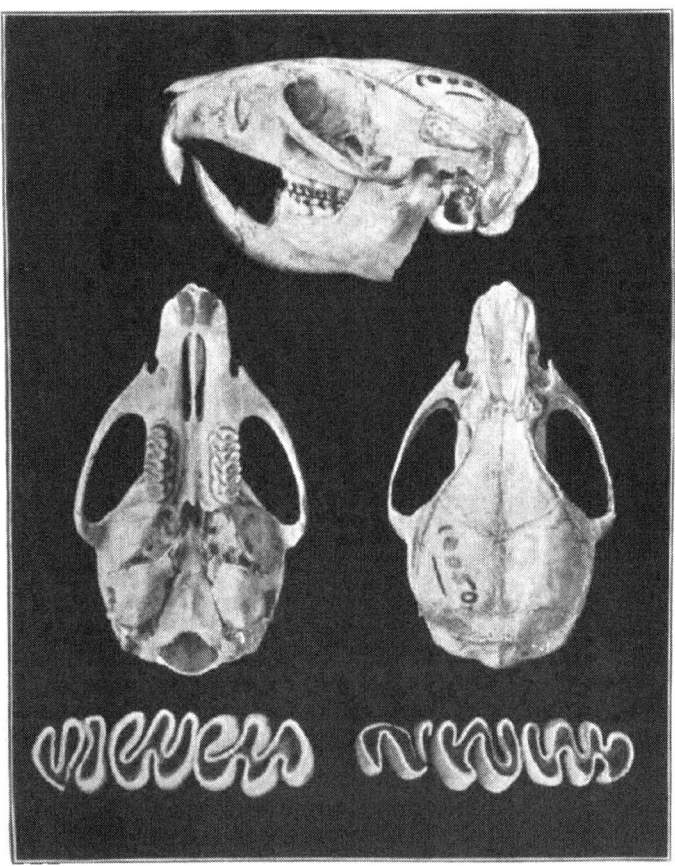

FIG. 38. SIGMODON H. BORUCÆ.
No. 10050 Am. Mus. Nat. Hist. Coll.   Twice nat. size.

UPPER TOOTH ROW.	LOWER TOOTH ROW
Enlarged 7 times.	Enlarged 7 times.

201. alleni (*Sigmodon*), Bailey, Proc. Biol. Soc. Wash., xv, 1902, p. 112.
ALLEN'S COTTON RAT.

Type locality. San Sebastian, Mascota, State of Jalisco, Mexico.

Geogr. Distr. Western part of State of Jalisco, and southern part of Province of Tepic, Mexico.

Genl. Char. Hands, feet, and tail slender; colors bright.

Color. Above dull tawny; beneath white or buffy; hands and feet yellowish brown; tail above black, beneath brownish.

Measurements. "Average of 8 adults: total length, 244; tail, 112; hind foot, 31.6. Skull of type: basal length, 29.3; nasals, 13; zygomatic breadth, 19.3; mastoid breadth, 13; alveolar length of upper molar series, 6.4." (Bailey, l. c.)

hispidus borucæ.

  borucæ (*Sigmodon*), Allen, Bull. Amer. Mus. Nat. Hist., 1897, p. 40.

BORUCA COTTON RAT.

Type locality. Boruca, Costa Rica.

Geogr. Distr. Costa Rica, Central America.

Color. Above yellowish brown, tinged with chestnut and lined with black, paler on the sides; under parts grayish white; limbs like the body; feet grayish brown; tail naked, blackish above, paler beneath.

Measurements. Total length, 275; tail, 115; hind foot, 30. Skull: basal length, 29; zygomatic breadth, 18; length of nasals, 12; length of upper tooth row, 5.5.

hispidus chiriquensis.

borucœ chiriquensis (*Sigmodon*), Allen, Bull. Am. Mus. Nat. Hist., 1904, p. 68.

BOQUERON COTTON RAT.

Type locality. Boqueron, Chiriqui, Panama.

Genl. Char. Similar to *S. h. saturatus*, but bullæ more pyriform and postpalatal opening much broader.

Color. Like *S. h. borucœ* but darker; upper parts dark yellowish brown varied with black; under parts buffy; nose and sides of upper lip ochraceous buff.

Measurements. Total length, 280; tail vertebræ, 105; hind foot, 32; ear, 20.

hispidus saturatus (*Sigmodon*), Bailey, Proc. Biol. Soc. Wash., xv, 1902, p. 111.

TEAPA COTTON RAT.

Type locality. Teapa, State of Tabasco, Mexico.

Geogr. Distr. States of Chiapas, Tabasco, and southern Vera Cruz, Mexico.

Genl. Char. Size of *S. h. berlandieri*; tail shorter.

Color. Above dark reddish brown; beneath dull cinnamon brown, sometimes whitish; hands and feet yellowish brown; tail above black, lighter below.

Measurements. Total length, 250; tail vertebræ, 103; hind foot, 31.2. Skull of type: basal length, 29.5; nasals, 12.7; zygomatic breadth, 19; mastoid breadth, 13.5; alveolar length of upper molar series, 6. (Bailey, l. c.)

hispidus furvus Bangs, Bull. Mus. Comp. Zoöl., 1903, XXXIX, p. 158.

CEIBA COTTON RAT.

Type locality. Ceiba, Honduras.

Genl. Char. Similar to *S. h. saturatus*, but darker. Skull with audital bullæ wider and flatter, basioccipital longer and narrower.

Color. Upper parts between mummy brown and burnt umber, varied with brownish, black-tipped hairs, redder on rump; under parts ochraceous; feet and hands dark brown; tail black, the under side slightly paler; ears blackish.

Measurements. Total length, 265; tail vertebræ, 105; hind foot, with claw, 32; ear from notch, 18. Skull: basal length, 31.4;

occipito-nasal length, 35.8; zygomatic width, 20.4; length of palate to palatal notch, 16.2; length of upper molar series, 6; length of single half mandible, 20.

hispidus toltecus.

>*toltecus (Hesperomys)*, Sauss., Rev. Mag. Zoöl., 2me Sér., 1860, p. 98.

TOLTEC COTTON RAT.

Type locality. "Cordilleras of Vera Cruz," Mexico.

Geogr. Distr. Alta Mira, State of Tamaulipas, southward to Orizaba, State of Vera Cruz, Mexico.

Genl. Char. Tail rarely equaling head and body; similar to *S. hispidus*; soles and tail naked.

Color. Above mixed black and yellowish brown; flanks paler, the brown color predominating; chin and under parts grayish white; tail bicolor, above blackish, paler beneath; feet grayish brown.

Measurements. Total length, 207–235; tail, 100–146; hind foot. 17–28. Skull: basal length, 32; zygomatic breadth, 21; mastoid breadth, 15; length of nasals, 14; length of upper tooth row, 6.7.

hispidus microdon (Sigmodon), Bailey, Proc. Biol. Soc. Wash., xv, 1902, p. 111.

SMALL-TOOTHED COTTON RAT.

Type locality. Puerto Morelos, Yucatan.

Geogr. Distr. Northern Yucatan, and State of Campeche, Mexico.

Genl. Char. Size small; teeth small; tail short.

Color. Above dark dull brown; beneath grayish white or buffy; tail above black, beneath brownish.

Measurements. Total length, 243; tail, 96; hind foot, 32. Skull of type: basal length, 28.4; nasals, 12.5; zygomatic breadth, 18; mastoid breadth, 13; alveolar length of upper molar series, 5.5. (Bailey, l. c.)

hispidus baileyi.

>*baileyi (Sigmodon)*, Allen, Bull. Am. Mus. Nat. Hist., 1903, p. 601.

BAILEY'S COTTON RAT.

Type locality. La Cienega de las Vacas, State of Durango, Mexico. Altitude, 8,500 feet.

Genl. Char. Similar to *S. h. major*, but smaller.

Color. Upper parts gray brown, sides tinged with buff, with the long hairs tipped with white or black; under parts white, basal part of hairs plumbeous; side of nose ochraceous buff; base of tail suffused with cinnamon buff; feet pale buffy gray; tail blackish brown above, pale buffy gray beneath; ears dark gray.

Measurements. Total length, 198; tail, 90; hind foot, 25; ear from notch, 18. Skull: total length, 31.5; Hensel, 27.3; length of nasals, 12.5; zygomatic width, 18.3; mastoid width, 13.3; alveolar length of upper molar series, 5.6.

hispidus mascotensis.
  mascotensis (*Sigmodon*), Allen, Bull. Amer. Mus. Nat. Hist., 1897, p. 54.
  colimæ Allen, Bull. Am. Mus. Nat. Hist., 1897, p. 55.
MASCOTA COTTON RAT.
  Type locality. Mineral San Sebastian, Mascota, State of Jalisco, Mexico.
  Geogr. Distr. Western portion of State of Jalisco to southern part of State of Oaxaca, Mexico.
  Genl. Char. Larger than *S. h. berlandieri*, tail longer, hind foot, larger.
  Color. Above cinnamon brown; sides paler; belly white; hands and feet yellowish gray; tail above dark brown, sides and beneath paler.
  Measurements. Total length, 272; tail vertebræ, 117; hind foot, 32. Skull: total length, 34.5; basal length, 30.4; zygomatic breadth, 19.5; mastoid breadth, 13.2; interorbital constriction, 4.8; length of nasals, 13.5; length of upper molar series, 6.7.

hispidus eremicus (*Sigmodon*), Mearns, Proc. U. S. Nat. Mus., 1897, p. 504.
WESTERN DESERT COTTON RAT.
  Type locality. Cienega Well, Colorado River, State of Sonora, Mexico, thirty miles south of Monument No. 204, Mexican boundary line.
  Geogr. Distr. Western Desert Tract, Lower Colorado River, State of Sonora, Mexico.
  Genl. Char. Nasals spatulate at extremity; coloring yellowish instead of grayish.
  Color. Yellowish gray, the sides and rump tinged with ochraceous; under surface white; feet grayish white; tail inclined to blackish above.
  Measurements. Total length, 280; tail vertebræ, 128; hind foot, 34; ear, 15. Skull: basal length, 30.6; zygomatic breadth, 20.3; mastoid breadth, 14; length of nasals, 13.3; length of upper molar series, 6.6.

hispidus berlandieri.
  berlandieri (*Sigmodon*), Baird, Proc. Acad. Nat. Scien. Phila., VII, 1855, p. 333.

hispidus pallidus Mearns, Proc. U. S. Nat. Mus., 1898, p. 504.
    Elliot, Syn. N. Am. Mamm., 1901, p. 145.
BERLANDIER'S COTTON RAT.
    Type locality. Rio Nazas, State of Coahuilla, Mexico.
    Geogr. Distr. Brownsville to El Paso, Texas, and Carlsbad, New
Mexico; south to southern part of State of Jalisco, Mexico.
    Genl. Char. Smaller than *S. hispidus texensis*; ears larger, color
paler.
    Color. Above buffy gray, mixed with black; under parts white;
feet grayish white; tail dusky brownish above, grayish white beneath.
    Measurements. Total length, 242; tail vertebræ, 103; hind foot,
30; ear, 14. Skull: basal length, 30.5; zygomatic breadth, 19; mastoid breadth, 13.4; length of nasals, 12.3; length of upper molar
series, 6.5.

hispidus arizonæ (*Sigmodon*), Mearns, Bull. Am. Mus. Nat. Hist.,
    1890, p. 287. Elliot, Syn. N. Am. Mamm., 1901, p. 144.
ARIZONA COTTON RAT.
    Type locality. Fort Verde, Yavapai County, Arizona.
    Geogr. Distr. Arizona into State of Sonora and Lower California,
Mexico.
    Genl. Char. Larger than *S. hispidus*; ears larger; tail longer,
colors paler.
    Color. Above light yellowish brown, mixed with ashy and lined
with black; under parts white; tail dusky above, whitish below.
    Measurements. Average total length, 320; tail vertebræ, 121;
hind foot, 35-36. Skull: total length, 40; zygomatic breadth, 23;
length of nasals, 15.9; length of upper molar series, 7.1.

hispidus major (*Sigmodon*), Bailey, Proc. Biol. Soc. Wash., xv,
    1902, p. 109.
LARGE COTTON RAT.
    Type locality. Sierra de Choix, 50 miles northeast of Choix, State
of Sinaloa, Mexico.
    Geogr. Distr. Province of Tepic to southern part of State of
Sonora, Mexico.
    Genl. Char. Very large; feet stout. Skull heavily ridged in adults;
interparietal strap-shaped, ends rounded; nasals notched posteriorly;
audital bullæ large, elongated.
    Color. Above light brownish gray; nose yellowish; beneath
white; hands and feet light gray; tail above blackish, beneath dark
gray.
    Measurements. Total length, 365; tail, 156; hind foot, 40.5.
Skull: basal length, 36; nasals, 16; zygomatic breadth, 23.5; mastoid

breadth, 16.4; alveolar length of upper molar series, 7.3. (Bailey, l. c.)

hispidus tonalensis (*Sigmodon*), Bailey, Proc. Biol. Soc. Wash., xv, 1902, p. 109.

TONALA COTTON RAT.

Type locality. Tonala, State of Chiapas, Mexico.

Geogr. Distr. Eastern part of State of Oaxaca, into western part of State of Chiapas.

Genl. Char. Similar to *S. mascotensis*, but larger.

Color. Above yellowish brown; beneath creamy white; feet and tail dull brownish gray.

Measurements. Total length, 350; tail, 166; hind foot, 41. Skull: basal length, 34.5; nasals, 15; zygomatic breadth, 22; mastoid breadth, 14.5; alveolar length of upper molar series, 6.8.

FIG. XXXVIII. SIGMODON H. INEXORATUS. OCOTLAN COTTON RAT.

hispidus inexoratus (*Sigmodon*), Elliot, Pub. Field Columb. Mus., III, 1903, p. 144. Zoölogy.

OCOTLAN COTTON RAT.

Type locality. Ocotlan, State of Jalisco, Mexico.

Genl. Char. Size large. Skull with infraorbital foramina oblong and narrow; frontal region broad.

Color. Above mixed creamy buff and black; sides paler cream buff, as are also the arms, hands, and thighs; hind feet gray; under parts grayish white; tail nearly naked, blackish above, whitish beneath; ears dark brown or blackish.

Measurements. Total length, type, 310; tail vertebræ, 146; hind foot, 36. Skull: total length, 37; Hensel, 31; zygomatic width, 20; mastoid width, 15; median palatal length, 8; length of incisive foramen, 7; upper tooth row, 7; lower tooth row, 7.

202. minimus (*Sigmodon*), Mearns, Proc. U. S. Nat. Mus., XVII,
1894, p. 130. Elliot, Syn. N. Am. Mamm., 1901, p. 146.
LEAST COTTON RAT.

Type locality. Northern boundary line between New Mexico
and Mexico, 100 miles west of the Initial Monument in Grant County,
on the west bank of the Rio Grande.

Geogr. Distr. Northern Mexico and southern New Mexico and
Arizona.

Genl. Char. Darker in color and hair more bristly than *S. h.
texensis*; ears, feet, and tail thickly covered with hair; tail indistinctly bicolor.

Color. Above mixed brown, gray, and black; beneath pale buff;
tail brownish black; feet yellowish gray.

Measurements. Total length, 223; tail, 94; hind foot, 28; ear
above crown, 14. Skull of type: basal length, 28.5; nasals, 11.3;
zygomatic breadth, 19; mastoid breadth, 14; alveolar length of upper
molar series, 5.9.

203. leucotis (*Sigmodon*), Bailey, Proc. Biol. Soc. Wash., XV, 1902.
p. 115.
WHITE-EARED COTTON RAT.

Type locality. Valparaiso Mountains, State of Zacatecas, Mexico.
Altitude, 8,700 feet.

Geogr. Distr. Known only from the type locality.

Genl. Char. Size medium; tail short, hairy. Skull heavily ridged
along sides; interparietal narrow with a median suture; no median
ridge on supraoccipital; nasals short; bullæ oval.

Color. Above dull brownish gray; under parts whitish; hands
and feet brownish gray; tail black, brownish at base beneath; ears
whitish gray.

Measurements. Average of 8 adult topotypes: total length, 234;
tail vertebræ, 91; hind foot, 29.4. Skull of type: basal length, 31;
nasals, 11.4; zygomatic breadth, 20.5; mastoid width, 15; alveolar
length of upper molar series, 6. (Bailey, l. c.)

204. ochrognathus (*Sigmodon*), Bailey, Proc. Biol. Soc. Wash., XV,
1902, p. 115.
OCHRACEOUS-FACED COTTON RAT.

Type locality. Chisos Mountains, Brewster County, Texas. Altitude, 8,000 feet.

Geogr. Distr. Chisos Mountains, Texas, to Parral, State of
Chihuahua, Mexico.

Genl. Char. Size small; tail long, hairy. Skull: interparietal with
a posterior indentation; nasals short, wide, and truncate posteriorly.

Color. Above yellowish gray, tinged about ears, face, and rump with ochraceous; nose, orbital ring, and base of tail bright ochraceous beneath white; hands and feet buffy gray; tail above blackish, beneath buffy gray.

Measurements. Total length, 260; tail, 117; hind foot, 29. Skull of type: basal length, 28; nasals, 11.6; zygomatic breadth, 19; mastoid breadth, 13; alveolar length of upper molar series, 5.5. (Bailey, l. c.)

205. alticola (*Sigmodon*), Bailey, Proc. Biol. Soc. Wash., xv, 1902, p. 116.
ALPINE COTTON RAT.

Type locality. Cerro San Felipe, State of Oaxaca, Mexico. Altitude, 10,000 feet.

Geogr. Distr. Mountains of State of Oaxaca, Mexico.

Genl. Char. Size medium; tail hairy; pelage long and soft. Skull: similar to that of *S. leucotis*; lateral pits of palate very shallow.

Color. Above dark ochraceous or light umber brown; beneath pale cinnamon brown; hands and feet yellowish gray; tail black, yellowish brown at base beneath; ears gray.

Measurements. Total length, 230; tail, 101; hind foot, 28.5. Skull of type: basal length, 28.5; nasals, 11; zygomatic breadth, 17.8; mastoid breadth, 13.5; alveolar length of upper molar series, 6.3. (Bailey, l. c.)

a.—amoles (*Sigmodon*), Bailey, Proc. Biol. Soc. Wash., xv, 1902, p. 116.
AMOLES COTTON RAT.

Type locality. Pinal de Amoles, State of Queretaro, Mexico.

Geogr. Distr. Known only from type locality.

Genl. Char. Similar to *S. alticola.* Skull wider, smaller bullæ, and lateral pits of palate deeper, and interparietal without median division.

Color. Like *S. alticola*, but upper parts duller and less tawny.

Measurements. Total length, 252; tail vertebræ, 105; hind foot, 29.5. Skull of type: basal length, 29.5; nasals, 5.7; zygomatic breadth, 19.6; mastoid breadth, 14; alveolar length of upper molar series, 6.3. (Bailey, l. c.)

206. austerulus (*Sigmodon*), Bangs, Bull. Mus. Comp. Zoöl., xxxix, 1902, p. 32.
VOLCAN DE CHIRIQUI COTTON RAT.

Type locality. Volcan de Chiriqui, Chiriqui, Panama. Altitude, 10,000 feet

Genl. Char. Similar to *S. h. boruca*, tail longer, well haired, color paler.

Color. Above cinnamon brown, dorsal region darker; rump shaded with russet; under parts white washed with pale buff; tail above dusky, beneath gray; hands and feet yellowish gray; ears gray.

Measurements. Total length, 260; tail vertebræ, 120; hind foot, 32; ear, 17. Skull: mastoid width, 14.8; length of upper molar series, 6; length of mandible, 19.2.

207. fulviventer (*Sigmodon*), Allen, Bull. Amer. Mus. Nat. Hist., 1889, p. 180.

FULVOUS-BELLIED COTTON RAT.

Type locality. Zacatecas, State of Zacatecas, Mexico.

Geogr. Distr. States of Zacatecas and Durango, Mexico.

Genl. Char. Similar to *S. hispidus*, but paler, and under parts ochraceous buff instead of white.

Color. Above yellowish brown mixed with black, darkest on median line; under parts ochraceous buff; arms and legs buffy; hands and feet yellowish gray; tail blackish, mixed with gray.

Measurements. Total length, 270; tail vertebræ, 108; hind foot, 33. (ex Type.) Skull of type; base broken: total length over incisors, 35.5; nasals, 13; zygomatic breadth, 21.5; alveolar length of upper molar series, 6.3.

208. melanotis (*Sigmodon*), Bailey, Proc. Biol. Soc. Wash., xv, 1902, p. 114.

BLACK-EARED COTTON RAT.

Type locality. Patzcuaro, State of Michoacan, Mexico. Altitude, 7,000 feet.

Geogr. Distr. Known from type locality only.

Color. Above dark ochraceous, lined with black; beneath dark rusty ochraceous; hands and feet yellowish brown; tail black, yellowish beneath at base; ears black.

Measurements. Type: total length, 275; tail, 100; hind foot, 31. Skull: basal length, 31.4; nasals, 12.2; zygomatic breadth, 20.5; mastoid breadth, 14.5; alveolar length of upper molar series, 6.3. (Bailey, l. c.)

The Rice Rats of the genus ORYZOMYS are, in a number of the species, among the largest of the Muridæ, with long, scantily haired tails and long hind feet. They are mostly dwellers of tropical regions, and in the United States are found only in the coast region of the eastern and southeastern portions; but in Mexico some are met with at 10,000 feet elevation, and these forms are provided with a woolly covering to protect them from the cold of such high altitudes.

48. Oryzomys. Rice Rats.

$$I.\frac{1-1}{1-1};\ M.\frac{3-3}{3-3} = 16.$$

C. H. Merriam. *Synopsis of the Rice Rats (Genus Oryzomys) of the United States and Mexico*, Proc. Wash. Acad. Scien., III, 1901, p. 273.

Oryzomys Baird, N. Am. Mamm., 1857, p. 458. Type *Mus palustris* Harlan.

Oligoryzomys Bangs, Proc. N. Eng. Zoöl. Club, I, 1900, p. 94, pl. I, fig. 2.

"Form rat-like; ears nearly buried in the fur; hairs of body coarse; tail longer than head and body; the hairs longest on the

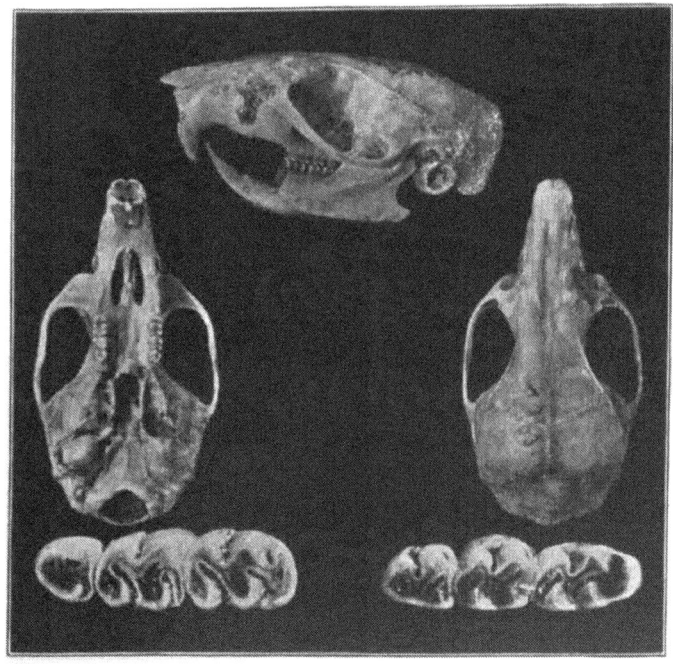

FIG. 39. ORYZOMYS COSTARICENSIS.
No. 7963 Am. Mus. Nat. Hist. Coll.   Twice nat. size.

UPPER TOOTH ROW.	LOWER TOOTH ROW.
Enlarged 10 times.	Enlarged 10 times.

under surface; hind feet very long; soles naked, with six tubercles,
all very small except the posterior, which is very long and narrow;
upper margin of the orbit raised into a compressed crest, as in *Sigmodon*." (Baird, l. c.)

KEY TO THE SPECIES.

A. Fur coarse; tail longer than head and body;
    hind feet very long. "Upper margin of
    orbit raised into a compressed crest."

  a. Under parts white.   PAGE

    a.' Above yellowish fulvous and black.........*O. nelsoni* 235

    b.' Above ochraceous and black.

       a." Tail above dusky, whitish below.......*O. albiventer* 236

       b." Tail above brown, paler brown

           beneath......................*O. crinitus aztecus* 245

    c.' Above ochraceous fulvous...........*O. yucatanensis* 236

  b. Under parts grayish white.

    a.' Above tawny ochraceous lined with

       black .... *O. panamensis* 241

    b.' Above mottled blackish brown.........*O. talamancæ* 241

    c.' Above rusty brown and black..............*O. alfari* 242

  c. Under parts whitish, or tinged with salmon
    or buff.

    a.' Above pale grayish.....................*O. peninsulæ* 236

    b.' Above reddish brown and black.............*O. couesi* 236

    c.' Above mixed deep rufous and black......*O. palatinus* 237

    d.' Above dark rufous brown................*O. hylocetes* 237

    e.' Above grizzled fulvous and black.........*O. rhabdops* 237

    f.' Above dark fulvous and black, small.....*O. chapmani* 238

    g.' Above dark fulvous and black, large....*O. c. caudatus* 238

    h.' Above fulvous and black, paler..........*O. c. dilutior* 238

    i.' Above blackish and fulvous...........*O. c. saturatior* 239

    j.' Above chestnut brown.....................*O. bulleri* 239

    k.' Above "deep dull fulvous"................*O. rufus* 239

    l.' Above bright fulvous; beneath tinged

       with fawn...............................*O. fulgens* 240

    m.' Above dark fulvous; under parts grayish

       white.................................. *O. molestus* 240

    n.' Above pale fulvous and black.........*O. zygomaticus* 241

    o.' Above grayish bistre...................*O. cozumelæ* 241

  d. Under parts buffy white.

    a.' Above fulvous brown lined with black...*O. mexicanus* 242

209. nelsoni (*Oryzomys*), Merr., Proc. Biol. Soc. Wash., XII, 1898, p. 15.

MARIA MADRE ISLAND RICE RAT.

Type locality. Maria Madre Island, Tres Marias Islands, State of Jalisco, Mexico.

Geogr. Distr. Tres Marias Islands, State of Jalisco, Mexico.

Genl. Char. Larger than *O. mexicanus*; tail very long, naked.

Color. Above yellowish fulvous, lined with black on head and back; flanks and thighs buffy ochraceous; under parts white; tail dark brown, yellowish beneath at base.

Measurements. Total length, 342; tail vertebræ, 190; hind foot, 38.

210. albiventer (*Oryzomys*), Merr., Proc. Wash. Acad. Scien., III, 1901, p. 279.
WHITE-BELLIED RICE RAT.

Type locality. Ameca, State of Jalisco, Mexico.
Geogr. Distr. State of Jalisco, Mexico; extent of range unknown.
Genl. Char. Size large; ear short; under parts white; molars large.
Color. Above ochraceous, lined with black; under parts white; tail dusky above, whitish below.
Measurements. Total length, 295; tail vertebræ, 162; hind foot, 37.3.

211. yucatanensis (*Oryzomys*), Merr., Proc. Wash. Acad. Scien., III, 1901, p. 294.
CHICHEN ITZA RICE RAT.

Type locality. Chichen Itza, Yucatan, Mexico.
Geogr. Distr. Yucatan, Mexico.
Genl. Char. Similar to *O. melanotis*, but without white cheek patch. Skull: superciliary beads distinct; incisive foramina short; molars small.
Color. Above ochraceous fulvous, extending to and including sides of face to nose; head and back lined with black; beneath white; tail yellowish beneath, dusky above; ears fulvous brown.
Measurements. Total length, 235; tail vertebræ, 119; hind foot, 32.

212. peninsulæ (*Oryzomys*), Thomas, Ann. Mag. Nat. Hist., 6th Ser., XX, 1897, p. 548.
LOWER CALIFORNIA RICE RAT.

Type locality. Santa Anita, Lower California, Mexico.
Geogr. Distr. Cape Region, Lower California, Mexico.
Genl. Char. Size large; fur woolly; tail thinly haired.
Color. Head and back anteriorly pale grayish, grading into dull fulvous on rump; under parts whitish; hands and feet silvery white; tail brown above, whitish below; ear pale brown.
Measurements. Total length, 298; tail vertebræ, 150; hind foot, 34; ear, 18. Skull: basilar length, 27.3; greatest breadth, 18.7; length of nasals, 13; width, 3.9; length of upper molar series, 5; length of palatine foramen, 7; width, 3.

213. couesi (*Hesperomys*), Alston, Proc. Zoöl. Soc., 1876, p. 756.
couesi (*Oryzomys*), Thomas, Ann. Mag. Nat. Hist., 6th Ser., 1893, p. 403.
COUES' RICE RAT.

Type locality. Coban, Guatemala.
Geogr. Distr. State of Chiapas into Guatemala, Central America.

Genl. Char. Ears small, rounded; hind feet large; tail long, scaly; pelage harsh.

Color. Above reddish brown mixed with black; flanks pale rufous; under parts whitish; breast washed with rufous; tail almost unicolor, paler beneath; hands and feet grayish.

Measurements. Total length, 276; tail vertebræ, 149; hind foot, 28.

214. palatinus (*Oryzomys*), Merr., Proc. Wash. Acad. Scien., III, 1901, p. 290.
TABASCAN RICE RAT.

Type locality. Teapa, State of Tabasco, Mexico.

Genl. Char. Size small; similar to *O. chapmani*, but more reddish generally. Skull light, slender; supraorbital bead barely perceptible; outer sides of zygomata parallel; incisive foramina separated by a broad septum.

Color. Above mixed deep rufous and black; sides and cheeks pale rufous; beneath whitish; hind feet whitish; tail dark above, paler beneath.

Measurements. Total length, 209; tail vertebræ, 106; hind foot, 25.

215. hylocetes (*Oryzomys*), Merr., Proc. Wash. Acad. Scien., III, 1901, p. 291.
MARSH RICE RAT.

Type locality. Chicharras, State of Chiapas, Mexico.

Genl. Char. Small; ears large; similar to *O. palatinus*. Skull: nasals flat, truncate anteriorly.

Color. Above dark rufous brown, beneath whitish; hind feet dark; ears and tail blackish.

Measurements. Total length, 217; tail vertebræ, 118; hind foot, 27.

216. rhabdops (*Oryzomys*), Merr., Proc. Wash. Acad. Scien., III, 1901, p. 291.
STRIPED-FACE RICE RAT.

Type locality. Calel, Guatemala. Altitude, 10,000 feet.

Geogr. Distr. State of Chiapas, Mexico, into Guatemala, Central America.

Genl. Char. Medium size; ears large. Skull: zygomata "squarish," spreading.

Color. Above grizzled fulvous and black; sides and cheeks paler; beneath whitish; side of nose pale fulvous; streak from nose to eye blackish; tail dusky brown above, yellowish beneath; ears blackish.

Measurements. Total length, 255; tail vertebræ, 141; hind foot, 29.5.

217. chapmani (*Oryzomys*), Thomas, Ann. Mag. Nat. Hist., 7th
        Ser., 1, 1898, p. 179.
CHAPMAN'S RICE RAT.
        Type locality. Jalapa, State of Vera Cruz, Mexico.
        Geogr. Distr. State of Vera Cruz, Mexico.
        Genl. Char. Size small. Skull small; nasals narrow posteriorly;
palatal foramina reaching nearly to anterior margin of the first upper
molar.
        Color. Upper parts dark fulvous and black, blacker on back;
beneath whitish, strongly suffused with slate; ears shining black;
wrists and ankles suffused with smoky brown; tail bicolor, blackish
above, paler below.
        Measurements. Total length, 121; tail vertebræ, 116; hind foot,
24; ear, 19. Skull: basal length, 21.6; basilar length, 19.8; greatest
breadth, 13.8; interorbital constriction, 4.3; length of nasals, 10.4;
breadth across squamosals, 11.4; palatal length, 11.5; length of
upper molar series, 3.7.

a.—caudatus (*Oryzomys*), Merr., Proc. Wash. Acad. Scien., III,
        1901, p. 289.
LONG-TAILED RICE RAT.
        Type locality. Comaltepec, State of Oaxaca, Mexico.
        Genl. Char. "Similar to *O. chapmani*, but larger, tail much
longer; color slightly darker. Skull larger and heavier, with longer
rostrum and broader nasals." (Merr., l. c.)
        Color. Above dark fulvous and black, darkest on dorsal region;
sides paler; under parts grayish white; tail above blackish brown,
beneath paler, becoming grayish towards base; hands and feet flesh
color; ears black.
        Measurements. Total length, 257; tail vertebræ, 141; hind
foot, 30.

b.—dilutior (*Oryzomys*), Merr., Proc. Wash. Acad. Scien., III,
        1901, p. 290.
PALE RICE RAT.
        Type locality. Huauchinango, State of Puebla, Mexico.
        Genl. Char. "Similar to *O. chapmani*, but slightly paler; hind
foot longer. Rostrum and nasals broader; rostrum more swollen at
base; anterior root of zygoma heavier."
        Color. Above rufous and black, darkest on dorsal line; sides
inclining to yellowish brown; under parts white, tinged with buff;

tail black above, pale brown beneath; hands and feet flesh color, covered sparsely with white hairs; ears black.

Measurements. Total length, 223; tail vertebræ, 117; hind foot, 28.

c.—saturatior (*Oryzomys*), Merr., Proc. Wash. Acad. Scien., III, 1901, p. 290.

TUMBALA RICE RAT.

Type locality. Tumbala, State of Chiapas, Mexico.

Genl. Char. Similar to *O. chapmani*, but decidedly darker, particularly on top of head and middle of back, which are blackish, slightly "peppered" with fine points of fulvous; under parts soiled buffy, in some specimens salmon; cheeks fulvous; ears, hind feet, and tail blackish.

Measurements. Total length, 218; tail vertebræ, 120; hind foot, 25.5.

218. bulleri (*Oryzomys*), Allen, Bull. Amer. Mus. Nat. Hist., 1897, p. 53.

BULLER'S RICE RAT.

Type locality. Valle de Banderas, Territorio de Tepic, State of Jalisco, Mexico.

Geogr. Distr. States of Jalisco and Colima, Mexico; limits unknown.

Genl. Char. Similar to *O. couesi*, but darker. Rostrum slender.

Color. Upper parts chestnut brown mixed with black; side ochraceous; under parts buffy white; ears brown; tail above brownish, yellowish below; hands and feet whitish.

Measurements. Total length, 242; tail vertebræ, 127; hind foot, 27; ear, 11. Skull: total length, 27.8; basal length, 23; zygomatic breadth, 15; interorbital constriction, 4.5; length of nasals, 10.2; palatal length, 11; width of braincase, 12.5; length of upper tooth row, 4.6.

219. rufus (*Oryzomys*), Merr., Proc. Wash. Acad. Scien., III, 1901, p. 287.

RUFOUS RICE RAT.

Type locality. Santiago, Territorio de Tepic, Mexico.

Genl. Char. Similar to *O. bulleri*, but smaller; pelage more red.

Color. Above deep fulvous, extending to head, mixed with scattering black hairs; under parts soiled white; tail dusky above, paler below.

Measurements. Total length, 250; tail vertebræ, 136; hind foot, 28.

220. fulgens (*Oryzomys*), Thomas, Ann. Mag. Nat. Hist., 6th Ser.,
1893, p. 403.
SHINING RICE RAT.

Type locality. Unknown. "Mexico."

Genl. Char. Size large; fur coarse, woolly; ears small; tail long.

Color. Above bright fulvous; under parts whitish, tinged with
fawn; outer side of limbs like back, inner side whitish; hands and
feet silvery fawn; tail blackish above, yellowish below.

Measurements. Total length, 301; tail, 151; hind foot, 37.5; ear
from notch, 13.3. Skull: zygomatic breadth, 17.8; length of nasals,
13.2; interorbital constriction, 4.8; length of palatine foramen, 7.3;
length of upper molar series, 5.2. (Thomas, l. c.)

FIG. XXXIX. ORYZOMYS MOLESTUS. OCOTLAN RICE RAT.

221. molestus (*Oryzomys*), Elliot, Pub. Field Columb. Mus., II,
1903, p. 145. Zoölogy.
OCOTLAN RICE RAT.

Type locality. Ocotlan, State of Jalisco, Mexico.

Genl. Char. Size large; ears small; tail very long; color beneath
uniform. Skull about half as broad as long; supraorbital beads diverging posteriorly from least interorbital width in almost straight lines,
unlike those of *O. fulgens*; palatal arch with an azygos central point;
palatine foramina very long and broad.

Color. Above dark fulvous; flanks pale buff; forehead darker
than back; under parts grayish white, with a nearly pure white
pectoral spot; limbs like back; hands and feet grayish; tail nearly
naked, pale brown above, lighter beneath; ears pale brown.

Measurements. Total length, 325; tail vertebræ, 170; hind foot,
38. Skull: greatest length, 35; Hensel, 28; zygomatic width, 18;
mastoid width, 14; length of nasals, 14; width of rostrum, 6; interorbital constriction, 6; palatal length, 7; length of upper tooth row, 6.

222. zygomaticus (*Oryzomys*), Merr., Proc. Wash. Acad. Scien., III, 1901, p. 285.
GUATEMALAN RICE RAT.
Type locality. Nenton, Guatemala.
Geogr. Distr. Guatemala.
Genl. Char. Size medium; hind feet large. Skull: superciliary ridges strongly developed and everted; zygomata curving downward below level of posterior root.
Color. Above pale fulvous, lined with black on dorsal portion; under parts white, suffused with buff.
Measurements. Total length, 290; tail vertebræ, 152; hind foot, 33.

223. cozumelæ (*Oryzomys*), Merr., Proc. Biol. Soc. Wash., XIV, 1901, p. 103.
COZUMEL ISLAND RICE RAT.
Type locality. Cozumel Island, Yucatan, Mexico.
Genl. Char. Size large, similar to *O. aquaticus.* Skull: braincase broad posteriorly, carrying lateral beads outwards.
Color. Above grayish bistre; sides and rump suffused with pale fulvous; beneath whitish to pale salmon; tail dusky above, paler beneath; ears dark brown.
Measurements. Total length, 332; tail vertebræ, 182; hind foot, 35.

224. panamensis (*Oryzomys*), Thomas, Ann. Mag. Nat. Hist., 7th Ser., VIII, 1901, p. 252.
PANAMA RICE RAT.
Type locality. Near city of Panama, Panama.
Genl. Char. Size large; tail about equal to head and body; fur soft. Skull: supraorbital ridges without heavy beads; molars small, narrow.
Color. Upper parts tawny ochraceous, lined with black; sides lighter; under parts grayish; outer sides of arms and legs gray; tail brown above, white below; hands and feet whitish; ears brown.
Measurements. Total length, 261; tail, 130; hind foot, 28.5; ear, 18. Skull: tip of nasals to front of interparietal, 28.5; zygomatic breadth, 16.4; interorbital constriction, 5; length of nasals, 12.5; length of upper molar series, 4.2.

225. talamancæ (*Oryzomys*), Allen, Proc. U. S. Nat. Mus., XIV, 1891, p. 193.
TALAMANCA RICE RAT.
Type locality. Talamanca, Costa Rica.

Geogr. Distr. Costa Rica, Central America.

Genl. Char. Ears large; soles naked; tubercles, 6; tail long as head and body.

Color. Above mixed russet and blackish brown; cheeks, sides of head, and flanks yellow brown; under parts grayish white; tail naked, above blackish, beneath dark brown; hands and feet pale yellowish gray; ears naked, blackish.

Measurements. Total length, 228.6; tail vertebræ, 114.3; hind foot, 30.8; ear from crown, 13.7. Skull: basal length, 28.5; total length, 31.2; greatest breadth, 15.8; interorbital constriction, 5.3; length of nasals, 12.7; palatal length, 16; length of upper molar series, 51.

226. alfari (*Hesperomys*), Allen, Bull. Amer. Mus. Nat. Hist., 1891, p. 214.

ALFARO'S RICE RAT.

Type locality. San Carlos, Costa Rica, Central America.

Geogr. Distr. Costa Rica, Central America.

Genl. Char. Tail longer than head and body; ears rather long; hind foot long; soles naked.

Color. Above rusty brown and black; sides of head and body more rufous; beneath ashy white; tail naked, blackish brown above, paler beneath; feet yellowish.

Measurements. Total length, 184; tail, 88.9; hind foot, 23.4; ear from crown, 10.2. Skull: occipito-nasal length, 24.5; Hensel, 18.5; zygomatic width, 13; interorbital constriction, 5; length of nasals, 10; palatal length, 10; length of upper tooth row, 3.2; length of mandible, 8; length of lower tooth row, 4.

227. mexicanus (*Oryzomys*), Allen, Bull. Am. Mus. Nat. Hist., 1897, p. 52.

TONILA RICE RAT.

Type locality. Hacienda San Marcos, Tonila, State of Jalisco, Mexico.

Geogr. Distr. State of Sinaloa, southern part, south to Isthmus of Tehuantepec, Mexico.

Genl. Char. Similar to *O. palustris*; size medium; hind feet large. Postpalatal border of nares V-shaped.

Color. Above fulvous brown, lined with black, sides paler; under parts grayish white, sometimes suffused with buff; tail naked, dusky above, paler below.

Measurements. Total length, 279; tail vertebræ, 142; hind foot, 30. Skull: total length, 27.8; basal length, 23; zygomatic width, 15; interorbital constriction, 4.5; width of braincase, 12.5; length of nasals, 10.2; palatal length, 11; length of upper tooth row, 4.6.

a.—*peragrus* (*Oryzomys*), Merr., Proc. Wash. Acad. Scien., III, 1901, p. 283.

WANDERING RICE RAT.

Type locality. Rio Verde, State of San Luis Potosi, Mexico.

Genl. Char. Similar to *O. mexicanus*, but grayer and tail longer.

Color. Above grayer than *O. mexicanus*, heavily lined with black; under parts buffy, deeper than in *O. mexicanus*; tail bicolor.

Measurements. Total length, 294; tail vertebræ, 167; hind foot, 35.

228. melanotis (*Oryzomys*), Thomas, Ann. Mag. Nat. Hist., 6th Ser., XI, 1893, p. 404.

BLACK-EARED RICE RAT.

Type locality. Mineral San Sebastian, State of Jalisco, Mexico.

Geogr. Distr. State of Jalisco, Mexico.

Genl. Char. Size small, slender; tail long, scantily haired.

Color. Upper parts grizzled rufous, brightest on rump and sides; whitish cheek patch between eye and mouth; under parts buffy white; ears black; hands and feet white; tail above blackish, beneath yellowish white.

Measurements. Total length, 224; tail, 127; hind foot, 28; ear from notch, 18. Skull: basal length, 25.1; zygomatic width, 15.2; length of nasals, 12; interorbital constriction, 5.1; width of brain-case, 12.8; length of interparietal, 3.4; breadth of interparietal, 10; palatal length, 15.5; length of palatine foramen, 5.8; length of upper molar series, 4.3.

229. rostratus (*Oryzomys*), Merr., Proc. Wash. Acad. Scien., III, 1901, p. 293.

BROAD-NOSED RICE RAT.

Type locality. Metlaltoyuca, State of Puebla, Mexico.

Geogr. Distr. Eastern Mexico, from State of Puebla to Isthmus of Tehuantepec.

Genl. Char. Similar to *O. melanotis*; pelage coarse. Skull large, long, flat; rostrum long, swollen at base; nasals broad, flat; superciliary bead moderate.

Color. Above ochraceous fulvous, lined with black; beneath buffy white; tail above dark brown, under side yellowish; ears dark brown.

Measurements. Total length, 277; tail vertebræ, 141; hind foot, 32.5.

a.—*megadon* (*Oryzomys*), Merr., Proc. Wash. Acad. Scien., III, 1901, p. 294.

TEAPA RICE RAT.

Type locality. Teapa, State of Tabasco, Mexico.

Geogr. Distr. State of Tabasco westward probably to State of Puebla.

Genl. Char. Similar to *O. rostratus*, but smaller and more red: and is distinguishable from *O. melanotis* by its darker color and larger hind foot.

Color. Above deep fulvous lined with black, darkest on median line; sides yellowish brown; under parts grayish white, with a buff tinge; tail black above, whitish beneath: thighs plumbeous; hands and feet flesh color; ears dark brown.

Measurements. Total length, 272; tail vertebræ, 140; hind foot, 31.

23'. victus (*Oryzomys*), Thomas, Ann. Mag. Nat. Hist., 7th Ser., I, 1898, p. 177.

ST. VINCENT RICE RAT.

Type locality. Island of St. Vincent, Lesser Antilles.

Genl. Char. Molars large and stout; palate ending close to posterior edge of third upper molar; braincase lengthened; parietal ridges developed.

Color. Above dark rufous; beneath buffy white; no orbital ring; tail brown above, paler below, nearly naked; hands and feet silvery white; ears brown.

Measurements. Total length, 217; tail, 121; hind foot, 26.7; ear, 14. Skull: basilar length, 21.4; basal length, 23.8; greatest breadth, 15.1; length of nasals, 11.2; interorbital constriction, 4.5; palatal length from henselion, 12.3; length of upper molar series, 4.1.

231. devius (*Oryzomys*), Bangs, Bull. Mus. Comp. Zoöl., XXXIX, 1902, p. 34.

LONELY RICE RAT.

Type locality. Boquete, Chiriqui, Panama. Altitude, 5,000 feet.

Genl. Char. Like *O. childi*, ex Colombia, but under parts white and fulvous.

Color. Upper parts russet brown; dorsal region darker; orbital region black; under side of head and neck grayish white: pectoral and ventral patch white; tail above dusky, beneath grayish; hands and feet yellowish white; ears large, black.

Measurements. Total length, 335-360; tail vertebræ, 165-195; hind foot, 33-36; ear, 22-23. Skull: basal length, 31.6; occipitonasal length, 36; zygomatic width, 18; mastoid width, 13.8; interorbital width, 5.6; length of nasals, 14.2; width of nasals, 3.8; length of palatal slits, 5.6; width of palatal slits, 2.6; length of palate to palatal notch, 15.6; upper molar series, 5.4; length of single half mandible, 20. (Bangs, l. c.)

232. crinitus (*Oryzomys*), Merr., Proc. Wash. Acad. Nat. Scien., III, 1901, p. 281.

LONG-HAIRED RICE RAT.

Type locality. Tlalpam, Federal District, Mexico.

Genl. Char. Large; hind feet long; ears short. Skull resembling that of *O. c. aztecus* Merr.

Color. Above buffy fulvous, darkest on rump, and lined with black; chin, throat, and forelegs whitish; rest of under parts buffy; tail dusky above, yellowish below.

Measurements. Total length, 307; tail vertebræ, 161; hind foot, 37.

a.—aztecus (*Oryzomys*), Merr., Proc. Wash. Acad. Scien., III, 1901, p. 282.

AZTEC RICE RAT.

Type locality. Yautepec, State of Morelos, Mexico.

Geogr. Distr. Southeastern Mexico, from States of Morelos and Puebla to Isthmus of Tehuantepec.

Genl. Char. Skull: nasals and zygomata slender, the latter bowed downward; palatal slits broadly open posteriorly. Under parts white.

Color. Above ochraceous, sparsely lined with black; under parts white; tail brownish above, paler beneath.

Measurements. Total length, 290; tail vertebræ, 154; hind foot, 35.

233. tectus (*Oryzomys*), Thomas, Ann. Mag. Nat. Hist., VIII, 1901, p. 251.

BOGAVA RICE RAT.

Type locality. Bogava, Chiriqui, Panama. Altitude, 800 feet.

Genl. Char. Size large; tail longer than head and body. Skull has the supraorbital ridges overhanging and expanded; frontal region concave; muzzle heavy; palatal foramina not reaching level of molars.

Color. Above tawny rufous, slightly lined with black; sides inclining to ochraceous; upper lip fulvous; chin white; under parts buffy; hands and feet pale buff; tail brown above, white below; ears with ochraceous tuft at base.

Measurements. Total length, 282; tail, 142; hind foot, 29.5; ear, 18. Skull: tip of nasals to back of interparietal, 33; greatest breadth, 17; interorbital constriction, 6.5; length of nasals, 11.6; palatal length, 13.8; length of upper molar series, 4.9.

234. angusticeps (*Oryzomys*), Merr., Proc. Wash. Acad. Scien., III, 1901, p. 292.

SANTA MARIA VOLCANO RICE RAT.

Type locality. Volcan Santa Maria, Guatemala. Altitude, 9,000 feet.

Geogr. Distr. Guatemala, Central America.

Genl. Char. Similar to *O. rhabdops.* Ears large. Skull long and narrow; nasals very long; no superorbital bead; teeth small.

Color. Above mixed fulvous and black; beneath buffy; tail above blackish, paler on basal half beneath; ears blackish.

Measurements. Total length, 245; tail vertebræ, 134; hind foot. 29.

235. goldmani (*Oryzomys*), Merr., Proc. Wash. Acad. Scien.. III, 1901, p. 288.

GOLDMAN'S RICE RAT.

Type locality. Coatzacoalcos, State of Vera Cruz, Mexico.

Genl. Char. Size small; ears large; tail long. Skull narrow; superciliary bead slight; nasals broad; bullæ and molars large.

Color. Above mixed fulvous and black; beneath buff; tail above dusky, beneath paler.

Measurements. Total length, 233; tail vertebræ, 124; hind foot. 30.

236. jalapæ (*Oryzomys*), Allen & Chapman, Bull. Amer. Mus. Nat. Hist., 1897, p. 206.

JALAPA RICE RAT.

Type locality. Jalapa, State of Vera Cruz, Mexico.

Geogr. Distr. State of Vera Cruz and northwestern parts of State of Puebla, Mexico.

Genl. Char. Similar to *O. mexicanus,* but buff beneath.

Color. Above dark brown tinged with yellowish and lined with black; indistinct dorsal band; chin and throat grayish white; under parts varying from pale buff or whitish buff to deep buff; tail above dark brown, below paler; hands and feet pale grayish brown.

Measurements. Total length, 278; tail vertebræ, 140; hind foot. 30; ear, 18. Skull: total length, 32; basal length, 227; zygomatic width, 16; width of braincase, 12.4; interorbital constriction, 5; length of nasals, 13.4; palatal length, 13; length of upper molar series, 4.5. (ex Type.)

a.—apatelius (*Oryzomys*), Elliot, Pub. Field Columb. Mus., III, 1904. p. 266.

DECEITFUL RICE RAT.

Type locality. San Carlos, State of Vera Cruz, Mexico.

Gen. Char. Similar to *O. jalapæ* in color on upper parts, grayish buff beneath; tail shorter, hind foot longer. Skull with longer and narrower braincase, zygomatic width less, nasals longer; anterior palatine foramina longer, mastoid width less; mandible more slender. narrower between angle and condyle.

Color. Above yellowish brown lined with black, darkest on dorsal region; chin and throat pale gray; rest of under parts gray tinged with buff; tail distinctly bicolor, above black, beneath yellowish white, naked; hands and feet buffy white; ears dark brown.

Measurements. Total length, 259; tail vertebræ, 130; hind foot, 31. Skull: occipito-nasal length, 31.5; Hensel, 29.7; interorbital constriction, 4.5; greatest zygomatic width, 16; least zygomatic width anteriorly, 13; width of braincase above zygomata, 12; length of nasals, 12; palatal length, 13.5; length of incisive foramina, 11.4; length of upper tooth row, 5; length of mandible, angle to base of incisors, 15; height at condyle, 7; length of lower tooth row, 5.

b.—rufinus (*Oryzomys*), Merr., Proc. Wash. Acad. Scien., III, 1901, p. 285.

FULVOUS RICE RAT.

Type locality. Catemaco, State of Vera Cruz, Mexico.

Genl. Char. Incisors and molars larger than those of *O. jalapæ.*

Color. Similar to *O. jalapæ,* but back and rump deep fulvous, lined with black; tail dusky all around, paler beneath; under parts buffy; hands and feet flesh color, sparsely covered with white hairs; ears blackish brown.

Measurements. Total length, 270; tail, 139; hind foot, 32.

237. teapensis (*Oryzomys*), Merr., Proc. Wash. Acad. Scien., III, 1901, p. 286.

TEAPA RICE RAT.

Type locality. Teapa, State of Tabasco, Mexico.

Geogr. Distr. State of Tabasco; possibly also State of Chiapas, Mexico.

Genl. Char. Ears small; hind feet rather short; colors dark.

Color. Above grizzled bistre; sides yellowish, lined with black; chin, throat and arms whitish; rest of under parts buffy fulvous; tail dusky, base beneath yellowish.

Measurements. Total length, 259; tail vertebræ, 179; hind foot, 32.5.

238. antillarum (*Oryzomys*), Thomas, Ann. Mag. Nat. Hist., 7th Ser., I, 1898, p. 177.

JAMAICA RICE RAT.

Type locality. Jamaica.

Geogr. Distr. Type locality only.

Genl. Char. Size of *O. couesi;* tail as long as head and body, nearly naked. Skull: supraorbital ridges well defined; interparietal narrow; nasal passing posteriorly the premaxillæ; palatal foramina compressed; palate extending beyond third upper molar.

Color. Above rufous sparsely lined with black, brightest on the rump; head suffused with grayish; under parts yellowish, hairs gray at base; tail pale brown above, lighter beneath; hands and feet whitish; ears blackish outside, yellowish inside.

Measurements. Total length, 260; tail, 130; hind foot, 28; ear, 13. Skull: basal length, 26; basilar length, 24; greatest breadth, 17; interorbital constriction, 5.2; breadth across squamosals, 12.9; length of nasals, 12.6; palatal length, 14; length of upper molar series, 4.6.

239. richmondi (*Oryzomys*), Merr., Proc. Wash. Acad. Scien., III, 1901, p. 284.
ESCONDIDO RIVER RICE RAT.
Type locality. Escondido River, 50 miles above Bluefields, Nicaragua.
Genl. Char. Size large; ear short. Rostrum long, slender.
Color. Above grizzled yellowish fulvous, lined with black; rump sometimes rusty; under parts ochraceous fulvous, palest on chin and lips; tail dark, palest beneath.
Measurements. Total length, 295; tail vertebræ, 150; hind foot. 33.5.

240. fulvescens (*Hesperomys*), Sauss., Rev. Mag. Zoöl., 2me Sér., 1860. p. 102.
FULVOUS RICE RAT.
Type locality. "Mexique." Orizaba, State of Vera Cruz?
Geogr. Distr. Jalapa and vicinity, State of Vera Cruz, south to Yucatan, Mexico, and Guatemala, Central America.
Genl. Char. Tail longer than head and body; hind feet long, fur slightly harsh.
Color. Above yellowish fulvous, lined with black; flanks fulvous; under parts yellowish rufous, except chin and throat, which are whitish; tail naked, indistinctly bicolor, grayish above, paler beneath; feet pale buff or whitish.
Measurements. Total length, 173; tail, 100; hind foot, 20; ear, 14.

A. Oligoryzomys.

"Size very small; tail long; hind foot long and slender; fifth hind toe moderately long. Skull small, delicate; interorbital region narrow; outer edges of frontals squarish, but unbeaded; braincase smooth and unridged; zygomatic plate narrow and with but slight forward projection; a decided longitudinal depression or trough in middle of nasals; molar teeth small and delicate, but essentially like those of true *Oryzomys;* incisor teeth very narrow and delicate." (Bangs.)

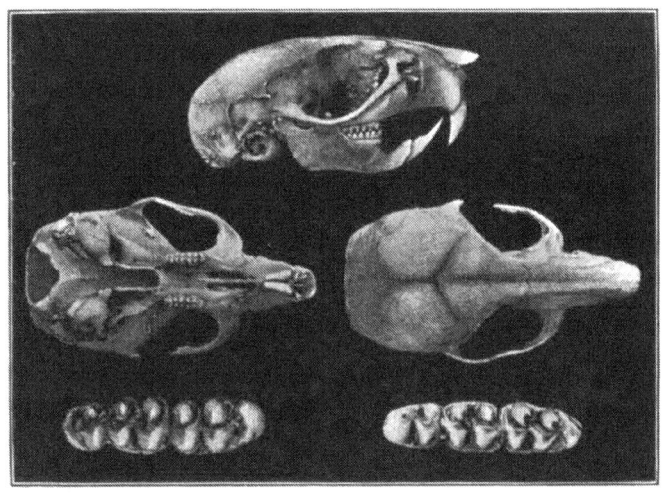

FIG. 40. ORYZOMYS VEGETUS.
Mus. Comp. Zoöl. Twice nat. size.

UPPER TOOTH ROW.
Enlarged 10 times.

LOWER TOOTH ROW.
Enlarged 10 times.

241. costaricen is (*Oryzomys*), Allen, Bull. Amer. Mus. Nat. Hist., 1893, p. 239.

EL GENERAL RICE RAT.

Type locality. El General, Costa Rica.

Geogr. Distr. Costa Rica, Central America. Altitude, from 2,150 to 10,342 feet.

Genl. Char. Size small; tail much longer than head and body. Skull with superciliary bead absent.

Color. Above varying from yellowish brown to yellowish chestnut, lined heavily with black; sides ochraceous buff; under parts deep buff; lateral line fulvous; tail naked, unicolor, pale brown; hands and feet buffy white; ears dusky brown on outside, yellowish inside.

Measurements. Total length, 196; tail vertebræ, 117; hind foot, 23; ear from crown, 7. Skull: total length, 21; basal length, 17.8; zygomatic width, 11.4; width of braincase, 10.1; interorbital constriction, 4.3; length of nasals, 8; length of upper molar series, 3. (ex Type.)

242. *vegetus (*Oryzomys*), Bangs, Bull. Mus. Comp. Zoöl., xxxix, 1902, p. 35.

_____

*This is considered by Dr. J. A. Allen to be the same as *O. costaricensis*. See Bull. Am. Mus. Nat. Hist., 1904, p. 69.

ACTIVE RICE RAT.

Type locality. Boquete, Chiriqui, Panama. Altitude, 4,000 feet.
Genl. Char. Like *O. costaricensis*, but larger; color darker. Incisors orange.

Color. Above bright yellowish red brown; dorsal region darker;
sides of head and body and rump orange rufous; chin and under sides
of neck whitish; rest of under parts ochraceous buff; hands and feet
yellowish white; tail above dusky, beneath grayish; ears dark brown.

Measurements. Total length, 205–235; tail vertebræ, 115–130.
hind foot, 24; ear, 13. Skull: basal length, 20.2; occipito-nasal
length, 24.4; zygomatic width, 12.6; mastoid width, 11; interorbital
width, 3.4; length of nasals, 9; upper molar series, 2.8; length of
single half of mandible, 12.4. (Bangs, l. c.)

49. Moschophoromys.

$$\text{I.}\frac{1-1}{1-1};\ \text{M.}\frac{3-3}{3-3}=16.$$

Moschophoromys, Elliot, Pub. Field Columb. Mus., III, 1904, p. 111.
Type *Mus desmaresti* Fischer.

Megalomys Trouess., Le Naturaliste, No. 45, 1881, p. 5. *Id.* Ann.
Scien. Nat., Zoöl., XIX, No. 5, 1885, p. 13, pl. 1. (nec Lauril-
lard, Paleont.)

Moschomys Trouess., Ann. Mag. Nat. Hist., 7th Ser., XI, 1903, p.
387. (nec Billberg, 1828, Microtinæ.)

Form rat-like; tail long, scaly; outer finger with a flat nail; hind
feet long, stout, without webs between the toes, nails curved; sole
tubercles, six. Skull very broad between orbits; supraorbital crest
well developed; molars of moderate size.

243. desmaresti (*Mus*), Fischer, Syn. Mamm., 1829, p. 316.

pilorides Desm., Dict. Scien. Nat., XLVI, 1826, p. 483. (nec
Pallas ex Ceylon.) *Id.* Trouess., Ann. Scien. Nat., Zoöl., XIX,
No. 5, 1885, p. 13, pl. 1. Ex Antilles francaises. (nec Pall.)

BLACK RICE RAT.

Type locality. Island of Martinique.

Genl. Char. Size and shape similar to those of *Mus norvegicus*.

Color. Shining black; chin, throat, belly, and base of tail white.

Measurements. Skull: occipito-nasal length, 62; length of nasals,
24; length of incisive foramina, 9.2; mastoid breadth, 21; length of
molar series, 9; posterior edge of foramen magnum to tip of nasals,
62; posterior edge of interparietal to tip of nasals, 61.5; width of brain-

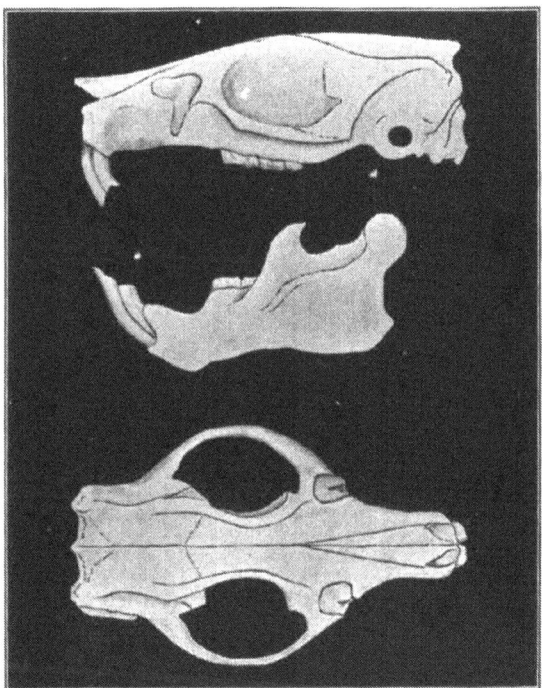

FIG. 41. MOSCHOPHOROMYS DESMARESTI.
Ex. Ann. Scien. Nat., Zoöl., xix, pl. i.   Nat. size.

case above squamosals, 21; breadth of interparietal, 11.5; length of lower molar series, 9.

244. luciæ (*Oryzomys*), Forsyth Major, Ann. Mag. Nat. Hist., 7th Ser., vii, 1901, p. 206.

ST. LUCIA RICE RAT.

Type locality. Island of St. Lucia, West Indies.

Genl. Char. Smaller than *M. desmaresti;* anterior margin of infraorbital foramen more convex anteriorly; belly brown.

Color. Brown, which hue also includes the belly.

Measurements. Skull: occipito-nasal length, 48.8; length of nasals, 19.4; length of incisive foramina, 8; mastoid breadth, 16; length of molar series, 7.5; from foramen magnum to front of incisors, 41; from posterior edge of foramen magnum to tip of nasals, 48.8; length of nasals, 19.4; Hensel, 12.8; length of incisive foramina, 8; greatest breadth of incisive foramina, 27; breadth of braincase at squamosals, 16; breadth of interparietal, 10.5; length of lower molar series, 8.

50. Zygodontomys.

$$I.\frac{1-1}{1-1}; M.\frac{3-3}{3-3} = 16.$$

Zygodontomys Allen, Bull. Am. Mus. Nat. Hist., ix, 1897, p. 38, pl. 1, figs. 1–7.   Type "*Oryzomys cherrii*" Allen.

Pelage soft; cusps of molars connected by median longitudinal ridge. Skull similar in general characters to *Oryzomys*.

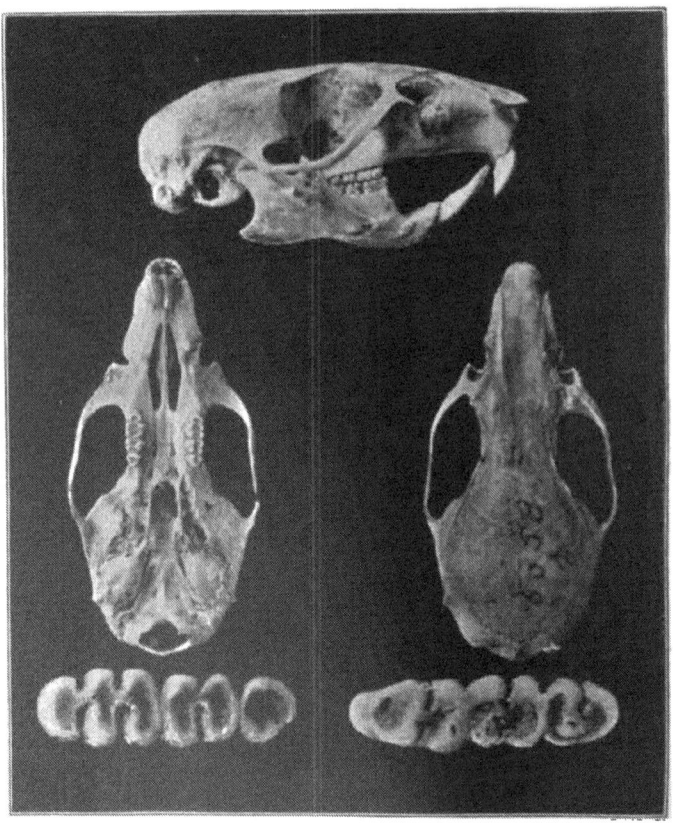

FIG. 42. ZYGODONTOMYS CHERRII.
No. 5358 Field Columbian Mus. Coll.   Twice nat. size.

UPPER TOOTH ROW.	LOWER TOOTH ROW.
Enlarged 10 times.	Enlarged 10 times.

KEY TO THE SPECIES.

FIG. XL. ZYGODONTOMYS CHERRII.
CHERRIE'S RICE RAT.

245. cherrii (*Oryzomys*), Allen, Bull. Am. Mus. Nat. Hist., 1895,
   p. 329.
CHERRIE'S RICE RAT.

Type locality. Boruca, Costa Rica.

Geogr. Distr. Costa Rica, Central America; range unknown.

Genl. Char. Size medium; tail less than length of head and
body; pelage coarse.

Color. Upper parts mixed yellowish brown and black, dorsal
line darkest; sides grayish brown, beneath grayish white; tail above
dusky brown, below grayish brown; feet and ears gray.

Measurements. Total length, 214; tail vertebræ, 92; hind foot,
23; ear from crown, 12. Skull: total length, 30; basal length, 28;
zygomatic breadth, 16; interorbital constriction, 6; breadth of braincase, 13. (ex Type.)

246. chrysomelas (*Oryzomys*), Allen, Bull. Amer. Mus. Nat. Hist.,
   1897, p. 37.

   caliginosus (*Hesperomys*), Allen, Bull. Amer. Mus. Nat. Hist.,
   1891, p. 210. (nec Tomes.)

SUERRE RICE RAT.

Type locality. Suerre, Costa Rica.

Geogr. Distr. Costa Rica, Central America.

Genl. Char. Size medium; tail half the length of head and body; ears small. Skull has the superciliary bead very broad; anterior palatal foramina not reaching anterior base of the first molar.

Color. Above mixed blackish brown and yellowish rufous; brighter on the sides; under parts yellowish brown tinged with gray; tail uniform black; hands and feet blackish brown; ears black.

Measurements. Total length, 187; tail, 90; hind foot, 25; ear, from notch, 10.5. Skull: total length, 28; basal length, 23.5; zygomatic breadth, 15; palatal length, 5.3; width of braincase, 12; length of nasals, 9.5; length of upper tooth row, 5.3. (ex Type.)

247. seorsus (*Zygodontomys*), Bangs, Am. Nat., XXXV, 1901, p. 642. ISOLATED RICE RAT.

Type locality. San Miguel Island, Bay of Panama.

Genl. Char. Similar to *Z. brevicauda* from Trinidad, but larger; tail scales coarse; skull larger, characters more pronounced.

Color. "Upper parts russet brown, shaded with dull ferrugineous, the latter more intense on the rump; whole dorsal region thickly set with brownish black-tipped hairs; sides paler, more yellowish; under parts dull buffy gray, strongly shaded with dull ferrugineous on anal region; line of demarcation between colors of upper and under parts indistinct; upper surface of hands and feet yellowish brown; soles naked; ears sparsely haired, dusky; tail very sparsely haired, coarsely scaly, indistinctly bicolor, blackish, rather paler towards base below."

Measurements. Total length, 280–320; tail vertebræ, 110–140; hind foot, 30–34; ear from notch, 15–18. Skull: basal length, 32.8; occipito-nasal length, 35.2; zygomatic width, 18.4; mastoid width, 12.8; length of nasals, 15; width of nasals, 4.4; interorbital constriction, 5.4; width across zygomatic plates, 11.2; length of incisive foramina, 7.4; width of incisive foramina, 3.2; length of palate, to palatal notch, 16; to end of pterygoid, 22; upper tooth row, 5.2; length of single half mandible, 21; lower tooth row, 5.2." (Bangs, l. c.)

51. Sigmodontomys.

Sigmodontomys Allen, Bull. Am. Mus. Nat. Hist., 1897, p. 38, pl. 1, figs. 8–14. Type *Sigmodontomys alfari* Allen.

Nasals narrowing posteriorly, pointed; anterior palatine foramina broad, not reaching front molars; palate reaching beyond posterior line of molars; palatal fossa broad, pterygoids parallel; base of zygoma broad with no anterior point; parietal large, exterior borders longest; first upper molar with two deep internal reëntrant angles, second molar with one, third none; first lower molar with two deep external reëntrant angles, and one on both the second and third molars.

248. *alfari (*Sigmodontomys*), Allen, Bull. Am. Mus. Nat. Hist., 1897, p. 39.

ALFARO'S RICE RAT.

Type locality. Jimenez, Costa Rica. Altitude, 700 feet.

Geogr. Distr. Costa Rica, Central America; range unknown.

Genl. Char. Similar to *Sigmodon*; fur long, soft, thick; tail longer than head and body; hind feet large, naked; tail naked; ears small.

Color. Upper parts yellowish brown, with dorsal region dusky brown; beneath grayish white; tail naked, dark brown; fore feet pale brown passing to grayish on toes; hind feet naked, grayish brown; ears blackish brown.

Measurements. Total length, 278; tail vertebræ, 155; hind foot, 37; ear, 14. Skull: total length, 35.2; basal length, 25.4; interorbital constriction, 12; width of braincase, 13.2; length of nasals, 14.3; palatal length, 15; length of upper tooth row, 5.3; length of mandible, 18. (ex Type.)

The genus RHITHRODONTOMYS contains the smallest species of the *Muridæ*, some indeed much smaller than certain species of shrews, in which family are found the least in size of American mammals. The Harvest Mice in general are about half the size of the house mouse and not unlike that animal in color, but usually have brighter flanks and a whiter under body. Although the Harvest Mice have been separated into various species and races, they are very difficult to distinguish, for there is a very great general resemblance among them all.

52. Rhithrodontomys. Harvest Mice.

$$I.\frac{1-1}{1-1}; \ M.\frac{3-3}{3-3} = 16.$$

J. A. Allen, *On the species of the genus Rheithrodontomys* (*sic*) Bull. Am. Mus. Nat. Hist., 1895, p. 107.

*The skull of the unique specimen of this species having been lost, it was not possible to illustrate the genus

Reithrodontomys (sic), Giglioli, Richer. intorn. alla Distrib. Geog. Gen., Roma, 1873, p. 160. Type *Mus lcconti.* Aud. & Bachman.

Reithrodon! Le Conte (nec Waterh.), Proc. Acad. Nat. Scien. Phil., 1853, p. 413.

Ochetodon Coues, Proc. Acad. Nat. Scien. Phil., 1874, p. 184.

Body slender; size very small; tail as long as the body without the head; anterior face of upper incisors with a deep longitudinal

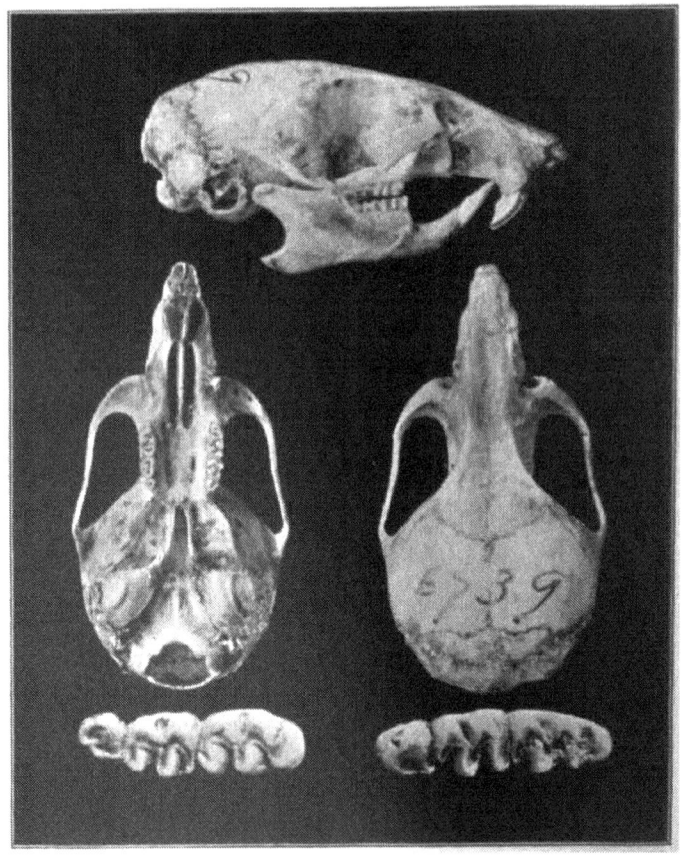

FIG. 43. RHITHRODONTOMYS MEGALOTIS.
No. 6739 Am. Mus. Nat. Hist. Coll. Enlarged 3 times.

UPPER TOOTH ROW.　　　　　LOWER TOOTH ROW.
Enlarged 12 times.　　　　　Enlarged 12 times.

groove, nearly as broad as the face of the tooth; lower incisors simple; anterior upper molar with four roots, three large, one very small. The lower half of the descending ramus is abruptly twisted inward nearly at a right angle to the lower border of the process; antorbital foramen is situated in the zygomatic portion of the maxillary and is almost circular above, contracting to a slit below; palate terminates opposite the posterior border of last molars as a transverse shelf; bullæ large, widely separated, but approximating from behind anteriorly.

KEY TO THE SPECIES AND SUBSPECIES.

249. australis (*Reithrodontomys!*), Allen, Bull. Am. Mus. Nat. Hist., 1895, p. 328.

IRAZÚ VOLCANO HARVEST MOUSE.

Type locality. Volcano of Irazú, Costa Rica.

Geogr. Distr. Costa Rica; range unknown.

Genl. Char. "Similar to *R. longicauda* in winter, but larger."

Color. Above yellowish brown, lined with black; sides lighter; under parts ashy plumbeous tinged with buff; tail above dusky brown, beneath whitish; hands and feet grayish; ears blackish.

Measurements. Total length, 158; tail vertebræ, 80; hind foot, 18; ear from crown, 10. Skull: total length, 23; basal length, 20; width of braincase, 11; zygomatic width, 10; interorbital constriction, 3.7; length of nasals, 8. (ex Type.)

a.—vulcanius (*Reithrodontomys!*), Bangs, Bull. Mus. Comp. Zoöl., XXXIX, 1902, p. 38.

CHIRIQUI HARVEST MOUSE.

Type locality. Volcan de Chiriqui, Chiriqui, Panama. Altitude, 10,300 feet.

Genl. Char. Similar to *R. australis*, but darker; pelage long, silky.

Color. Dorsal region sepia, shading on sides to isabella; top of head grayer; under parts isabella, tinged with cinnamon between arms and about vent; feet and hands grayish; tail dusky above, grayish beneath; ears sepia.

Measurements. Total length, 170; tail vertebræ, 96; hind foot, 19; ear, 17. Skull: basal length, 19.4; occipito-nasal length, 22.8; zygomatic width, 11.4; mastoid width, 11; interorbital width, 3.4; length of nasals, 8.2; width of nasals, 2.6; length of palate to palatal notch, 9.2; length of palatal slits, 5; width of palatal slits, 1.8; upper molar series, 3.2; length of single half mandible, 11.4. (Bangs, l. c.)

250. megalotis (*Reithrodon!*), Baird, N. Am. Mamm., 1857, p. 451.

aztecus, Allen, Bull. Am. Mus. Nat. Hist., 1893, p. 79.

deserti, Allen, Bull. Am. Mus. Nat. Hist., 1895, p. 127.

megalotis (*Rhithrodontomys*), Elliot, Syn. N. Am. Mamm., 1901, p. 151. *Id.* Suppl. p. 494.

BIG-EARED HARVEST MOUSE.

Type locality. Between Janos and San Luis Spring, State of Sonora, Mexico, near boundary of Grant County, New Mexico.

Geogr. Distr. States of Sonora and Chihuahua, Mexico, north to Nevada.

Genl. Char. Tail equals the body in length, without head.

Color. Upper parts mouse gray (sometimes tinged with reddish on back), and lined with black; rump washed with fulvous; under

FIG. XLI. RHITHRODONTOMYS MEGALOTIS.
BIG-EARED HARVEST MOUSE.

parts yellowish or grayish white; tail above like back; sides and beneath pale gray; hands and feet grayish white. Young animals have blackish ear spots and are grayer.

Measurements. Total length, 143; tail vertebræ, 70; hind foot, 19. Skull: occipito-nasal length, 20; Hensel, 14; zygomatic width, 10; interorbital constriction, 3; length of nasals, 8; palatal length, 7; length of upper molar series, 3.

a.—sestinensis (*Rhcithrodontomys!*), Allen, Bull. Am. Mus. Nat. Hist., 1903, p. 602.
RIO SESTIN MEADOW MOUSE.

Type locality. Rio Sestin, State of Durango, Mexico, Altitude, 7,500 feet.

Genl. Char. Similar to *R. megalotis*; tail longer.

Color. Like *R. megalotis*, "but upper parts more strongly varied with black and less fulvous, and with relatively longer tail." (Allen, l. c.)

Measurements. Total length, 130; tail vertebræ, 69; hind foot, 18; ear from notch, 14.

b.—zacatecæ (*Rcithrodontomys!*), Merr., Proc. Wash. Acad. Scien., III, 1901, p. 557.
VALPARAISO HARVEST MOUSE.

Type locality. Valparaiso Mountains, State of Zacatecas, Mexico.

Genl. Char. Similar to *R. megalotis*, but darker; nasals equal in length to premaxillæ.

Color. Above dull fulvous brown lined with black; sides buffy ochraceous; beneath plumbeous washed with buffy salmon; pectoral patch fulvous; tail dark brown above, gray beneath; hands and feet white; ears dark brown.

Measurements. Total length, 156; tail vertebræ, 87; hind foot, 17.5.

c.—*obscurus* (*Reithrodontomys!*), Merr., Proc. Wash. Acad. Scien., III, 1901, p. 558.

DUSKY HARVEST MOUSE.

Type locality. Sierra Madre, near Guadaloupe y Calvo, State of Chihuahua, Mexico.

Genl. Char. "Similar to *R. megalotis*, but everywhere much darker; upper parts conspicuously lined with and darkened by black hairs, under parts buffy salmon (instead of white), with pale fulvous pectoral patch; skull as in *R. megalotis*."

Color. Above mixed rufous and black, darkest on dorsal region; sides paler, a mixed gray and dark buff; under parts buff; no lateral line; hands and feet white; tail above dark brown, beneath whitish.

Measurements. Total length, 167; tail vertebræ, 90; hind foot, 19.

251. longicaudus (*Reithrodon!*), Baird, N. Amer. Mamm., 1857, p. 451.
longicaudus (*Rhithrodontomys*), Elliot, Syn. N. Am. Mamm., 1901, p. 151. *Id.* Suppl. p. 494.

LONG-TAILED HARVEST MOUSE.

Type locality. Petaluma, Sonoma County, California.

Geogr. Distr. Lower California, Mexico, to California, west to the Sierra Nevada, from coast region of Monterey County, north to Mendocino County (Eureka, Humboldt County, see Elliot, Field Columb. Mus., III, p. 186. Zoölogy), and in the interior from San Joaquin County, north to Tehama County. Probably further south irregularly in the coast and San Bernardino ranges of mountains. (Allen.)

Genl. Char. Small; colors darkish; tail long, more than half the length of head and body.

Color. Above yellowish brown and black, darkest on dorsal line; sides more yellowish, and with a fulvous lateral line from cheeks to rump; beneath grayish white, tinged often with yellow, and frequently with a fulvous spot on the breast; ears dusky, rusty brown tuft at the anterior base; feet whitish; tail dusky above, grayish white below.

Measurements. Average total length, 136.5; tail vertebræ, 72; hind foot, 17; ear, 11.2. Skull: occipito-nasal length, 19.5; Hensel, 14; zygomatic width, 8.5; interorbital constriction, 3; length of nasals, 7; palatal length, 7.3; length of upper tooth row, 3.

a.—pallidus (*Reithrodontomys!*), Rhoads, Amer. Nat., 1893, p. 835.
   pallidus (*Rhithrodontomys*), Elliot, Syn. N. Am. Mamm., 1901,
      p. 152. *Id.* Suppl. p. 494.
LOWER CALIFORNIA HARVEST MOUSE.
   Type locality. Santa Isabel, San Jacinto Mountains, San Diego County, California.
   Geogr. Distr. Southern California and northern Lower California, Mexico.
   Genl. Char. Larger than *R. longicaudus* and paler.
   Color. Above buffy gray, darker on dorsal line; face and lateral line ochraceous; under parts and feet white; spot between fore legs and on breast buff; tail bicolor.
   Measurements. Total length, 137; tail vertebræ, 73; hind foot, 16. Skull: occipito-nasal length, 19.5; Hensel, 14; zygomatic width, 10; interorbital constriction, 3; breadth of braincase, 9.3; length of nasals, 7; palatal length, 7.

252. saturatus (*Reithrodontomys!*), Allen & Chapman, Bull. Amer.
      Mus. Nat. Hist., 1897, p. 201.
LAS VIGAS HARVEST MOUSE.
   Type locality. Las Vigas, State of Vera Cruz, Mexico.
   Geogr. Distr. State of Vera Cruz, Mexico; range unknown.
   Genl. Char. Similar to *R. megalotis*, but larger and darker.
   Color. Upper parts dark brown, lined with black on median line and tinged with grayish fulvous; sides grayish fulvous; under parts whitish plumbeous; tail dark brown above, gray below; hands and feet grayish white.
   Measurements. Total length, 148; tail vertebræ, 74; hind foot, 18. Skull: total length, 21; basal length, 17; zygomatic width, 11; width of braincase, 10; interorbital constriction, 2.5; length of nasals, 8; palatal length, 8; length of upper tooth row, 3.3.

a.—cinereus (*Reithrodontomys!*), Merr., Proc. Wash. Acad. Scien.,
      1901, p. 556.
CINEREOUS HARVEST MOUSE.
   Type locality. Chalchicomula, State of Puebla, Mexico.
   Genl. Char. "Similar to *R. saturatus*, but very much paler and grayer, particularly the head and shoulders; tail shorter and more sharply bicolor; skull similar, but slightly smaller; rostrum shorter."

Color. Above yellowish brown lined with black, darkest on dorsal region and top of head; sides paler and with less black; under parts gray; hands and feet white; tail above dark brown, beneath whitish.

Measurements. Total length, 149; tail vertebræ, 73; hind foot, 19.

b.—alticola (*Reithrodontomys!*), Merr., Proc. Wash. Acad. Scien., III, 1901, p. 556.

MOUNTAIN HARVEST MOUSE.

Type locality. Cerro San Felipe, Oaxaca City, State of Oaxaca, Mexico. Altitude, 10,000 feet.

Genl. Char. "Similar to *R. saturatus*, but tail somewhat shorter; rostrum shorter; bullæ larger."

Color. Above yellowish brown, tinged with rufous and heavily lined with black; sides less black; under parts gray; plumbeous of under fur showing through; tail above dark brown, beneath whitish; hands and feet white; ears brown.

Measurements. Total length, 153; tail vertebræ, 75; hind foot, 19.

253. peninsulæ (*Rhithrodontomys*), Elliot, Pub. Field Columb. Mus., III, 1903, p. 164. Zoölogy.

PENINSULAR HARVEST MOUSE.

Type locality. San Quentin, Lower California, Mexico.

Genl. Char. Tail about half the total length; pelage more red than that of *R. longicaudus*, with longer tail and hind feet and much larger ears; skull larger, nasals longer, braincase broader, interorbital space of nearly equal width throughout, sides nearly parallel.

Color. Above, sides of head, and body, and on thighs reddish brown, deepest on thighs and rump and darkest on dorsal region, which part is mixed with black; under parts white or grayish white, the plumbeous under fur showing; a dark ochraceous spot on sides of chest and a paler one on middle of chest between the arms; tail dusky brown above, white beneath; hands and feet white, ears dark brown.

Measurements. Total length, 170; tail vertebræ, 89; hind foot, 17.5; ear, 16. Skull: occipito-nasal length, 27; Hensel, 20; zygomatic breadth, 10; interorbital constriction, 3; palatal length, 8.5; length of nasals, 13.5.

254. difficilis (*Reithrodontomys!*), Merr., Proc. Wash. Acad. Scien., III, 1901, p. 556.

CAPTIOUS HARVEST MOUSE.

Type locality. Orizaba, State of Vera Cruz, Mexico. Altitude, 4,500 feet.

Genl. Char. "Skull similar in general to that of *R. saturatus*, but incisive foramina more broadly open; molar series slightly shorter."

Color. Above fulvous and black, beneath pale fulvous; chin, throat, hands, and feet whitish; tail above dusky, beneath whitish.

Measurements. Total length, 177; tail vertebræ, 100; hind foot, 19.5.

255. mexicanus (*Reithrodon!*), Sauss., Rev. Mag. Zoöl., 2me Sér., XII, 1860, p. 109, pl. IX, fig. 1.
  sumichrasti, Sauss., Rev. Mag. Zoöl., 2me Sér., 1861, p. 3.
MEXICAN HARVEST MOUSE.

Type locality. Mountains in State of Vera Cruz, Mexico.

Geogr. Distr. State of Vera Cruz, Mexico.

Genl. Char. Size large; tail longer than head and body; feet large.

Color. Above rufous brown; sides orange brown; under parts white, tinged with buff on throat and breast; ears brown; tail blackish, almost naked; hands white; feet dusky.

Measurements. Total length, 153-169; tail vertebræ, 80-100; hind foot, 18-20; ear, 15-17.

a.—intermedius (*Reithrodontomys!*), Allen, Bull. Amer. Mus. Nat. Hist., 1895, p. 136.
  intermedius (*Rhithrodontomys*), Elliot, Syn. N. Am. Mamm., 1901, p. 153. *Id.* Suppl. p. 494.
BROWNSVILLE HARVEST MOUSE.

Type locality. Brownsville, Cameron County, Texas.

Geogr. Distr. Northern Mexico, into Texas.

Genl. Char. Smaller than *R. mexicanus* and paler.

Color. Above grayish brown, washed with pale yellowish, mixed with dark hairs on middle of back, sides lighter; lateral line yellowish, beneath white; ears brown; apical third of inner side rufous; feet whitish; tail dusky, nearly unicolor.

Measurements. Total length, 194; tail vertebræ, 108; hind foot, 21; ear, 13.

b.—fulvescens (*Reithrodontomys!*), Allen, Bull. Amer. Mus. Nat. Hist., 1894, p. 319.
OPOSURA HARVEST MOUSE.

Type locality. Oposura, State of Sonora, Mexico.

Geogr. Distr. State of Sonora, Mexico; range unknown.

Genl. Char. Similar to *R. mexicanus*, more yellowish.

Color. Above yellowish brown, lined with black; lateral line fulvous; under parts whitish; tail above pale brown, beneath lighter; hands and feet whitish.

Measurements. Total length, 183; tail vertebræ, 102; hind foot, 19; ear, 14. Skull: occipito-nasal length, 21; Hensel, 15; zygomatic width, 10; interorbital constriction, 3; length of nasals, 7; palatal length, 8; length of upper molar series, 3; length of mandible, 9.5; length of lower molar series, 4.

c.—gracilis (*Reithrodontomys!*), Allen & Chapman, Bull. Amer. Mus. Nat. Hist., 1897, p. 9.

CHICHEN ITZA HARVEST MOUSE.

Type locality. Chichen Itza, Yucatan, Mexico.
Geogr. Distr. Yucatan, Mexico.
Genl. Char. Similar to *R. mexicanus*, but smaller and paler.
Color. Above yellowish brown, lined with blackish on median line; sides fulvous brown; under parts white; ears brownish; tail unicolor, brown; hands and feet whitish.
Measurements. Total length, 165; tail vertebræ, 98; hind foot, 16; ear, 13. Skull: total length, about 20; width of braincase, 10; interorbital constriction, 3; length of nasals, 7; length of upper molar series, 3. (ex Type.)

256. tenuis (*Reithrodontomys!*), Allen, Bull. Am. Mus. Nat. Hist., 1899, p. 15.

SLENDER HARVEST MOUSE.

Type locality. Rosario, State of Sinaloa, Mexico.
Geogr. Distr. State of Sinaloa, Mexico; range unknown.
Genl. Char. Similar to *R. m. fulvescens*, but smaller.
Color. Upper parts yellowish brown, lined with black; lateral line from before the eyes and along flanks fulvous; under parts grayish white; base of hairs pale plumbeous; ears dusky, internally reddish, tail grayish brown; feet whitish.
Measurements. Total length, 152–170; tail vertebræ, 82–90; hind foot, 20; ear, 15.

257. chrysopsis (*Reithrodontomys!*), Merr., Proc. Biol. Soc. Wash., XIII, 1900, p. 152.

MOUNT POPOCATEPETL HARVEST MOUSE.

Type locality. Mt. Popocatepetl, Mexico.
Geogr. Distr. State of Mexico, Mexico; range unknown.
Genl. Char. Size large; ears large; tail very long, hairy. Skull: no superorbital beads; zygoma notched by antorbital slits; bullæ small; rostrum narrow.
Color. Upper parts golden yellow, mixed with black; under parts white, tinged with salmon fulvous; tail above dusky, beneath white; ears and ankles dusky; hands and feet white.

Measurements. Total length, 194; tail vertebræ, 108; hind foot, 21.

a.—toluca (*Reithrodontomys!*), Merr., Proc. Wash. Acad. Scien., 1901, p. 549.

VOLCAN TOLUCA HARVEST MOUSE.

Type locality. North slope of the Volcan Toluca, State of Mexico, Mexico. Altitude, 11,500 feet.

Genl. Char. Similar to *R. chrysopsis*, but darker; skull smaller.

Color. Above yellowish brown and black, forming a perceptible dorsal band, sides lighter; under parts grayish plumbeous; hands and feet brownish; tail above dark brown, beneath whitish; ears blackish brown.

Measurements. Total length, 180; tail vertebræ, 98; hind foot. 21.

258. perotensis (*Reithrodontomys!*), Merr., Proc. Wash. Acad. Scien., III, 1901, p. 550.

COFRE DE PEROTE HARVEST MOUSE.

Type locality. Cofre de Perote, State of Vera Cruz, Mexico. Altitude, 9,500 feet.

Genl. Char. Similar to *R. chrysopsis*, but colors duller; ears large; skull smaller; nasals equal to premaxillæ.

Color. Above yellowish fulvous, lined with black, forming a dorsal band and grading to fulvous on rump; top of head blackish; under parts buffy salmon; tail dark brown above, paler beneath; hands and feet white; wrists and ankles brownish; ears dark brown.

Measurements. Total length, 176; tail vertebræ, 102; hind foot. 19.

259. orizabæ (*Reithrodontomys!*), Merr., Proc. Wash. Acad. Scien., III, 1901, p. 550.

ORIZABA HARVEST MOUSE.

Type locality. Mount Orizaba, State of Puebla, Mexico. Altitude, 9,500 feet.

Genl. Char. Similar to *R. chrysopsis*; tail long.

Color. Above golden fulvous, darkest toward rump; beneath buffy salmon; lips and chin grayish; tail dark above, paler beneath; hands and feet whitish; wrists and ankles dark brown.

Measurements. Total length, 182; tail vertebræ, 105; hind foot, 20.

260. colimæ (*Reithrodontomys!*), Merr., Proc. Wash. Acad. Scien., III, 1901, p. 551.

ALPINE HARVEST MOUSE.

Type locality. Sierra Nevada de Colima, State of Jalisco, Mexico. Altitude, 12,000 feet.

Genl. Char. Size small; ears large; tail short; nasals equal to premaxillæ.

Color. Above buffy fulvous to golden fulvous, lined with black; beneath buffy salmon; tail dusky above, whitish below; hands whitish, feet brownish, whitish towards toes; wrists and ankles dark, nearly black; ears blackish.

Measurements. Total length, 165; tail vertebræ, 90; hind foot, 20.

a.—nerterus (*Reithrodontomys!*), Merr., Proc. Wash. Acad. Scien., III, 1901, p. 551.

COLIMA HARVEST MOUSE.

Type locality. Sierra Nevada de Colima, State of Jalisco, Mexico. Altitude, 6,500 feet.

Genl. Char. Similar to *R. colimæ*; tail longer, skull and molars smaller.

Color. Above golden fulvous, lined with black, darkest on dorsal line; side golden fulvous; under parts salmon buff; hands and feet brown; tail above blackish brown, paler beneath; ears dark brown.

Measurements. Total length, 190; tail vertebræ, 110; hind foot, 20.

261. costaricensis (*Reithrodontomys!*), Allen, Bull. Am. Mus. Nat. Hist., 1895, p. 139.

COSTA RICA HARVEST MOUSE.

Type locality. La Carpintera, Costa Rica. Altitude, 6,000 feet.
Geogr. Distr. Costa Rica, Central America.
Genl. Char. Size large; similar in color to *Peromyscus cherrii.*

Color. Upper parts ferrugineous brown, lined with black; sides orange rufous; under parts white, washed with yellow; fulvous patch sometimes on breast; tail nearly naked, dusky brown, unicolor; hands and feet whitish, with a median dusky stripe on feet; ears brown.

Measurements. Total length, 197; tail vertebræ, 111; hind foot, 20.5; ear, 12. Skull: occipito-nasal length, 22.5; Hensel, 17; zygomatic width, 12; interorbital constriction, 3.5; length of nasals, 8; palatal length to incisive foramina, 4; length of upper molar series, 3.

a.—jalapæ (*Reithrodontomys!*), Merr., Proc. Wash. Acad. Scien., III, 1901, p. 532.

JALAPA HARVEST MOUSE.

Type locality. Jalapa, State of Vera Cruz, Mexico. Altitude, 4,000 feet.

Genl. Char. Similar to *R. costaricensis*, but paler; nasals truncate posteriorly.

Color. Above pale ferrugineous brown mixed with black; nose, upper lip, and under parts white; hands whitish brown; feet grayish brown; tail above dark brown, paler beneath.

Measurements. Total length, 197; tail vertebræ, 119; hind foot, 21.

262. goldmani (*Reithrodontomys!*), Merr., Proc. Wash. Acad. Scien., III, 1901, p. 552.
GOLDMAN'S HARVEST MOUSE.

Type locality. Metlaltoyuca, State of Puebla, Mexico. Altitude, 800 feet.

Genl. Char. Size small; similar in color to *R. costaricensis*, paler.

Color. Above ochraceous fulvous, darkest on crown; side of nose, upper lip, chin, and under parts white; tail dusky; hands whitish; ankles and feet brownish; ears light brown.

Measurements. Total length, 190; tail, 109; hind foot, 21.5.

263. rufescens (*Reithrodontomys!*), Allen & Chapman, Bull. Amer. Mus. Nat. Hist., 1897, p. 199.
RUFOUS HARVEST MOUSE.

Type locality. Jalapa, State of Vera Cruz, Mexico.

Geogr. Distr. State of Vera Cruz, Mexico.

Genl. Char. Similar to *R. mexicanus*, but larger; ears black.

Color. Upper parts rufous mixed with black, darkest on median line; sides orange rufous; chin, throat, and inside of arms whitish; rest of under parts orange buff to buffy gray; ears black or blackish; tail above blackish, paler beneath; hands and feet dusky.

Measurements. Total length, 177; tail vertebræ, 99; hind foot, 20. Skull: total length, 23; basal length, 18.7; zygomatic width, 11.3; interorbital constriction, 3.5; length of nasals, 9; palatal length, 9; length of upper tooth row, 3.6. (ex Type.)

264. tenuirostris (*Reithrodontomys!*), Merr., Proc. Wash. Acad. Scien., III, 1901, p. 547.
SLENDER-NOSED HARVEST MOUSE.

Type locality. Todos Santos, Guatemala. Altitude, 10,000 feet.

Genl. Char. Size large; tail long; braincase contracted anteriorly; rostrum and nasals long, slender, latter equal in length to premaxillæ.

Color. Above deep fulvous, grading into ferrugineous on lower back; beneath salmon fulvous; chin whitish; tail dusky; hands and wrists, ankles, and feet dark brown; ears blackish brown.

Measurements. Total length, 210; tail vertebræ, 124; hind foot, 23. Skull: basal length, 20; zygomatic breadth posteriorly, 13; palatal length, 11.5; length of molar series. 4.5. (ex Type.)

a.—aureus (*Reithrodontomys!*), Merr., Proc. Wash. Acad. Scien., III, 1901, p. 548.

CALEL HARVEST MOUSE.

Type locality. Calel, Guatemala. Altitude, 10,200 feet.

Genl. Char. Similar to *R. teniurostris*, but paler; nasals narrow and exceeding premaxillæ in length; incisive foramina passing plane of first molars.

Color. Above yellowish brown, lined with black, with a reddish tinge on rump; under parts salmon fulvous; tail dark brown; hands and feet dark brown, becoming white towards toes; ears blackish.

Measurements. Total length, 196; tail vertebræ, 112; hind foot, 22.5; Skull: basal length, 20; zygomatic breadth, 13; palatal length, 11.5; length of molar series, 4.5. (ex Type.)

265. microdon (*Reithrodontomys!*), Merr., Proc. Wash. Acad. Scien., III, 1901, p. 548.

TODOS SANTOS HARVEST MOUSE.

Type locality. Todos Santos, Guatemala. Altitude, 10,000 feet.

Genl. Char. Small; ears and tail long; miniature of *R. tenuirostris*.

Color. Above reddish fulvous, deepest on rump; sides bright fulvous; beneath salmon fulvous; orbital ring dark, tail dusky; hands white; ankles and feet dark brown; toes white; ears dark brown.

Measurements. Total length, 185; tail vertebræ, 113; hind foot, 21. Skull: basal length, 17; zygomatic breadth, 11.5; palatal length, 10; length of molar series, 3. (ex Type.)

a.—albilabris (*Reithrodontomys!*), Merr., Proc. Wash. Acad. Scien., III, 1901, p. 549.

WHITE-LIPPED HARVEST MOUSE.

Type locality. Cerro San Felipe, State of Oaxaca, Mexico. Altitude, 10,000 feet.

Genl. Char. Similar to *R. microdon*, but paler; skull with certain comparative variations.

Color. Above fulvous, darkest on dorsal region; lips, nose, and under parts pure white; tail dark brown, paler beneath; wrists dusky; feet and ankles blackish brown; ears brown.

Measurements. Total length, 187; tail vertebræ, 117; hind foot, 20. Skull: basal length, 17; zygomatic breadth posteriorly, 11.5; length of palate, 10; length of molar series, 3. (ex Type.)

266. hirsutus (*Reithrodontomys!*), Merr., Proc. Wash. Acad. Scien., III, 1901, p. 553.

HISPID HARVEST MOUSE.

Type locality. Ameca, State of Jalisco, Mexico. Altitude, 5,500 feet.

Genl. Char. Size large; tail long; pelage coarse. Skull large; braincase flattened; nasals not so long as premaxillæ; bullæ small; length of upper molar series, 3.7.

Color. Above yellowish fulvous lined with black; sides brighter fulvous; beneath white, as are also the sides of nose and upper lip; tail above dusky, whitish beneath; hands buffy; feet grayish.

Measurements. Total length, 233; tail vertebræ, 143; hind foot, 22.

267. griseiflavus *(Reithrodontomys!)*, Merr., Proc. Wash. Acad. Scien., III, 1901, p. 553.

AMECA HARVEST MOUSE.

Type locality. Ameca, State of Jalisco, Mexico. Altitude, 4,000 feet.

Genl. Char. Similar to *R. hirsutus*, but smaller.

Color. Above buffy gray, grading into buffy fulvous on hind back, and lined with black; sides pale golden fulvous; beneath white, sometimes tinged with buffy; pectoral spot and axilla fulvous; lips and chin, hands and feet whitish; ankles dark; tail above brownish, beneath whitish.

Measurements. Total length, 169; tail vertebræ, 94; hind foot, 22.

a.—helvolus (Reithrodontomys!), Merr., Proc. Wash. Acad. Scien., III, 1901, p. 554.

OAXACA HARVEST MOUSE.

Type locality. Oaxaca City, State of Oaxaca, Mexico.

Genl. Char. Similar to *R. grisciflavus*, but redder.

Color. Above on back and sides orange fulvous, sparsely lined with black, but making a dorsal line to middle of back; throat and lips white, under parts gray; tail brown above, whitish beneath; hands and feet whitish; ears brown.

Measurements. Total length, 188; tail vertebræ, 111; hind foot, 20.

268. levipes *(Reithrodontomys!)*, Merr., Proc. Wash. Acad. Scien., III, 1901, p. 554.

SAN SEBASTIAN HARVEST MOUSE.

Type locality. San Sebastian, State of Jalisco, Mexico. Altitude, 3,000 feet.

Genl. Char. Size medium; tail rather long; hair coarse.

Color. Above fulvous lined with black; sides bright oranze fulvous; beneath pale fulvous; hands and feet whitish; tail dark brown above, paler beneath; ears brown.

Measurements. Total length, 188; tail vertebræ, 110; hind foot, 21.

a.—*otus* (*Reithrodontomys!*), Merr., Proc. Wash. Acad. Scien., III, 1901, p. 555.
LARGE-EARED HARVEST MOUSE.

Type locality. Sierra Nevada de Colima, State of Jalisco, Mexico. Altitude, 6,500 feet.

Genl. Char. "Size large, similar to *R. levipes*; ears decidedly larger; tail longer. Skull similar, but rostrum slightly longer; bullæ averaging slightly larger."

Color. Above fulvous and black, darkest on dorsal region; under parts pale fulvous; tail dark brown above, whitish beneath; hands pale brown; feet whitish; ears dark brown.

Measurements. Total length, 202; tail vertebræ, 120; hind foot, 22.

b.—*tolteus* (*Reithrodontomys!*), Merr., Proc. Wash. Acad. Scien., III, 1901, p. 555.
TLALPAM HARVEST MOUSE.

Type locality. Tlalpam, Federal District, Mexico.

Genl. Char. "Similar to *R. levipes*, but larger; color more yellowish (less fulvous), and much more heavily lined with black hairs. Skull narrower; zygomata less widely spreading anteriorly; nasals longer and ending with premaxillæ; bullæ slightly larger."

Color. Above yellowish brown lined with black, with a conspicuous dorsal line; sides pale golden brown; under parts grayish white; tail above pale brown, beneath whitish; hands and feet white; ears pale brown.

Measurements. Total length, 196; tail vertebræ, 106; hind foot, 21.

269. Inexpectatus (*Rhithrodontomys*), Elliot, Pub. Field Columb. Mus., III, 1903, p. 145. Zoölogy.
PATZCUARO HARVEST MOUSE.

Type locality. Patzcuaro, State of Michoacan, Mexico.

Genl. Char. Similar to *R. levipes*, but whitish on under parts instead of pale fulvous, and tip of tail white; ears large; tail long.

Color. Upper parts mixed black and tawny ochraceous; sides ochraceous buff; chin, upper part of throat, hands, and feet white; under parts grayish white, slightly tinged with buff; limbs like sides; tail above blackish brown, tip white, beneath whitish; ears naked, dark brown, with a slight edging of pale brown and a tuft of ochraceous hairs at base.

Measurements. Total length, 180, tail vertebræ, 113, hind foot, 21. Skull: length from alveolus of incisors to posterior margin of palate, 9; palatal length, 5, length of nasals, 9, interorbital constriction, 3; length of molar series, 4. (ex Type)

270. dorsalis (*Reithrodontomys*), Merr., Proc. Wash. Acad. Scien., III, 1901, p. 557.
GRAY-BACKED HARVEST MOUSE.
Type locality. Calel, Guatemala
Genl. Char. Size large; molars small; rostrum long.
Color. Above grayish washed with buffy, and lined with black, dorsal area blackish; sides ochraceous, beneath buffy white; lips, chin, hands, and feet whitish; tail above dark, beneath whitish.
Measurements. Total length, 171, tail vertebræ, 92, hind foot, 1.

271. creper (*Reithrodontomys*), Bangs, Bull. Mus. Comp. Zool. XXXIX, 1902, p. 39
DUSKY HARVEST MOUSE.
Type locality. Volcan de Chiriqui, Chiriqui, Panama. Altitude 11,000 feet.
Genl. Char. Hind foot large, tail long; pelage long, silky.
Color. Dorsal region bistre, sides raw umber, under parts dark cinnamon; hands and feet brownish; toes whitish; tail, basal two-thirds uniform dusky, terminal third white; ears dusky.
Measurements. Total length, 215; tail vertebræ, 130; hind foot 23; ear, 15. Skull: basal length, 21.4; occipito-nasal length, 25.4; mastoid width, 11.6; length of nasals, 8.8, length of palate, to palatal notch, 10; length of palatal slits, 4.8, upper molar series, 4.2; single half of mandible, 13.6. (Bangs, l. c.)

The next genus is peculiarly a South American one, where, with the exceptions mentioned below, all of its species are found.

53. Acodon.

$$I.\frac{1-1}{1-1};\ M.\frac{3-3}{3-3}=16.$$

Akodon Meyen, Nova Acta Phys. Med. Acad. Caes. Leop. Carol., XVI, 1833, p. 599, tab. XLIII, fig. 1.   Type *Akodon boliviensis* Meyen.
Abrothrix (*sic*) Waterh., Proc. Zoöl. Soc., 1837, p. 21.

Fur long, soft; tail short, hairy; ears hairy; thumb with a short nail. First molar with two indentations on both sides, the second molar one, and the third molar with one on outer side only.

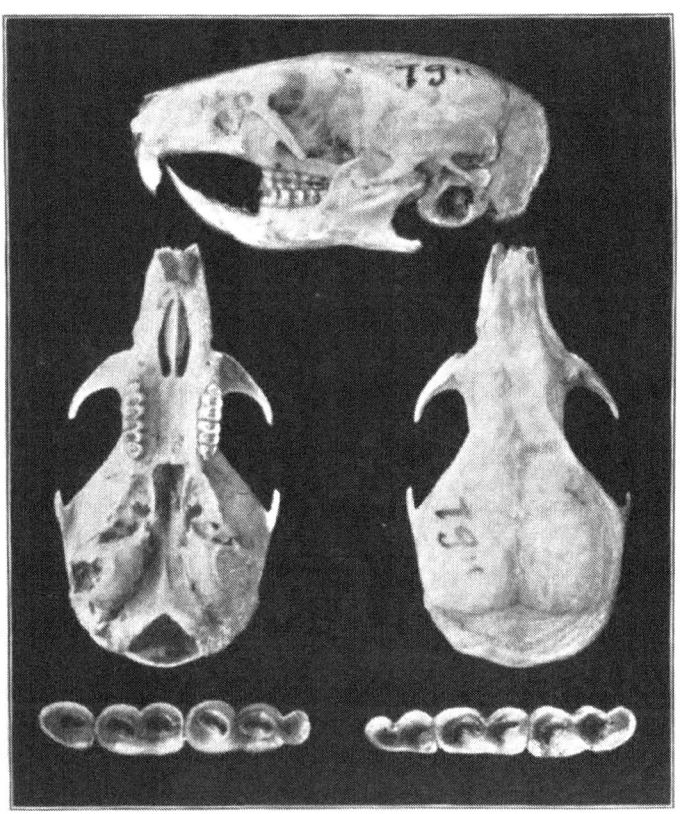

FIG. 44. ACODON TEGUINA.
No. 7911 Am. Mus. Nat. Hist. Coll.   Enlarged 3 times.

UPPER TOOTH ROW.   LOWER TOOTH ROW.
Enlarged 10 times.   Enlarged 10 times.

KEY TO THE SPECIES AND SUBSPECIES.

272. teguina (*Hesperomys*), Alston, Proc. Zoöl. Soc., 1876, p. 755.
ALSTON'S MOUSE.

Type locality.   Coban, Guatemala.
Geogr. Distr.   Range unknown.

Genl. Char.   Tail about as long as body without head, hairy; ears moderate, rounded, sparsely covered with hair.

Color.   Above reddish brown, lined with black; chin, throat, and breast grayish reddish brown; belly deep fawn; tail, feet, and ears dusky.

FIG. XLII. ACODON TEGUINA. ALSTON'S MOUSE.

Measurements.   Total length, 134; tail, 52; hind foot, 42; ear, 10. Skull: occipito-nasal length, 20.5; Hensel, 16; zygomatic width, 11.5; interorbital constriction, 4; length of nasals, 6; palatal length, 8.5; length of upper tooth row, 3.5.

a —*apricus* (*Akodon*), Bangs, Bull. Mus. Comp. Zoöl., xxxix, 1902, p. 40.

BOQUETE MOUSE.

Type locality.   Boquete, Chiriqui, Panama.   Altitude, 4,000–5,000 feet.

Genl. Char.   Similar to *A. teguina*, but paler.

Color.   Above Vandyke brown; dorsal region and top of head dusky; beneath dull cinnamon rufous; tail, hands, feet, and ears blackish.

Measurements.   Total length, 125–142; tail vertebræ, 50–58; hind foot, 18; ear, 13.   Skull: basal length, 20.2; occipito-nasal length, 23; zygomatic width, 12; mastoid width, 10.8; interorbital width, 4.6; length of nasals, 9; width of nasals, 2.8; length of palate, to palatal notch, 9.6; upper molar series, 4; length of single half of mandible, 12.8.   (Bangs, l. c.)

273. irazu (*Akodon*), Allen, Bull. Am. Mus. Nat. Hist., 1904, p. 46.

VOLCANO OF IRAZU MOUSE.

Type locality.   Volcan de Irazu, Costa Rica.

Genl. Char.   Smaller than *A. teguina*, with smaller ears.

Color. Upper parts yellowish brown; under parts buffy brown.

Measurements. Total length, 125; tail vertebræ, 50; hind foot, 17. Skull: occipito-nasal length, 22; Hensel, 17; zygomatic breadth, 11.5; mastoid breadth, 10; interorbital constriction, 4; length of nasals, 8.3; palatal length, 8; length of upper tooth row, 4; length of mandible, condyle to base of incisors, 11.

274. xerampelinus (*Akodon*), Bangs, Bull. Mus. Comp. Zoöl., xxxix, 1902, p. 41.

BANGS' RED MOUSE.

Type locality. Volcan de Chiriqui, Chiriqui, Panama. Altitude, 10,300 feet.

Genl. Char. Similar to *A. teguina*, but paler.

Color. Above yellowish brown; beneath broccoli brown; hands, feet, tail, and ears blackish.

Measurements. Total length, 145; tail vertebræ, 65; hind foot, 17; ear, 14. Skull: basal length, 19.2; occipito-nasal length, 22.6; zygomatic width, 11.6; mastoid width, 10.8; interorbital width, 4.2; length of nasals, 2.6; length of palate, to palatal notch, 9.6; upper molar series, 4; length of single half mandible, 13. (Bangs, l. c.)

The Wood Rats are among the handsomest species of the *Muridæ* in North America. They are large in size, with a velvety fur, in some as soft as spun silk, with the upper parts mouse color, brown, or even a reddish hue, and white under parts, hands, and feet. One of the genera, NEOTOMA, presents the usual rat tail, scaly, long, and naked; but another, TEONOMA (north of United States and Mexican boundary), has a tail similar to a squirrel's, hairy, at times almost bushy, and the members of this genus are very handsome animals indeed. Being almost strictly nocturnal, the Wood Rat is not often seen by day, but as soon as darkness falls, if the cabin of the woodsman is near its haunts, it will be overrun with these animals, and they will be seen on floors and shelves and other parts of the home diligently seeking food. In such places they become very bold. climbing upon the bed and racing over the body of its sleeping occupant. The naked-tailed species seems to have the majority of numbers, but in appearance are excelled by the bushy-tailed. The subfamily has various genera, the members of which are more or less closely allied to each other, but none equal in size and general appearance those of NEOTOMA and TEONOMA.

Subfam. II. **Neotominæ. Wood Rats.**

54. Neotoma.

$$I.\frac{1-1}{1-1};\ M.\frac{3-3}{3-3}=16.$$

C. H. Merriam, *The Neotominæ, with a description of a new genus and species and a synopsis of the known forms.* Proc. Acad. Nat. Scien. Phil., 1894, p. 225.

Neotoma Say & Ord, Journ. Acad. Nat. Scien. Phil., 1825, p. 345, pls. xxi, xxii. Type *Neotoma floridana* Ord.

Skull long; zygomatic width equals half the length of skull; edge of maxilla bounding antorbital foramen, rounded; the foramen broad and open above, compressed into a somewhat narrow slit below; intermaxillæ reaching back to interorbital region; nasals much shorter; palate terminating with a concave border posteriorly between last two molars; audital bullæ small, their axes oblique to that of the skull; occipital plane of skull perpendicular, at right angles to the superior surface; process of jugals descending downward and backward,

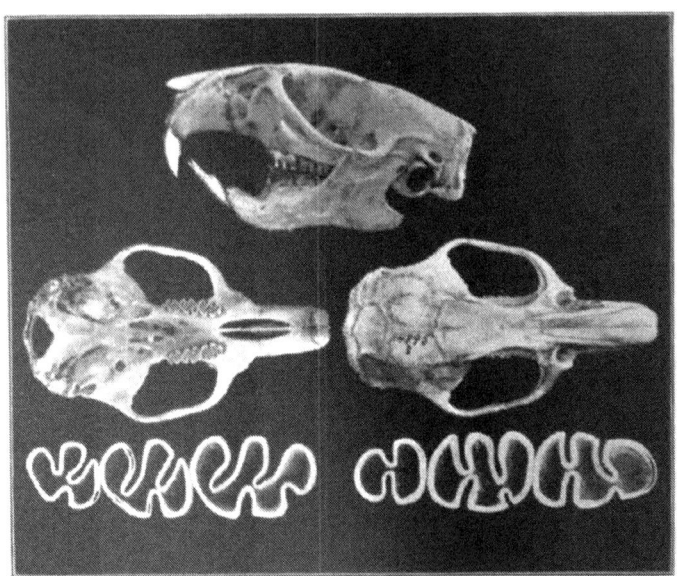

FIG. 45. NEOTOMA MICROPUS.
No. 4895 Field Columbian Mus. Coll. Nat. size.

UPPER TOOTH ROW.   LOWER TOOTH ROW.
Enlarged 5 times.   Enlarged 5 times.

that of squamosal joining it at almost a right angle. Mandible with long, acute, coronoid process, higher than condyle; roots of lower incisors causing protuberances on each side of the jaw; upper molar teeth with usually one internal and two external reëntrant loops; first and second lower molars with two external and two internal loops; last molar with only one of each; upper molars three-rooted; lower with but two roots.

A. Neotoma. Naked-tailed Wood Rats.

Tail long, naked, round.

KEY TO SPECIES AND SUBSPECIES.

275. distincta (*Neotoma*), Bangs, Proc. Biol. Soc. Wash., XVI, 1903, p. 89. TEXOLO WOOD RAT.

Type locality. Texolo, State of Vera Cruz, Mexico.

Genl. Char. Tail black; rostrum swollen over roots of incisors.

Color. Upper parts mummy brown; back sprinkled with hairs tipped with brownish black; sides paler; sides of nose and upper lips dull grayish brown; chin, upper throat and narrow belly stripe dull gray; pectoral collar ochraceous buff; rest of under parts yellowish white; hands and feet dull grayish brown; fingers and toes white; tail black; ear naked, dusky.

Measurements. Total length, 370–417; tail vertebræ, 165–206. hind foot, 40–41; ear, 20–26. Skull: basal length, 43; occipito-nasal length, 48; zygomatic width, 25; mastoid width, 19.4; interorbital constriction, 5.2; length of nasals, 5.6; length of palate, to palatal notch, 22.2; length of palatal slits, 10.6; length of upper molar series, 9.4; length of single half mandible, 29.6.

276. fuscipes (*Neotoma*), Baird, N. Amer. Mam., 1857, p. 495. Elliot. Syn. N. Am. Mamm., 1901, p. 158.

monochrura Rhoads, Amer. Nat., XXVIII, 1894, p. 67.

splendens True, Proc. U. S Nat. Mus., 1894, p. 353.

DUSKY-FOOTED WOOD RAT.

Type locality.   Petaluma, Sonoma County, California.

Geogr. Distr.   Coast region of California and Oregon, from Monterey Bay to the Columbia River. State of Durango, Mexico? (Sclater.)

Genl. Char.   Tail nearly as long as head and body; size large; hind feet short; soles naked.

Color.   Above mixed reddish brown and black; sides reddish brown; dorsal region darkest; limbs outside to wrists and ankles like the back; feet dusky; under parts yellowish white; tail dusky all around.

Measurements.   Total length, 410; tail vertebræ, 198; hind foot, 41. Skull: occipito-nasal length, 54; Hensel, 44; zygomatic width, 27; interorbital constriction, 6; length of nasals, 19; palatal length, 24; length of upper tooth row, 9; length of mandible, 26.

a.—*macrotis* (*Neotoma*), Thomas, Ann. Mag. Nat. Hist., 6th Ser., XII, 1893, p. 234. Elliot, Syn. N. Am. Mamm., 1901, p. 159.

simplex True, Proc. U. S. Nat. Mus., 1894, p. 354.

LARGE-EARED WOOD RAT.

Type locality.   San Diego, San Diego County, California.

Geogr. Distr.   Lower California, Mexico, and southern California.

Genl. Char.   Similar to *N. fuscipes*, but smaller, and feet white; *hairs on central portion of under parts white to the roots.* Tail short.

Color.   Above mixed reddish brown and black; lighter on sides; dorsal area darkest; feet and entire under parts white; tail above brownish black, beneath pale brown.

Young specimens have none, or very little, of the reddish brown color, but are a pale yellowish brown, and the belly is often tinged with buff.   This pelage is the most common and may also represent that of winter.

Measurements.   Total length, 404; tail vertebræ, 195; hind foot, 41; ear, 32.5. Skull: occipito-nasal length, 43; Hensel, 34.5; zygomatic width, 22; interorbital constriction, 6; length of nasals, 16; palatal length, 19; length of upper tooth row, 8; length of mandible, 21.

277. torquata (*Neotoma*), Ward, Amer. Nat., XXV, 1891, p. 160.

COLLARED WOOD RAT.

Type locality.   Between Tetela del Volcan and Zacualpan Amilpas, State of Morelos, Mexico.

Geogr. Distr.   State of Morelos, Mexico.   Type locality only.

Genl. Char.   Breast collar pale Vandyke brown; hairs of belly at base, gray; tail bicolor; soles naked.

Color. Above pale Vandyke brown, hairs tipped with black; darkest on dorsal line; under parts white with a yellow tinge; breast crossed by a pale brown collar; chin white; hands and feet white; tail covered with short hairs, above clove brown, sides and beneath whitish; ears seal brown.

Measurements. Total length, 498; tail vertebræ, 160; hind foot, 35; ear, 21. Skull: total length, 45; zygomatic width, 23; length of upper molar series, 9; length of mandible, 25; length of lower molar series, 9.

278. cumulator (*Neotoma*), Mearns, Proc. U. S. Nat. Mus., 1898, p.
  503. Elliot, Syn. N. Am. Mamm., 1901, p. 154.
COLLECTOR WOOD RAT.
 Type locality. Old Fort Yuma, San Diego County, California.
 Geogr. Distr. Lower California and State of Sonora, Mexico, southern California.
 Genl. Char. Size large; color similar to that of *N. intermedia.* but paler; nasals broad anteriorly.
 Color. Above grayish fulvous, lined with black; gray on limbs and ochraceous buff on sides; feet and under parts white; tail black above, white beneath; ears mixed gray and black.
 Measurements. Total length, 406; tail vertebræ, 188; hind foot, 37; ear, 30.5. Skull: greatest length, 47.5; greatest width, 23.

279. bryanti (*Neotoma*), Merr., Amer. Nat., XXI, 1887, p. 191.
CERROS ISLAND WOOD RAT.
 Type locality. Cerros or Cedros Island, Lower California, Mexico.
 Geogr. Distr. Known only from type locality.
 Genl. Char. Size large; tail naked; specimen imperfect.
 Color. General hue dark slate, on both upper and under parts; fulvous patch behind ear; hands and feet white.
 Measurements. Total length, 372; tail vertebræ, 165; hind foot, 38; ear, 26. Skull: total length, 48; basal length, 44; zygomatic width, 25; mastoid width, 20; interorbital constriction, 5; length of nasals, 17.5; breadth of nasals anteriorly, 5; palatal length, 21; length of upper molar series, 8.

280. ferruginea (*Neotoma*), Tomes, Proc. Zoöl. Soc., 1861, p. 282.
RUSTY WOOD RAT.
 Type locality. Dueñas, Guatemala.
 Geogr. Distr. Tehauntepec, State of Oaxaca, Mexico, south into Guatemala.
 Genl. Char. Tail about as long as head and body; similar to *N. fuscipes*, but smaller and redder.

Color. Above rusty red, somewhat golden on the sides, and mixed with black on the back; outside of arms and legs to wrists and ankles, dusky; inner side of arms white, of legs grayish; sides of lips and under parts white; tail above dusky, beneath paler.

Measurements. Total length, 343–353; tail vertebræ, 165–177; hind foot, 33–35. Skull: occipito-nasal length, 46; zygomatic width, 28; length of nasals, 20; length of upper molar series, 10.

FIG. XLIII. NEOTOMA MICROPUS. SMALL-FOOTED WOOD RAT.

281. micropus (*Neotoma*), Baird, Proc. Acad. Nat. Scien. Phil., 1855, p. 333. Elliot, Syn. N. Am. Mamm., 1901, p. 155.

canescens Allen, Bull. Am. Mus. Nat. Hist., 1891, p. 285.

SMALL-FOOTED WOOD RAT.

Type locality. Charco Escondido, State of Tamaulipas, Mexico.

Geogr. Distr. State of Tamaulipas, Mexico, northward to Oklahoma and New Mexico.

Genl. Char. Tail short, hardly two-thirds the length of the body; ears large; feet small, soles naked.

Color. Above grayish lead color, lined with dark brown; sides paler; shoulders and flanks occasionally tinged with yellowish brown; under parts and feet white; tail above dusky, beneath grayish white.

Measurements. Total length, 359; tail vertebræ, 185; hind foot, 36; ear, 30. Skull: occipito-nasal length, 49; Hensel, 36; zygomatic width, 24; interorbital constriction, 6; length of nasals, 6.5; palatal length, 19; length of upper tooth row, 10; length of mandible, condyle to alveolus of incisor, 25.

282. leucodon (*Neotoma*), Merr., Proc. Biol. Soc. Wash., IX, 1894, p. 120. WHITE-TOOTHED WOOD RAT.

Type locality. San Luis Potosi, State of San Luis Potosi, Mexico.

Geogr. Distr. States of Zacatecas, San Luis Potosi, and Vera Cruz, Mexico.

Genl. Char. Similar to *N. micropus* in size; postpalatal notch narrow; jugals short; molars white; upper first molar with two internal salient angles.

Color. Above ochraceous buff, lined with black; face grayish; under parts white, as are also the hands and feet; tail above blackish, beneath white.

Measurements. Total length, 358; tail vertebræ, 164; hind foot, 39.

283. latifrons (*Neotoma*), Merr., Proc. Biol. Soc. Wash., IX, 1894, p. 121.

QUERENDARO WOOD RAT.

Type locality. Querendaro, State of Michoacan, Mexico.

Geogr. Distr. State of Michoacan, Mexico.

Genl. Char. Similar to *N. leucodon*, smaller; longer hind feet; shorter tail.

Color. Above like *N. leucodon*; cheeks and sides tinged with fulvous; under parts, hands, and feet white; tail above dusky, beneath whitish.

Measurements. Total length, 350; tail vertebræ, 149; hind foot, 42.

284. mexicana (*Neotoma*), Baird, Proc. Acad. Nat. Scien. Phil., 1855, p. 333. Elliot, Syn. N. Am. Mamm., 1901, p. 158.

MEXICAN WOOD RAT.

Type locality. Mountains near Chihuahua, State of Chihuahua, Mexico.

Geogr. Distr. State of Chihuahua, Mexico, north into New Mexico and Texas.

Genl. Char. Ears large; feet small; soles naked. Nasal bones end on the same line posteriorly, and do not extend backward to the anterior extremity of orbits; highest point of condyle higher than coronoid process.

Color. Above pale yellowish brown, lined with dark brown; sides yellowish brown; outside of legs, feet, and under parts white; tail dusky above, whitish beneath.

Measurements. Total length, 340; tail vertebræ, 151; hind foot, 37; ear, 32. Skull: occipito-nasal length, 44.5; Hensel, 35; zygomatic width, 23; interorbital constriction, 6; length of nasals, 17; palatal length, 19; length of upper tooth row, 7; length of mandible, 22.

285. navus (*Neotoma*), Merr., Proc. Biol. Soc. Wash., XVI, 1903, p. 47.

ACTIVE WOOD RAT.

Type locality. Sierre Guadalupe, State of Coahuila, Mexico.

Genl. Char. Size medium; tail long. Skull: frontals expanded posteriorly, forming supraorbital shelves; bullæ small; teeth slender, with anterior lobe of first upper molar having a deep notch on inner side, as in *N. mexicana.*

Color. Upper parts and sides of face buffy ochraceous, lined with black; head grayish; under parts white, the plumbeous under fur showing; axillæ salmon; tail dusky above, white below; hands and feet from wrists and ankles white.

Measurements. Total length, 350; tail vertebræ, 164; hind foot, 34. Skull: basal length, 37; zygomatic width, 21.5; palatal length, 21.5; length of upper molar series, alveolar border, 8.5.

286. sinaloæ (*Neotoma*), Allen, Bull. Amer. Mus. Nat. Hist., 1898, p. 149. SINALOA WOOD RAT.

Type locality. Tatameles, State of Sinaloa, Mexico.

Geogr. Distr. State of Sinaloa, Mexico.

Genl. Char. Similar to *N. mexicana*; tail longer; bullæ small.

Color. Upper parts dark fulvous brown mixed with black; under parts, hands and feet white; tail above blackish brown, beneath lighter.

Measurements. Total length, 315–332; tail vertebræ, 155–160; hind foot, 31–32. Skull: total length, 41.5; basal length, 37; zygomatic width, 21.5; mastoid width, 15.2; interorbital constriction, 4.8; length of nasals, 15.4; palatal length, 17; length of upper tooth row, 7.

287. arenacea (*Neotoma*), Allen, Bull. Amer. Mus. Nat. Hist., 1898, p. 150.

LOWER CALIFORNIA WOOD RAT.

Type locality. San José del Cabo, Lower California, Mexico.

Geogr. Distr. Cape Region, Lower California, Mexico.

Genl. Char. Similar to *N. f. macrotis*, but smaller and paler; intermaxillæ extending beyond nasals; supraorbital bead on frontals.

Color. Upper parts grayish brown mixed with black, tinged with fulvous; outside of forearm and leg blackish; under parts, hands, and feet white; hairs on median band white to roots; rest plumbeous at base; tail above blackish brown, beneath grayish white.

Measurements. Total length, 349; tail vertebræ, 164–167; hind foot, 35–36. Skull: total length, 46.2; basal length, 41; zygomatic width, 23.5; mastoid width, 17.2; interorbital constriction, 5.3; length of nasals, 18.2; palatal length, 15; length of upper molar series, 7.

288. anthonyi (*Neotoma*), Allen, Bull. Amer. Mus. Nat. Hist., 1898, p. 151.

Todos Santos Island Wood Rat.

Type locality. Todos Santos Island, Lower California, Mexico.

Geogr. Distr. Known from type locality only.

Genl. Char. Nasals rounded anteriorly; slight supraorbital bead; first upper molar with sulcus on anterior internal border. Size small.

Color. Upper parts grayish brown, slightly mixed with black, and tinged with fulvous; forearm above externally and outer side of leg blackish; under parts, hands, and feet white to roots of hairs; tail above blackish brown, beneath grayish white.

Measurements. Total length, 330–345; tail vertebræ, 132–146; hind foot, 34–36; ear, 23–25. Skull: total length, 46; basal length, 42; zygomatic width, 25; mastoid width, 18.2; interorbital constriction, 4.8; length of nasals, 18; palatal length, 18; length of upper molar series, 8.

289. intermedia (*Neotoma*), Rhoads, Am. Nat., XXVIII, 1894, p. 69.
　　Elliot, Syn. N. Am. Mamm., 1901, p. 161.
californica Price, Proc. Calif. Acad. Scien., 1894, p. 154, pl. XI.
venusta True, Proc. U. S. Nat. Mus. 1894, p. 247, Sept.

Rhoads' Wood Rat.

Type locality. Dulzuras, San Diego County, California.

Geogr. Distr. Lower California, Mexico, and southern California, south of the Bay of Monterey.

Genl. Char. Size small; tail slender, short, bicolor; ears large; soles naked.

Color. Above light brownish gray, lined with black; chin, center of breast, inside of legs, and feet, white; rest of under parts soiled grayish buff; tail above sooty blackish, beneath white.

Measurements. Total length, 318; tail vertebræ, 160; hind foot, 35; ear, 28. Skull: occipito-nasal length, 42; Hensel, 33; zygomatic width, 20; interorbital constriction, 5.5; length of nasals, 16; palatal length, 18; length of upper molar series, 8.5; length of mandible, 19.

a.—melanura (*Neotoma*), Merr., Proc. Biol. Soc. Wash., IX, 1894, p. 126.

Black-tailed Wood Rat.

Type locality. Ortiz, State of Sonora, Mexico.

Geogr Distr. State of Sonora, Mexico.

Genl. Char. Size small; first upper molar has the "anterior loop partly divided by antero-internal sulcus."

Color. Winter Pelage. Upper parts mixed black and ochraceous buff; sides ochraceous; under parts, hands, and feet white to roots of hairs, except on sides of belly, where the bases are plumbeous at roots; ankles blackish; tail black above, white beneath.

Measurements. Total length, 333; tail vertebræ, 170; hind foot, 34; ear, 25.

b.—angusticeps (*Neotoma*), Merr., Proc. Biol. Soc. Wash., IX, 1894, p. 127. Elliot, Syn. N. Am. Mamm., 1901, p. 162.
NARROW-HEADED WOOD RAT.

Type locality. Southwest corner Grant County, New Mexico.

Geogr. Distr. Northern Mexico (State of Chihuahua), and southwestern New Mexico.

Genl. Char. Similar to *N. i. albigula*; ears smaller; color more fulvous.

Color. Above fulvous; ochraceous buff on head, lined with black; feet and under parts creamy white to roots of hair, except on belly, where the base of hairs is plumbeous; tail grayish brown above, white beneath.

Measurements. Total length, 335; tail vertebræ, 150; hind foot, 33; ear, 25. Skull: basal length, 42; Hensel, 39.5; zygomatic width, 24; interorbital constriction, 6.

c.—albigula (*Neotoma*), Hartley, Proc. Calif. Acad. Scien., 2d Ser., 1894, p. 157, pl. XII. Elliot, Syn. N. Am. Mamm., 1901, p. 162.
WHITE-THROATED WOOD RAT.

Type locality. Vicinity of Fort Lowell, near Tucson, Pima County, Arizona.

Geogr. Distr. State of Sonora, Mexico, north into Arizona.

Genl. Char. Similar to both *N. intermedia* and *N. mexicana*, but the yellow ground color of upper parts and sides is deeper and the general color is lighter.

Color. Above pale yellowish brown mixed with black; central line on back darker; sides pale yellow mixed with a pale brown; feet dusky white; under parts grayish white, except throat, which is pure white to base of hairs; tail blackish brown above, soiled white beneath; ears light brown on naked part.

Measurements. Total length, 342; tail vertebræ, 153; hind foot, 32. Skull: occipito-nasal length, 44; Hensel, 37; zygomatic width, 23; interorbital constriction, 5; length of nasals, 16; palatal length, 20; length of upper tooth row, alveolar border, 6.5; length of mandible, angle to alveolus of incisor, 20.5.

d.—durangæ (*Neotoma*), Allen, Bull. Am. Mus. Nat. Hist., 1903, p. 602.
DURANGO WOOD RAT.

Type locality. San Gabriel, State of Durango, Mexico. Altitude, 7,000 feet.

Genl. Char. "Externally similar to *N. i. albigula*, but averaging rather larger, with a shorter and broader skull and much heavier dentition." (Allen, l. c.)

Measurements. Total length, 356; tail vertebræ, 159; hind foot, 32; ear from notch, 30. Skull: total length, 45; Hensel, 38; length. of nasals, 18; zygomatic width, 24; width of braincase above zygomata, 18.5; mastoid width, 18; interorbital constriction, 6; length of upper tooth row, 9.

290. orizabæ (*Neotoma*), Merr., Proc. Biol. Soc. Wash., IX, 1894, p. 122.

ORIZABA WOOD RAT.

Type locality. Mt. Orizaba, State of Puebla, Mexico.

Geogr. Distr. States of Puebla, Tlaxcala and Vera Cruz, Mexico.

Genl. Char. Similar to *N. fulvienter*, but more buffy ochraceous above, and white on belly; feet shorter.

Color. Above ochraceous buff, mixed with black on back; head grayish; under parts, hands, and feet white; on each side of breast a spot of salmon color; tail above dusky, beneath whitish.

Measurements. Total length, 356; tail vertebræ, 163; hind foot, 33.

291. tenuicauda (*Neotoma*), Merr., Proc. Biol. Soc. Wash., VII, 1892, p. 169.

SLENDER-TAILED WOOD RAT.

Type locality. North slope of the Sierra Nevada de Colima, State of Colima, Mexico. Altitude, 12,000 feet.

Geogr. Distr. State of Colima, Mexico.

Genl. Char. Size small; tail and ears almost naked.

Color. Above dark brown, tinged with yellowish fulvous; darker on flanks; under parts whitish; salmon color at arm pits; hands and feet whitish; tail above dusky, beneath whitish.

Measurements. Total length, 340; tail vertebræ, 160; hind foot, 31.

bella felipensis Elliot, Pub. Field Columb. Mus., 1903, III, p. 217. Zoölogy.

SAN FELIPE DESERT RAT.

Type locality. San Felipe, Gulf of California, Lower California, Mexico.

Genl. Char. Largest of the pale colored desert rats, nearest to *N. bella*; feet and ears larger; tail longer. Skull with shorter and broader nasals, shorter pterygoid fossa, wider basioccipital and basisphenoid; bullæ much smaller; braincase narrower posteriorly.

Color. Upper parts cream buff, lined with black; sides cream color; lips, hands, feet, lower portion of sides, and under parts pure white; basal part of hairs on side and under parts plumbeous, except on chin, center of breast, and a line down through the center of the abdomen to anal region, which have the hairs white to the roots; tail dusky above, whitish beneath; ears pale brown.

Measurements. Total length, 335; tail vertebræ, 158; hind foot, 34; ear, 34. Skull: occipito-nasal length, 41; Hensel, 34; zygomatic breadth, 21; interorbital constriction, 5; width of braincase at root of zygomata, 18; posterior width, 14; palatal arch to alveoli of incisors, 19; postpalatal length, 15; median length of nasals, 15; posterior width of nasals, 2.5; anterior width of nasals, 4.5; palatal arch to hamular processes of pterygoids, 7; width of basioccipital anteriorly, 8; length of upper molar series, 7; length of mandible, angle to alveolus of incisor, 23; height at condyle, 11; at coronoid process, 12.5; length of lower molar series, 8.

291a. *picta (*Neotoma*), Goldman, Proc. Biol. Soc. Wash., XVII, 1904, p. 79.
PAINTED WOOD RAT.

Type locality. Chilpancingo, State of Guerrero, Mexico.

Genl. Char. Size medium; tail long, slender.

Color. Upper parts orange rufous, sprinkled with black on head and back; beneath white, plumbeous under fur showing; axillæ orange rufous; tail above dusky, beneath paler; hands yellowish white; feet dusky; toes white; ears dusky.

Measurements. Total length, 368; tail vertebræ, 180; hind foot, 37. Skull: greatest length, 43.3; Hensel, 35; zygomatic width, 23; interorbital constriction, 5; length of nasals, 17.4; palatal length, 8.6; length of upper molar series, alveolar border, 8.7.

291b. isthmica (*Neotoma*), Goldman, Proc. Biol. Soc. Wash., XVII, 1904, p. 80.
ISTHMIAN WOOD RAT.

Type locality. Tehuantepec, State of Oaxaca, Mexico.

Genl. Char. Size large; tail long, stout; ears medium.

Color. Above between orange rufous and ferrugineous; outer side of forearms and hind legs grayish fulvous; top of head and back sprinkled with black; under parts and inner sides of arms and legs soiled white; tail above brownish, beneath paler; hands white; feet dusky; toes white.

Measurements. Total length, 395; tail vertebræ, 198; hind foot, 38. Skull: greatest length, 48.4; Hensel, 38.4; zygomatic breadth, 23.7; interorbital constriction, 6.2; length of nasals, 19; palatal length, 8.3; upper molar series, alveolar border, 9.

---

*Descriptions of the four following Wood Rats were published too late to be included in the regular numerical order.

291c. parvidens (*Neotoma*), Goldman, Proc. Biol. Soc. Wash., XVII, 1904, p. 81.

JUQUILA WOOD RAT.

Type locality. Juquila, State of Oaxaca, Mexico.

Genl. Char. Size small; tail short.

Color. Upper parts ferrugineous; outer side of arms and legs brownish fulvous; under parts white; axillæ orange rufous; tail above dusky, beneath paler; hands yellowish white; feet dusky; toes yellowish white.

Measurements. Total length, 295; tail vertebræ, 141; hind foot, 31. Skull: greatest length, 40.5; Hensel, 32.5; zygomatic breadth, 20.7; interorbital constriction, 5.3; length of nasals, 15.2; palatal length, 7.3; upper molar series, alveolar border, 7.4.

291d. tropicalis (*Neotoma*), Goldman, Proc. Biol. Soc. Wash., XVII, 1904, p. 81.

TROPICAL WOOD RAT.

Type locality. Totontepec, State of Oaxaca, Mexico.

Genl. Char. Size small; tail short, slender, thinly haired.

Color. Upper parts dark brown; cheeks, shoulders, and sides fulvous; pectoral band salmon; under parts dull whitish; median line on belly white; hands and feet dusky; toes whitish; tail above dusky, beneath paler.

Measurements. Total length, 325; tail vertebræ, 156; hind foot, 34. Skull: greatest length, 41.3; Hensel, 33.5; zygomatic breadth, 22.2; interorbital constriction, 5.8; length of nasals, 16.5; palatal length, 7.9; length of upper molar series, alveolar border, 8.3.

292. goldmani (*Neotoma*), Merr., Proc. Biol. Soc. Wash., 1903, p. 48.

GOLDMAN'S RAT.

Type locality. Saltillo, State of Coahuila, Mexico.

Genl. Char. Size small; tail short; ears large. Skull small and rounded; frontals as in *N. mexicana*, but broader and flatter; nasals wedge-shaped, truncate posteriorly; premaxillæ reaching beyond nasals.

Color. Upper parts buffy grayish, lined with black; flanks buffy ochraceous; head gray; cheeks buffy ochraceous; under parts, hands and feet white; tail above dark brown, beneath whitish.

Measurements. "Average of four; total length, 279; tail vertebræ, 128; hind foot, 30. Skull: basal length, 33; zygomatic width, 19; palatal length, 18.2; interorbital breadth, 5.5; upper molar series, 7."

293. fulviventer (*Neotoma*), Merr., Proc. Biol. Soc. Wash., IX, 1894, p. 121.

FULVOUS-BELLIED WOOD RAT.

Type locality. Toluca Valley, State of Mexico, Mexico.

Geogr. Distr. State of Mexico, Mexico.

Genl. Char. Similar to *N. tenuicauda*, but larger; ears and feet small; tail slender.

Color. Above fulvous, dusky on middle of back; under parts pale fulvous; hands and feet white; tail above blackish, beneath whitish.

Measurements. Total length, 356; tail vertebræ, 163; hind foot, 33.

55. Nelsonia.

$$I.\tfrac{1-1}{1-1}; \ M.\tfrac{3-3}{3-3} = 16.$$

Nelsonia Merr., Proc. Biol. Soc. Wash., XI, 1897, p. 277, figs. 14–15.

Type *Nelsonia neotomodon* Merriam.

Skull similar to that of a large *Peromyscus*, but flatter; antorbital slits faintly notching upper surface of maxillary root of zygoma; interior angle of antorbital slits protrudes forwards as processes; teeth large; crowns flat, with deep reëntrant angles; third upper molar with one deep external reëntrant angle separating the crown into

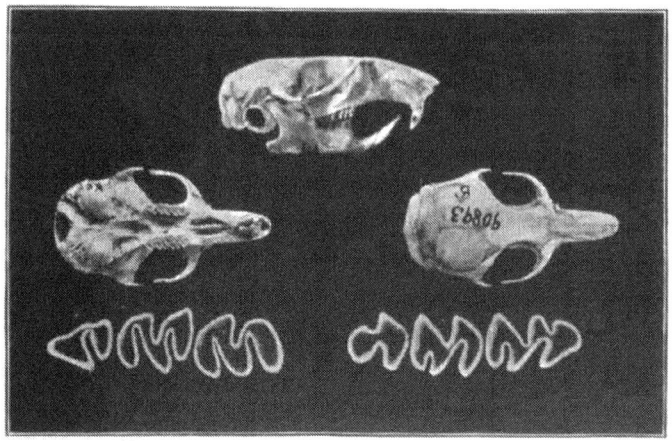

FIG. 46. NELSONIA NEOTOMODON.
No. 90893 Am. Mus. Nat. Hist. Coll. Nat. size.
UPPER TOOTH ROW.   LOWER TOOTH ROW.
Enlarged 6 times.   Enlarged 6 times.

two unequal lobes; second lower molar with one external and one internal reëntrant angle, separating the crown into two transverse loops; the posterior occasionally exhibiting a second reëntrant angle; third lower molar with one internal reëntrant angle, and a slight external projection. The other molar teeth with patterns like those of *Neotoma* and other allied genera.

KEY TO THE SPECIES.

294. neotomodon (*Nelsonia*), Merr., Proc. Biol. Soc. Wash., XI, 1897, p. 278.

ZACATECAS WOOD RAT.

Type locality. Mountains near Plateado, State of Zacatecas, Mexico. Altitude, 8,200 feet.

Geogr. Distr. State of Zacatecas, Mexico.

Genl. Char. Tail well haired; ears large, nearly naked.

Color. Above grayish brown, lined with black on lower back; sides pale fulvous; under parts white; orbital ring black; arms and legs dusky to ankles exteriorly; upper lip and side of nose white; tail above dusky, beneath white, tip white; hands and feet white.

Measurements. Total length, 247; tail vertebræ, 121; hind foot, 29. Skull: occipito-nasal length, 33; Hensel, 36; zygomatic width, 17; interorbital constriction, 5; length of nasals, 7; palatal length, 9; length of upper tooth row, 6; length of mandible, 10.

295. goldmani (*Nelsonia*), Merr., Proc. Biol. Soc. Wash., XVI, 1903, p. 80.

MT. TANCITARO WOOD RAT.

Type locality. Mt. Tancitaro, State of Michoacan, Mexico.

Genl. Char. Similar to *N. neotomodon*, but darker and grayer; tail well haired. Skull more angular, flatter between orbits; vertical lamella on anterior base of zygoma forming a spine; nasals narrower; nostrils more constricted at base.

Color. Above dark slate gray, washed with pale ochraceous; beneath white, plumbeous under fur showing; tail above dusky, beneath paler. Young dark slate.

Measurements. Total length, 248; tail vertebræ, 122; hind foot, 20.

56. Xenomys.

$$I.\frac{1-1}{1-1}; \ M.\frac{3-3}{3-3} = 16.$$

Xenomys Merr., Proc. Biol. Soc. Wash., VII, 1892, p. 160. Type *Xenomys nelsoni* Merriam.

Bullæ large, elongated, and parallel to axis of skull; vertical ridge on anterior border of squamosal; this last terminates between posterior root of zygoma and occiput; paroccipital processes long and

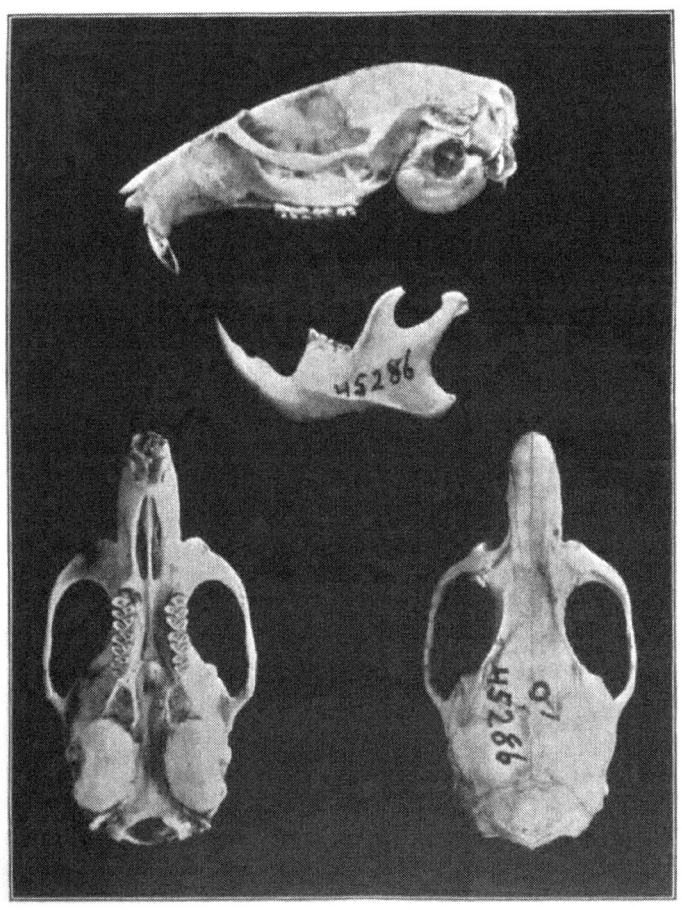

FIG. 47. XENOMYS NELSONI.
No. 45286 U. S. Nat. Mus. Coll. Enlarged ½

stout; upper molars with three roots, lower with two; the series large, with flat crowns, and with rounded alternating closed triangles; first upper molar with one anterior and one posterior closed loop, and one external and two internal lateral closed triangles; second and third upper molars with one anterior and one posterior closed loop and one lateral closed triangle on each side; last lower molar deeply incised on inner face by a reëntrant angle; on outer side is an anterior and posterior loop and a nearly closed triangle about the middle of the tooth; supraorbital beads well developed; lachrymals and interparietal large.

296. nelsoni (*Xenomys*), Merr., Proc. Biol. Soc. Wash., VII, 1892, p. 161.

NELSON'S WOOD RAT.

Type locality. Hacienda Magdalena, between City of Colima and Manzanillo, State of Colima, Mexico.

Geogr. Distr. State of Colima, Mexico.

Genl. Char. Size moderate; tail shorter than head and body. Skull resembling that of *Neotoma*, but differing in the characters given above.

Color. Above fulvous, mixed with black on the back; orbital ring dusky; a white spot above the eye and one below root of ear; lips and fore part of cheeks white, rest of face fulvous; under parts creamy white; tail dark umber brown, unicolor; hands and feet whitish.

Measurements. Total length, 300; tail vertebræ, 143; hind foot, 30; ear from crown, 18; from anterior root, 22. (Ear lengths from dried skin, others taken in the flesh.) Skull: occipito-nasal length, 40.6; Hensel, 33.3; zygomatic width, 32; interorbital constriction, 6; length of nasals, 14; palatal length, 19.3; length of upper tooth row, 8.6.

57. Neotomodon.

$$I.\frac{1-1}{1-1}; \ M.\frac{3-3}{3-3} = 16.$$

Neotomodon Merr., Proc. Biol. Soc. Wash., XII, 1898, p. 127. Type *Neotomodon alstoni* Merriam.

Skull broad; braincase short; molars rooted, large, crowns flat; upper first and second molars with three external salient loops and two reëntrant angles; and two internal salient loops and one reëntrant angle; third upper molar small, rounded; lower first and second molars with three salient loops and two reëntrant angles externally and internally; third lower molar with two loops and one reëntrant angle externally, and one internal anterior loop and one reëntrant angle.

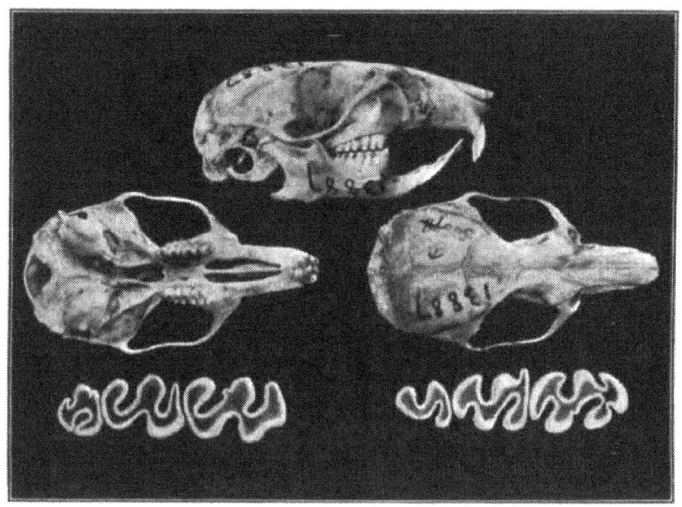

FIG. 48. NEOTOMODON ALSTONI.
No. 13887 Am. Mus. Nat. Hist. Coll. Enlarged ½.

UPPER TOOTH ROW.	LOWER TOOTH ROW.
Enlarged 7 times.	Enlarged 7 times.

KEY TO THE SPECIES.

A. Ears large, nearly naked; tail short, fur soft, plantar tubercles, 6; mammæ, 6.
  a. Breast buffy.
    a.' Under parts white.   PAGE
      a." Large; total length, 212 mm..........*N. perotensis* 293
      b." Small; total length, 194 mm.............*N. orizabæ* 294
    b.' Under parts plumbeous, washed with
      white.................................*N. alstoni* 294

297. perotensis (*Neotomodon*), Merr., Proc. Biol. Soc. Wash., XII, 1898, p. 129.
PEROTE WOOD RAT.
  Type locality. Cofre de Perote, State of Vera Cruz, Mexico. Altitude, 9,500 feet.
  Geogr. Distr. State of Vera Cruz, Mexico.
  Genl. Char. Similar to *N. alstoni*, but ears and tail shorter.

Color. Summer Pelage. Above grayish buff, dorsal band blackish; sides buffy; under parts white, breast tinged with buffy; tail above dusky, beneath white; hands and feet white.

Measurements. Total length, 212; tail vertebræ, 91; hind foot, 24.

298. orizabæ (*Neotomodon*), Merr., Proc. Biol. Soc. Wash., XII, 1898, p. 129.

MOUNTAIN WOOD RAT.

Type locality. Mt. Orizaba, State of Puebla, Mexico. Altitude, 9,500 feet.

Geogr. Distr. State of Puebla, Mexico.

Genl. Char. Smaller than the other species, and with shorter tail.

Color. Above buffy gray; under parts white, chest buffy; tail above brownish, beneath white; hands and feet whitish.

Measurements. Total length, 194; tail vertebræ, 81; hind foot, 24.

299. alstoni (*Neotomodon*), Merr., Proc. Biol. Soc. Wash., XII, 1898, p. 127.

ALSTON'S WOOD RAT.

Type locality. Nahuatzin, State of Michoacan, Mexico.

Geogr. Distr. States of Michoacan, Morelos, and Mexico, Mexico. Altitude, 8,500 feet.

Genl. Char. Size moderate; ears large; tail shorter than head and body; hind feet large.

Color. Above grayish to fulvous brown, darkest on back; under parts plumbeous washed with white; breast tinged with buff; tail above dusky, beneath whitish; hands and feet white.

Measurements. Total length, 220; tail vertebræ, 101; hind foot. 26.5. Skull: occipito-nasal length, 33; Hensel, 25.5; zygomatic width, 17; interorbital constriction, 4.5; length of nasals, 12; palatal length, 14; length of upper tooth row, 6.5; length of mandible, 14.5.

58. Teanopus.

$$I.\frac{1-1}{1-1}; \ M.\frac{3-3}{3-3} = 16.$$

Teanopus Merr., Proc. Biol. Soc. Wash., 1903, p. 81. Type *Teanopus phenax* Merriam.

Ears large, nearly naked; tail long, thickly haired; soles of feet covered with small tubercles. Skull: audital bullæ greatly inflated, like those of *Xenomys*; antorbital slits large and broadly open; sphenoid vacuities open; braincase without temporal shield; angle of jaw elongate, with lower border inflected and upturned, forming a shallow trough as in *Teonoma;* infracondylar notch deep; last lower molar with reëntrant loop on inner side, passing in front of its mate on outer side.　(Merr., l. c.)

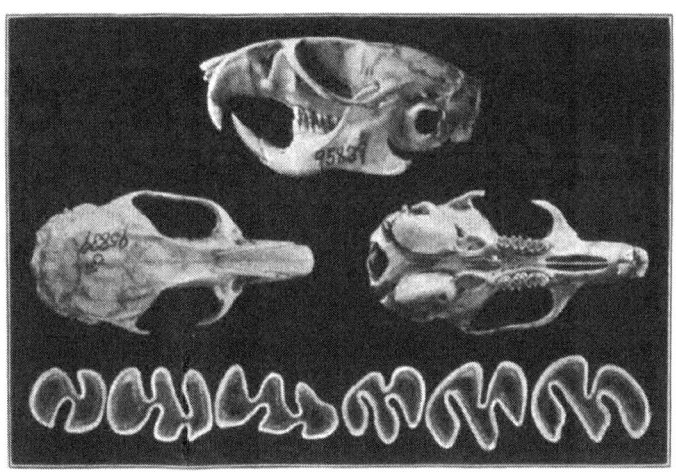

FIG. 49. TEANOPUS PHENAX.
No. 95839 U. S. Nat. Mus.　Nat. size.　Tooth rows enlarged 6 times.

300. phenax (*Teanopus*), Merr., Proc. Biol. Soc. Wash., XVI, 1903, p. 81.

CHEATING WOOD RAT.

Type locality.　Camoa, Rio Mayo, State of Sonora, Mexico.

Genl. Char.　Similar to *Hodomys vetulus* in size and appearance.

Color.　Above buffy gray; beneath yellowish white, plumbeous under fur showing on posterior half; cheeks pale gray; outer side of fore leg grayish dusky, inner side white; hind foot whitish; ankles dusky, bordered beneath with white.

Measurements.　Total length, 352; tail vertebræ, 172; hind foot, 37.5. Skull: occipito-nasal length, 42.5; Hensel, 35; zygomatic width, 22; interorbital constriction, 5; length of nasals, 16; palatal length, 18; length of upper molar series, 7; length of mandible, 23; length of lower molar series, 8.

59. Hodomys.

$$I.\frac{1-1}{1-1}; \ M.\frac{3-3}{3-3} = 16.$$

Hodomys Merr., Proc. Acad. Nat. Scien. Phil., 1894, p. 232. Type *Neotoma alleni* Merriam.

Cranium long, narrow; squamosal and supraoccipital articulating; bullæ small, narrow anteriorly, shorter than molar series; pterygoid fossa longer than broad; basioccipital broad; spheno-palatine vacuities closed; mandible broadly expanded posteriorly; first and second upper molars with four roots each, and the enamel fold on inner side divides the middle transverse loop; three roots on third upper molar; third lower molar with two salient and one reëntrant angle on each side, and with or without an antero-external vertical sulcus.

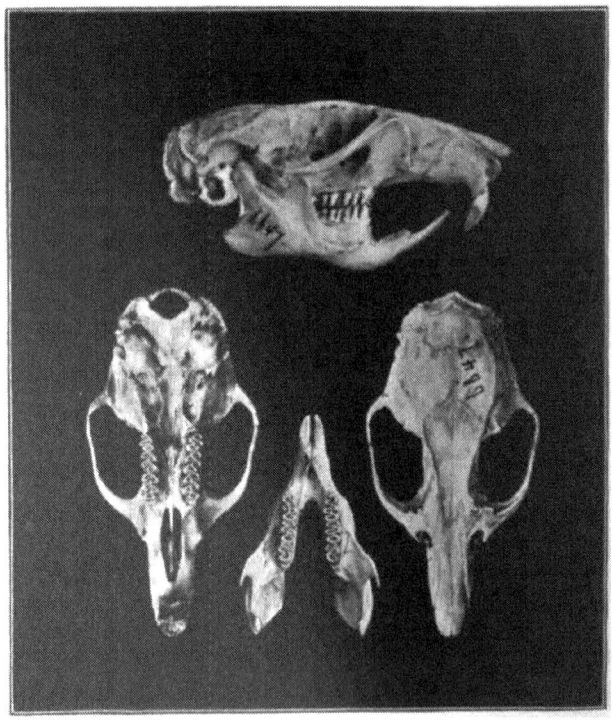

FIG. 50. HODOMYS ALLENI.
No. 5021 Field Columbian Mus. Coll. Nat. size

KEY TO THE SPECIES.

FIG. XLIV. HODOMYB ALLENI. ALLEN'S WOOD RAT.

301. alleni (*Neotoma*), Merr., Proc. Biol. Soc. Wash., VII, 1892, p. 168.

ALLEN'S WOOD RAT.

Type locality. Manzanillo, State of Colima, Mexico.

Geogr. Distr. State of Colima, Mexico.

Genl. Char. Size large; tail shorter than head and body, annulations visible; molar series large; first and second upper molars with a lateral triangle on each side; last lower molar S-shaped, and with a reëntrant angle on outer side.

Color. Above tawny ferrugineous; sides of face mouse or bluish gray; under parts white, basal portion of hairs plumbeous; tail blackish, unicolor; hands and feet whitish.

Measurements. Total length, 472; tail vertebræ, 225; hind foot, 46; ear, 29. Skull: occipito-nasal length, 52; Hensel, 45; zygomatic width, 26; interorbital constriction, 5; length of nasals, 19; palatal length, 23; length of upper molar series, 10; length of mandible, 26.

302. vetulus (*Hodomys*), Merr., Proc. Acad. Nat. Scien. Phil., 1894, p. 236.

GRAY-FACED WOOD RAT.

Type locality. Tehuacan, State of Puebla, Mexico.

Geogr. Distr. Southern Mexico, State of Puebla.

Genl. Char. Smaller than *H. alleni;* tail bicolor; feet white; cranial characters comparatively on smaller scale, but palate proportionately longer; frontals broader and less upturned at margins of orbits; third lower molar without antero-external sulcus.

Color. Above dull fulvous, mixed with black; face gray; under parts whitish and washed with fulvous, the latter sometimes restricted to sides of belly; tail above blackish, beneath whitish; fore and hind feet white.

Measurements. Total length, 380; tail vertebræ, 166; hind foot. 38; ear, 29. Skull: total length, 47; basal length, 41; Hensel, 39; zygomatic width, 25; length of crowns of upper molar series, 9.

The Subfamily MICROTINÆ contains the Meadow Mice of North America. In general, these troublesome creatures (for they prove to be great pests to the agriculturalist) inhabit low, swampy meadows. near streams, along the banks of which their narrow runways can readily be seen amid the grass; but others again are found in lofty mountainous districts, and still others on thirsty plains. Small in size, dark of pelage, and quick of movement, they are difficult to see in the usually thick grass amid which they live, as their rather stubby forms pass quickly before the observer. They make their nests in burrows and are very prolific. From the true mouse they are distinguishable by a short tail and legs, short, blunt muzzle. and ears buried in the fur. There are numerous species and races, some of the latter separated on such fine lines as to be practically indistinguishable, and the entire group is divided into several subgeneric sections, based mainly on the differences in the structure of the teeth. These animals are the representatives in North America of the voles of Europe. The species on the American Continent are most numerous north of the United States and Mexican boundary line.

Subfam. III. **Microtinæ. Meadow Mice, Voles, etc.**

60. Microtus.

$$I.\frac{1-1}{1-1};\ M.\frac{3-3}{3-3}=16.$$

G. S. Miller. *Genera and Subgenera of Voles and Lemmings.* N. Am. Faun., 1896, No. 12.

V. Bailey. *Revision of the American Voles of the genus Microtus.* N. Am. Faun., 1900, No. 17.

Microtus Schrank, Faun. Boica, I, 1st Abth., 1798, p. 72.   Type *Mus arvalis* Pallas.

Arvicola Lacép., Mém. l'Instit., 1801, III, p. 495.

Mynomes Rafin., Am. Month. Mag., 1817, II, p. 45.

Psammomys LeConte, Ann. N. Y. Lyc. Nat. Hist., 1830, p. 132. (nec Cretzschmer.)

Pitymys McMurtr., Am. ed. Cuv., Anim. King., I, App., 1831, p. 434 (footnote).

Ammomys Bon., Sagg. Dist. Met. degli. Anim. Vert., 1831, p. 20 (footnote).

Pinemys Less., Hist. Nat. Mamm. et Ois. decouv. depuis, 1788; Ouvre de Buff., V, 1836, p. 436.

Hemiotomys Selys Longchamps, Essai, Mon. Campagn. Envir. Liege, 1836, p. 7, pl. 1.

Lagurus Glog., Hand-u-Hilfsb. Naturgesch., 1841, p. 97.

Neodon Hodg., Ann. Mag. Nat. Hist., 2d Ser., III, 1849, p. 203.

Agricola Blas., Faun. Wirbelt. Deutsch., 1857, pp. 334-335, 368-374, figs. 202-206.

Chilotus Baird, N. Am. Mamm., 1857, p. 516.

Paludicola Blas., Faun. Wirbelt Deutsch., 1857, pp. 333-334, 343-368, figs. 183-201.

Pedomys Baird, N. Am. Mamm., 1857, p. 517.

Sylvicola Fatio., Les Campagn., Bass. Léman, Ass. Zoöl. Léman, 1867, pp. 63-72, pl. I, figs. 18-25, pl. VI.

Ochetomys Fitzin., Sitzungb., K. Akad. Wiss. Wien., LVI, 1867, p. 47.

Praticola Fatio, Les Campagn. Bass. Léman, Ass. Zoöl. Léman., 1867, p. 36. (Part.)

Terricola Fatio, Les Campagn. Bass. Léman, Ass. Zoöl. Léman, 1867, p. 36.

Micrurus Forsyth-Major, Alt. dell. Soc. Tosc. Scien. Nat., Pisa, III, 1876, p. 126.

Eremiomys Palyakoff Mem. Acad. Imp. Sci. St. Petersb., XXXIX, Suppl., 1881, p. 35.

Neofiber True, Science, 1884, p. 34.

Campicola Schulze, Schrift. Natur. Ver. Harz. Wernig., V, 1890, pp. 24, 25.

Tetramerodon Rhoads, Proc. Acad. Nat. Scien. Phil., 1894, p. 282.

Aulacomys Rhoads, Am. Nat., 1894, p. 182.

Orthriomys Merr., Proc. Biol. Soc. Wash., XII, 1898, p. 106.

Herpetomys Merr., Proc. Biol. Soc. Wash., XII, 1898, p. 107.

Molars not rooted; mandibular molars without closed triangles on outer side; upper incisors not grooved; root of lower incisor extends back to third molar, displacing base of that tooth and terminating above the dental foramina of the ascending ramus; tail terete, longer than hind foot; posterior border of bony palate very variable; middle part of zygoma only slightly expanded; postorbital process of squamosal shelf-like.

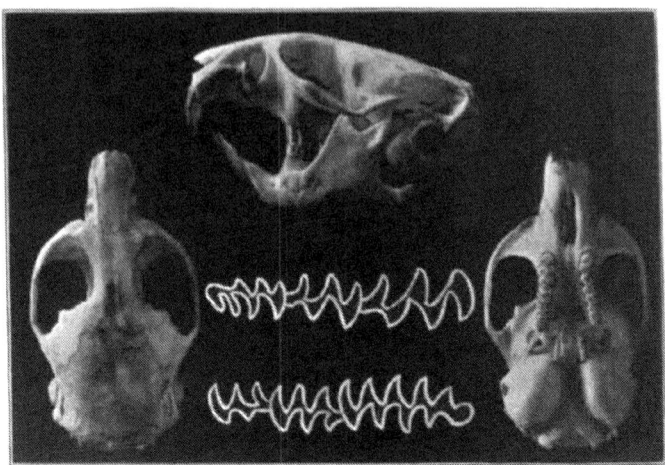

FIG. 51. MICROTUS CALIFORNICUS HYPERYTHRUS.
No. 10758 Field Columbian Mus. Coll. Enlarged ½.
Tooth rows enlarged 6 times.

KEY TO SUBGENERA.

A. Microtus.

Palate normal; lower third molar usually with three closed triangles; lower first molar normally with five closed triangles and nine
salient angles; upper third molar normally with three closed triangles and seven or eight salient angles; upper incisors not grooved;

mammæ four to eight, two or four pectoral and two or four inguinal; normal number eight; plantar tubercles six; soles moderately hairy.

KEY TO SPECIES AND SUBSPECIES OF THE SUBGENUS.

A. Size small. (Mammæ four.)
    a. Above dark brown and black; beneath chest-   <small>PAGE</small>
      nut fulvous.............................*M. fulviventer* 301
    b. Above mixed cinnamon; beneath buffy.....*M. mexicanus* 301
    c. Above bistre and black; beneath plumbeous,
      tinged with drab........................*M. m. phæus* 302
B. Size large. (Mammæ eight.)
    a. Above dark tawny..................*M. c. hyperythrus* 302

303. fulviventer (*Microtus*), Merr., Proc. Biol. Soc. Wash., XII,
    1898, p. 106.

FULVOUS-BELLIED MEADOW VOLE.

Type locality. Cerro San Felipe, State of Oaxaca, Mexico. Altitude, 10,200 feet.

Geogr. Distr. State of Oaxaca, Mexico.

Genl. Char. Similar to *M. mexicanus*, but redder.

Color. Above dark brown and black; under parts chestnut fulvous; tail blackish above, pale fulvous beneath.

Measurements. Total length, 154; tail vertebræ, 38; hind foot, 20. Skull: basal length, 25.4; length of nasals, 7.4; zygomatic width, 15.5; mastoid breadth, 12.4; length of upper molar series, alveolar border, 6.5.

304. mexicanus (*Hemiotomys*), Sauss., Rev. Mag. Zoöl., 2me Sér.,
    1861, p. 3.

MEXICAN MEADOW VOLE.

Type locality. Mount Orizaba, State of Puebla, Mexico.

Geogr. Distr. Eastern part of State of Puebla, Mexico.

Genl. Char. Size small; tail short; ears large. Skull broad; incisive foramina wide and short; first lower molar with six interior salient angles; mammæ four, two inguinal, two pectoral.

Color. Above mixed cinnamon and black, paler on sides; beneath buffy; sides of nose cinnamon; tail above dusky, gray below; feet gray. The above is properly the winter pelage, that of summer being darker.

Measurements. Average of 10 adults: Total length, 138; tail vertebræ, 29; hind foot, 19.35; maximum, 148; 30; 20. Skull: basal length, 24.5; zygomatic breadth, 15.3; mastoid width, 11.6; length of nasals, 7.4; length of upper tooth row, alveolar border, 6.6. (Bailey, N. Am. Faun., No. 17.)

a.—phæus (*Arvicola*), Merr., Proc. Biol. Soc. Wash., VII, 1892, p. 171.

DARK MEADOW VOLE.

Type locality. North slope of Sierra Nevada de Colima, State of Jalisco, Mexico. Altitude, 10,000 feet.

Geogr. Distr. State of Jalisco, Mexico. High mountains.

Genl. Char. Size medium. Skull similar to that of *M. mogollonensis.*

Color. Above bistre and black; under parts plumbeous, tinged with drab; tail above sooty, beneath paler.

Measurements. Total length, 155; tail vertebræ, 34; hind foot, 20.5; ear, 14. Skull: basal length, 25.2; zygomatic width, 15.5; mastoid breadth, 12; length of nasals, 7.3; length of upper molar series, alveolar border, 6.08.

FIG. XLV. MICROTUS C. HYPERYTHRUS.
REDDISH MEADOW VOLE.

californicus hyperythrus (*Microtus*), Elliot, Pub. Field Columb. Mus., III, 1903, p. 161. Zoölogy.

REDDISH MEADOW VOLE.

Type locality. San Quentin, Lower California.

Geogr. Distr. Sea-coast in the vicinity of San Quentin, up to 8,000 feet elevation in the San Pedro Martir Mountains, Lower California, Mexico.

Genl. Char. More reddish in color than *M. californicus*; hind foot longer, and tail about equal to that of *M. californicus*. Skull much larger.

Color. Upper parts dark tawny, slightly lined with black; sides paler; entire under parts plumbeous faintly washed with white; hands and feet grayish buff; tail above dusky, beneath paler.

Measurements. Total length, 203; tail vertebræ, 52; hind foot, 24.5; ear, 16. Skull: occipito-nasal length, 31; Hensel, 26; zygomatic breadth, 17.5; interorbital constriction, 3; length of nasals, 8.5; palatal length, 14.5; mastoid breadth, 14; width of braincase above auditory meatus, 10; length of upper tooth row, alveolar border, 6.5.

B. Pitymys.

Palate normal; lower third molar without closed triangles; lower first molar with five closed triangles and nine salient angles; upper third molar with two or three closed triangles and six salient angles; mammæ, four inguinal; plantar tubercles, five; soles hairy.

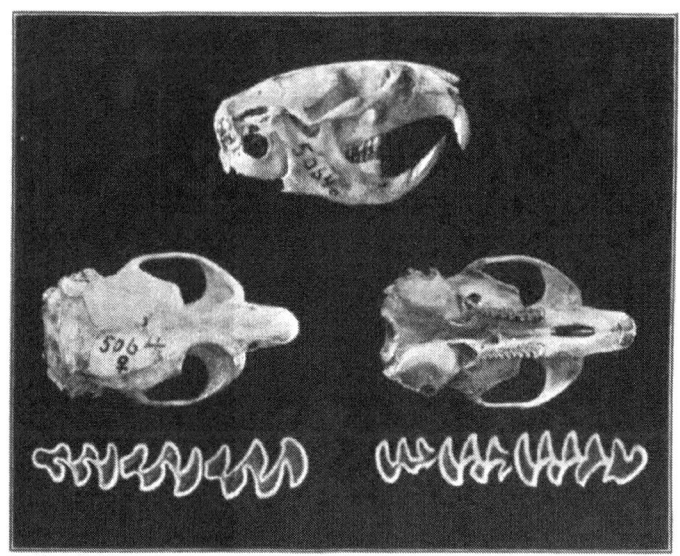

FIG. 52. MICROTUS (PITYMYS) QUASIATER.
No. 5064 Field Columbian Mus. Coll. Enlarged ⅔.
UPPER TOOTH ROW.   LOWER TOOTH ROW.
Enlarged 7 times.   Enlarged 7 times.

305. quasiater (*Arvicola*), Coues, Proc. Acad. Nat. Scien. Phil., 1874, p. 191.

COUES' MEADOW VOLE.

Type locality. Jalapa, State of Vera Cruz, Mexico.

Geogr. Distr. State of Vera Cruz, Mexico.

Genl. Char. Similar to *M. pinetorum*, but darker.

Color. Upper parts dark chestnut brown; beneath blackish ash, hoary in certain lights; tail above like the back, possibly a little darker; hands and feet brownish.

Measurements. Total length, 124–130; tail vertebræ, 16–18; hind foot, 16.5–17; ear, 13–14. Skull: occipito-nasal length, 24.5; Hensel, 22; zygomatic width, 19; interorbital constriction, 3.5; length of nasals, 7.5; palatal length, 11.5; length of upper tooth row, 6.

C. Orthriomys.

First lower molar with one external and two internal closed triangles, and two open triangles; third lower molar with one external and one internal closed triangle and two internal transverse loops; one reëntrant angle between the two closed triangles; third upper molar with one external and one internal closed triangle; anterior loop in second and third upper molars pyriform.

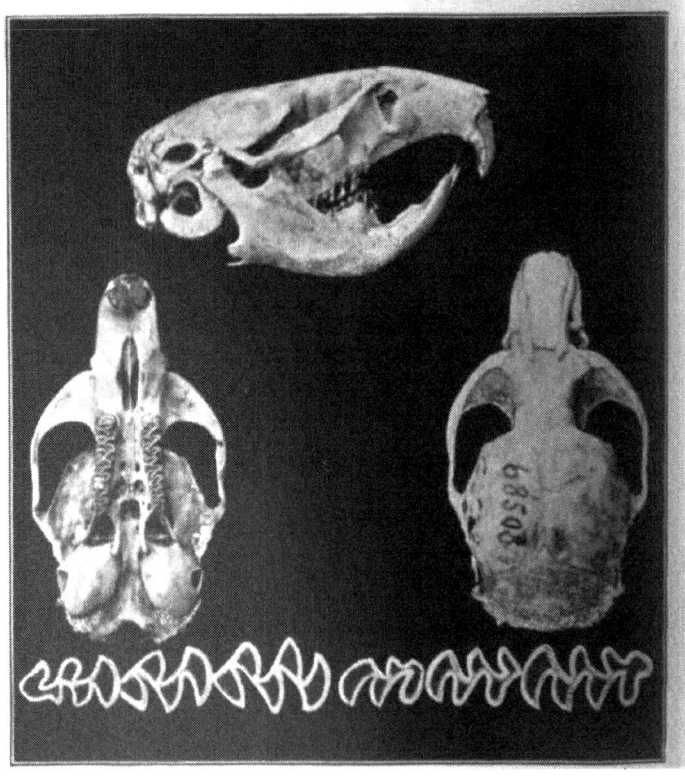

Fig. 53. Microtus (Orthriomys) umbrosus.
No. 68508 U. S. Nat. Mus. Coll. Twice nat. size.

Upper tooth row.	Lower tooth row.
Enlarged 7 times.	Enlarged 7 times.

306. umbrosus (*Microtus*), Merr., Proc. Biol. Soc. Wash., XII, 1898, p. 107.

MOUNT ZEMPOALTEPEC MEADOW VOLE.

Type locality. Mt. Zempoaltepec, State of Oaxaca, Mexico. Altitude, 8,200 feet.

Geogr. Distr. State of Oaxaca, Mexico.

Genl. Char. Size rather large; tail long; ear short; color dark.

Color. Above dusky, mixed with brown; under parts slate, tinged with fulvous.

Measurements. Total length, 184; tail vertebræ, 65; hind foot, 23. Skull: occipito-nasal length, 28; Hensel, 24.5; zygomatic width,

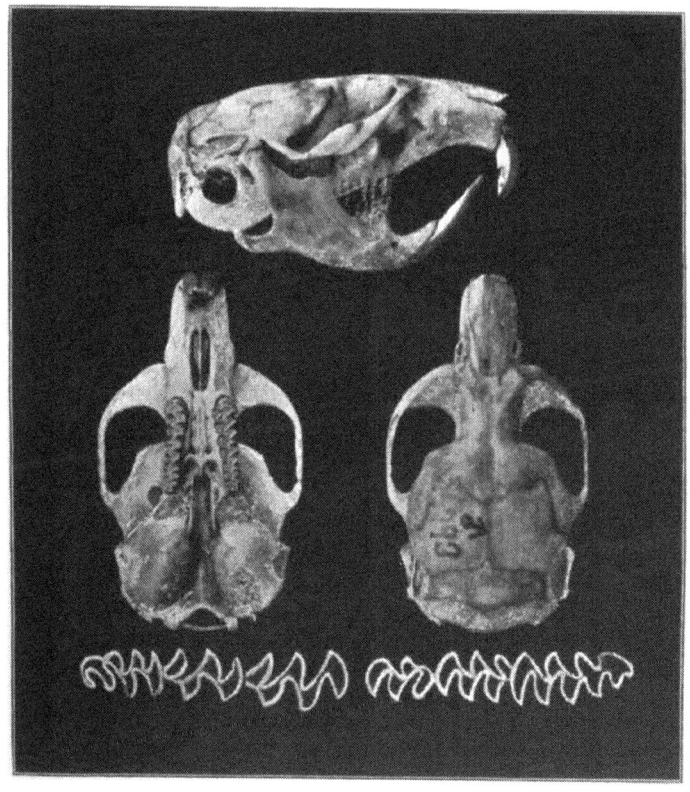

FIG. 54. MICROTUS (HERPETOMYS) GUATEMALENSIS.
No. 76793 U. S. Nat. Mus. Coll. Twice nat. size.

UPPER TOOTH ROW.	LOWER TOOTH ROW.
Enlarged 6 times.	Enlarged 6 times.

15.5; interorbital constriction, 4.5; length of nasals, 7; palatal length, 13; length of upper tooth row, 7.

D. Herpetomys.

Third upper molar with two external and one internal closed triangles and posterior crescentic loop, forming two internal salient angles; first lower molar with one external and two internal closed triangles, and two open triangles; third lower molar with one external and one internal closed triangle, and one anterior and one posterior obliquely transverse internal loop; plantar tubercles, 5; mammæ, 6, 4 pectoral, 2 inguinal.

307. guatemalensis (*Microtus*), Merr., Proc. Biol. Soc. Wash., XII, 1898, p. 108.

GUATEMALAN MEADOW VOLE.

Type locality. Todos Santos, State of Huehuetenango, Guatemala. Altitude, 10,000 feet.

Geogr. Distr. Guatemala, Central America.

Genl. Char. Size medium; color dark; tail short, with few hairs; ears nearly hidden in fur. Skull: bullæ large, swollen; incisive foramina rectangular, truncate anteriorly and posteriorly; root of zygoma anteriorly notched; jugals nearly parallel.

Color. Upper parts mixed black and golden fulvous; under parts slaty black; lips white; tail blackish, unicolor.

Measurements. Total length, 155; tail vertebræ, 40; hind foot, 21. Skull: occipito-nasal length, 26.5; Hensel, 23.5; zygomatic width, 15; interorbital constriction, 4.5; length of nasals, 7.5; palatal length, 13.5; length of upper molar series, 7.

The next genus FIBER contains the Muskrats, the species most familiar to man probably of all the Muridæ, save those of the genus *Mus*, which includes those species commonly called "house rats and mice." The habits of the Muskrat resemble in a considerable degree those of the Beaver, and their large-domed houses, formed of sticks, roots, and grasses are often seen rising from the surface of a pond or lake. The general plan of these structures is very similar to a beaver's dwelling, and the entrance is beneath the water, with the nest or sleeping apartment toward the roof so as to be, if possible, above any sudden rise of the water. Holes in the banks by the side of streams are often made, in which the Muskrat lives, and these excavations sometimes cause the banks to cave in and a large portion of ground to disappear beneath the stream. Muskrats are shy and watchful, and are not often seen by day, but towards evening they become active and swim about in the vicinity of their homes, disappearing with a loud splash beneath the surface if alarmed.   Immense numbers of muskrat skins are sold every year, and made into clothing or linings for garments.

61. Fiber.

$$I.\frac{1-1}{1-1}; \ M.\frac{3-3}{3-3} = 16.$$

Fiber G. Cuv., Leçons d'Anat. Comp., 1, 1800, tab. 1.   Type *Castor zibethicus* Linnæus.

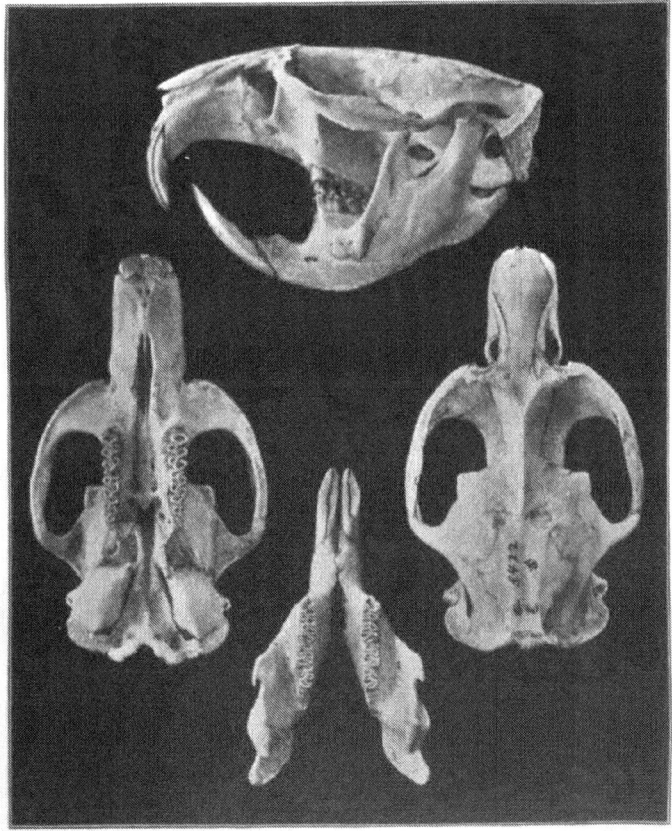

FIG. 55. FIBER ZIBETHICUS PALLIDUS.
No. 5422 Field Columbian Mus. Coll. Nat. size.

Ondatra Lacép., Tab. Mamm., 1799, p. 9. Less. Man., 1827, p. 286. (nec Link, 1795.)

Moschomys Billberg, Syn. Faun. Scandinav., 1, 1828, Mamm., Conspectus A.

Size large; hind feet oblique to the leg; tail flattened sideways for nearly its entire length and fringed with stiff hairs; ears very small, deeply buried in fur; muzzle furry, except nasal pads, which are naked. Palms and soles naked, fringed with hairs, five-tubercled; dentition and skull arvicoline; squamosals much expanded; parietals reduced; interparietal nearly as long as broad; upper incisors almost a circle in shape within and without the jaw; lower incisors enter jaw to root of the condylar process; descending process of condyle hamular and much twisted; palate terminates opposite middle of last molar and has a median azygos protuberance; pterygoid fossa wide and deep; nasals narrow posteriorly, widening rapidly anteriorly, tumid, and terminating behind the incisors; interorbital constriction excessive; processes of squamosal and maxilla have their ends in contact; the jugal being merely a splint, not necessary for the continuity of the zygomatic arch.

zibethicus pallidus (*Fiber*), Mearns, Bull. Am. Mus. Nat. Hist., 1890, p. 280. Elliot, Syn. N. Am. Mamm., 1901, p. 213.

PALE MUSK RAT.

Type locality. Fort Verde, Yavapai County, Arizona.

Geogr. Distr. Lower California? and State of Sonora? north, probably, to Montana.

Genl. Char. Size small, two-thirds that of the typical form. Skull like that of the eastern muskrat, but smaller.

Color. Rusty brown, paler beneath; scattered hairs on tail liver brown.

Measurements. Total length, 500; tail vertebræ, 203; hind foot, 69. Skull: occipito-nasal length, 56; Hensel, 52; zygomatic width, 37; interorbital constriction, 6; length of nasals, 19; palatal length, 31; length of upper molar series, 15; length of mandible, 37.

The Gophers, or Pouched Rats, as the mole-like creatures which compose the next family are called, are stout, shapeless animals, whose powerful shoulders and fore legs with enormous claws on the front toes, suitable for digging, blunt head, minute eyes and small ears, admirably fit them for a life under ground. In the localities in which they abound their long tunnels ramify the soil in all directions and are indicated by the earth raised above the surrounding level, in the same manner as is witnessed in the case of moles. So completely do they live in the ground that one is rarely seen upon it. These animals are provided with cheek pouches, some of enormous size reaching even to the shoulders, and these are convenient receptacles for food and afford a means for transporting it from place to place. In some localities Gophers are veritable pests, undermining the soil with their endless galleries, and flinging the earth excavated from these burrows on every side. In size these animals vary considerably, some being as large as a full grown rat, and others again not half that bigness. The pelage is exceedingly soft, even silky. The skull is heavy, its muscles large and powerful, and the cutting teeth strong and effective, adze-shaped. The family is divided into two chief genera, *Geomys* and *Thomomys*, distinguished by the presence or absence of median grooves on the incisors.

Fam. IV. **Geomyidæ. Pouched Rats.**

C. H. Merriam, *Monographic Revision of the Pocket Gophers*, N. Am. Faun., No. 8, 1895.

Large, fur-lined cheek pouches present, opening outside the mouth. Squamosals expanded; jugal extending to lachrymal; palate sloping below level of zygomata, which are strong and flaring; molars rootless; lower jaw strong; form arvicoline; fore feet fossorial; eyes and ears minute.

KEY TO THE GENERA.

A. Skull large, flat, rather massive; upper incisors grooved; jugal extending to lachrymal; mandible powerful.
  a. Upper premolar with three enamel plates, the
    posterior absent.   PAGE
    a.′ Upper incisor bisulcate.....................*Geomys* 310
    b.′ Upper incisor unisulcate.
      a.″ First and second upper molars with one
        enamel plate each, posterior absent.
        a.‴ Squamosals not greatly expanded
          laterally; orbitosphenoids articulat-
          ing anteriorly with alisphenoids.....*Cratogeomys* 311
        b.‴ Squamosals greatly expanded later-
          ally; orbitosphenoids not articulat-
          ing anteriorly with alisphenoids.....*Platygeomys* 316
      b.″ First and second upper molars with two
        enamel plates each.

62. Geomys. Pocket Gophers.

$$I.\frac{1-1}{1-1}; \ P.\frac{1-1}{1-1}; \ M.\frac{3-3}{3-3} = 20.$$

Geomys Rafin., Am. Month. Mag., II, No. 1, 1817, p. 45. Type
Geomys pinetis Raf. = *Mus tuza* Ord, 1815.
Saccophorus Kuhl, Beitr. Zoöl. und vergl. Anat., 1820, p. 65.
Pseudostoma Say, Long's Exped. Rocky Mts., 1823, I, p. 406.
Ascomys Licht., Abhand. K. Acad. Wiss. Berl., 1825, p. 20. fig. 2.

Upper incisors grooved along the middle; first and second upper
molars with two enamel plates each, posterior one complete; root of
inferior incisor slightly protuberant; zygomata widest anteriorly;
posteriorly but little greater than mastoid breadth; parietals ridged;
audital bullæ elongate, somewhat acute anteriorly, not greatly inflated;
basioccipital very broad posteriorly; pterygoid fossa wide; fore claws
immensely developed; mammæ, three pair, two inguinal, one pectoral.

308. arenarius (*Geomys*), Merr., Mon. Geom., N. Am. Faun., No. 8,
1895, p. 139. Elliot, Syn. N. Am. Mamm., 1901, p. 219.
SAND-LOVING POCKET GOPHER.

Type locality. El Paso, El Paso County, Texas.

Geogr. Distr. From Juarez, State of Chihuahua, Mexico, north to
Las Cruces, and Valley of Upper Rio Grande from El Paso, and west
to Deming, New Mexico.

Genl. Char. Size medium, tail long, well haired, except tip; color
pale. Skull: no sagittal crest; prominent knob at distal end of
squamosal arm of zygoma; interparietal truncate posteriorly on plane
of lambdoid suture; occiput moderately bulging.

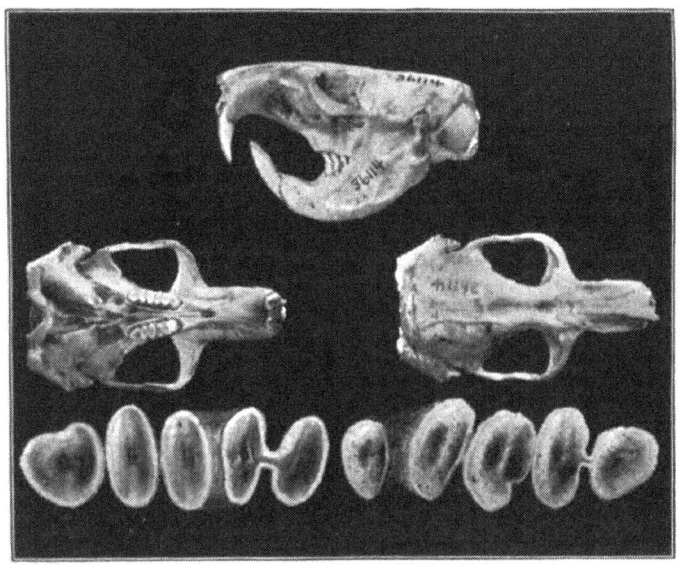

FIG. 56. GEOMYS ARENARIUS.
No. 36114 U. S. Nat. Mus. Coll. Nat. size.

UPPER TOOTH ROW.   LOWER TOOTH ROW.
Enlarged 7 times.   Enlarged 7 times.

Color. Above drab brown, lined with black; under parts and feet white.

Measurements. Total length, 258; tail vertebræ, 88; hind foot, 33. Skull: basal length, 37.5–40.5; Hensel, 34–37; zygomatic width, 24–28; interorbital breadth, 6–7; length of upper molar series, 7.5–8; length of single half of mandible, 27–28.5.

63. Cratogeomys. Powerful Pocket Gophers.

Cratogeomys Merr., Mon. Geom., N. Am. Faun., No. 8, 1895, p. 150, pls. and figs. Type *Geomys merriami* Thomas.

Upper incisor with a single groove, usually open; upper premolar with three enamel plates, posterior absent; shaft convex forward; upper and lower premolars subequal in length. First and second upper molar with one enamel plate each, posterior absent. Last upper molar with a deep sulcus on outer side, none on inner. Skull: a depression extends obliquely across squamosals from root of zygoma to occiput near median line; breadth of cranium posteriorly less than zygomatic breadth; zygomata broad and heavy.

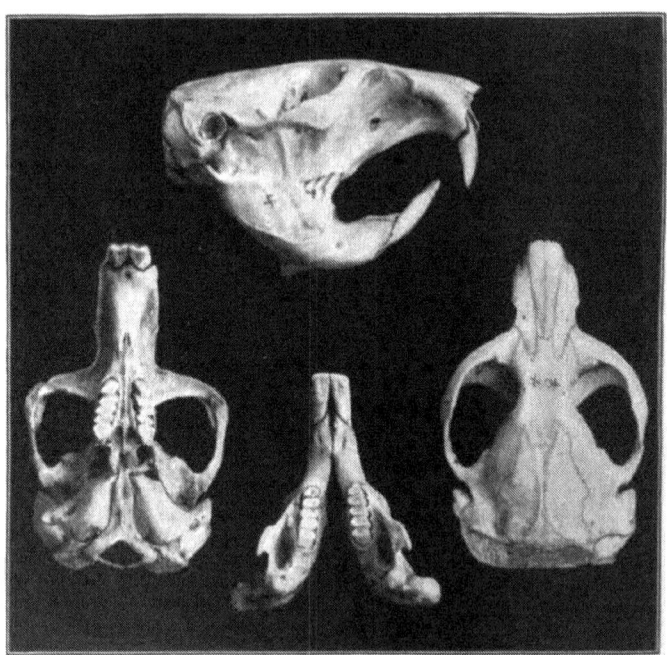

FIG. 57. CRATOGEOMYS CASTANOPS.
No. 4 Field Columbian Mus. Coll. Nat. size.

KEY TO SPECIES AND SUBSPECIES.

*A. Basioccipital truncate, wedge-shaped. (Sides approximating anteriorly.)

  a. Sagittal crest well developed.

    a.' Lower incisor strongly beveled on outer   PAGE
      side.................................*C. merriami* 313

    b.' Lower incisor not beveled on outer side.

      a." Top of skull strongly convex in profile...*C. fulvescens* 316

      b." Top of skull nearly flat in profile.

        a."' Nasals normal, rather long and nar-
          row............................*C. perotensis* 313

        b."' Nasals short, narrow posteriorly,
          broad anteriorly ....*C. estor* 314

* Merr. Mon. *Geomyidæ.* p. 151.

b. No sagittal crest.

309. merriami (*Geomys*), Thomas, Ann. Mag. Nat. Hist., 6th Ser., xii, 1893, p. 271, pls. and figs.

MERRIAM'S POCKET GOPHER.

Type locality. "Southern Mexico."

Geogr. Distr. Valley of Mexico and Toluca Valley, State of Mexico, and States of Puebla and Hidalgo. "South end of Valley of Mexico to an altitude of 10,000 or 11,000 feet; east to Atlixco, State of Puebla, north to Irolo, State of Hidalgo, and west to Lermain, Toluca Valley." (Merr.)

Genl. Char. Size large; tail moderately haired. Skull massive; incisors very large; outer edge of enamel on lower incisors forming a bead; squamosals covering parietals and meeting in a median crest.

Color. Upper parts varying from mixed chestnut brown and black to slate black (melanistic); under parts paler; dusky patch around and behind ears.

Measurements. Total length, 380; tail vertebræ, 112; hind foot, 50. Skull: basal length, 51–70.5; Hensel, 46.5–64; zygomatic breadth, 35–49; interorbital width, 7.5–9.5; length of upper molar series, 11.5–15.5; length of single half of mandible, 36.5–52.

310. perotensis (*Cratogeomys*), Merr., Mon. Geomyidæ, N. Am. Faun., No. 8, 1895, p. 154, pl. 8, fig. 6.

PEROTE POCKET GOPHER.

Type locality. Cofre de Perote, State of Vera Cruz, Mexico. Altitude, 9,500 to 12,000 feet.

Geogr. Distr. State of Vera Cruz, Mexico.

Genl. Char. Size smaller than *C. merriami*; hind feet and tail hairy; no naked nose pad.

Color. Upper parts mixed fulvous and black; usually a white patch at base of tail; dusky patch behind ear; under parts plumbeous washed with fulvous; hind feet white basally, usually dark for remaining portion, sometimes all white; tail dusky and white.

Measurements. Total length, 300; tail vertebræ, 79; hind foot, 40. Skull: basal length, 51.5–55; Hensel, 47–51; zygomatic breadth, 37–39.5; interorbital width, 7–7.5; length of upper molar series, 10.5–12; length of mandible, 37.5–40.5.

310a. estor (*Cratogeomys*), Merr., Mon. Geomyidæ, N. Am. Faun., No.
    8, 1895, p. 155, pl. 8, figs. 4, 5.
LAS VIGAS POCKET GOPHER.
    Type locality. Las Vigas, State of Vera Cruz, Mexico. Altitude,
8,000 feet.
    Geogr. Distr. Northeastern foothills of Cofre de Perote and hills
to the north, State of Vera Cruz, Mexico.
    Genl. Char. Smaller than *C. perotensis;* small naked nasal pad;
hind feet and tail hairy. Skull: superior outline nearly straight.
    Color. Apparently there is no difference in the coloration of
this form and that of *C. perotensis*, the same description acting
equally well for both.
    Measurements. Total length, 315; tail vertebræ, 94; hind foot,
41. Skull: basal length, 52; Hensel, 47-51; zygomatic width, 38-38.5;
interorbital width, 7-8; length of upper molar series, 11; length of
mandible, 37.5-42.

311. oreocetes (*Cratogeomys*), Merr., Mon. Geomyidæ, N. Am. Faun.,
    No. 8, 1895, p. 156, pl. 8, figs. 1, 2.
MOUNTAIN GOPHER.
    Type locality. Mount Popocatepetl, State of Mexico, Mexico.
Altitude, 11,500 feet.
    Geogr. Distr. Higher slopes of Mount Popocatepetl, Mexico.
    Genl. Char. Size large; small nasal pad; tail nearly naked. Skull:
temporal ridges developed; lambdoid crest present; bullæ sub-
globular; groove on upper incisors on inner side wide; outer side of
tooth beveled.
    Color. Above dusky, head and middle of back darkest, washed
with pale brown; brown spot beneath eyes; fore feet dusky; hind
feet white.
    Measurements. Total length, 318; tail vertebræ, 92; hind foot,
43. Skull: basal length, 51; Hensel, 47; zygomatic width, 32.5;
interorbital width, 8; length of upper tooth row, 10.5; length of single
half of mandible. 36.

312. peregrinus (*Cratogeomys*), Merr., Mon. Geomyidæ, N. Am. Faun.,
    No. 8, 1895, p. 158, pl. 8, fig. 3.
MOUNT IZTACCIHUATL POCKET GOPHER.
    Type locality. Mount Iztaccihuatl, State of Mexico, Mexico.
Altitude, 11,500 feet.
    Geogr. Distr. Higher slopes of Mount Iztaccihuatl, Mexico.
    Genl. Char. Size medium; fore foot large, claws nearly equaling
hind foot and claws; nasal pad small. Skull: zygomata broad and
bowed outward, rostrum short; nasals broad; premaxillæ broad,

reaching the plane of the orbit; cranium broad posteriorly, the squamosals expanding laterally; audital bullae short; a single broad groove on upper incisor.

Color. Above mixed dusky and whitish; throat and sides of face darker; under parts paler; hind feet whitish; fore feet similar to but darker than upper parts; tail dusky.

Measurements. Total length, 304; tail vertebræ, 87; hind foot, 42. Skull: basal length, 52; Hensel, 47.5; zygomatic width, 35; interorbital width, 7.5; length of upper molar series, 11.5; length of single half of mandible, 37.

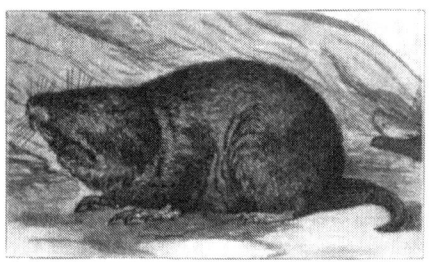

FIG. XLVI. CRATOGEOMYS CASTANOPS.
CHESTNUT-FACED POCKET GOPHER.

313. castanops (*Pseudostoma*), Baird, Rep. Stansb. Exped. to Great Salt Lake, 1852, p. 313.

clarkii, Baird, Proc. Acad. Nat. Scien. Phil., 1855, p. 332.

castanops (*Cratogeomys*), Elliot, Syn. N. Am. Mamm., 1901, p. 220. CHESTNUT-FACED POCKET GOPHER.

Type locality. Prairie Road to Bent's Fort, near the present town of Las Animas, Bent County, Colorado, on the Arkansas River.

Geogr. Distr. States of Chihuahua and Coahuila, Mexico, north to Colorado.

Genl. Char. Size medium; tail medium. Skull broad, heavy; basioccipital rectangular; sides parallel; rostrum and braincase long; superior profile convex; end of maxillary root of zygoma greatly expanded, forming a plate.

Color. Above yellowish brown mixed with black; beneath buffy, plumbeous base of hairs visible; fore feet black; hind feet blackish brown; tail hairs blackish.

Measurements. Total length, 295; tail vertebræ, 77; hind foot, 33. Another specimen from Paladura Cañon, Texas, measured in total length, 280; tail vertebræ, 80; hind foot, 39, taken in the flesh. Skull: basal length, 47.5–56.5; Hensel, 40.5–49; zygomatic width,

30–38; interorbital width, 6.5–7.5; length of upper tooth row, 9.5–10.5; length of single half of mandible, 31.5–38.

a.—*goldmani* (*Cratogeomys*), Merr., Mon. Geomyidæ, N. Am. Faun., No. 8, 1895, p. 160.
GOLDMAN'S POCKET GOPHER.
    Type locality. Cañitas, State of Zacatecas, Mexico.
    Geogr. Distr. State of Zacatecas, Mexico.
    Genl. Char. Similar to *C. castanops* in coloration. Skull: rostrum shorter than that of *C. castanops*; braincase broader; basioccipital somewhat larger.
    Color. Above mixed black and buffy ochraceous; under parts paler.
    Measurements. Total length, 270; tail vertebræ, 83; hind foot, 34. Skull: basal length, 44.5–46; Hensel, 41–42.5; zygomatic width, 32–32.5; interorbital width, 7.5; length of upper molar series, 9–9.5; length of single half of mandible, 31–32.5.

314. fulvescens (*Cratogeomys*), Merr., Mon. Geomyidæ, N. Am. Faun., No. 8, 1895, p. 161, pl. 12, fig. 2.
FULVOUS POCKET GOPHER.
    Type locality. Chalchicomula, State of Puebla, Mexico.
    Geogr. Distr. States of Puebla and Tlaxcala, from "Esperanza north to Perote and west to the northeast base of Mount Malinche, in Tlaxcala," Mexico.
    Genl. Char. Similar to *C. castanops*, but larger and darker; superior outline of skull convex; fronto maxillary suture reaching anteriorly the plane of the front of the zygoma.
    Color. Above mixed yellowish brown and blackish; under parts ochraceous buff.
    Measurements. Total length, 318; tail vertebræ, 102; hind foot, 43. Skull: basal length, 49–55; Hensel, 45–50.5; zygomatic width, 34–40; interorbital width, 6.5–8; length of upper molar series, 10–12; length of single half of mandible, 35–38.

64. Platygeomys. Broad-headed Pocket Gophers.

$$I.\frac{1-1}{1-1}; \ P.\frac{1-1}{1-1}; \ M.\frac{3-3}{3-3} = 20.$$

Platygeomys Merr., Mon. Geomyidæ, N. Am. Faun., No. 8, 1895, p. 162, pls. and figs. Type *Geomys gymnurus* Merriam.

    Skull large, massive, flat; squamosals expanded laterally, giving considerable width to the occipital region, concealing the postglenoid notch, and increasing the glenoid fossa; zygomata heavy and widely spread; jugal large and broad; endoturbinals forming an elongated oblique plate; anterior border of mesethmoid rounded above; upper premolar with three enamel plates; first and second upper molars with only one each; a single sulcus on upper incisor near median line.

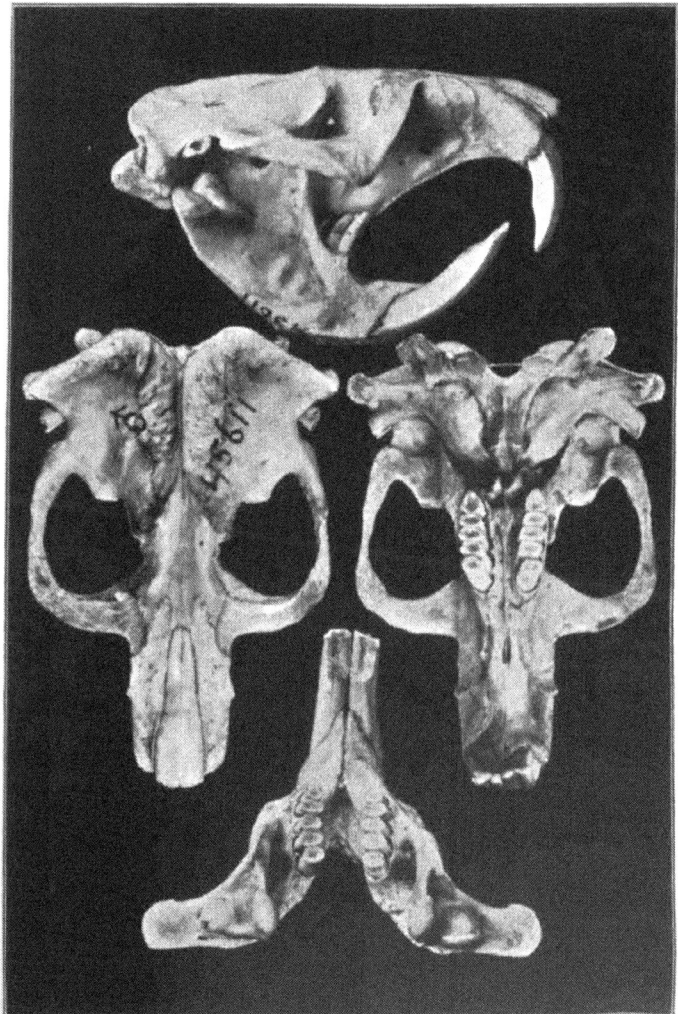

FIG. 58. PLATYGEOMYS GYMNURUS.
No. 45611 U. S. Nat. Mus. Coll. Nat. size

KEY TO SPECIES AND SUBSPECIES.

A. Nasals wedge-shaped; hind feet nearly naked.   PAGE
   a. Above sooty, washed with reddish brown......*P. fumosus*   318
   b. Above chestnut or slate black..............*P. gymnurus*   318
B. Nasals truncate; hind feet hairy, whitish.
   a. Above liver brown......................*P. tylorhinus*   319
   b. Above pale fulvous.................*P. t. angustirostris*   319
   c. Above chestnut..........................*P. planiceps*   319
   d. Above pale, dull chestnut.................*P. neglectus*   319

315. fumosus (*Geomys*), Merr., Proc. Biol. Soc. Wash., VII, 1892, p. 165.

   fumosus (*Platygeomys*), Merr., Mon. Geomyidæ, N. Am. Faun., No. 8, 1895, p. 170, pl. 11, fig. 4; pl. 14, fig. 8.

SOOTY POCKET GOPHER.

   Type locality.   Colima City, State of Colima, Mexico.
   Geogr. Distr.   State of Colima, Mexico.
   Genl. Char.   Size medium; tail and hind feet nearly naked. Skull: posterior portion of cranium broad; squamosals expanded; zygomata rounded anteriorly, greatest width at median portion; nasals wedge-shaped.
   Color.   Above sooty washed with reddish brown; under parts pale plumbeous washed with light brown.
   Measurements.   Total length, 287; tail vertebræ, 82; hind foot, 42. Skull: basal length, 49–55.5; Hensel, 44–51; zygomatic width, 35.5–39; interorbital width, 8–9.5; length of upper molar series, 11.5–13; length of single half of mandible, 40.5–46.5.

316. gymnurus (*Geomys*), Merr., Proc. Biol. Soc. Wash., VII, 1892, p. 166.

   gymnurus (*Platygeomys*), Merr., Mon. Geomyidæ, N. Am. Faun., No. 8, 1895, p. 164, pls. and figs.

NAKED-TAILED POCKET GOPHER.

   Type locality.   Zapotlan, State of Jalisco, Mexico.
   Geogr. Distr.   Valley of Zapotlan and slopes of the Sierra Nevada de Colima, State of Jalisco, Mexico.
   Genl. Char.   Size large; pad on nose and tail naked; feet nearly hairless.
   Color.   Upper parts chestnut, sometimes slate black; under parts paler; young slate black; sides and rump with whitish bristles.
   Measurements.   Total length, 352; tail vertebræ, 105; hind foot, 53. Skull: basal length, 57–62.5; Hensel, 53–57.5; zygomatic width, 42–46.5; interorbital width, 9–10; length of upper molar series, 13–14.5; length of single half of mandible, 41–45.

317. tylorhinus (*Platygeomys*), Merr., Mon. Geomyidæ, N. Am. Faun., No. 8, 1895, p. 167, pl. 13, fig. 1.

TOUGH-SKINNED POCKET GOPHER.

Type locality. Tula, State of Hidalgo, Mexico.

Geogr. Distr. States of Hidalgo and Michoacan, on north slope of the Sierra Madre, Mexico.

Genl. Char. Size large, similar to *P. gymnurus*, but smaller; skull lighter and smaller; nasals broad posteriorly and truncate.

Color. Upper parts liver brown; under parts paler, plumbeous base of hairs showing; hind feet white.

Measurements. Total length, 345; tail vertebræ, 100; hind foot, 45. Skull: basal length, 48–60; Hensel, 45–57; zygomatic width, 40–46; interorbital width, 7.5–9; length of upper molar series, 12–13; length of mandible, 46–52.

a.—*angustirostris* (*Platygeomys*), Merr., Proc. Biol. Soc. Wash., 1903, p. 81.

SLENDER-NOSED POCKET GOPHER.

Type locality. Patamban, State of Michoacan, Mexico.

Genl. Char. Similar to *P. tylorhinus*, but paler. Skull smaller; rostrum, nasals, and incisors narrower and arched posteriorly.

Color. Above pale fulvous grizzled with black; beneath pale slaty plumbeous washed with pale fulvous.

Measurements. Body measurements not given. Skull: basal length, 53.5; zygomatic width, anteriorly, 38; breadth of rostrum, anteriorly, 10; breadth of nasals, anteriorly, 6; at middle, 3.

318. planiceps (*Platygeomys*), Merr., Mon. Geomyidæ, N. Am. Faun., No. 8, 1895, p. 168, pl. 13, fig. 3; pl. 14, fig. 9.

FLAT-HEADED POCKET GOPHER.

Type locality. North slope of the Volcano Toluca, State of Mexico, Mexico. Altitude, 8,600 feet.

Geogr. Distr. Volcano Toluca to City of Toluca, State of Mexico.

Genl. Char. Similar to *P. tylorhinus*; tail longer. Jugal narrow and slightly expanded; upper incisor with single sulcus.

Color. Above chestnut; under parts paler; hind feet whitish; black spot around ear.

Measurements. Total length, 372; tail vertebræ, 121; hind foot, 46. Skull: basal length, 52.5–59; Hensel, 49–55; zygomatic width, 38–42.5; interorbital width, 7.5–8; length of upper molar series, 12–13; length of single half of mandible, 46–51.

319. neglectus (*Platygeomys*), Merr., Proc. Biol. Soc. Wash., XV, 1902, p. 68.

NEGLECTED POCKET GOPHER.

Type locality. Mount Cerro de la Calentura, eight miles northwest of Pinal de Amoles, State of Queretaro, Mexico. Altitude, 9,000 feet.

Genl. Char. "Size small for a *Platygeomys*; general appearance and characters as in *P. planiceps*, but color paler and duller (less chestnut); size smaller; rostrum and nasals shorter; frontal flat interorbitally (not elevated on each side over the orbits); zygomatic arches parallel (instead of strongly divergent anteriorly); jugal light and slender, its faces not strongly developed."

Measurements. Type. "Total length, 310; tail vertebræ, 96; hind foot, 42." (Merr., l. c.)

65. Pappogeomys. Ancient Pocket Gophers.

$$\text{I.}\frac{1-1}{1-1}; \text{ P.}\frac{1-1}{1-1}; \text{ M.}\frac{3-3}{3-3} = 20.$$

Pappogeomys Merr., Mon. Geomyidæ, N. Am. Faun., No. 8, 1895, p. 145, pls. and figs. Type *Geomys bulleri* Thomas.

Molars with enamel pattern of *Geomys*, and the incisors unisulcate, as in some other genera, i. e., *Cratogeomys*, etc. Sphenoid fossa shortened by the orbitosphenoids; mesethmoid higher than long, and with the lower edge extending between wings of vomer posteriorly; zygomata slender; upper premolar with but three enamel plates; first and second upper molars with two enamel plates each; last upper molar has a single exterior sulcus; upper incisor with a single deep median sulcus.

KEY TO SPECIES.   PAGE

A. Size small; tail naked; occiput extending considerably beyond lambdoidal suture.
   a. Above rusty chestnut, beneath paler...........*P. bulleri* 320
   b. Above pale plumbeous tinged with chestnut;
      beneath paler ....*P. albinasus* 321

320. bulleri (*Geomys*), Thomas, Ann. Mag. Nat. Hist., 6th Ser., x, 1892, p. 196, August.

nelsoni Merr., Proc. Biol. Soc. Wash., 1892, p. 164, September.

BULLER'S POCKET GOPHER.

Type locality. Near Talpa, west slope of Sierra de Mascota, State of Jalisco, Mexico. Altitude, 8,500 feet.

Geogr. Distr. State of Jalisco, Mexico.

Genl. Char. Size small; tail naked; occiput extending considerably behind lambdoidal suture; nasals narrow; premaxillæ short, rounded posteriorly, and just reaching plane of orbits.

Color. Above rusty chestnut; under parts paler.

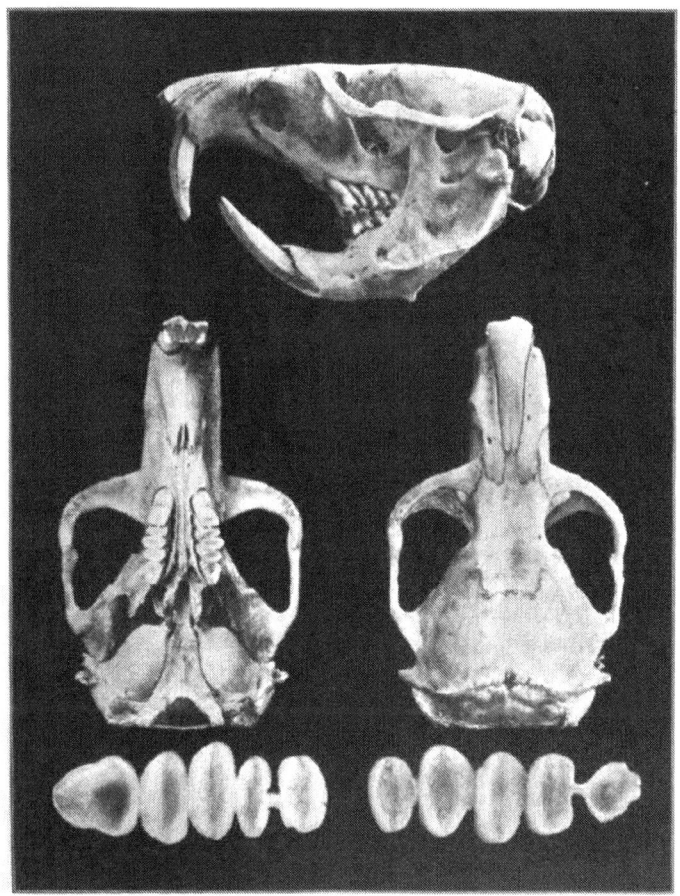

FIG. 59. PAPPOGEOMYS BULLERI.
No. 8359 Field Columbian Mus. Coll. Enlarged ½.

UPPER TOOTH ROW.   LOWER TOOTH ROW.
Enlarged 6 times   Enlarged 6 times.

Measurements. Total length, 238; tail vertebræ, 83; hind foot, 33. Skull: basal length, 35–38; Hensel, 32–35; zygomatic width, 23–25.5; interorbital width, 7–8; length of upper molar series, 8.5–9; length of single half of mandible, 25–28.

321. albinasus (*Pappogeomys*), Merr., Mon. Geomyidæ, N. Am. Faun., No. 8, 1895, p. 149.

WHITE-NOSED POCKET GOPHER.

Type locality. Atemajac, a suburb of Guadalajara, State of Jalisco, Mexico.

Geogr. Distr. State of Jalisco, Mexico. Altitude, 4,000 to 5,100 feet.

Genl. Char. Size small; nasal pad and tail naked. Skull similar to that of *P. bulleri,* but generally larger.

Color. General color of whole body pale plumbeous tinged with chestnut; palest on under parts; white patch on nose reaching nearly to eyes.

Measurements. Total length, 226; tail vertebræ, 68; hind foot, 31. Skull: basal length, 36.5; Hensel, 33; zygomatic width, 25.5; interorbital width, 7.5; length of upper molar series, 9.5; length of single half of mandible, 26.5.

66. Orthogeomys. Straight-headed Pocket Gophers.

$$I.\frac{1-1}{1-1}; \ P.\frac{1-1}{1-1}; \ M.\frac{3-3}{3-3} = 20.$$

Orthogeomys Merr., Mon. Geomyidæ, N. Am. Faun., No. 8, p. 172, pl. and figs. Type *Geomys scalops* Thomas.

Skull elongated, flat; frontal broad; orbital plates of frontal separated by orbitosphenoids; braincase subcylindrical; interorbital constriction lacking; upper premolar with three or four enamel plates; upper incisor with one open sulcus.

KEY TO THE SPECIES.

A. Pelage not bristly; nasal pad large.   PAGE
   a. Nasals long, broad posteriorly.................*O. grandis*   322
   b. Nasals long, narrow posteriorly.
      a.' Premaxillæ very broad, inclined to a point
        posteriorly, on a line with the orbits.........*O. nelsoni*   324
      b.' Premaxillæ narrower, truncate posteriorly,
        not reaching the orbits....................*O. scalops*   324
B. Pelage bristly; nasal pad small or absent........*O. latifrons*   324

322. grandis (*Geomys*), Thomas, Ann. Mag. Nat. Hist., 6th Ser., xii, 1893, p. 270.

GIANT POCKET GOPHER.

Type locality. Duenas, Guatemala.

Genl. Char. Size large; fur coarse. Skull large, heavy; premaxillæ longer than nasals posteriorly; interorbital space broad, its edges rounded anteriorly and inflated; incisors pale yellow, with a deep groove; molars large.

Color. Smoky chocolate brown; "muzzle, cheeks and chin whitish brown; tail naked; hands and feet thinly covered with whitish hairs."

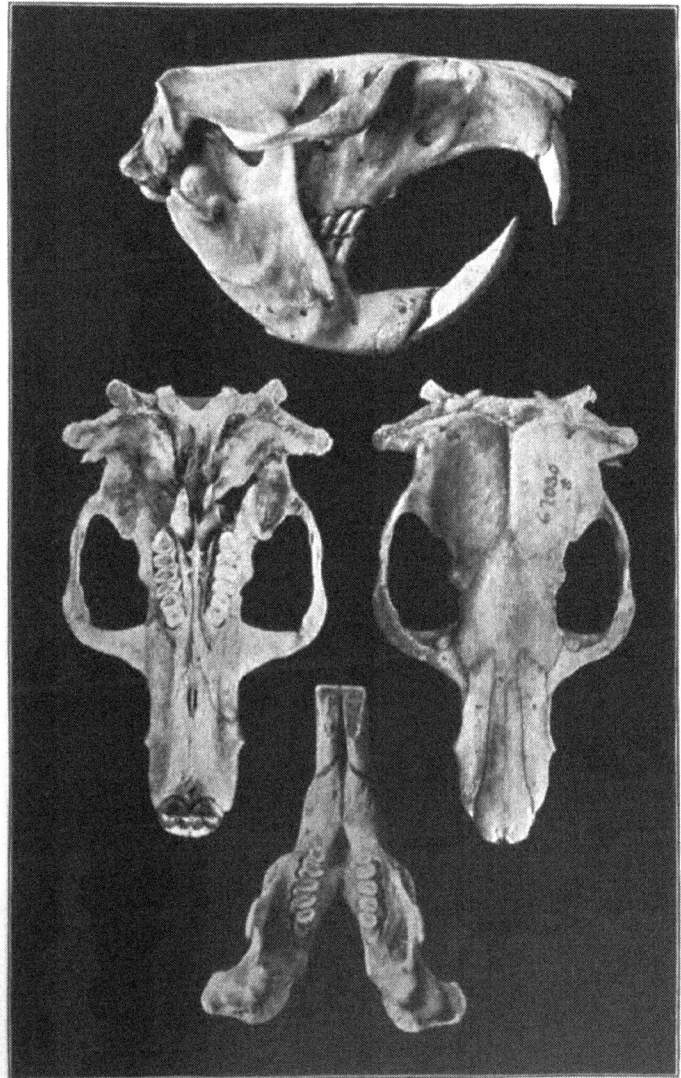

FIG. 60. ORTHOGEOMYS SCALOPS.
No. 6730 U. S. Nat. Mus. Coll. Nat. size.

Measurements. Total length, 320; tail, 135; hind foot, 57. Skull: basal length, 68; Hensel, 62; zygomatic width, 43.8; interorbital width, 15; length of upper molar series, 16.5; length of single half of mandible, 49.

323. nelsoni (*Orthogeomys*), Merr., Mon. Geomyidæ, N. Am. Faun., No. 8, 1895, p. 176, fig. 63.
NELSON'S POCKET GOPHER.
Type locality. Mount Zempoaltepec, State of Oaxaca, Mexico. Altitude, 8,000 feet.
Geogr. Distr. Mount Zempoaltepec and adjacent region, State of Oaxaca, Mexico.
Genl. Char. Size large; ears large; tail, except base, naked. Skull very large, long; nasals very narrow posteriorly; premaxillæ reaching far beyond posterior end of nasals; interorbital constriction slight; frontal narrow, short.
Color. General hue dark brown; slightly paler below.
Measurements. Total length, 397; tail vertebræ, 123; hind foot, 53. Skull: basal length, 70; Hensel, 64; zygomatic width, 44–45; interorbital width, 16–17; length of upper molar series, 15; length of single half of mandible, 50–51.

324. scalops (*Geomys*), Thomas, Ann. Mag. Nat. Hist., 6th Ser., XIII, 1894, p. 437.
DIGGER POCKET GOPHER.
Type locality. Tehuantepec, State of Oaxaca, Mexico.
Geogr. Distr. State of Oaxaca, and possibly in State of Chiapas, Mexico.
Genl. Char. Size large; nasal pad, tail, and hind feet naked.
Color. General color dark seal brown, nearly black.
Measurements. Total length, 369; tail vertebræ, 103.5; hind foot, 50. Skull: basal length, 60.5–64; Hensel, 55.5–59.5; zygomatic width, 37–42; interorbital width, 14.2–16; length of upper molar series, 13–14.5; length of single half of mandible, 43.5–46.5.

325. latifrons (*Orthogeomys*), Merr., N. Am. Faun., No. 8, 1895, p. 178, pl. 11, and figs. 5, 6; text fig. 64.
BRISTLED POCKET GOPHER.
Type locality. Guatemala. Exact locality unknown.
Geogr. Distr. Guatemala. Range unknown.
Genl. Char. Pelage bristly; tail long, naked; feet scantily haired; incisors with nearly median groove; zygomata narrow, slender, broadest posteriorly; nasals short, narrow.
Color. Uniform dull sooty brown above and below.
Measurements. Total length, 320; tail, 100; hind foot, 39.

67. Heterogeomys. Distinct Pocket Gophers.

$$I.\frac{1-1}{1-1}; \ P.\frac{1-1}{1-1}; \ M.\frac{3-3}{3-3} = 20.$$

Heterogeomys Merr., Mon. Geomyidæ, N. Am. Faun., No. 8, 1895, p. 179, pls. and figs. Type *Geomys hispidus* Le Conte.

Skull high, narrow; zygomata moderately wide; frontal broad and flat; orbital plate of frontal usually perforated by a foramen above apex of sphenoidal fissure; nasals much arched anteriorly; squamous part of occipital plane high above mastoid bullæ; orbito-sphenoids narrow and long, not articulating with alisphenoids; upper

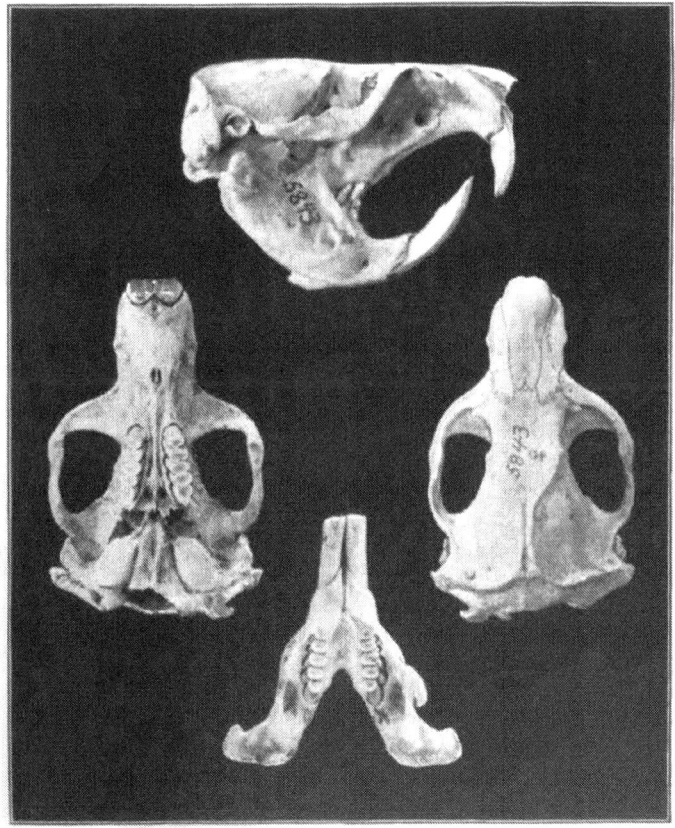

FIG. 61. HETEROGEOMYS TORRIDUS.
No. 5843 Field Columbian Mus. Coll. Nat. size.

part of optic foramen disappearing in advanced life; squamosal expansion slight; fronto-maxillary suture reaching orbit in front of lachrymal. (ex Merr., l. c.)

KEY TO THE SPECIES.

326. hispidus (*Heterogeomys*), Le Conte, Proc. Acad. Nat. Scien. Phil., 1852, p. 158, pls. and figs.

HARSH-COATED POCKET GOPHER.

Type locality.   Near Jalapa, State of Vera Cruz, Mexico.

Geogr. Distr.   State of Vera Cruz, Mexico, at 4,000 to 4,500 feet altitude.

Genl. Char.   Size large; upper incisors with one deep sulcus each on inner side; nose pad and tail naked.

Color.   General hue above dark seal brown; under parts slightly paler.

Measurements.   Total length, 345; tail vertebræ, 92; hind foot, 47.3.   Skull: basal length, 55–57.5; Hensel, 51–53; zygomatic width, 36–45; interorbital width, 10–11.5; length of upper molar series, 13.5–14; length of single half of mandible, 38–42.

327. torridus (*Heterogeomys*), Merr., Mon. Geomyidæ, N. Am. Faun. No. 8, 1895, p. 183, pls. and figs.

TROPICAL POCKET GOPHER.   *Tultusia* in Guatemala.

Type locality.   Chichicaxtle, State of Vera Cruz, Mexico.

Geogr. Distr.   Lowlands, State of Vera Cruz, through States of Oaxaca and Chiapas, Mexico, into Guatemala.

Genl. Char.   Similar to *H. hispidus*; size large.   Skull broad and heavy.

Color.   Dark seal brown, slightly paler below.

Measurements.   Total length, 323; tail vertebræ, 88; hind foot, 52.   Skull: basal length, 54–60; Hensel, 50–55.5; zygomatic width, 37–43.5; interorbital breadth, 10.5–11; length of upper molar series, 13–14.5; length of single half of mandible, 38–42.

68. Macrogeomys.   Large Size Pocket Gophers.

$$I.\frac{1-1}{1-1}; \ P.\frac{1-1}{1-1}; \ M.\frac{3-3}{3-3} = 20.$$

Macrogeomys Merr., N. Am. Faun., No. 8, 1895, p. 185, pls. and figs.   Type *Geomys heterodus* Peters.

Size large; large naked nasal pad; tail naked; pelage silky.   Skull:

frontal flat, depressed medianly, broad; postorbital processes large; palato-pterygoids short, broad, truncate posteriorly; braincase high above root of zygoma; lambdoid crest straight; occipital plane flat, sloping forward; upper premolar with four enamel plates; first upper and second lower molars with two enamel plates each; last upper molar with lengthened heel and deep outer sulcus; upper incisor with only one narrow deep sulcus on inner third of face.

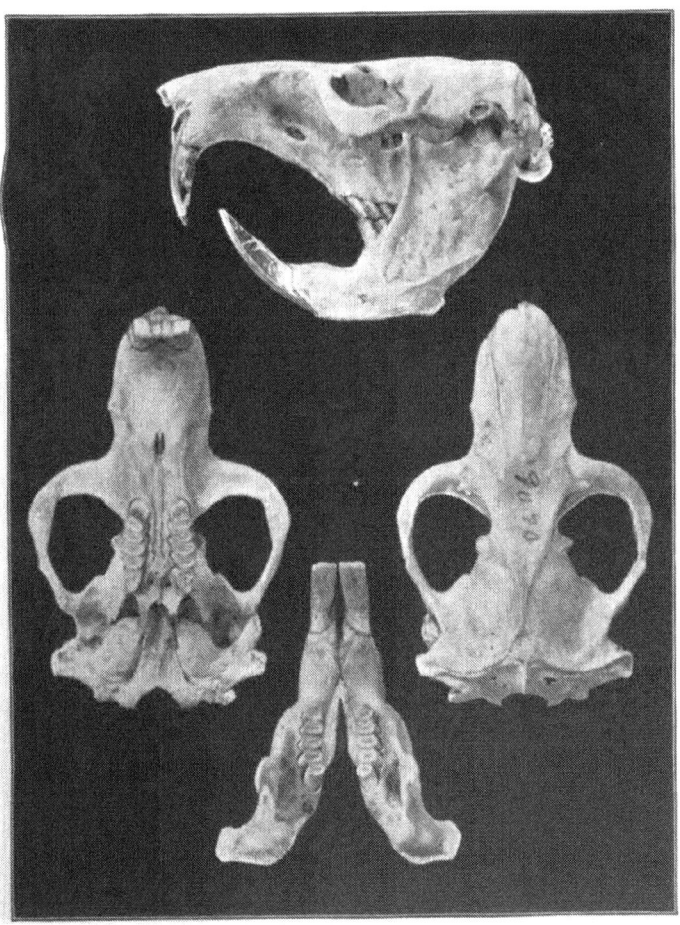

FIG. 62. MACROGEOMYS CHERRII.
No. 9070 Am. Mus. Nat. Hist. Coll. Nat. size.

KEY TO THE SPECIES.

A. Audital bulla rounded on outer side; no white
   patch on head.   PAGE
   a. Upper parts sepia brown; skull short, broad...*M. heterodus*   328
   b. Upper parts chocolate brown anteriorly, buffy
      posteriorly.........................*M. dolichocephalus*   328
   c. Upper parts dark seal brown, nearly black......*M. cavator*   329
   d. Upper parts dusky chocolate brown, uniform ....*M. pansa*   329
B. Audital bulla flattened on outer side; white
   patch on head.
   a. Upper parts dark brown, beneath whitish;
      roots of maxilla and squamosal meeting above
      jugal.................................*M. costaricensis*   330
   b. Upper parts plumbeous, beneath paler; roots
      of maxilla and squamosals not meeting above
      jugal].....................................*M. cherrii*   330

328. heterodus (*Geomys*), Peters, Monatsb. K. Preuss. Akad. Wiss.
      Berl., 1864, p. 177, pls. and figs.
IRAZU POCKET GOPHER.

Type locality. Costa Rica. Locality unknown.

Geogr. Distr. Irazu Range, and possibly other parts of Costa Rica.

Genl. Char. Size large; upper incisor with a deep narrow groove
on inner side of median line; large naked nasal pad; tail naked, as
are also the feet, except a few hairs on toes; no external ears; incisors in front orange. Skull large, short; frontal broad, flat, concave
between orbits, with deep notches in front of postorbital processes,
which are large; zygomata wide; jugal broad; nasals wedge-shaped;
pterygoids U-shaped, truncate posteriorly.

Color. Above hair brown; remaining portions of body and the
muzzle soiled gray, this hue reaching well up on side of rump, and
also covering base of tail.

Measurements. Total length, 325; tail, 65; hind foot, 41. Skull:
basal length, 58 60; Hensel, 51.2-55; zygomatic width, 42.5; interorbital width, 10-14; length of upper molar series, 14-15; length of
mandible, 44 45.

329. dolichocephalus (*Macrogeomys*), Merr., N. Am. Faun., No. 8,
      1895, p. 189, pls. and figs.
NARROW-HEADED POCKET GOPHER.

Type locality. San José, Costa Rica.

Geogr. Distr. Range unknown.

Genl. Char. Size large, similar to *M. heterodus*, but darker.
Skull long and narrow; zygomata narrow, the breadth but slightly

greater than the mastoid breadth, and the jugal, which is short and broadest in the center, is overlapped by the maxillary and squamosal arms of the zygoma; braincase subcylindrical; nasals short, terminating in front of the zygomatic arches; pterygoids broad and short; mandible long, narrow.

Color. Above dull chocolate brown; muzzle and lower part of rump buffy; under parts pale chocolate brown.

Measurements. Total length, 380; tail, 75; hind foot, 45. Skull: basal length, 52–65; Hensel, 48–60; zygomatic width, 33–40.5; interorbital breadth, 9.5; length of upper molar series, 13–15.5; length of single half of mandible, 38.5–48.5.

330. cavator (*Macrogeomys*), Bangs, Bull. Mus. Comp. Zoöl., xxxɪx, 1902, p. 42.

BOQUETE POCKET GOPHER.

Type locality. Boquete, Chiriqui, Panama. Altitude, 4,800 feet.

Genl. Char. Similar to *M. dolichocephalus*, but with comparative difference in crania; colors darker.

Color. Above dark seal brown, almost black; beneath similar but grizzled, a white patch beneath chin and on under sides of wrists; feet and hands naked, yellowish brown; tail naked, yellowish brown, end black.

Measurements. Total length, 360–410; tail vertebræ, 108–125; hind foot, 47–54; ear, 7–8. Skull: type, basal length, 64; occipito-nasal length, 67.8; zygomatic width, 45.8; mastoid width, 33.4; interorbital width, 11; length of nasals, 25.4; length of palate to palatal notch, 44.6; upper molar series, 15.4; length of single half of mandible, 51.2. (Bangs, l. c.)

331. pansa (*Macrogeomys*), Bangs, Bull. Mus. Comp. Zoöl., xxxɪx, 1902, p. 44.

BROAD-FOOTED POCKET GOPHER.

Type locality. Bogaba, Chiriqui, Panama. Altitude, 600 feet.

Genl. Char. Smaller than *M. cavator*; hind foot proportionately larger.

Color. Above dull, dusky, chocolate brown; under parts grizzled, belly whitish; feet and hands naked, yellowish brown; tail naked, yellowish brown, tip dusky.

Measurements. Total length, 320–330; tail vertebræ, 110; hind foot, 48–52; ear, 5–7. Skull: basal length, 54; occipito-nasal length, 57.6; zygomatic width, 36; mastoid width, 27.8; interorbital width, 11.8; length of nasals, 23; length of palate to palatal notch, 37; upper molar series, 13; length of single half of mandible, 41. (Bangs, l. c.)

332. costaricensis (*Macrogeomys*), Merr., N. Am. Faun., No. 8, 1895, p. 192, pls. and figs.
PACUARE POCKET GOPHER.

Type locality. Pacuare, Costa Rica.

Geogr. Distr. Range unknown.

Genl. Char. Unique specimen immature. Skull similar to that of *M. dolichocephalus*, but with broader nasals; zygomata standing at nearly right angles to axis of skull, the anterior angle abruptly rounded; jugal narrower; pterygoids shorter and broader; audital bullæ disk-shaped and separated by a groove from the bullæ.

Color. Above dark brown; under parts whitish; a large white patch on top of head between eyes and ears.

Measurements. Total length, 330; tail about 80; hind foot, 33, without claw. Skull: basal length, 48.5; Hensel, 44.5; zygomatic width, 33; interorbital width, 9; length of upper molar series, 13; length of single half of mandible, 37.

333. cherrii (*Geomys*), Allen, Bull. Am. Mus. Nat. Hist., 1893, p. 337.
CHERRIE'S POCKET GOPHER.

Type locality. Santa Clara, Costa Rica.

Geogr. Distr. Costa Rica; range unknown.

Genl. Char. Nasal pad, tail, and hind feet naked; white head patch. Skull differs from that of *M. costaricensis* in certain particulars, such as the jugal, which is large and long and is not covered by the maxillary arm of the zygoma; in the horizontal part of the zygomatic arch not being strongly convex upward, and has not the constriction between the orbital and temporal fossa; and the orbitotemporal fossæ are broadest at the middle instead of being narrowest.

Color. Above chocolate or sooty brown, except a large white patch on crown; throat, breast, and inner side of limbs dusky grayish; rest of under parts grayish buff; tail and feet naked, reddish.

Measurements. Total length, 275; tail, 80; hind foot, 40. Skull: basal length, 47.5; Hensel, 44; zygomatic width, 34; interorbital width, 9.5; length of upper molar series, 12; length of single half of mandible, 37.

69. Zygogeomys. Zygomata Pocket Gophers.

$$I.\frac{1-1}{1-1};\ P.\frac{1-1}{1-1};\ M.\frac{3-3}{3-3} = 20.$$

Zygogeomys Merr., Mon. Geomyidæ, N. Am. Faun., No. 8, 1895, p. 195, pls. and figs. Type *Zygogeomys trichopus* Merriam.

Skull long and narrow; maxillary and squamosal branches of zygoma in contact above jugal, which is mainly external; rostrum long and narrow; orbitosphenoids large, and with the exception of a foramen at apex, close the upper part of sphenoidal fissure, and ankylosed with the alisphenoid; mesethmoid quadrangular; upper incisors bisulcate, chief sulcus on median line, minor on inner convexity. First and second upper molars with two enamel plates

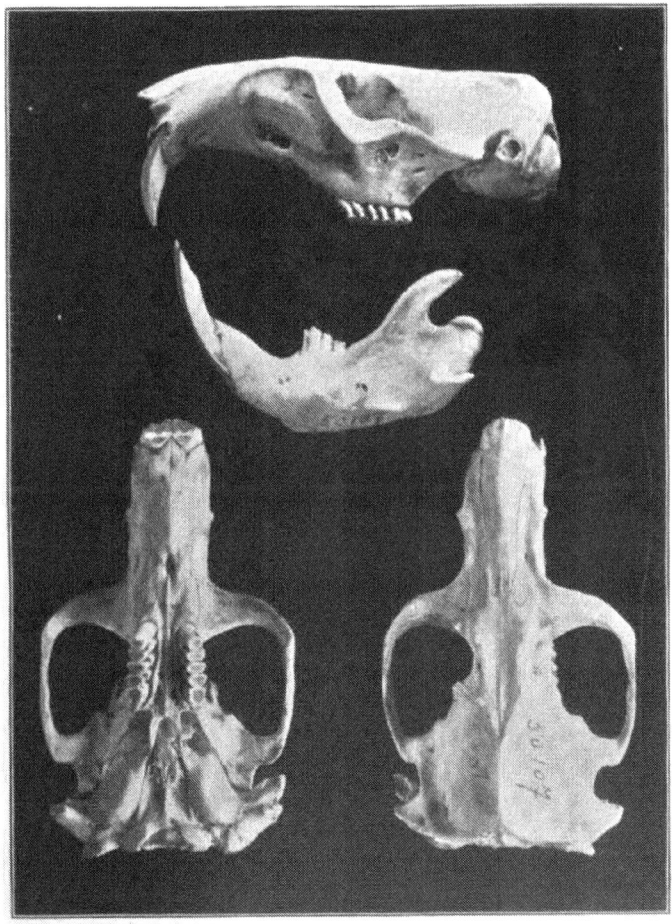

FIG. 63. ZYGOGEOMYS TRICHOPUS.
No. 50107 U. S. Nat. Mus. Coll. Nat. size.

each; upper premolar with four enamel plates; third upper molar incomplete double prism, crown longer than broad; sulcus on middle of outer side only, inner enamel plate covering two-thirds or three-fourths of inner side of tooth, reaching end of heel posteriorly; outer enamel plate covering about half of outer side of tooth, its anterior bent strongly outward; upper incisors bisulcate, principal sulcus on inner side of median line.   (ex Merr., l. c.)

334 trichopus (*Zygogeomys*), Merr., Mon. Geomyidæ, N. Am. Faun., No. 8, 1895, p. 196, pls. and figs.
PINE ZONE POCKET GOPHER.

Type locality.   Nahuatzin, State of Michoacan, Mexico.

Geogr. Distr.   Sierra Nevada, State of Michoacan, Mexico.   Altitude, 6,800 to 9,500 feet.

Genl. Char.   Size large; tail long, naked; naked pad on nose; feet covered with hair; other characters as in genus.

Color.   Dark slate to seal brown, washed with ferrugineous on upper parts; under parts plumbeous, tinged with fulvous; white patch on throat; hind feet whitish or dark gray.

Measurements.   Total length, 346; tail vertebræ, 115; hind foot, 46.   Skull: basal length, 46.5–58.5; Hensel, 43–54; zygomatic width, 35–39; interorbital width, 8–9.5; length of upper molar series, 10.5–12; length of single half of mandible, 33–42.

The genus THOMOMYS is distinguished from *Geomys* and the allied genera by the absence of median grooves on the incisors, by the smaller and more feeble claws, and by even larger cheek pouches. The habits of the species are the same, and they often prove as great pests to the agriculturalist as do their relatives of the other genera. While usually of smaller size than the species of Geomys, there is one, *T. bulbivorus*, from the northwestern portion of the United States, which equals the largest member of that genus.   The Gophers feed upon all kinds of herbage and bulbous roots.   They are irascible, bite severely, are very quick in their movements, alert to each danger that threatens, and fertile in methods for avoiding it.

70. Thomomys.   Pocket Gophers.

$$I.\frac{1-1}{1-1};\ P.\frac{1-1}{1-1};\ M.\frac{3-3}{3-3} = 20.$$

Thomomys Wied, Nov. Act. Phys. Med. Akad. Caes. Leop. Carol., XIX, 1839, p. 377.   Type *Thomomys rufescens* Wied.

Diplostoma Rich., Faun. Bor. Amer., I, 1829, p. 206. (nec Rafin.)

Oryctomys Ed. & Gerv., Mag. Zoöl., VI, 1836, p. 23.   (Part.)

Upper incisors without median sulcus, but with a fine marginal groove, not, however, always present; root of inferior incisors forming a protuberance on outside of condylar ramus; zygomata wider posteriorly than mastoid diameter of skull; audital bullæ inflated; basioccipital narrower in the middle than bullæ at same place. pterygoid fossa rather short, and wide posteriorly; upper molars with exterior edge of crown acute; lower molars with interior edge acute; fore claws moderately developed.

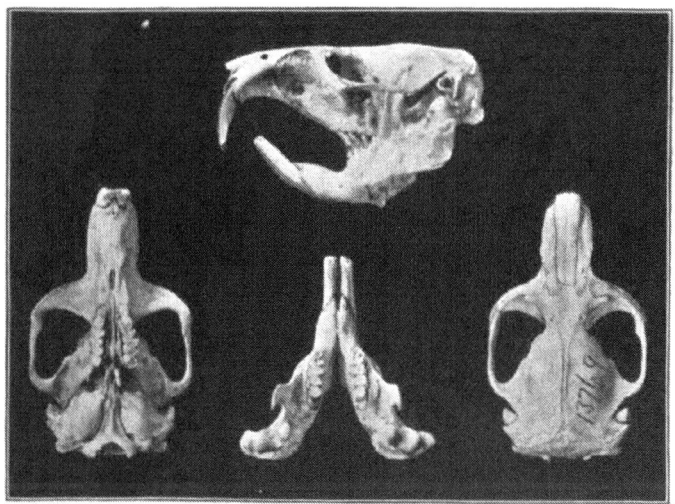

FIG. 64. THOMOMYS ATROVARIUS.
No. 13769 Am. Mus. Nat. Hist. Coll.   Nat. size.

KEY TO THE SPECIES AND SUBSPECIES.

A. Upper incisors without median grooves; fore claws weak.

  a. Upper parts yellowish brown or fulvous.

FIG. XLVII. THOMOMYS FULVUS.
REDDISH BROWN POCKET GOPHER.

335 fulvus (*Geomys*), Woodh., Proc. Acad. Nat. Scien. Phil., 1852,
    p. 201. Elliot, Syn. N. Am. Mamm., 1901, p. 227.
    umbrinus. Baird, N. Am. Mamm., 1857, p. 399.
REDDISH BROWN POCKET GOPHER.

    Type locality. San Francisco Mountains, Coconino County, Arizona.

    Geogr. Distr. State of Sonora, Mexico, north into New Mexico, and "Arizona to Central California." Lower California?

Genl. Char. Size medium; tail half as long as body and head; claws large, long.

Color. Above reddish brown, darkest on dorsal region; sides and under parts yellowish white, tinged with rufous on abdomen; lips and ears dusky; inside of pouches whitish; tail with basal part like back.

Measurements. Total length, 239; tail vertebræ, 79; hind foot, 29; ear, 8. Skull: basal length, 37; Hensel, 35; zygomatic width, 25; interorbital breadth, 7; length of upper molar series, 8; length of mandible, 25.

a.—anitæ (*Thomomys*), Allen, Bull. Amer. Mus. Nat. Hist., 1898, p. 146.

SANTA ANITA POCKET GOPHER.

Type locality. Santa Anita, Lower California, Mexico.

Geogr. Distr. Cape Region, Lower California, Mexico.

Genl. Char. Similar to *T. fulvus;* nasals gradually broaden postero-anteriorly.

Color. Above yellowish brown, darkest on median line; under parts buff; about mouth, cheek pouches and ears blackish.

Measurements. Total length, 250; tail vertebræ, 62; hind foot, 35; ear, 8.5. Skull: total length, 42; basal length, 38.7; zygomatic width, 26.5; mastoid width, 21; interorbital constriction, 8; length of nasals, 14.6; width of nasals posteriorly, 2.5.

b.—alticola (*Thomomys*), Allen, Bull. Am. Mus. Nat. Hist., 1899, p. 13.

SIERRA LAGUNA POCKET GOPHER.

Type locality. Sierra Laguna, Lower California, Mexico. Altitude, 7,000 feet.

Geogr. Distr. High altitudes in the Sierra Laguna, Lower California, Mexico.

Genl. Char. Similar to *T. f. anitæ*, but darker. Skull similar in size and characters.

Color. Above yellowish brown and black, darkest on dorsal region; nose and edges of cheek pouches blackish; spot at base of ear dusky; under parts reddish fulvous; inside of cheek pouches and anal region white; tail and feet whitish.

Measurements. Total length, 225; tail, 61; hind foot, 30.

c.—nigricans (*Thomomys*), Rhoads, Proc. Acad. Nat. Scien. Phil., 1895, p. 36.

BLACKISH POCKET GOPHER.

Type locality. Witch Creek, San Diego County, California.

Geogr. Distr. Lower California. Mexico, north into southern California.

Genl. Char. Size large; claws short, thick; Skull massive, angular; dentition weak; interparietal longer than wide.

Color. Above tawny brown; rump lighter; head blackish; ears and aural patch sooty; beneath tawny ash; feet and lower surface of limbs ash.

Measurements. Total length, 260; tail vertebræ, 89; hind foot, 33.5. Skull: total length, 39; zygomatic breadth, 24.4; basilar length, 35.5; nasals, 11.9; interorbital constriction, 6.9; length of mandible, 25.

d.—intermedius (*Thomomys*), Mearns, Proc. U. S. Nat. Mus., 1897, XIX, p. 719. Elliot, Syn. N. Am. Mamm., 1901, p. 228.
PLATEAU POCKET GOPHER.

Type locality. Huachuca Mountains, southern Arizona. Altitude, 9,000 feet.

Geogr. Distr. Mountains connecting Colorado Plateau with that of Mexico.

Genl. Char. Smaller than *T. fulvus.*

Color. General hue mars brown; dorsal area plumbeous black; feet and tail soiled white; base of under fur plumbeous.

Measurements. Total length, 220; tail vertebræ, 66; hind foot, 24. Skull: 31.5×22. (Mearns, l. c.)

336 martirensis (*Thomomys*), Allen, Bull. Amer. Mus. Nat. Hist., 1898, p. 147.
SAN PEDRO MARTIR POCKET GOPHER.

Type locality. San Pedro Martir Mountains. Lower California, Mexico. Altitude, 8,200 feet.

Geogr. Distr. San Pedro Martir Mountains, Lower California, Mexico. Altitude, 7,000 to 8,200 feet.

Genl. Char. Larger than *T. fulvus;* color less fulvous.

Color. Upper parts mixed grayish brown and blackish. darkest on median line; side washed with pale fulvous; under parts grayish white; hairs plumbeous at base; around mouth and entrance to cheek pouches dusky; ear patch blackish; tail and feet grayish white.

Measurements. Total length, 248; tail vertebræ, 67; hind foot, 31. Skull: total length, 44.5; basal length, 41; zygomatic width, 25; mastoid width, 21; interorbital constriction, 9; length of nasals, 16; width of nasals anteriorly, 5.

337. aphrastus (*Thomomys*), Elliot, Pub. Field Columb. Mus., 1903. p. 219. Zoölogy.

TROUBLESOME POCKET GOPHER.

Type locality. San Tomas, Lower California, Mexico.

Genl. Char. About the size of *T. fulvus*, but darker and grayer. Nasals short, broad anteriorly; pterygoid almost touching the bullæ.

Color. (March.) Upper parts mixed broccoli brown and ochraceous, the slate at base of hairs showing through occasionally and giving a grayish tint to the pelage; sides bright ochraceous buff; spot behind ear, nose and openings of pouches black; under parts ochraceous buff, plumbeous of under fur showing through; hands and feet grayish white; tail ochraceous buff above, yellowish beneath. Another specimen in July from San Quentin is lavender gray, darkest on dorsal region, with top of head ochraceous buff and black, and sides of head pinkish buff. This individual is beginning to change to the darker pelage of winter.

Measurements. Total length, 222; tail vertebræ, 69; hind foot, 29; ear, 7.5. Skull: occipito-nasal length, 38; Hensel, 35; interorbital constriction, 6; zygomatic width, 25; length of nasals, 12.5; anterior width of nasals, 4.5; palatal length, 23; length of upper molar series, 7.5; length of mandible, 22; length of lower molar series, 7; height at coronoid process, 15.

338. peregrinus (*Thomomys*), Merr., Proc. Biol. Soc. Wash., VIII, 1893, p. 146.

WANDERING POCKET GOPHER.

Type locality. Salazar, State of Mexico, Mexico.

Geogr. Distr. State of Mexico. Range unknown.

Genl. Char. Similar to *T. fulvus*, but darker; incisors curving forward; rostrum narrower than interorbital width.

Color. Above dark umber brown; sooty black on head and on median line; sides dull fulvous; under parts buffy; end of nose, around mouth and ear patch blackish; fore feet umber brown to base of toes; hind feet plumbeous (under fur showing through).

Measurements. Total length, 207; tail vertebræ, 72; hind foot, 28.5.

339. orizabæ (*Thomomys*), Merr., Proc. Biol. Soc. Wash., VIII, 1893, p. 145.

ORIZABA POCKET GOPHER.

Type locality. Mt. Orizaba, State of Puebla, Mexico. Altitude, 9,500 feet.

Geogr. Distr. State of Puebla, Mexico. Range unknown.

Genl. Char. Size medium; color similar to that of *T. fulvus;* tail long; feet well haired; muzzle of skull longer and broader than in *T. peregrinus*, and broader than interorbital width.

Color. Upper parts dark umber brown; sides fulvous; under parts buffy fulvous; under side of face blackish; feet and apical third of tail white. Melanistic phase, which the type represents, is slate black all over, except ends of fore and hind feet and the apical third of tail and inside of cheek pouches, which are white.

Measurements. Total length, 217; tail vertebræ, 68; hind foot, 30.

340. atrovarius (*Thomomys*), Allen, Bull. Amer. Mus. Nat. Hist., 1898, p. 148.

TATAMELES POCKET GOPHER.

Type locality. Tatameles, State of Sinaloa, Mexico.

Geogr. Distr. State of Sinaloa, Mexico. Range unknown.

Genl. Char. Size medium; nasals narrow; similar in color to *T. orizabæ.*

Color. General hue plumbeous, tinged with brown above and with gray on the under parts, the tips of the hairs being of that color; tail hairs plumbeous and gray.

Measurements. Total length, 210; tail vertebræ, 65; hind foot, 28; ear, 7. Skull: total length, 40; zygomatic width, 24.5; mastoid width, 19.5; interorbital width, 7; length of nasals, 9; width of nasals anteriorly, 3.7.

341. toltecus (*Thomomys*), Allen, Bull. Amer. Mus. Nat. Hist., 1893, p. 52, pl. i, fig. 13.

JUAREZ POCKET GOPHER.

Type locality. Juarez, northern part of State of Chihuahua, Mexico.

Geogr. Distr. State of Chihuahua, Mexico. Extent of range unknown.

Genl. Char. Larger than *T. fulvus.* Skull heavy, broad; premaxillæ much longer posteriorly than nasals; interparietal very small; upper incisors slightly grooved.

Color. Upper parts grayish rufescent brown, lined with black on median line; under parts pale grayish buff; tail and feet grayish buff; edge of cheek pouches white.

Measurements. Total length, 230; tail vertebræ, 60; hind foot, 27. Skull: total length, 43; basilar length, 40; greatest zygomatic breadth, 27; interorbital constriction, 7; length of nasals, 14.

342. perditus (*Thomomys*), Merr., Proc. Biol. Soc. Wash., xiv, 1901, p. 108.

LOST POCKET GOPHER.

Type locality. Lampazos, State of Nuevo Leon, Mexico.

Geogr. Distr. Range unknown.

Genl. Char. Small, similar to *T. toltecus*, but grayer and smaller. Nasals notched behind and terminating on plane of premaxillæ; interparietal subquadrate; braincase swollen.

Color. Above mixed drab gray and black; sides buffy; rump and sides of shoulders buffy fulvous; under parts, fore legs, and feet buffy salmon; hind feet whitish; space around mouth dusky.

Measurements. Total length, 195; tail vertebræ, 59; hind foot, 26.5.

343. sinaloæ (*Thomomys*), Merr., Proc. Biol. Soc. Wash., xiv, 1901, p. 108.

Sinaloa Pocket Gopher.

Type locality. Altata, State of Sinaloa, Mexico.

Geogr. Distr. Known only from the type locality.

Genl. Char. Similar to *T. cervinus*, but darker. Skull has "strongly spreading, depressed, and sharply angular zygomata."

Color. Above pale chestnut brown; under parts pale chestnut fulvous; space around mouth pale dusky.

Measurements. Total length, 233; tail vertebræ, 73; hind foot, 31.

344. goldmani (*Thomomys*), Merr., Proc. Biol. Soc. Wash., xiv, 1901, p. 108.

Goldman's Pocket Gopher.

Type locality. Mapimi, State of Durango, Mexico.

Geogr. Distr. Range unknown.

Genl. Char. Very small, allied to *T. perditus.* Nasals shorter than premaxillæ.

Color. Above "bright rusty fulvous, mixed with dark-tipped hairs; under parts white; nose and region around mouth dusky."

Measurements. Total length, 208; tail vertebræ, 68; hind foot, 30.

345. nelsoni (*Thomomys*), Merr., Proc. Biol. Soc. Wash., xiv, 1901, p. 109.

Nelson's Pocket Gopher.

Type locality. Parral, State of Chihuahua, Mexico.

Genl. Char. Size medium. Skull: "Zygomata strongly spreading, broader behind than in front, with well developed anterior angle; temporal impressions marked; interparietal subquadrate, becoming subtriangular in old age; nasals narrowly cuneate, notched behind, and falling well short of premaxillæ; bullæ medium; under jaw very long, the postcoronoid notch narrow and completely covered by coronoid process."

Color. Above "dull chestnut brown, mixed with black-tipped hairs on middle of back; under parts same color, but paler; nose and region around mouth dusky; feet whitish, but brown of hind leg coming well down over ankle and covering part of foot." (Merr. l. c.)

Measurements. Total length, 196; tail vertebræ, 60; hind foot, 28.

346. cervinus (*Thomomys*), Allen, Bull. Amer. Mus. Nat Hist., 1895, p. 203, fig. 1.   Elliot, Syn. N. Am. Mamm., 1901, p. 230.
CERVINE POCKET GOPHER.

Type locality.   Phœnix, Maricopa County, Arizona.

Geogr. Distr.   State of Sonora, Mexico, to southern Arizona.

Genl. Char.   Size large; color pale; rostrum broad, heavy; skull large.

Color.   Above fawn, obscured on dorsal region with dusky; beneath gray, base of hair plumbeous; blackish area about ears; pouches inside white; feet whitish; tail above grayish fawn, paler beneath.

Measurements.   Total length, 228; tail vertebræ, 63; hind foot. 28.   Skull: occipito-nasal length, 44; Hensel, 38; zygomatic width. 26.5; interorbital width, 7; mastoid breadth, 20; median length of nasals, 13; lateral length of nasals, 10; width of nasals anteriorly, 5.

347. perpallidus (*Thomomys*), Merr., Scien., VIII, 1886. p. 588. Elliot, Syn. N. Am. Mamm., 1901, p. 229.
PALE POCKET GOPHER.

Type locality.   Colorado Desert, San Diego County, southern California.

Geogr. Distr.   Lower California, Mexico, Colorado Desert, southern California, and northeastward to the Painted Desert, Arizona.

Genl. Char.   Similar to *T. clusius* (Coues, Proc. Acad. Nat. Scien. Phil., 1875, p. 138. Ex. Bridger's Pass., Rocky Mts.), but tail longer, half the length of head and body; color pale.

Color.   Above pale brownish yellow; sides yellowish white; beneath white; feet white; tail white for two-thirds the length, tip blackish.

Measurements.   Total length, 228; tail vertebræ, 76; hind foot. 52.   Skull: occipito-nasal length, 38; Hensel, 33; zygomatic width. 23; interorbital width, 7; mastoid width, 19.5; median length of · nasals, 12; length of upper molar series, 7; length of mandible, 25.

The Kangaroo Rats, as their name implies, are remarkable for the great length of their hind legs and tail, and they progress by long leaps exactly similar to those of the animal from which they derive their trivial name. They possess cheek pouches, and these, like those of the members of the GEOMYIDÆ, are external, causing these families to differ from all other Mammals in this respect. These pouches are covered with hair on the inside as well as on the outside. The skull of these animals is peculiarly shaped and very light, being not thicker than a sheet of paper. The Kangaroo Rats are divided into two genera, one with four toes on the hind feet, the other with five; but in the latter the first digit, although possessing a claw, is rudimentary. These handsome little creatures live in burrows which they excavate in sandy soil, and they are preyed upon by the spotted skunks, which are numerous in the localities they frequent. They live in colonies and are very industrious, digging their burrows, some of which are several inches in diameter, in the shifting sand. They are hardy and do not seem to heed the severest weather, and run about on the snow when the thermometer registers below zero. At this time they feed on the seeds of cockle and sand burrs, and in summer on the seeds of the prickly pear and other hardy desert plants. They prefer the sandy districts near rivers, where vegetation is scanty, and on sandy portions of the prairies, and avoid those places where the soil is rich. They are nocturnal, passing the day in sleep, and their tracks in the sand each morning exhibit their activity and the extent of their night wanderings. They are exceedingly pretty creatures, with a pleasing coloring and fur as soft as silk, and with large, soft, expressive eyes.

Fam. V. **Heteromyidæ. Kangaroo Rats. Pocket Mice.**

Incisors narrow; molars rootless; mastoids enormously developed, appearing on top of the skull; hind feet long; digits four or five; pelage soft.

Subfam. IV. **Dipodomyinæ.**

Anterior molar without lobe to the prism. Skull two-thirds as wide as long, occipital plane emarginate; zygomatic plate of maxillary nearly roofing the orbit; pit on inner side of jaw near molars.

71. **Dipodomys. Four-toed Kangaroo Rats.**

$$I.\frac{1-1}{1-1}; \; M.\frac{3-3}{3-3} = 16.$$

Dipodomys Gray, Ann. Mag. Nat. Hist., VII. 1841. p. 521. Type *Dipodomys phillipsi* Gray.
Macrocolus Wagn., Archiv. fur Natürg., 1846, I, p. 172.

Skull light, depressed, smooth, thin, broad posteriorly, tapering anteriorly; anterior outline emarginate; rostrum extending beyond incisors, acuminate; zygomata delicate, straight, abutting against tympanics, which are greatly inflated and possess a nontubular orifice of meatus; mastoids enormous; squamosals reduced; parietal triangle-shaped; interparietal small, situated between forks of occipital; bullæ in contact below the basisphenoid; palate terminating posteriorly with a sharp median somewhat lengthened spur; external to this is a fossa with two small anterior foramina, and a large one behind pterygoids, with a hamular termination abutting the bullæ; basioccipital narrow, acuminate, reduced, separated by a fissure for its entire length from the bullæ; a similar fissure divides the last named bones from the alisphenoid and squamosal. The interorbital foramen is placed low down, midway on the side of the rostrum; incisive foramina minute slits between incisors and molars. In the rear of the skull the occipital bones appear as a rim to the foramen magnum; upper incisors sulcate, pointing backward; molars rootless; mandible small, thick, with a conspicuous acute lamina twisting obliquely outward and upward; mental foramen outside, near incisors; hind legs elongate; tail longer than head and body, penicillate; soles

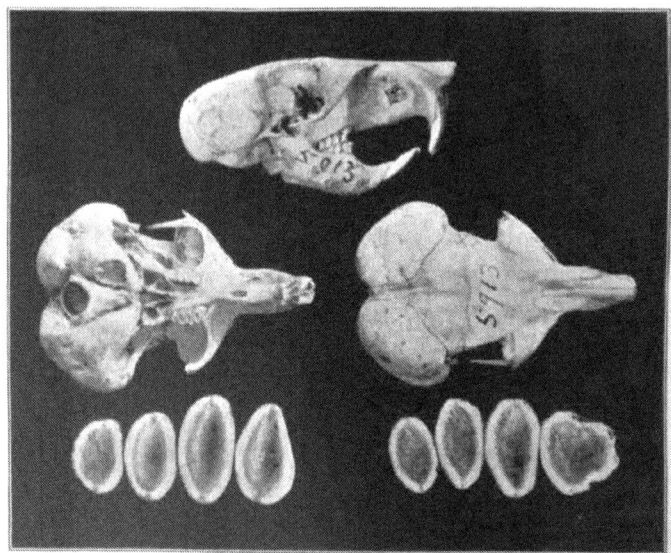

FIG. 65. DIPODOMYS PHILLIPSI.
No. 5913 Am. Mus. Nat. Hist. Coll. Nat. size.
UPPER TOOTH ROW.   LOWER TOOTH ROW.
Enlarged 7 times.   Enlarged 7 times.

hairy; hind foot with four toes only; cheek pouches large; fur of velvet softness. All the species of both genera have facial crescentic lines more or less distinctly marked; side of mouth white; white spot over the eyes and another behind the ear, and a white stripe across the thighs, usually reaching the tail; digits four.

KEY TO THE SPECIES.

A. Hind foot with four toes.
  a. Under parts white.   PAGE

348. phillipsi (*Dipodomys*), Gray, Ann. Mag. Nat. Hist., VII, 1841, p. 522.

PHILLIPS' KANGAROO RAT.

Type locality. Real del Monte, north end of Valley of Mexico.

Geogr. Distr. Valley of Mexico, State of Mexico, and States of Puebla, eastern Tlaxcala and western Vera Cruz, Mexico.

Genl. Char. Size small; colors dark; tip of tail white. Skull: interorbital constriction very considerable; mastoids relatively small and separated on top of skull; superior outline arched.

Color. Above sepia brown and ochraceous mixed with black; sides of nose, spot at base of ear, stripe across thigh, and under parts white; stripe across nose from base of whiskers and base of tail black; tail black above and below, sides and tip white.

Measurements. Total length, 270; tail vertebræ, 168; hind foot, 41. Skull: occipito-nasal length, 42; Hensel, 29.5; zygomatic width, 22; mastoid width, 28; width of parietal, 19.5; posterior width of basioccipital between bullæ, 5.5; palatal length, 17; postpalatal length, 12; length of bullæ, 11.5; length of upper tooth row, alveolar border, 5; length of mandible, 18.5; length of lower tooth row, 5.

349. ornatus (*Dipodomys*), Merr., Proc. Biol. Soc. Wash., ix, 1894, p. 110.

ORNAMENTED KANGAROO RAT.

Type locality. Berriozabal, State of Zacatecas, Mexico.

Geogr. Distr. State of Zacatecas, Mexico.

Genl. Char. Similar to *D. phillipsi*; hind foot shorter. Upper premolar without the antero-internal lobe.

Color. Above golden clay color; head and median line of back darker; orbital ring, facial crescents, and inner sides of hind legs black; stripes on thighs and under parts white; tail black above and below, sides and tip white.

Measurements. Total length, 274; tail vertebræ, 167; hind foot, 39.

350. perotensis (*Dipodomys*), Merr., Proc. Biol. Soc. Wash., ix, 1894, p. 111.

PEROTE KANGAROO RAT.

Type locality. Perote, State of Vera Cruz, Mexico.

Geogr. Distr. State of Vera Cruz, Mexico.

Genl. Char. Similar to the preceding species in color and cranial characters, the differences being mainly comparative, viz., skull narrower on top, more arched; breadth of supra-occipital between mastoids greater; angle of mandible larger.

Color. Above brownish clay mixed with black on head and back; sides ochraceous buff; facial crescents and inner side of hind leg black; tail black, with white side stripes and tip.

Measurements. Total length, 271; tail vertebræ, 162; hind foot, 40.

351. mitchelli (*Dipodomys*), Mearns, Proc. U. S. Nat. Mus., 1897, xix, p. 719.

TIBURON ISLAND KANGAROO RAT.

Type locality. Tiburon Island, Gulf of California, Mexico.

Geogr. Distr. Known only from type locality.

Genl. Char. Similar to *D. m. simiolus* Rhoads, but smaller.

Color. Above yellowish brown mixed with black on back; sides ochraceous buff; under parts white; tail black, with white side stripes, pencil grayish; ears black inside, yellowish outside; stripe on under side of hind foot black.

Measurements. Total length, 110; tail vertebræ, 130; hind foot, 38.5. (ex Type.) Skull: total length, 36.5; breadth of skull, 22.5; length of nasals, 13.3. (Mearns, l. c.)

352. deserti (*Dipodomys*), Steph., Am. Nat., xxi, 1887, p. 42, pl. v.
Elliot, Syn. N. Am. Mamm., 1901, p. 235.
DESERT KANGAROO RAT.

Type locality. Mojave River, near San Bernardino County, California.

Geogr. Distr. State of Sonora and Lower California, Mexico, north to Mojave and Colorado deserts, southeastern California.

Genl. Char. Large; colors pale; tail longer than head and body; mastoids greatly developed.

Color. Above pale yellowish brown, hairs plumbeous at base; legs, feet, and under parts white; tail white, with a pale brown dorsal stripe, growing darker towards end, and extending from near base to white tip; white spot over eye and behind ear; white thigh stripe.

Measurements. Total length, 133; tail vertebræ, 204; hind foot, 52. Skull: total length, 42; Hensel, 36; zygomatic width, 21; interorbital constriction, 13; mastoid width, 28.5; length of nasals, 15; width of parietal anteriorly, 17.5; palatal length to incisive foramina, 7; length of upper molar series, 5; length of mandible, 18.

353. spectabilis (*Dipodomys*), Merr., N. Am. Faun., No. 4, 1890, p. 46. Elliot, Syn. N. Am. Mamm., 1901, p. 235.
HANDSOME KANGAROO RAT.

Type locality. Dos Cabezos, Cochise County, Arizona.

Geogr. Distr. States of Sonora and Chihuahua, Mexico, and portions of Texas and Arizona, eastward to Sierra Blanca, Texas.

Genl. Char. Size large; tail nearly twice the length of head and body; mastoids not meeting behind parietals; maxillary bridge of orbit broad.

Color. Above ochraceous buff, lined with black, sides paler; top of head and back darkest; facial crescent crossing nose black; hip patch ochraceous; hind leg white above, dusky below; dorsal stripe of tail dusky for basal half, followed by a broad black band, terminating in a long white brush; ventral stripe dusky, the two meeting about two-thirds the length of tail from base, the lateral white stripes disappearing.

Measurements. Total length, 355; tail vertebræ, 211; pencil, 30; hind foot, 56; ear, 17.5. Skull: total length, 46; Hensel, 31; zygomatic width, 25; interorbital constriction, 15; mastoid breadth, 28; length of upper molar series, 5.5; length of nasals, 10; length of mandible, 21.

354. merriami (*Dipodomys*), Mearns, Bull. Amer. Mus. Nat. Hist., 1890, p. 290. Elliot, Syn. N. Am. Mamm., 1901, p. 232.

FIG. XLVIII. DIPODOMYS MERRIAMI.
MERRIAM'S KANGAROO RAT.

MERRIAM'S KANGAROO RAT.

Type locality. New River, between Phœnix and Prescott, Arizona.

Geogr. Distr. States of Sonora and Chihuahua, Mexico, north into Arizona.

Genl. Char. Limbs and tail slender, latter longer than head and body; skull small.

Color. Above mouse gray, tinged with pinkish buff; sides sandy; nose and sides of face black nearly to eyes; aural spot and under parts white; tail drab gray, white band on each side not reaching extremity.

Measurements. Total length, 281; tail vertebræ, 149; hind foot, 36. Skull: total length, 37; Hensel, 22; zygomatic width, 17; inter-orbital constriction, 13; mastoid width, 23; length of nasals, 14; length of upper molar series, 4; length of mandible, 18.5.

a.—ambiguus (Dipodomys), Merr., N. Am. Faun., No. 4, 1890, p. 42. Elliot, Syn. N. Am. Mamm., 1901, p. 234.

DOUBTFUL KANGAROO RAT.

Type locality. El Paso, El Paso County, Texas.

Geogr. Distr. State of Chihuahua, Mexico, north to north-western Texas.

Genl. Char. Tail one-third longer than head and body; size me-dium; body slender.

Color. Above buffy drab; sides tinged with pale buff, and lined everywhere with black; beneath white; upper and lower tail stripes dusky to tip; lateral white stripes terminating with the vertebræ.

Measurements. Total length, 233; tail vertebrae, 133; hind foot, 37; ear, 7.

b.—parvus (Dipodomys), Rhoads, Am. Nat., XXVIII, 1894, p. 70. Elliot, Syn. N. Am. Mamm., 1901, p. 234.

SMALL KANGAROO RAT.

Type locality. San Bernardino, San Bernardino County, California.

Geogr. Distr. Lower California, Mexico, into southern California.
Genl. Char. Similar to *D. merriami*, smaller, tail longer and without black markings on face and nose.

Color. Above buffy gray, sides buff; under parts white; usual white eye and ear spot, and stripe on thighs; dark dorsal and ventral tail stripes brownish black; pencil sooty black; lateral stripes white; orbital ring black; under surface of hind foot brownish.

Measurements. Total length, 248; tail vertebræ, 154; hind foot, 35; ear, 10. Skull: basilar length, 21; mastoid breadth, 22.5; interorbital constriction, 13; length of nasals, 13; length of upper molar series on crowns, 3.6; length of mandible, 13.9; height of coronoid process from angle, 5.1.

c.—simiolus (*Dipodomys*), Rhoads, Proc. Acad. Nat. Scien. Phil., 1893, p. 410. Elliot, Syn. N. Am. Mamm., 1901, p. 234.
   similis Rhoads, Proc. Acad. Nat. Scien. Phil., 1893, p. 411. Elliot, Syn. N. Am. Mamm., 1901, p. 234.

ALLIED KANGAROO RAT.

Type locality. Agua Caliente, Mojave Desert, California.
Geogr. Distr. Lower California and State of Sonora, Mexico, into California.
Genl. Char. Similar to *D. deserti*, but smaller; pencil not white.
Color. Above pale yellowish brown like *D. deserti*, inclining to cinnamon on rump; beneath white; tail, upper fourth dark ashy extending to tip, brownish black on middle third of under side; feet white, soles slightly darker.

Measurements. Total length, 241; tail vertebræ, 149; hind foot, 38; ear from crown, 9; pencil, 35. Skull: basilar length, 21.8; mastoid breadth, 24; interorbital constriction, 14.5; length of nasals, 13; length of mandible, angle to base of incisors, 13.8; coronoid process from angle, 5.6.

d.—arenivagus Elliot, Pub. Field Columb. Mus., III, 1903, p. 249.
   Zoölogy.

Type locality. San Felipe, Lower California, Mexico.
Genl. Char. Size small, similar to *D. m. simiolus*, but paler; ear larger; hind foot shorter. Skull narrower across mastoids and parietal; nasals shorter.
Color. Upper parts pinkish buff, palest on the head and darkest on rump, the plumbeous under fur showing in places; no black streaks on face; white spots behind ears and above eyes; upper parts of sides from eye to rump like color of rump; nose, sides of face, lower part of flanks, entire under parts and limbs, pure white; a narrow line of pinkish buff across thighs; hands yellowish white.

feet white; tail with a bushy pencil, the upper parts to tip pale drab, sides and beneath white; ears naked, yellowish.

Measurements. Type. Total length, 225; tail vertebræ, 134; hind foot, 36; ear, 15. Average of ten specimens: Total length, 234.7; tail, 137.3; hind foot, 36.7; ear, 14.1. Skull: total length, posterior line of mastoids to anterior end of nasals, 34; Hensel, 20; zygomatic width, 15; width of mastoids, 22; greatest width of parietal, 15; length of nasals, 12; greatest width of rostrum, 5; palatal length, 11; length of upper tooth row, 3; length of mandible, condyle to tip of incisors, 16; length of lower tooth row, 3.

e.—atronasus (*Dipodomys*), Merr., Proc. Biol. Soc. Wash., IX, 1894, p. 113.

BLACK-NOSED KANGAROO RAT.

Type locality. Hacienda La Parada, 25 miles northwest of the City of San Luis Potosi, State of San Luis Potosi, Mexico.

Geogr. Distr. Known only from type locality, Mexico.

Genl. Char. Similar to *D. merriami*, but darker.

Color. Above dark clay color; sides ochraceous buffy; nose from tip to eyes blackish; usual black stripes on face and thighs; tail black above and below, white on sides to middle third of the length.

Measurements Total length, 267; tail vertebræ, 162; hind foot, 40.

f.—melanurus (*Dipodomys*), Merr., Proc. Calif. Acad. Scien., 2d Ser., 1893, p. 345.

BLACK-TAILED KANGAROO RAT.

Type locality. San Jose del Cabo, Lower California, Mexico.

Geogr. Distr. Cape region of Lower California, Mexico.

Genl. Char. Like *D. merriami*; terminal third of the tail blackish.

Color. Above mixed pale ochraceous buff and black; face and supra-orbital spot white; under parts white; penicillate part of tail blackish, rest like *D. merriami*.

Measurements. Total length, 240; tail vertebræ, 141; hind foot, 36.5.

The description given of the Kangaroo Rats of the genus *Dipodomys* will answer perfectly for the members of PERODIPUS. They closely resemble each other in general appearance, and the presence of a fifth toe on the hind foot was not suspected, so minute is it, for a considerable time after the animals were known to naturalists.

72. Perodipus. Five-toed Kangaroo Rats.

$$I.\frac{1-1}{1-1}; \ \overset{\bullet}{M}.\frac{3-3}{3-3} = 16.$$

Perodipus Fitzin., Sitz. Math.-Natur. Classe, K. Akad. Wiss. Wien.,
1867, LVI, p. 126. Type *Dipidomys agilis* Gambel.
Dipodops Merr., N. Am. Faun., 1890, No. 3, p. 72.

Skull similar to that of *Dipodomus*, with greatly developed mastoids and thread-like zygomata; audital bullæ in contact below basisphenoid; hind feet with five toes, the first digit rudimentary, but having a claw, and reaching to the end of the metatarsal bones of the other digits.

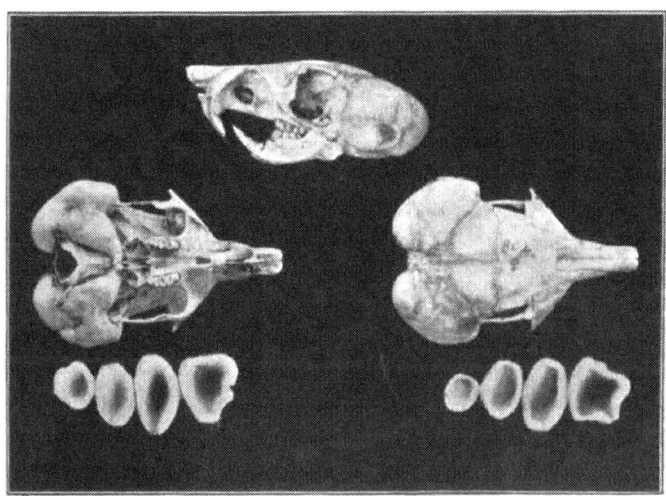

FIG. 66. PERODIPUS CHAPMANI.
No. 802 Field Columbian Mus. Coll. Nat. size
UPPER TOOTH ROW.   LOWER TOOTH ROW.
Enlarged 7 times.   Enlarged 7 times.

KEY TO THE SPECIES.

A. Hind feet with five toes.   PAGE

355. ordi (*Dipodomys*), Woodh., Proc. Acad. Nat. Scien. Phil., 1853,
    p. 235.   Elliot, Syn. N. Am. Mamm., 1901, p. 238.
ORD'S KANGAROO RAT.
    Type locality.   El Paso, El Paso County, Texas.
    Geogr. Distr.   State of Chihuahua, Mexico, north to Snake Plains.
    Genl. Char.   Size medium; tail little shorter than head and body.
    Color.   Above ochraceous buff, darkest on back, mixed with
black on rump; sides paler; side of nose, spot behind the ear, stripe
across thigh, and under parts, white; tail with dorsal and ventral
stripes dusky, pencil dusky, base of hairs white.
    Measurements.   Total length, 240; tail vertebræ, 134; hind foot,
38; ear, 7.

a.—palmeri (*Dipodops*), Allen, Bull. Amer. Mus. Nat. Hist., 1891,
    p. 276.
PALMER'S KANGAROO RAT.
    Type locality.   San Luis Potosi, State of San Luis Potosi, Mexico.
    Geogr. Distr.   State of San Luis Potosi, Mexico.
    Genl. Char.   Similar to *P. ordi*, but darker.
    Color.   Upper parts brownish ochraceous, mixed with black;
under parts, arms (excepting a patch of buff), band across thigh
and hind feet white; tail dusky brown, with narrow lateral white
stripes, tip dusky brown; ears whitish, with a dusky patch on outer
anterior border, and another on inner lower posterior border.
    Measurements.   Total length, 249; tail vertebræ, 141; hind foot,
35; ear, 11.4.   Skull: total length, 38; basal length, 25.4; mastoid
width, 24; interorbital constriction, 13.2; length of nasals, 11.4;
length of mandible, condyle to tips of incisors, 17.8; height at coronoid
process, 5.8.

356. chapmani (*Dipodomys*), Mearns, Bull. Am. Mus. Nat. Hist.,
    1890, p. 291.   Elliot, Syn. N. Am. Mamm., 1901, p. 237.
CHAPMAN'S KANGAROO RAT.
    Type locality.   Forte Verde, Yavapai County, Arizona, Mexico.
    Geogr. Distr.   State of Sonora, Mexico, north into Arizona.
    Genl. Char.   More slender than *C. ordi*, tail longer, and color
darker.   Skull lighter, smaller.
    Color.   Above mouse gray, mixed with black and buff; sides
sandy buff; usual line on side of nose, ear, and eye spots, and thigh
stripe white; under parts white; dorsal and ventral stripes on tail
drab gray; lateral white stripes extending to near end of vertebræ.
    Measurements.   Total length, 280; tail vertebræ, 148; hind foot,
38.   Skull: total length, 37.2; zygomatic width, 17; Hensel, 23.5;

interorbital constriction, 13; length of nasals, 13.5; mastoid width, 23; length of upper molar series, 5; length of mandible, 18.3.

357. obscurus (*Perodipus*), Allen, Bull. Am. Mus. Nat. Hist., 1903, p. 603.

DUSKY KANGAROO RAT.

Type locality. Rio Sestin, State of Durango, Mexico.

Genl. Char. Color very dark; size moderate.

Color. Dorsal area gray brown suffused with fulvous; flanks more fulvous; lower half of cheeks, sides of neck, small spot above eye, postauricular patch, band on thighs, fore legs, inner side of hind legs and entire under parts white; tip of nose and bar at base of whiskers black; ear buffy white; tail with dorsal and ventral stripes blackish; sides and basal ring white.

Measurements. Total length, 232; tail vertebræ, 130; hind foot, 35; ear from notch, 12.7. Skull: total length, 36; greatest mastoid breadth, 23; length of nasals, 13.

358. agilis (*Dipodomys*), Gambel, Proc. Acad. Nat. Scien. Phil., 1848, p. 77. Elliot, Syn. N. Am. Mamm., 1901, p. 236.

hermanni & wagneri Le Conte, Proc. Acad. Nat. Scien. Phil., 1848, p. 77.

NIMBLE KANGAROO RAT.

Type locality. Los Angeles, Los Angeles County, California.

Geogr. Distr. Lower California, Mexico to southern and middle California.

Genl. Char. Size small; color dark; body rather stout.

Color. Above ashy brown, heavily lined with black, especially on top and sides of head and lower back; sides yellowish brown; spot over eye and behind ear, side of snout, stripe on thighs and under parts white; facial crescent black and broad, and meeting on bridge of nose; tail blackish and with a blackish tip, lateral white stripes terminating at base of pencil; under part of leg and base of foot dusky; rest white.

Measurements. Total length, 280; tail vertebræ, 170; hind leg, 39.5; ear, 14.5. Skull: total length, 38; Hensel, 24; zygomatic width, 17.5; interorbital constriction, 12; mastoid breadth, 24; palatal length to incisive foramina, 7; median length of nasals, 7.5; length of upper molar series, 3.5; length of mandible, condyle to tips of incisors, 20.

The Pocket Mice, while related to the Kangaroo Rats, have not their long hind limbs; indeed, these members scarcely exceed the fore legs in length, and the pelage, instead of being soft and silky, is usually harsh and coarse. The skull, however, is of a papery construction, and the tail is long, and often tufted, and the general appearance of the animals, especially in some of the species, is not unlike that of the Kangaroo Rats. The cheek pouches have rather narrow openings, but extend back nearly to the ears. The Pocket Mice differ somewhat in their habits from the *Dipodomyinæ*, and apparently hibernate, for they are rarely seen during the winter in localities where the temperature goes below zero. They are prairie dwellers, and make their burrows amid the buffalo grass, sinking them perpendicularly for five or six inches, and the excavated earth is piled up in little mounds near the opening. These mice are generally small in size, some species being indeed almost minute. They are divided into two subgenera, distinguished from each other by variations in the skulls.

Subfam. V. **Heteromyinæ.**

C. H. Merriam. *Revision of the North American Pocket Mice*, N. Am. Faun., No. 1, 1889.

W. H. Osgood. *Revision of the Pocket Mice of the genus Perognathus*, N. Am. Faun., No. 18, 1900.

73. **Perognathus.** Pocket Mice.

$$I.\tfrac{1-1}{1-1}; \; P.\tfrac{1-1}{1-1}; \; M.\tfrac{3-3}{3-3} = 20.$$

Perognathus Wied. Nov. Act. Phys. Med. Acad. Caes. Leop. Carol., XIX, 1839, pp. 368–373, pl. XXXIV. Type *Perognathus fasciatus* Wied.

Cricetodipus Peale, Rep. Mamm. & Ornith. U. S. Expl. Exped., 1848, p. 53, pl. 13, fig. 2.

Abromys (*sic*) Gray, Proc. Zoöl. Soc., 1868, p. 202.

Otognosis Coues, Proc. Acad. Nat. Scien. Phil., 1875, p. 305.

Chætodipus Merr., N. Am. Faun., No. 1, 1889, p. 5, pl. III, fig. 15.

Skull depressed and flat above; nasals lengthened, projecting beyond incisors; mastoids less developed than those of the species of *Dipidomyinæ*, and in certain species not projecting beyond plane of occiput; zygomata much as in ordinary rodents; occiput not emarginate; molars rooted; no pit between last lower molar and coronoid process; tail moderate; soles naked or sparsely haired; hind limbs scarcely exceeding the fore limbs in length.

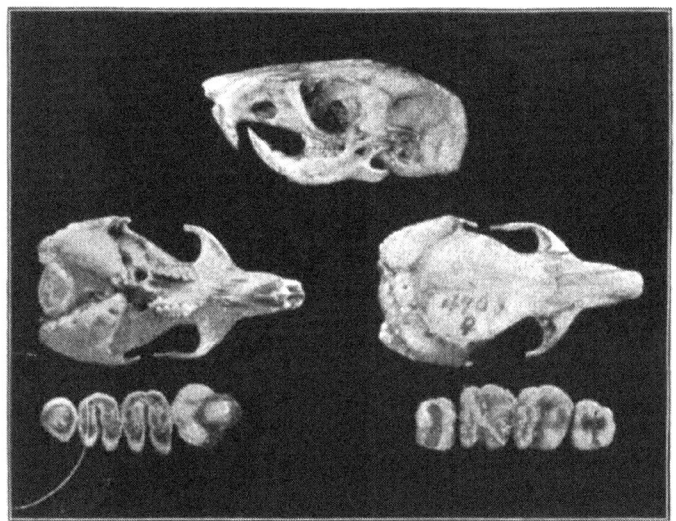

FIG. 67. PEROGNATHUS MERRIAMI.
No. 4963 Field Columbian Mus. Coll. Twice nat. size
UPPER TOOTH ROW.   LOWER TOOTH ROW.
Enlarged 10 times.   Enlarged 10 times.

A. Perognathus.

Mastoids well developed, extending beyond occipital line; inter-parietal pentagonal; mastoid side of parietal longest; audital bullæ nearly contiguous below basisphenoid.

KEY TO THE SPECIES AND SUBSPECIES OF THE SUBGENUS.

A. Size small; hind foot less than 18 mm.
   a. Tail about 60 mm.   PAGE
      a.' Above ochraceous and black............*P. merriami*   353
      b.' Above pale fulvous or buff and black.
         a." Hind foot, 15 mm......................*P. flavus*   354
         b." Hind foot, 17 mm................*P. f. mexicanus*   355
         c." Hind foot, 17.4 mm..............*P. p. brevinasus*   355
      c.' Above pale ochraceous drab and black....*P. pacificus*   355
B. Size large; hind foot over 18 mm.........*P. a. melanotis*   356

359. merriami (*Perognathus*), Allen, Bull. Am. Mus. Nat. Hist., 1892, p. 45, pl. III, figs. 1–6. Elliot, Syn. N. Am. Mamm., 1901, p. 243.

mearnsi Allen, Bull. Am. Mus. Nat. Hist., 1896, p. 237.

FIG. XLIX. PEROGNATHUS MERRIAMI.
MERRIAM'S POCKET MOUSE.

MERRIAM'S POCKET MOUSE.

Type locality. Brownsville, Cameron County, Texas.

Geogr. Distr. Alta Mira, State of Tamaulipas, Mexico, into Texas and New Mexico.

Genl. Char. Size of *P. flavus*, color more yellow, sides golden.

Color. Autumn Pelage. Above yellow, heavily lined with black; sides golden, sparsely mixed with black; beneath white; thighs golden; feet white; tail pale brownish yellow above, becoming blackish at tip; beneath paler.

Summer Pelage. Above ochraceous, lined with black; lateral stripe ochraceous; under parts white; ears dusky; aural spot buff; tail pale grayish brown, lightest beneath. This phase of pelage represents *P. mearnsi* Allen.

Measurements. Total length, 118; tail vertebræ, 55; hind foot, 16; ear, 4. Skull: basilar length of Hensel, 14.8; occipito-nasal length, 20.4; greatest mastoid breadth, 11.2; length of interparietal, 2.3; greatest width of interparietal, 3.6.

360. flavus (*Perognathus*), Baird, Proc. Acad. Nat. Scien. Phil., 1855, p. 332. Elliot, Syn. N. Am. Mamm., 1901, p. 244.
BAIRD'S POCKET MOUSE.

Type locality. El Paso, El Paso County, Texas.

Geogr. Distr. State of Chihuahua, Mexico, into southeastern Arizona and Texas.

Genl. Char. Size very small; ear without antitragal lobe; tail less than head and body; audital bullæ meeting below basisphenoids.

Color. Above pale fulvous, lined with black, no lateral stripe; feet and under parts white; buff patch behind ear, and white one on lower margin; orbital ring pale fulvous; tail pale brownish or olive gray, of nearly the same hue above and below.

Measurements. Total length, 117; tail vertebræ, 71; hind foot, 17; ear, 4. Skull: length of Hensel, 14.6; occipito-nasal length, 21; greatest mastoid breadth, 12; length of interparietal, 2.6; greatest width of interparietal, 2.9.

a.—mexicanus (Perognathus), Merr., Proc. Acad. Nat. Scien. Phil., 1894, p. 265.
MEXICAN POCKET MOUSE.

Type locality. Tlalpam, Federal District, Mexico.

Geogr. Distr. States of Zacatecas and San Luis Potosi to the State of Jalisco on the west and State of Hidalgo on the east, and south to the State of Mexico.

Genl. Char. Similar to *P. flavus*, but darker and larger.

Color. Above buff, mixed with black; spot behind ear, and the lateral line ochraceous; under parts white; tail dusky above, paler beneath.

Measurements. Total length, 115.7; tail, 53.7; hind foot, 17.4.

panamintinus brevinasus (Perognathus), Osgood, N. Am. Faun., No. 18, 1900, p. 30. Elliot, Syn. N. Am. Mamm., 1901, p. 246.
SHORT-NOSED POCKET MOUSE.

Type locality. San Bernardino, San Bernardino County, California.

Geogr. Distr. Lower California north to southern California.

Genl. Char. Similar to *P. panamintinus*, but darker. Skull smaller; interparietal smaller, nasals shorter; lower premolar larger than last molar.

Color. Above pinkish buff, lined with black; lateral line pinkish buff; ears dusky, orbital ring buffy; tail buffy white, faintly dusky above.

Measurements. Total length, 124; tail vertebræ, 66; hind foot, 17.4. Skull: length of Hensel, 14.9; occipito-nasal length, 21.4; mastoid breadth, 11.9; length of interparietal, 2.5; greatest width of interparietal, 3.5; interorbital constriction, 5.2; length of nasals, 8.3.

361. pacificus *(Perognathus)*, Mearns, Bull. Am. Mus. Nat. Hist., 1898, p. 299. Elliot, Syn. N. Am. Mamm., 1901, p. 246.
PACIFIC POCKET MOUSE.

Type locality. Mouth of Tijuana River, Mexican boundary Monument, No. 258, shore of Pacific Ocean, San Diego County, California.

Geogr. Distr. Edge of Pacific Ocean, on a flat at mouth of Tijuana River, and probably into Lower California, Mexico.

Genl. Char. Similar to *P. bimaculatus*, but smaller; tail more hairy; lower premolar more nearly quadrate; audital bullæ more separated.

Color. Above pale ochraceous drab, lined with black; ears and spot at base of whiskers blackish; orbital area pale buff; spot behind eye and also the lateral line on body pale buff; feet and under parts white; tail hoary at base, tip dusky.

Measurements. Total length, 109; tail vertebræ, 53; hind foot, 15.5; ear from crown, 4.7. Skull: length of Hensel, 13; occipito-nasal length, 19; greatest mastoid breadth, 11; length of inter-parietal, 2.2; greatest width of interparietal, 3.4; interorbital con-striction, 4.9; length of nasals, 7.

apache melanotis (*Perognathus*), Osgood, N. Am. Faun., No. 18, 1900, p. 27.

BLACK-EARED POCKET MOUSE.

Type locality. Casas Grandes, State of Chihuahua, Mexico.

Geogr. Distr. Known only from type locality.

Genl. Char. Similar to *P. apache*, but darker, smaller.

Color. Above rich buff, lined with black; under parts pure white; orbital region buff; ears, edges, and spot at base white; inside black; tail above dusky, beneath buffy white.

Measurements. Total length, 133; tail vertebræ, 65; hind foot, 19.5. Skull: length of Hensel, 15.3; occipito-nasal length, 21.5; greatest mastoid width, 11.8; length of interparietal, 2.8; greatest interparietal width, 4.3; interorbital constriction, 5; length of nasals, 8.

B. Chætodipus.

"Mastoids moderately developed, not projecting behind plane of occiput; interparietal broadly pentagonal, or strap-shaped; mas-

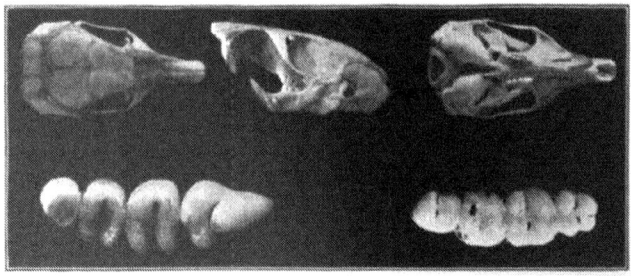

FIG. 68. PEROGNATHUS (CHÆTODIPUS) FEMORALIS MESOPOLIUS.
No. 10374 Field Columbian Mus. Nat. size.
Tooth rows enlarged 9 times.

toid side of parietal not longest; audital bullæ separated anteriorly by full width or nearly full width of basisphenoid."

penicillatus arenarius (*Perognathus*), Merr., Proc. Cal. Acad. Scien.,
        2d Ser., IV, 1894, p. 461.
LITTLE DESERT POCKET MOUSE.
   Type locality. San Jorge, near Comandu, Lower California,
Mexico.
   Geogr. Distr. East coast of Lower California.
   Genl. Char. Size very small; tail exceeding head and body; no
bristles. Similar to *P. penicillatus.*
   Color. Above buffy drab, mixed with black; sides paler; no
lateral line; white spot on lower margin of ears; under parts white;
tail brownish above, white below.
   Measurements. Total length, 136; tail vertebræ, 70; hind foot,
20. Skull: length of Hensel, 15.3; occipito-nasal length, 23; greatest
mastoid breadth, 12; length of interparietal, 3.5; greatest width of
interparietal, 6.4; interorbital constriction, 6.2; length of nasals, 8.8.

penicillatus angustirostris (*Perognathus*), Osgood, N. Am. Faun.,
        No. 18, 1900, p. 47. Elliot, Syn. N. Am. Mamm., 1901, p. 252.
SLENDER-NOSED POCKET MOUSE.
   Type locality. Carriso Creek, Colorado Desert, San Diego
County, California.
   Geogr. Distr. Northern Lower California, east to the Colorado
Desert and southwestern Arizona, north to Colorado Desert, California.
   Genl. Char. Similar to *P. penicillatus*, but smaller; nasals and
ascending premaxillæ long and narrow.
   Color. Above vinaceous buff, sprinkled with black; under parts,
fore legs, and feet white; lateral stripe indistinct, pale fulvous; tail
above dusky extending over pencil, beneath white.
   Measurements. Total length, 191; tail vertebræ, 105; hind foot,
24.4. Skull: length of Hensel, 18; occipito-nasal length, 26; greatest
mastoid breadth, 13; length of interparietal, 3.1; greatest width of
interparietal, 7; interorbital constriction, 6.4; length of nasals, 10.3.

penicillatus pricii (*Perognathus*), Allen, Bull. Am. Mus. Nat. Hist.,
        1894, p. 318. Elliot, Syn. N. Am. Mamm., 1901, p. 253.
   obscurus Allen, Bull. Am. Mus. Nat. Hist., 1895, p. 216.
PRICE'S POCKET MOUSE.
   Type locality. Oposura, State of Sonora, Mexico.

Geogr. Distr. Northwestern Mexico west of the Sierra Madre, north into south central Arizona.

Genl. Char. Similar to *P. penicillatus*, but smaller; no spines on rump; pelage harsh; skull short.

Color. Above and on sides vinaceous buff, lined with black; lateral stripe indistinct; under parts white; tail above dusky, beneath white; ears like back.

Measurements. Total length, 172; tail vertebræ, 90; pencil, 10; hind foot, 23; ear, 7.5. Skull: occipito-nasal length, 26; Hensel, 18.4; greatest mastoid breadth, 13; length of interparietal, 6.7; greatest width of interparietal, 6.2; length of nasals, 9.4.

penicillatus eremicus (*Perognathus*), Mearns, Bull. Amer. Mus. Nat. Hist., 1898, p. 300. Elliot, Syn. N. Am. Mamm., 1901, p. 253. EASTERN DESERT POCKET MOUSE.

Type locality. Fort Hancock, El Paso County, Texas.

Geogr. Distr. North central Mexico east of the Sierra Madre to La Ventura, State of Coahuila, and north into western Texas.

Genl. Char. Small; colors pale; skull rather heavy.

Color. Above whitish drab, tinged with fawn and lined with dusky; lateral stripe faint; feet and lower parts white; tail above and tip pale brown, beneath white.

Measurements. Average total length of six, 163; tail vertebræ, 83; hind foot, 22.1; ear from base, 9.1. Skull: Hensel, 17.5; greatest mastoid breadth, 12.6; length of interparietal, 3; greatest width of interparietal, 7; interorbital constriction, 6.4; length of nasals, 9.3.

362. pernix (*Perognathus*), Allen, Bull. Amer. Mus. Nat. Hist., 1898, p. 149. SINALOA POCKET MOUSE.

Type locality. Rosario, State of Sinaloa, Mexico.

Geogr. Distr. States of Sinaloa and Jalisco, Mexico.

Genl. Char. Size small; colors dark; no spines; tail long.

Color. Upper parts hair brown; lateral line pinkish ochraceous buff; under parts whitish; tail brownish above, whitish beneath; ears dusky with white dot on inferior margins.

Measurements. Total length, 175; tail vertebræ, 97; hind foot, 22.3. Skull: occipito-nasal length, 24.4; Hensel, 17.4; greatest mastoid breadth, 12.2; length of interparietal, 3.3; greatest width of interparietal, 7.2; interorbital constriction, 5.4; length of nasals, 8.6.

a.—rostratus (*Perognathus*), Osgood, N. Am. Faun., No. 18, 1900, p. 51. BROAD-NOSED POCKET MOUSE.

Type locality. Camoa, Rio Mayo, State of Sonora, Mexico.

Geogr. Distr. Southern part State of Sonora and northern portion State of Sinaloa, plains of the coast, Mexico.

Genl. Char. Similar to *P. pernix*, but skull shorter and wider.

Color. Above broccoli brown; lateral line pinkish buff; under parts whitish; generally paler than *P. pernix*.

Measurements. Total length, 162; tail vertebræ, 94; hind foot, 23.5. Skull: occipito-nasal length, 22.7; Hensel, 16.5; greatest mastoid width, 11.7; length of interparietal, 3.4; greatest width of interparietal, 7; interorbital constriction, 5.5; length of nasals, 8.6.

363. helleri (*Perognathus*), Elliot, Pub. Field Columb. Mus., III, 1903, p. 166. Zoölogy.

HELLER'S POCKET MOUSE.

Type locality. San Quentin, Lower California, Mexico.

Genl. Char. Size similar to *P. p. arenarius*, color very different, conspicuous lateral line. Skull with shorter nasals, broader rostrum, wider interorbital space, braincase broader, mastoids less prominent, bullæ smaller and more pointed anteriorly.

Color. Above mixed black and dark buff, giving a dark yellowish brown appearance to the upper parts; distinct bright buff lateral line from nose to rump; under parts pure white; tail above dark brown, almost dusky, beneath pure white, pencil-like upper part dusky; hands and feet gray; ears dark brown.

Measurements. Total length, 159; tail vertebræ, 83; hind foot, 20.5; ear, 8. Skull: occipito-nasal length, 23; Hensel, 14; zygomatic breadth, 11.5; interorbital constriction, 6; mastoid breadth, 11.5; greatest width of braincase, 10.5; palatal length, 8.5; length of nasals, 7.5; width of rostrum, 4; length of upper tooth row, alveolar border, 3.

364. hispidus (*Perognathus*), Baird, N. Am. Mamm., 1857, p. 421. Elliot, Syn. N. Am. Mamm., 1901, p. 251.

HISPID POCKET MOUSE.

Type locality. Charco Escondido, State of Tamaulipas, Mexico.

Geogr. Distr. States of Nuevo Leon and Tamaulipas, Mexico, north into Texas and Oklahoma.

Genl. Char. Size large; tail about equal to head and body; no spines; antitragus lobed; hind foot broad, short.

Color. Above brownish black and cinnamon; lateral stripe bright fulvous; under parts, fore legs, and feet white; tail black above, white below; no tuft.

Measurements. Total length, 208; tail vertebræ, 102; hind foot, 24. Skull: length of Hensel, 20.2; occipto-nasal length, 28; greatest mastoid breadth, 13.8; length of interparietal, 4.1; greatest width of interparietal, 7.2; interorbital constriction, 7; length of nasals, 10.6.

a.—zacatecæ (*Perognathus*), Osgood, N. Am. Faun., No. 18, 1900, p. 45.
ZACATECAS POCKET MOUSE.

Type locality. Valparaiso, State of Zacatecas, Mexico.

Geogr. Distr. States of Zacatecas and Guanajuato, Mexico.

Genl. Char. Similar to *P. hispidus*, but larger and darker.

Color. Upper parts olive brown; lateral line ochraceous; under parts white; spots on side of whiskers black; tail black above, white beneath.

Measurements. Total length, 211; tail vertebræ, 105; hind foot, 27.5. Skull: occipito-nasal length, 30.2; Hensel, 22.5; greatest mastoid breadth, 15; length of interparietal, 4; greatest width of inter-parietal, 8; interorbital constriction, 7; length of nasals, 12.

b.—paradoxus (*Perognathus*), Merr., N. Am. Faun., No. 1, 1889, p. 24, pl. III, fig. 18. Elliot, Syn. N. Am. Mamm., 1901, p. 252.
STRANGE POCKET MOUSE.

Type locality. Banner, Trego County, Kansas.

Geogr. Distr. State of Sonora, north to northern and western Kansas.

Genl. Char. Large; ears large, with antitragal lobes; soles naked.

Color. Above yellowish brown, lined with black; sides fulvous; fore legs fulvous outside; feet and under parts white; tail above fuliginous, beneath white, tinged with fulvous.

Measurements. Total length, 205–242 (Merr. measurements, l. c. 100 mm. error); tail vertebræ, 113; hind foot, 26; ears, 5. Skull: occipito-nasal length, 32; Hensel, 24; greatest mastoid breadth, 15; length of interparietal, 4.7; greatest width of interparietal, 8; inter-orbital constriction, 7.5; length of nasals, 13.2.

365. baileyi (*Perognathus*), Merr., Proc. Acad. Nat. Scien. Phil., 1894, p. 262, fig. 1. Elliot, Syn. N. Am. Mamm., 1901, p. 251.
BAILEY'S POCKET MOUSE.

Type locality. Magdalena, State of Sonora, Mexico.

Geogr. Distr. States of Sonora and Lower California, Mexico, into Arizona.

Genl. Char. Large; tail very much longer than head and body; mastoids considerably developed; bullæ nearly meeting below basisphenoid; interparietal pentagonal, broadest anteriorly.

Color. Above drab brown, lined with black; beneath white; pale lateral stripe; tail above dusky, beneath white.

Measurements. Total length, 210; tail vertebræ, 122; hind foot, 27. Skull: occipito-nasal length, 30; Hensel, 21.5; greatest mastoid breadth, 15.6; length of interparietal, 4.2; greatest width of inter-parietal, 6.8; interorbital constriction, 6.8; length of nasals, 12.2.

a.—rhydinorhis (*Perognathus*), Elliot, Pub. Field Columb. Mus., III, 1903, p. 167. Zoölogy.

SAN QUENTIN POCKET MOUSE.

Type locality. San Quentin, Lower California, Mexico.

Geogr. Distr. West coast of Lower California in the vicinity of San Quentin.

Genl. Char. Similar to *P. baileyi,* but darker, with very slender nasals, larger mastoids, and more slender upper incisors.

Color. Upper parts and sides pale buff, finely lined with black, darkest on head; sides of head, the nose, cheeks, and line above eyes yellowish buff, lined with black; ochraceous lateral line from lips to thighs; under parts pure white; tail above and pencil dark brown, beneath yellowish white; hands and feet grayish white.

Measurements. Total length, 232; tail vertebræ, 128; hind foot, 27; ear, 11.5. Skull: occipito-nasal length, 31; Hensel, 22; zygomatic width, 16; interorbital constriction, 6.5; width of interparietal, 6.5; length of nasals, 10; palatal length, 12; mastoid breadth, 15; length of upper tooth row, 4.

366. cnecus (*Perognathus*), Elliot, Pub. Field Columb. Mus., III, 1903, p. 169. Zoölogy.

BUFF-COLORED POCKET MOUSE.

Type locality. Rosarito, San Pedro Martir Mountains, Lower California, Mexico.

Genl. Char. Size very large; color of under parts deep cream buff; skull large, with very long, slender nasals, broadening at tip.

Color. Upper parts ochraceous buff, lined with black; top of nose and whiskers blackish; spot in front of eye, and broad lateral band from lip to posterior part of thigh ochraceous buff; sides above lateral line grayish buff; rest of under parts, arms, legs, hands, and thighs creamy buff, and a pure white spot with hairs white to the roots on chest between arms, extending in a narrow line towards abdomen; tail hairy, above blackish, beneath cream buff; ears dark brown.

Measurements. Total length, 228; tail vertebræ, 126; hind foot, 28; ear, 9.5. Skull: occipito-nasal length, 32; Hensel, 22; zygomatic width, 16.5; greatest mastoid breadth, 16; greatest width of inter-parietal, 8; interorbital constriction, 7; median length of nasals, 13; lateral length, 14; posterior width of nasals, 2; anterior width, 3;

greatest width of rostrum, 5; palatal length, 13; length of tooth row, alveolar border, 4; length of mandible, tip of angle to alveolus of incisor, 14.2.

367. margaritæ (*Perognathus*), Merr., Proc. Cal. Acad. Scien., 2d
  Ser., 1894, p. 459.
MARGARITA POCKET MOUSE.

Type locality. Santa Margarita Island, Lower California, Mexico.
Geogr. Distr. Type locality only.
Genl. Char. Size medium; tail longer than head and body. Skull: mastoids small; occiput not projecting posteriorly.
Color. Upper parts pale fawn mixed with black; lateral line indistinct; subauricular spot small; under parts and feet whitish; tail above dusky, beneath whitish.
Measurements. Total length, 180; tail vertebræ, 102; hind foot, 22.5. Skull: occipito-nasal length, 25.9; Hensel, 18; greatest mastoid breadth, 12; length of interparietal, 12; greatest width of interparietal, 8; interorbital constriction, 6.5; length of nasals, 10.3.

368. spinatus (*Perognathus*), Merr., N. Am. Faun., No. 1, 1889,
  p. 21. Elliot, Syn. N. Am. Mamm., 1901, p. 255.
SPINY POCKET MOUSE.

Type locality. Twenty-five miles below the Needles, Colorado River, San Bernardino County, California.
Geogr. Distr. Northern Lower California, Mexico, and southern California.
Genl. Char. Size medium; tail longer than head and body, crested; antitragal lobe large; soles naked; rump spinous.
Color. Above drab gray, lined with black; no lateral stripe; fore legs, feet, and under parts white; tail above dusky; below white; pencil dusky.
Measurements. Total length, 179; tail vertebræ, 104; pencil, 15; hind foot, 21; ear from crown, 3.5. Skull: occipito-nasal length, 28; Hensel, 19.8; greatest mastoid breadth, 13.1; length of interparietal, 4; greatest width of interparietal, 8.3; interorbital constriction, 6.5; length of nasals, 11.2.

a.—peninsulæ (*Perognathus*), Merr., Proc. Cal. Acad. Scien., 2d
  Ser., 1894, p. 460.
CAPE ST. LUCAS POCKET MOUSE.

Type locality. San José del Cabo, Lower California, Mexico.
Geogr. Distr. Cape region of Lower California, Mexico.
Genl. Char. Similar to *P. spinatus*, but larger; tail shorter; ears large, rounded; skull with comparative differences.

Color. Exactly like that of *P. spinatus.*

Measurements. Total length, average, 188; tail vertebræ, 101; hind foot, 24. Skull: occipito-nasal length, 26.5; Hensel, 18; greatest mastoid breadth, 13; length of interparietal, 3.7; greatest width of interparietal, 7.6; least interorbital width, 6.6; length of nasals, 9.8.

b.—bryanti (*Perognathus*), Merr., Proc. Cal. Acad. Scien., 2d Ser., 1894, p. 458.
BRYANT'S POCKET MOUSE.

Type locality. San José Island, Lower California, Mexico.
Geogr. Distr. Type locality only.
Genl. Char. Similar to *P. s. peninsulæ;* larger, tail longer.
Color. Like *P. s. peninsulæ.*
Measurements. Total length, 216; tail vertebræ, 127; hind foot, 25. Skull: occipito-nasal length, 27.3; Hensel, 18.9; greatest mastoid breadth, 13.1; length of interparietal, 3.5; greatest width of interparietal, 8.1; interorbital constriction, 6.8; length of nasals, 10.3.

369. intermedius (*Perognathus*), Merr., N. Am. Faun., No. 1, 1889, p. 18, pl. 11, fig. 13. Elliot, Syn. N. Am. Mamm., 1901, p. 253.
INTERMEDIATE POCKET MOUSE.

Type locality. Mud Spring, Mohave County, Arizona.
Geogr. Distr. State of Chihuahua, Mexico, into Texas and Arizona.
Genl. Char. Tail much longer than head and body; antitragal lobe large; soles naked.
Color. Above drab gray, tinged with pale fulvous and lined with black; lateral line pale fulvous; fore legs, feet, and under parts white; tail above sooty brown, beneath whitish.
Measurements. Total length, 183; tail vertebræ, 106; pencil, 18; hind foot, 21; ear, 4.5. Skull: occipito-nasal length, 24.5; Hensel, 17; greatest mastoid breadth, 13.5; length of interparietal, 3; greatest width of interparietal, 8; interorbital constriction, 6.3; length of nasals, 9.4.

370. nelsoni (*Perognathus*), Merr., Proc. Acad. Nat. Scien. Phil., 1894, p. 266.
NELSON'S POCKET MOUSE.

Type locality. Hacienda La Parada, 25 miles northwest of the city of San Luis Potosi, State of San Luis Potosi, Mexico.
Geogr. Distr. States of San Luis Potosi and Durango, south into State of Jalisco, and east into State of Tamaulipas, Mexico.
Genl. Char. Similar to *P. intermedius,* but larger and darker; bristles on rump; tail crested.

Color. Upper parts and sides hair brown; lateral line fawn; under parts soiled white; tail black above, whitish beneath; ears dusky.

Measurements. Total length, 182; tail vertebræ, 104; hind foot, 23; ear, 8. Skull: occipito-nasal length, 26; Hensel, 18; greatest mastoid breadth, 13.8; length of parietal, 3.5; greatest width of interparietal, 7.6; interorbital constriction, 6.7; length of nasals, 10.

a.—canescens (*Perognathus*), Merr., Proc. Acad. Nat. Scien. Phil., 1894, p. 267.
JARAL POCKET MOUSE.

Type locality. Jaral, State of Coahuila, Mexico.

Geogr. Distr. Known only from type locality.

Genl. Char. Similar to *P. intermedius*; larger and more grayish.

Color. Upper parts drab gray; lateral line pinkish buff; under parts white; tail mouse gray above, white beneath.

Measurements. Total length, 193; tail vertebræ, 117; hind foot, 22. Skull: occipito-nasal length, 25; Hensel, 17.5; greatest mastoid breadth, 13.5; length of interparietal, 37; greatest width of interparietal, 7.2; interorbital constriction, 6.1; length of nasals, 9.3.

371. goldmani (*Perognathus*), Osgood, N. Am. Faun., No. 18, 1900, p. 54.
GOLDMAN'S POCKET MOUSE.

Type locality. Sinaloa, State of Sinaloa, Mexico.

Geogr. Distr. States of Sonora and Sinaloa, Mexico; coast plains.

Genl. Char. Size large; ears large; similar in color to *P. nelsoni.*

Color. Upper parts broccoli brown, darkened on anterior half with black; lateral line pinkish buff; ear whitish exteriorly for apical half; under parts white; tail blackish above, white below.

Measurements. Total length, 202; tail vertebræ, 108; hind foot, 28. Skull: occipito-nasal length, 27.7; Hensel, 20.6; greatest mastoid width, 14.3; length of interparietal, 3.8; greatest width of interparietal, 6.5; interorbital constriction, 6.5; length of nasals, 11.1.

a.—artus (*Perognathus*), Osgood, N. Am. Faun., No. 18, 1900, p. 55.
BATOPILAS POCKET MOUSE.

Type locality. Batopilas, State of Chihuahua, Mexico.

Geogr. Distr. States of Chihuahua, Durango, and Sinaloa, Mexico.

Genl. Char. Similar to *P. goldmani*; rump bristles undeveloped; slight cranial differences.

Color. Indistinguishable from *P. goldmani.* The less prominent rump bristles seem chiefly to distinguish this from the preceding one.

Measurements. Total length, 191; tail vertebræ, 106; hind foot, 24.6. Skull: occipito-nasal length, 25.4; Hensel, 18.8; greatest mastoid breadth, 12.4; length of interparietal, 3.3; greatest width of interparietal, 7.1; interorbital constriction, 6.1; length of nasals, 9.7.

372. fallax (*Perognathus*), Merr., N. Am. Faun., No. 1, 1889, p. 19, pl. III, fig. 14. Elliot, Syn. N. Am. Mamm., 1901, p. 254.
SHORT-EARED CALIFORNIA POCKET MOUSE.

Type locality. Reche Cañon, three miles southeast of Colton, San Bernardino County, California.

Geogr. Distr. Northern Lower California, Mexico, into southern California.

Genl. Char. Large; tail crested, longer than head and body; antitragal lobe higher than broad; soles naked.

Color. Above dark grizzled yellowish brown lined with black; lateral line pale fulvous, covering upper surface of fore leg; beneath white; tail above sooty brown, beneath white.

Measurements. Total length, 183–211; tail vertebræ, 104–126; hind foot, 24–25; ear, 6. Skull: occipito-nasal length, 26; Hensel, 18; greatest mastoid breadth, 14; length of interparietal, 3.8; greatest width of interparietal, 7.8; interorbital constriction, 6.6; length of nasals, 10.

a.—pallidus (*Perognathus*), Mearns, Proc. Biol. Soc. Wash., XIV, 1901, p. 135.
PALLID POCKET MOUSE.

Type locality. East slope of the Coast Range Mountains, on the Mexican boundary line, San Diego County, California.

Geogr. Distr. Lower California, Mexico, north to Riverside County, California.

Genl. Char. Similar to *P. fallax,* but paler.

Color. Above pale broccoli brown, slightly mixed with black; under parts creamy white; lateral line pale pinkish buff; tail drab; feet creamy white; ears sparsely covered with white hairs.

Measurements. "Average of six specimens: total length, 195 (188–206); tail vertebræ, 107 (98–112); hind foot, 24.2 (23.7–25); ear from crown, 6.9 (6.5–7)." (Mearns, l. c.)

373. anthonyi (*Perognathus*), Osgood, N. Am. Faun., No. 18, 1900, p. 56.
ANTHONY'S POCKET MOUSE.

Type locality. South Bay, Cerros or Cedros Island, Lower California, Mexico.

Geogr. Distr. Known only from type locality.

Genl. Char. Similar to *P. fallax*, but smaller, and comparative differences in skull.

Color. Upper parts grayish fawn, mixed with black; lateral line faint, brownish fawn; white spots under eye; tail above dusky, beneath whitish.

Measurements. Total length, 168; tail vertebræ, 92; hind foot, 23.5. Skull: occipito-nasal length, 25.4; Hensel, 17.4; greatest mastoid breadth, 12.9; length of interparietal, 2.6; greatest width of interparietal, 5.8; interorbital constriction, 6; length of nasals, 10.2.

374. femoralis (*Perognathus*), Allen, Bull. Am. Mus. Nat. Hist., 1891, p. 281. Elliot, Syn. N. Am. Mamm., 1901, p. 254.
GREAT CALIFORNIA POCKET MOUSE.

Type locality. Dulzura, San Diego County, California.

Geogr. Distr. Northern Lower California, Mexico, and adjoining portions of southern California.

Genl. Char. Tail much longer than head and body; size large.

Color. Above grayish yellowish brown, lined with black; lateral line fulvous; fore legs, feet, and under parts white; tail above sooty brown, including tip, beneath white.

Measurements. Total length, 198–241; tail vertebræ, 112–133; hind foot, 25–27. Skull: occipito-nasal length, 29.6; Hensel, 20.3; greatest mastoid breadth, 14.3; length of interparietal, 4; greatest width of interparietal, 8.1; length of nasals, 11.4; interorbital constriction, 7.1.

a.—mesopolius (*Perognathus*), Elliot, Pub. Field Columb. Mus., III, 1903, p. 168. Zoölogy.
GRAY POCKET MOUSE.

Type locality. Piñon, San Pedro Martir Mountains, Lower California, Mexico; 5,000 feet elevation.

Genl. Char. Size large; ear large; hind foot and tail long; similar in color to *P. femoralis*, but grayer and without the bistre hue. Skull with greater interorbital constriction and less mastoid breadth.

Color. Above mixed pale gray and light buff, lined sparingly with black; sides grayer than upper parts; lateral line bright buff; under parts, hands, and feet pure white; tail and pencil dusky above, white beneath; ears light brown.

Measurements. Total length, 232; tail vertebræ, 136; hind foot, 27; ear, 14. Skull: occipito-nasal length, 27; Hensel, 18; zygomatic width, 13; mastoid breadth, 13; interorbital constriction, 6; palatal length, 10; length of nasals, 10; posterior width of nasals 2; anterior width of nasals, 3; greatest width of rostrum, 5.

The harsh pelage exhibited in *Perognathus* and *Chætodipus* of the Pocket Mice, is in the members of HETEROMYS carried farther, even to the presence of small flattened spines in the fur. They are animals of moderate size, with tails about equaling the head and body in length, of various colorations, with a heavier skull, and incisors smooth in front. Only one species is found north of the Mexican boundary, but a number are natives of Mexico, Central America, and the West Indies. They have a rather long, slender body, and moderately short legs.

74. Heteromys. Spiny Pouched Rats.

$$I.\frac{1-1}{1-1};\ P.\frac{1-1}{1-1};\ M.\frac{3-3}{3-3} = 20.$$

Heteromys Desm., Nouv. Dict. Hist. Nat. Mamm., 1817, p. 313.
    Type *Mus anomalus* Thompson.
Dasynotus Wagl., Nat. Syst. Amphib., 1830, p. 21.
Saccomys F. Cuv., Mem. Mus. Hist. Nat., Paris, x, 1823, p. 419.
Xylomys Merr., Proc. Biol. Soc. Wash., xv, 1902, p. 43.
Liomys Merr., Proc. Biol. Soc. Wash., xv, 1902, p. 44.

Molars rooted; flattened spines mingled with the fur; skull almost flat above, slightly arched in middle of superior outline; nasals projecting beyond incisors; mastoids level with plane of occiput; zygomata depressed, but not dipping to level of palate; occiput

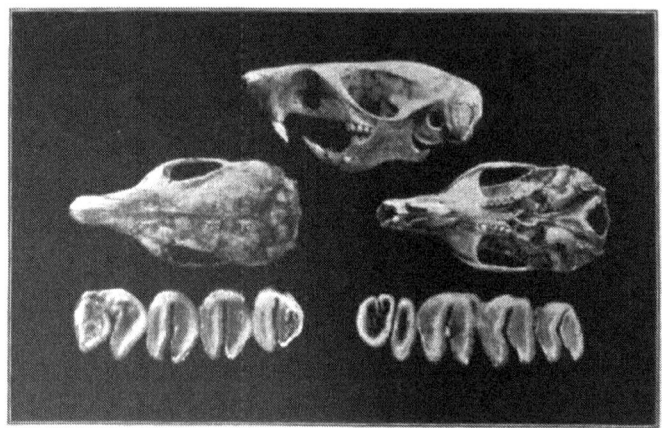

FIG. 69. HETEROMYS GAUMERI.
No. 5844 Field Columbian Mus. Coll. Nat. size.
UPPER TOOTH ROW.   LOWER TOOTH ROW.
Enlarged 8 times.   Enlarged 8 times.

horizontal, flat; tail long, exceeding body and head; soles hairy, with five tubercles; pouches covered with hair; tail scaly, short hairs from between the scales.

A. *Heteromys.

KEY TO THE SPECIES AND SUBSPECIES OF THE SUBGENUS.

A. Pelage harsh; flattened spines present.
  a. Soles hairy, 5-tuberculate.   PAGE

375. irroratus (*Heteromys*), Gray, Proc. Zoöl. Soc., 1868, p. 205.
LA PARADA SPINY RAT.

Type locality. La Parada, State of Oaxaca, Mexico.

Geogr. Distr. State of Oaxaca, Mexico.

Genl. Char. Tail short, hairy; lateral line present.

Color. Upper parts grayish yellow, mixed with black; lips and under parts, inner side of legs, and feet white; faint lateral line pale fawn; tail beneath white.

Measurements. Head and body, 118; tail, imperfect ? 106; hind foot and claws, 32.5. (ex mounted specimen in Brit. Mus. O. Thomas in litt.)

376. bulleri (*Heteromys*), Thomas, Ann. Mag. Nat. Hist., 6th Ser., XI, 1893, p. 330.

---

*The arrangement of the members of HETEROMYS under its two subgenera *Heteromys* and *Liomys* cannot be satisfactorily accomplished, as the teeth of the adults in various instances have no distinguishing characters; hence the present separation of the species and races in the keys given may be regarded only as tentative.

BULLER'S SPINY RAT.

Type locality. La Laguna, Sierra de Juanacatlan, State of Jalisco, Mexico. Altitude, 7,000 feet.

Geogr. Distr. High elevations, State of Jalisco, Mexico.

Genl. Char. Size between *H. alleni* and *H. irroratus*; skull stout.

Color. Above grizzled smoky gray and yellow; lateral line yellowish; under parts pure white; hands and feet white; arms white to elbow; hind leg dark gray on outer side, white on inner; tail brown above, white beneath; ears without white edges.

Measurements. Total length, 234; tail, 120; hind foot, 28.5; ear from notch, 15. Skull: basal length, 29.5; greatest length, 34.5; greatest breadth, 16.8; length of nasals, 13.5; interorbital constriction, 8.5; length of interparietal, 4; breadth of interparietal, 6.4; palatal length, 21; length of upper molar series, 5.3.

377. salvini (*Heteromys*), Thomas, Ann. Mag. Nat. Hist., 6th Ser., XI, 1893, p. 331.

longicaudatus, Alst., Biol. Cent. Am., Mamm., I, 1880, p. 167, pl. XVII, fig. 2. (nec Gray.)

SALVIN'S SPINY RAT.

Type locality. Dueñas, Gautemala.

Genl. Char. Size equal to that of *H. bulleri*; feet shorter; fur spiny.

Color. General hue grizzled black and yellowish above; under parts pure white; outer sides of forearm slaty gray; tail brown above, whitish below.

Measurements. Total length, about 210; tail, 95; hind foot, 26.5. Skull: basal length, 28; greatest length, 33.6; greatest breadth, 15.2; length of nasals, 13.1; interorbital constriction, 7.6; length of interparietal, 4.5; breadth of interparietal, 10.2; palatal length, 19.2; length of upper molar series, 4.4.

a.—nigrescens (*Heteromys*), Thomas, Ann. Mag. Nat. Hist., 6th Ser., XII, 1893, p. 234.

BLACK SPINY RAT.

Type locality. Costa Rica.

Genl. Char. Similar to *H. salvini*, but dorsal region uniform in color.

Color. Above dark uniform smoky brown; a few yellow-tipped hairs on sides; lateral line absent; limbs dark gray; under parts white; tail above brown, beneath white.

Measurements. Length of head and body, 127; tail incomplete; hind foot without claws, 25. "Skull: greatest length, 32.7; greatest breadth, 15.4; length of nasals, 3.8; interorbital constriction, 6.8;

breadth of interparietal, 8; length of interparietal, 4.1; length of upper molar series, 4.7."

378. gaumeri (*Heteromys*), Allen & Chapman, Bull. Amer. Mus. Nat. Hist., 1897, p. 9.

GAUMER'S SPINY RAT.

Type locality. Chichen Itza, Yucatan, Mexico.

Geogr. Distr. Yucatan, Mexico.

Genl. Char. Size large; premaxillæ and nasals terminating equally.

Color. Above dark smoke gray; sides lighter, fulvous; lateral line and outer surface of fore legs orange ochraceous; outer surface of hind legs like back; under parts and inner surface of limbs white; ears dusky with white edges; tail crested and tufted, dusky above, grayish white below; middle of dorsal surface covered with spines which are whitish at base, black apically, and mixed in with orange ochraceous hairs.

Measurements. Total length, 292; tail vertebræ, 162; hind foot, and claw, 32; ear from notch, 14.5. Skull: basal length, 31; greatest length, 37; greatest breadth, 16; interorbital breadth, 10.5; length of nasals, 16; length of interparietal, 11; palatal length, 20; upper tooth row, 5.

379. annectens (*Heteromys*), Merr., Proc. Biol. Soc. Wash., xv, 1902, p. 43.

ALLIED SPINY RAT.

Type locality. Pluma, State of Oaxaca, Mexico.

Genl. Char. Size medium; tail long; skull similar to that of *H. gaumeri*, but "smaller, narrower interorbitally, with more abruptly spreading zygomata and smaller interparietal." Hind foot 6-tuberculate.

Color. Upper parts blackish brown; lateral line fulvous; under parts, hands, and feet white, the latter sometimes "clouded"; tail above dusky, beneath whitish, except tip, which is all dark.

Measurements. Total length, 300; tail vertebræ, 165; hind foot, 33.

380. hispidus (*Heteromys*), Allen, Bull. Amer. Mus. Nat. Hist., 1897, p. 56.

HISPID SPINY RAT.

Type locality. Compostella, Territorio de Tepic, Mexico.

Geogr. Distr. Territorio de Tepic, Mexico.

Genl. Char. Pelage soft; soles 6-tuberculate, hairy.

Color. Upper parts reddish brown, mixed with black-tipped bristles; under parts and feet white; lateral line reddish fulvous; tail above pale brown, beneath lighter.

Measurements. Total length, 220-230; tail vertebræ, 115-123; hind foot, 27-28; ear, 13.5. Skull: total length, 30; basal length, 24; greatest mastoid breadth, 13.5; interorbital constriction, 7; length of nasals, 12; palatal length, 12.

381. desmarestianus (*Heteromys*), Gray, Proc. Zoöl. Soc., 1868, p. 204. Alston, Biol. Centr. Amer., Mamm., I, 1880, p. 167, pl. 17, fig. 1.

COBAN SPINY RAT.

Type locality. Coban, Guatemala.

Geogr. Distr. Guatemala.

Genl. Char. Ears moderate; teeth small, "upper flat, lower keeled, narrow in front."

Color. "Chestnut brown; tip of nape, lips, chin, and under side of the body, hands, and feet and under side of tail white; spines of back white with chestnut tips." (Gray, l. c.)

Measurements. Total length, 247.5; tail, 118.5.

382. longicaudatus (*Heteromys*), Gray, Proc. Zoöl. Soc., 1868, p. 204.

LONG-TAILED SPINY RAT. *Tultusia* in Guatemala, also applied to *Heterogeomys torridus.*

Type locality. Mexico.

Geogr. Distr. State of Oaxaca, Mexico; range unknown.

Genl. Char. Tail long; no lateral line; soles 6-tuberculate.

Color. Above blackish brown, in some specimens with rufous intermixed; sides dark drab brown, uniform, or mixed with rufous; limbs uniform in color with sides; under parts, hands, and feet white; tail above blackish brown, beneath yellowish.

Measurements. Total length, 247.15; tail, 118.75.

383. repens (*Heteromys*), Bangs, Bull. Mus. Comp. Zoöl., XXXIX, 1902, p. 45.

BOQUETE SPINY RAT.

Type locality. Boquete, Chiriqui, Panama. Altitude, 4,000 feet.

Genl. Char. Size large; feet large; soles 6-tuberculate.

Color. Top of nose and face grayish dusky; upper parts of body mixed dusky, brown, and tawny ochraceous; no lateral line; upper lip, under parts, and inner sides of arms and legs white; upper surface of arms gray; outer surface of legs dusky; tail above dusky, beneath white, pencil whitish; ears dusky, bordered with whitish.

Measurements. Total length, 282-300; tail vertebræ, 145-155; hind foot, 32-33; ear, 14-15. Skull: type; basal length, 31.4; occipito-nasal length, 35.4; zygomatic width, 16.4; mastoid width,

14.8; interorbital width, 9.2; length of nasals, 14.8; width of nasals, 4.2; upper molar series, 4.8; length of single half mandible, 17.2; (Bangs, l. c.)

384. goldmani (*Heteromys*), Merr., Proc. Biol. Soc. Wash., xv, 1902, p. 41.
GOLDMAN'S SPINY RAT.

Type locality. Chicharras, State of Chiapas, Mexico.

Genl. Char. Size large; tail longer than head and body, naked; sole of hind foot with 6 pads.

Color. Above dusky gray, head and back darkest; upper lip, inner side of limbs, hands, and feet and under parts white; tail dusky above, paler below, tip whitish; ears without white edges.

Measurements. Total length, 347; tail vertebræ, 199; hind foot, 40.

a.—lepturus (*Heteromys*), Merr., Proc. Biol. Soc. Wash., xv, 1902, p. 42.
SHORT-TAILED SPINY RAT.

Type locality. Mountains near Santo Domingo, State of Oaxaca, Mexico.

Genl. Char. Similar to *H. goldmani*, but smaller and not so black.

Color. "Head and back grizzled with fulvous; nose and ankles dusky by contrast; hind feet and tail decidedly shorter; rostrum broader, broadening gradually to zygomata with much less of the usual notch."

Measurements. "Total length, 340; tail vertebræ, 191; hind foot, 39." (Merr., l. c.)

385. griseus (*Heteromys*), Merr., Proc. Biol. Soc. Wash., xv, 1902, p. 42.
GRAY SPINY RAT.

Type locality. Mountains near Tonala, State of Chiapas, Mexico.

Genl. Char. Size large; hind foot 6-tuberculate.

Color. Above drab, grizzled with black; upper lip, hands, feet, and under parts white; tail above dusky, beneath whitish, tip dark all around; faint buff lateral line on head and body; ears without white edging.

Measurements. Total length, 325; tail vertebræ, 186; hind foot, 38.

386. adspersus (*Heteromys*), Peters, Monatsb. K. Preuss. Akad. Wiss. Berl., 1874, p. 357.
SPOTTED SPINY RAT.

Type locality. Panama.

Genl. Char. Soles 6-tuberculate; tail half as long as head and body.

Color. Head dark gray; upper parts black, mixed with ochre yellow, base of all the hairs white; flat grooved spines, with black tips; the stiff hairs with a black ring and reddish yellow tip, these last distributed sparingly over the head and rump; tip of nose, lips, inner sides of arms and legs and under parts of body white; tail above black, beneath white; ears naked on outer side, black on inner.

Measurements. Total length, 240; tail, 95; hind foot and claw, 30.

B. Liomys.

$$I.\tfrac{1-1}{1-1};\ P.\tfrac{1-1}{1-1};\ M.\tfrac{3-3}{3-3} = 20.$$

Pelage harsh; tail well haired; molars with two parallel transverse loops, without "additional lobes or permanent enamel islands," uniting in old animals and forming a horseshoe as in *Heteromys*.

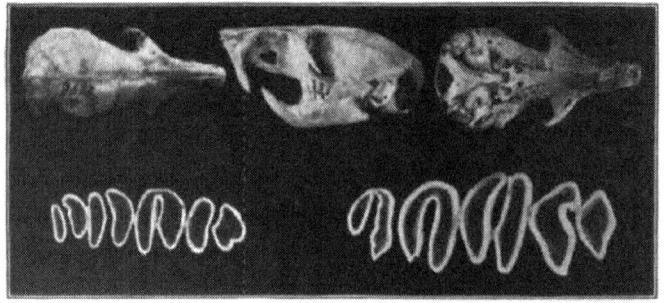

FIG. 70. HETEROMYS (LIOMYS) ALBOLIMBATUS.
No. 8673 Field Columbian Mus. Coll.   Nat. size.

LOWER TOOTH ROW.   UPPER TOOTH ROW.
Enlarged 6 times.   Enlarged 6 times.

KEY TO THE SPECIES AND SUBSPECIES OF THE SUBGENUS.

387. albolimbatus (*Heteromys*), Gray, Proc. Zoöl. Soc., 1868, p. 205. GRAY'S SPINY MOUSE.

Type locality. La Parada, State of Oaxaca, Mexico.

Geogr. Distr. State of Oaxaca, Mexico.

Genl. Char. Edges of ears and tip of tail white; tail long as body

Color. Above mixed gray, black, and yellow; lateral line buff saddle conspicuous; rump gray; under parts and feet white; tai above blackish brown, beneath white.

Measurements. Total length, 207–226; tail vertebræ, 108–113; hind foot, 25–28. Skull: occipito-nasal length, 29.5; Hensel, 20; zygomatic width, 14; interorbital constriction, 7.5; length of nasals, 10.5; length of upper molar series, 5; length of mandibles, 13; length of lower tooth row, 5.

388. canus (*Liomys*), Merr., Proc. Biol. Soc. Wash., xv, 1902, p. 44. HOARY SPINY MOUSE.

Type locality. Parral, State of Chihuahua, Mexico.

Genl. Char. Size large; hind foot 5-tuberculate. Skull similar to that of *H. alleni*, but larger and heavier and with broader interorbital space.

Color. Above gray, mixed with pale fulvous, white, and dark gray, lateral line indistinct; under parts, hands, and feet white; tail blackish above, white beneath, tip all dark.

Measurements. Total length, 276; tail vertebræ, 138; hind foot, 34.

FIG. L. HETEROMYS ALLENI. ALLEN'S SPINY MOUSE.

389. alleni (*Heteromys*), Coues, Bull. Mus. Comp. Zoöl., VIII, 1881,
p. 187.
ALLEN'S SPINY MOUSE.

Type locality. Hacienda Angostura, Rio Verde, State of San
Luis Potosi, Mexico.

Geogr. Distr. State of San Luis Potosi, Mexico.

Genl. Char. Soles hairy, 5-tuberculate; tail long; pelage coarse,
with flattened spines intermixed in the hairs; incisors yellow.

Color. Above blackish, the hairs having buff bases and tips
black, the lighter color showing amid the darker one; narrow lateral
line from nose to thighs bright buff; under parts white; upper part
of arms and thighs like back; hands and feet white; ears like back,
edged with white; lips white; tail bicolor, above black, beneath white.

Measurements. Total length, 238–260; tail, 115–136; hind foot,
28–30.

390. torridus (*Liomys*), Merr., Proc. Biol. Soc. Wash., XV, 1902,
p. 45.
TORRID SPINY MOUSE.

Type locality. Cuicatlan, State of Oaxaca, Mexico.

Genl. Char. Size small; arms whitish, hind foot 5-tuberculate.

Color. Above gray, grizzled with black, forming saddle on the
back; lateral line indistinct; arms, hands, feet, and under parts
whitish; tail above dark. beneath whitish, tip all dark.

Measurements. Total length, 222; tail vertebræ, 125; hind foot, 27.

a.—minor (*Liomys*), Merr., Proc. Biol. Soc. Wash., xv, 1902, p. 45.
SMALLEST SPINY MOUSE.
Type locality. Huajuapam, State of Oaxaca, Mexico.
Genl. Char. Like *H. torridus*, but smaller.
Color. Above grizzled gray, no saddle on back, remainder similar to *L. torridus*.
Measurements. Total length, 222; tail vertebræ, 125; hind foot, 27.

391 exiguus (*Heteromys*), Elliot, Pub. Field Columb. Mus., III, 1903, p. 146. Zoölogy.
LITTLE SPINY MOUSE.
Type locality. Puenta de Ixtla, State of Morelos, Mexico.
Genl. Char. Size small; above buff and dark brown, and with a "saddle" on back.
Color. Above mixed buff brownish black and dark gray, forming a dark "saddle" on middle of back; sides of head, neck, and body, rump, and thighs light gray; lips, under parts, and hands white; arms, legs, and feet yellowish white; tail above blackish brown, beneath white; no lateral line separating the gray of sides from the white under parts.
Measurements. Total length, 185; tail vertebræ, 106; hind foot, 24. (Type.)

392. plantinarensis (*Liomys*), Merr., Proc. Biol. Soc. Wash., xv, 1902, p. 46.
PLANTINAR SPINY MOUSE.
Type locality. Plantinar, State of Jalisco, Mexico.
Genl. Char. Size small; ears small; tail but slightly haired; rostrum and nasals curved, the latter notched posteriorly; hind foot 5-tuberculate.
Color. Above grizzled fulvous and black; lateral line fulvous, broad; under parts yellowish white; hands and feet white; tail drab above, beneath whitish; ear edged with whitish.
Measurements. Total length, 202; tail vertebræ, 102; hind foot, 26.

393. pictus (*Heteromys*), Thomas, Ann. Mag. Nat. Hist., 6th Ser., XII, 1893, p. 233.
PAINTED SPINY MOUSE.
Type locality. Mineral San Sebastian, State of Jalisco, Mexico. Altitude, 4,300 feet.

Geogr. Distr.   State of Jalisco, Mexico.

Genl. Char.   Size and dimensions of skull similar to those of *H. salvini* from Guatemala.

Color.   Above mixed grizzled rufous and orange; sides bright rufous; lateral line orange rufous; fore limbs white; hind limbs dusky on outer sides, white on inner; under parts white; tail blackish above, white beneath; ear black, edges white.

Measurements.   Total length, 217; tail, 113; hind foot, 24.8; ear from notch, 12.

a.—rostratus (*Liomys*), Merr., Proc. Biol. Soc. Wash., xv, 1902, p. 46. LONG-NOSED SPINY MOUSE.

Type locality.   Ometepec, State of Guerrero, Mexico.

Genl. Char.   Similar to *H. pictus*, but somewhat larger; pelage coarser.   Skull heavy; nasals long, truncate, or notched posteriorly.

Color.   Like that of *H. pictus*, but not so red.

Measurements.   Total length, 252; tail vertebræ, 133; hind foot, 29.

b.—isthmius (*Liomys*), Merr., Proc. Biol. Soc. Wash., xv, 1902, p. 46. ISTHMIAN SPINY MOUSE.

Type locality.   Tehuantepec, State of Oaxaca, Mexico.

Genl. Char.   Similar to *H. pictus*, but differing in color.

Color.   Similar to *H. pictus*, but with the upper parts very much paler and less red, and lateral line faint or absent; cheeks pale grayish brown.

Measurements.   Total length, 245; tail vertebræ, 130; hind foot, 30.

*393a. parviceps (*Liomys*) Goldman, Proc. Biol. Soc. Wash., xvii, 1904, p. 82. URUAPAN SPINY MOUSE.

Type locality.   Uruapan, State of Michoacan, Mexico.

Genl. Char.   Size very small, hind feet short, 6-tuberculate.

Color.   Upper parts grizzled brownish fulvous; under parts white; faint fulvous lateral line; tail above brownish, beneath whitish; hands and feet white; ears with white edges.

Measurements.   Total length, 202; tail vertebræ, 110; hind foot, 24.   Skull: greatest length, 28.3; Hensel, 20; zygomatic width, 13; interorbital constriction, 6.7; length of nasals, 11.5; interparietal, 3.2 × 8.3; length of upper molar series, alveolar border, 4.2.

---

*Description published too late to be included in the regular order of numerals.

394. sonorana (*Liomys*), Merr., Proc. Biol. Soc. Wash., xv, 1902, p. 47.

SONORA SPINY MOUSE.

Type locality. Alamos, State of Sonora, Mexico.

Genl. Char. Size medium; skull with long and slender rostrum, the sides parallel; nasals long, emarginate posteriorly; hind foot 6-tuberculate.

Color. Above grayish drab, grizzled with fulvous and dark hairs; lateral line fulvous; under parts, hands, and feet white; brown band across nose; tail above dusky; beneath whitish, grading into dusky on terminal third; ears edged with white.

Measurements. Total length, 262; tail vertebræ, 142; hind foot, 32.5.

395. veræcrucis (*Liomys*), Merr., Proc. Biol. Soc. Wash., xv, 1902, p. 47.

VERA CRUZ SPINY MOUSE.

Type locality. San Andreas Tuxtla, State of Vera Cruz, Mexico.

Genl. Char. Size medium; skull small; upper surface of anterior root of zygomata strongly depressed and rounded; nasals notched posteriorly; interparietal sub-triangular; hind foot 6-tuberculate.

Color. Above dark brown, mixed with black and fulvous; lateral line faint on sides of body; under parts white; tail dusky above, whitish beneath; ankles dusky.

Measurements. Total length, 220; tail vertebræ, 108; hind foot, 25.

396. obscurus (*Liomys*), Merr., Proc. Biol. Soc. Wash., xv, 1902, p. 48.

DUSKY SPINY MOUSE.

Type locality. Carrizal, State of Vera Cruz, Mexico.

Genl. Char. Size rather large; skull heavy, and with high brain-case; hind foot 6-tuberculate.

Color. Above blackish, grizzled with fulvous and whitish, and becoming grayish on the sides.

Measurements. Total length, 230; tail vertebræ, 124; hind foot, 31.

397. phæura (*Liomys*), Merr., Proc. Biol. Soc. Wash., xv, 1902, p. 48.

DARK-TAILED SPINY MOUSE.

Type locality. Pinotepa, State of Oaxaca, Mexico.

Genl. Char. Size small; ears large; hind foot 6-tuberculate; tail short, unicolor except basal third beneath. Skull broad, short, and flat; rostrum and nasals short, slender.

Color. Above drab brown, mixed with black and pale fulvous; lateral line fulvous; tail all dusky except basal third beneath which is paler.

Measurements. Total length, 204; tail vertebræ, 95 (broken); hind foot, 29.

398. orbitalis (*Liomys*), Merr., Proc. Biol. Soc. Wash., xv, 1902, p. 48.

CATEMACO SPINY MOUSE.

Type locality. Catemaco, State of Vera Cruz, Mexico.

Genl. Char. Size medium; skull broad, heavy; nasals suddenly expanding on anterior half; zygomata widely and squarely spread; rostrum short, sides parallel; superciliary beads developed, reaching to middle of parietals; hind foot 6-tuberculate.

Color. Above dark brown, slightly grizzled with fulvous, blackish on middle of back; lateral line indistinct; hands, feet, and under parts white; tail above dusky, beneath whitish, tip all dark.

Measurements. Total length, 225; tail vertebræ, 109; hind foot, 29.

399. crispus (*Liomys*), Merr., Proc. Biol. Soc. Wash., xv, 1902, p. 49.

CURLY SPINY MOUSE.

Type locality. Tonala, State of Chiapas, Mexico.

Genl. Char. Size small; tail shorter than head and body, scantily haired; hind foot 6-tuberculate.

Color. Above gray, sprinkled with white, darkest on dorsal region, the hairs with tips recurved; hands, feet, and under parts whitish; no lateral line; ears without white edges; tail dusky, paler beneath.

Measurements. Total length, 210; tail vertebræ, 99 (broken); hind foot, 27.5.

a.—setosus (*Liomys*), Merr., Proc. Biol. Soc. Wash., xv, 1902, p. 49.
HUEHUETAN SPINY MOUSE.

Type locality. Huehuetan, State of Chiapas, Mexico.

Genl. Char. Similar to *H. crispus*, but with coarser pelage; tail longer; skull larger, and with its characters correspondingly intensified; nasals more cuneate, truncate posteriorly.

Color. Like *H. crispus*, but darker and with fewer recurved hairs.

Measurements. Total length, 225; tail vertebræ, 110; hind foot, 29.

400. heterothrix (*Liomys*), Merr., Proc. Biol. Soc. Wash., xv, 1902, p. 50.

HONDURAS SPINY MOUSE.

Type locality. San Pedro Sula, Honduras.

Genl. Char. Size medium; tail nearly naked; skull similar to that of *H. c. setosus*, but larger and heavier; roots of zygomata broad; hind foot 6-tuberculate.

Color. Above dark drab brown, grizzled with ferrugineous and black; hands, feet, and under parts yellowish white; tail dark above, paler below.

Measurements. Total length, 255; tail vertebræ, 126; hind foot, 31.

401. paralius (*Heteromys*), Elliot, Pub. Field Columb. Mus., III, 1903, p. 233. Zoölogy.

LITTORAL SPINY MOUSE.

Type locality. San Carlos, State of Vera Cruz, Mexico.

Genl. Char. Similar to *H. texensis*; size large; tail long, usually with a white tip; grayish patch behind ears; ears large; skull with a greater occipito-nasal length and wider zygomatic arch; soles 6-tuberculate.

Color. Top of head and back behind shoulders blackish brown mixed with reddish, the base of hairs grayish; back of ears and sides of head and body grayish, with blackish brown hairs intermingled; lateral stripe bright buff, extending from nose to thighs; nose and upper lip, hands, and feet white; under parts yellowish white; a patch of orange buff on each side of root of tail; tail above black, beneath whitish, with an all-around white tip.

Measurements. Total length, 255; tail vertebræ, 136; hind foot, 29. Skull: occipito-nasal length, 33; Hensel, 23; interorbital constriction, 8; zygomatic width, 15; length of nasals, 12; palatal length, 13; length of upper molar series, 4; length of mandible, 13; length of lower molar series, 4.

O. Xylomys.

$$I.\tfrac{1-1}{1-1}; \ P.\tfrac{1-1}{1-1}; \ M.\tfrac{3-3}{3-3} = 20.$$

Xylomys Merr., Proc. Biol. Soc. Wash., XV, 1902, p. 43. Type *Heteromys nelsoni* Merriam.

Pelage without bristles. Skull has small superorbital beads and a high and rounded braincase; frontals elongate; maxillary root of zygomata large, rectangular; no tubercle over root of lower incisor; lower jaw broad, the angle everted slightly; last upper molar with two transverse loops and a posterior loop; posterior prism double.

402. nelsoni (*Heteromys*), Merr., Proc. Biol. Soc. Wash., XV, 1902, p. 43.

NELSON'S SPINY MOUSE.

Type locality. Pinabete, State of Chiapas, Mexico.

Genl. Char. Size very large; ears large. Skull long, slendei superorbital beads faint.

Color. Above mouse gray, darkest on top of head; upper lip, hands, feet, under parts, inner side of fore legs and streak on hind leg white; hind feet clouded above; ears without white edges.

Measurements. Total length, 356; tail vertebræ, 195; hind foot, 43.5.

Fam. VI. **Octodontidæ. The Octodonts.**

This Family has been divided into several subfamilies, only two of which, the CAPROMYINÆ and LONCHERINÆ, are necessary to be mentioned as coming within the scope of this work. Save one species, the Coypu (*Myocastor coypu*), yielding the "Nutria fur" of commerce, the members of this family are not generally known except to naturalists, and among them, in the New World and on certain of its islands, are found the curious Tree Rats of the genus CAPROMYS, from which the name of the first of the above-mentioned subfamilies is derived; the allied PLAGIODONTIA and various species of Spiny Rats of different genera compose the other subfamily LONCHERINÆ. Their trivial names are Hedge-hog or Spiny Rats, as many have variously shaped spines mingled with the fur. Some are of considerable size, and all have a more or less harsh fur, in some instances even bristly. The various species are arboreal, terrestial, or aquatic in habits, the Coypu having webbed hind feet and a cylindrical, tapering otter-like tail. The technical characters by which these subfamilies are separated exist chiefly in the skull and teeth.

Subfam. I. **Loncherinæ.**

In the succeeding genera the fur is usually mixed with flattened lancet-shaped spines, contracted at the base, and acutely pointed. Sometimes they are ridged, and never project beyond the hair. Some species are prettily marked in brown and white, but many have sombre hues only. Certain members of this subfamily are destitute of spines, and so the trivial names for these animals would be inappropriate for them, but it will answer well enough for the majority of the species found within the limits of this work.

75. Loncheres.

$$I.\frac{1-1}{1-1}; \ P.\frac{1-1}{1-1}; \ M.\frac{3-3}{3-3} = 20.$$

Loncheres Illig., Prodr. Syst. Mamm. Av., 1811, p. 90. Type *Myoxus chrysurus* Zimm.=*Echinomys cristatus* Desmarest.

Lonchetes Billberg, Syn. Faun. Scandinav., 1, 1828, Mamm., Conspectus A.

Palate long, narrow, V-shaped posteriorly; nasals long, broad anteriorly, decreasing in width to posterior end, which is truncate and on a line with the premaxillæ; rostrum deep; interorbital constriction moderate; braincase broad anteriorly, narrowing slightly towards occipital region; zygomata parallel with axis of skull, nearly straight, the jugal composing most of the arch; bullæ large, swollen slightly, oblique; interpterygoid fossa broad, widest posteriorly; hamular processes of pterygoids long, abutting the bullæ; upper

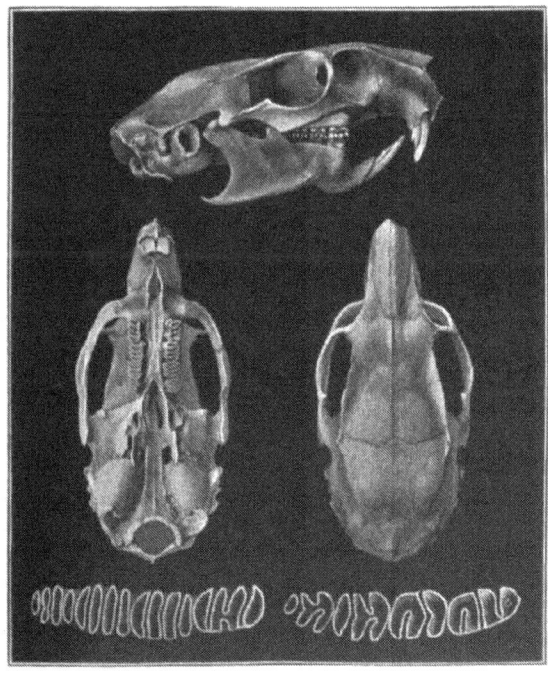

FIG. 71. LONCHERES LABILIS.
Mus. Comp. Zoöl. Coll. Nat. size.

UPPER TOOTH ROW.   LOWER TOOTH ROW.
Enlarged 3 times.   Enlarged 3 times.

molars with one internal and three external loops; lower molars with one external and two internal loops; mandible broad with conspicuous bead on inferior outline, the angle prolonged into a lengthened delicate spur; condyle broad, rounded at extremity; coronoid process short, pointed; symphyseal portion narrow, lower outline V-shaped.

403. labilis (*Loncheres*), Bangs, Amer. Nat., XXXV, 1901, p. 638. GLIDING SPINY RAT.

Type locality. San Miguel Island, Bay of Panama.
Genl. Char. Pelage long, stiff; spines wanting.

FIG. LI. LONCHERES LABILIS. GLIDING SPINY RAT.

Color. Top of head, nose, and cheeks mixed black and yellowish; patch of yellowish white at base of whiskers, also one above eye and behind ear; rest of upper parts bright ferrugineous, varied with black in certain specimens; chin grayish white; under parts buff or ferrugineous—in some individuals these colors show in patches; upper surface of hands and feet yellowish brown; nails white; tail at base like body, then black, tipped sometimes with yellowish white; ears blackish, nearly naked.

Measurements. Total length, 420–540; tail vertebræ, 175–240; hind foot, 42–48; ear from notch, 14–16. Skull: basal length, 47.8; occipito-nasal length, 56.6; zygomatic width, 27.4; mastoid width, 22.4; interorbital constriction, 12; length of nasals, 16; width of nasals, 7.2; palatal length, to palatal notch, 21.2; to end of pterygoid, 36.4; upper tooth row, 13; length of single half mandible, 34; lower tooth row, 13.2.

The next is a comparatively large genus of moderate sized Spiny Rats. One curious characteristic of these animals is the tendency to lose their tails, the separation taking place at the fifth caudal vertebræ at the posterior border of the pelvis. The fifth caudal is abnormal, the posterior half having apparently been lost by absorption. This interesting fact was ascertained by Mr. F. M. Chapman,* who also states that on skinning specimens the tail easily broke away at the fifth caudal vertebræ. So frequently does the loss of the tail occur, that in Trinidad the natives believe there are two species, those with tails, and the tailless.

76. †Proechinomys. Spiny Rats.

$$I.\frac{1-1}{1-1}; \ P.\frac{1-1}{1-1}; \ M.\frac{3-3}{3-3} = 20.$$

Proechimys (*sic*) Allen, Bull. Am. Mus. Nat. Hist., 1899, p. 264.
Type *Echimys! trinitatis* Allen & Chapman.
Echimys (*sic*) Geoff. St. Hil., Ann. Scien. Nat., 2me Sér., x, 1838, p. 125. (nec G. Cuvier, 1809.) *Id.* Mag. Zoöl., 2me Sér., 1840, p. 30.

Nasals very long, rounded anteriorly, narrowing posteriorly, and longer than premaxillæ; orbital constriction slight, greatest width of skull at fronto-parietal suture; bullæ large, converging to a point anteriorly; interpterygoid fossa broad, widest anteriorly; processes of the pterygoids long and broad, shaped somewhat like the head of a spear, pointed, the exterior half twisted outward and lying parallel with axis of skull; palate short, wide, posterior margin with V-shaped notch; incisive foramina broad, rather short; root of upper incisor curving backward to anterior base of zygoma; molar pattern rather simple, upper premolars with two external and two internal loops, the molars with two internal; lower molars with two external loops. Lower part of mandible broadly flattened and rounded outward, narrowing anteriorly to angle, which terminates in a short pointed process; condyle very broad, slightly rounded; coronoid process short, and pointed backward.

KEY TO SPECIES AND SUBSPECIES.
PAGE

---

* Bull. Amer. Mus. Nat. Hist., 1893, p. 226.

† εχινοσ, μῦσ=ECHINOMYS nec *Echimys*=PROECHINOMYS.

404. semispinosus (*Echimys!*), Tomes, Proc. Zoöl. Soc., 1860, p. 265.
SHORT-SPINED RAT.

Type locality. Ecuador.

Geogr. Distr. Nicaragua south to Ecuador.

Genl. Char. Size large; ears small; tail half as long as head and body; long and strong claws on hands and feet; spines short, flexible, confined to middle of back.

Color. Above mixed reddish brown and black; cheeks, sides of neck, and body paler; orbital ring black; under parts white; tail above black, beneath ashy brown; hands and feet ashy brown.

Measurements. Total length, 370–400; tail vertebræ, 140–169; hind foot, 47–55; ear, 18–19. Skull: total length, 57; zygomatic breadth, 29; upper tooth row, 16; length of nasals, 21.

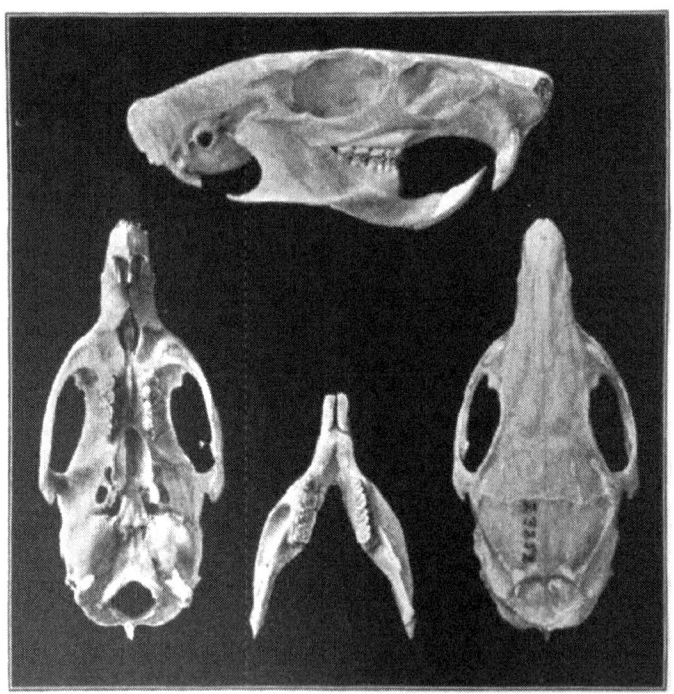

FIG. 72. PROECHINOMYS CENTRALIS.
No. 23252 U. S. Nat. Mus. Coll. Nat. size.

405. centralis (*Echinomys*), Thomas, Ann. Mag. Nat. Hist., 6th Ser.,
XVIII, 1896, p. 312.

Nicaraguan Spiny Rat.

Type locality. San Emilio, north end of Lake Nicaragua, Nicaragua.
Geogr. Distr. Costa Rica, Nicaragua.
Genl. Char. Similar to *P. semispinosus,* but brighter colored; nasals pointed posteriorly; zygomata broad; pterygoid processes broad, spatulate.

Color. Similar to *P. semispinosus,* "but brighter and richer, less heavily black-lined on the back, and with the spineless fur of the sides and rump much brighter rufous; hands and feet dull whitish above, the darker mark which runs along the outer side of the metatarsal in some species little marked." (Thomas, l. c.)

Measurements. Total length, 437; tail, 170; hind foot, 55 (dried skin). Skull: basal length, 47.5; basilar length, 41; greatest breadth, 26.5; length of nasals, 21.3; width of nasals, 6; interorbital constriction, 12.5; breadth of interparietal, 12.5; Hensel, 20; length of palatine foramina, 5; width, 2.5; length of upper molar series, 9.2; length of mandible, condyle to the incisor, 34.

a.—*chiriquinus* (*Proechimys!*), Thomas, Ann. Mag. Nat. Hist., 7th
    Ser., v, 1900, p. 220.

Bogava Spiny Rat.

Type locality. Bogava, Chiriqui, Panama. Altitude, 800 feet.
Geogr. Distr. Panama, Central America.
Genl. Char. Similar to *P. centralis,* but darker. Skull: muzzle broad, heavy; nasals short, broad; supraorbital ridges broad.

Color. Above dark rufous; sides of face grayish brown; feet brown.

Measurements. Total length, 450; tail, 150; hind foot, 55; ear, 14. Skull: greatest breadth, 31; length of nasals, 23.5; width of nasals, 7.5; interorbital constriction, 15.8; greatest breadth on ridges, 25; palatal length from henselion, 23; length of upper molar series, 9.2; length of palatal foramina, 6.5; width of palatal foramina, 3.7.

b.—*panamensis* (*Proechimys!*), Thomas, Ann. Mag. Nat. Hist.,
    7th Ser., v, 1900, p. 220.

Panama Spiny Rat.

Type locality. "Savanna near City of Panama," Panama.
Geogr. Distr. Panama, Central America.
Genl. Char. Similar to *P. centralis;* head and shoulders grayish brown.

Color. Above rufous; head and limbs grayish brown; cheek and sides of neck paler; hind feet brown, remainder like *P. centralis.*

Measurements. Total length, 475; tail, 178; hind foot, 54; ear, 26. Skull: basilar length, 43; greatest breadth, 29; length of nasals,

24; interorbital constriction, 13.2; greatest breadth on ridges, 23.6; length of upper molar series, 8.9.

406. burrus (*Proechimys!*), Bangs, Am. Nat., XXXV, 1901, p. 640. SAN MIGUEL SPINY RAT.

Type locality. San Miguel Island, Bay of Panama.

Genl. Char. Similar to *P. c. panamensis*, but larger and more red; nasals long, broad, and truncate posteriorly.

Color. Upper parts deep ferrugineous, varied with brownish black; top of nose, cheeks, and lower sides more yellowish; anal regions like back; rest of under parts pure white; tail above black, beneath grayish; hands and feet dusky brown; ears dusky, nearly naked.

Measurements. Total length, 410–490; tail vertebræ, 140–205; hind foot, 50–60; ear from notch, 20–22. Skull: basal length, 52; occipito-nasal length, 61.2; zygomatic width, 29; mastoid width, 22.2; interorbital constriction, 13.4; length of nasals, 24.2; width of nasals, 7.6; length of palate to palatal notch, 21; end of pterygoid, 34; length of palatal foramina, 5.2; width of palatal foramina, 3; upper tooth row, 9.8; length of single half of mandible, 33.8; lower tooth row, 10.2.

The subfamily CAPROMYINÆ comprises large arboreal rats, which are found only in some of the West India islands, the Bahamas, and certain of the islands in the Gulf of Mexico. They have comparatively short tails; in some species this member is very short, and one has a prehensile tail. All these are naked and scaly, only a few scattering hairs being observable. The food of these animals consists of leaves, twigs, and bark; and in one island at least, the most eastern of the Plana Keys, Mr. Ingraham observed the species that bears his name, associating together in considerable numbers. Not much is known of the habits of these singular creatures, but as their size makes them rather conspicuous and they are practically defenseless, while their flesh is said to be palatable, it is probable they will, ere long, become extinct in the limited localities in which they are now found. Indeed, the species inhabiting Jamaica is stated to be practically extinct already. When on the ground they greatly resemble the muskrat in their shape and movements.

Subfam. II. Capromyinæ.

F. M. Chapman. *A Revision of the genus Capromys.* Bull. Am. Mus. Nat. Hist., 1901, p. 313.

77. Capromys.

$$I.\frac{1-1}{1-1}; \ P.\frac{1-1}{1-1}; \ M.\frac{3-3}{3-3} = 20.$$

Capromys Desmarest, Mém. Soc. d'Hist. Nat., 1, 1822, p. 43. Type *Capromys fournieri* Desmarest = *Isodon pilorides* Say.

Geocapromys Chapman, Bull. Am. Mus. Nat. Hist., 1901, p. 314.

Procapromys Chapman, Bull. Am. Mus. Nat. Hist., 1901, p. 322.

Incisors moderate; upper molars with one internal and two external enamel folds; ear rather small; tail long; form slender; habits arboreal.

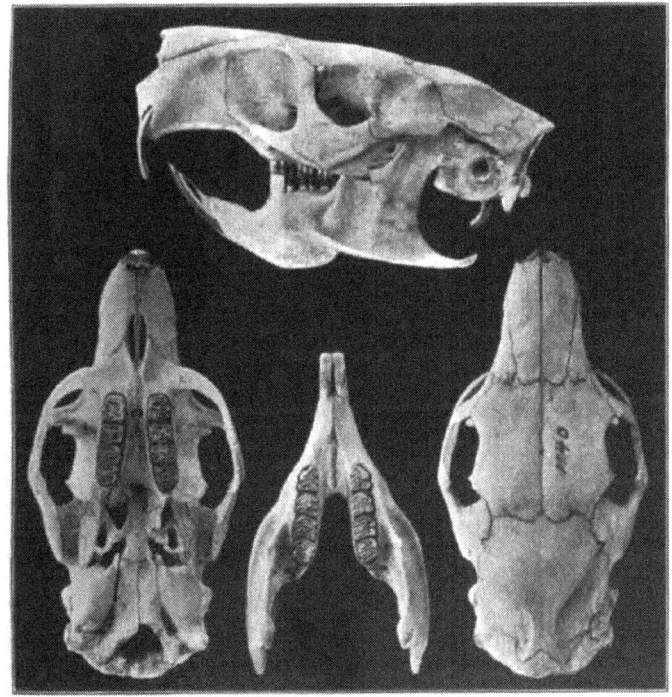

FIG. 73. CAPROMYS PILORIDES.
No. 1140 Field Columbian Mus. Coll. ½ nat. size.

KEY TO THE SPECIES AND SUBSPECIES.

A. Tail long.

 a. Above reddish chestnut and black, to reddish   PAGE
  brown and black; beneath yellowish white. *C. pilorides* 390

FIG. LII. CAPROMYS PILORIDES. HAIRY HUTIA.

A. Capromys.

407. pilorides (*Isodon*), Say, Journ. Acad. Nat. Scien. Phil., II,
    1822, p. 333, fig.

fournieri Desm., Mém. Soc. Hist. Nat., 1822, p. 43, pl. I.

quemi Fisch., Syn. Mamm., 1829, p. 312.

HAIRY HUTIA.

Type locality. "South America or one of the West Indian
Islands." Cuba?

Geogr. Distr. Cuba.

Genl. Char. Size large; tail long, sparsely haired; muzzle white.

Color. Variable; above reddish chestnut and black, the hairs
being black, tipped with reddish chestnut, darkest on median line
and paler on the sides where the hairs are tipped with yellowish
brown; nose and sides of head white, mixed with blackish and yel-

lowish brown; under parts mixed gray and yellowish brown; limbs, hands, and feet similar to back, but with less chestnut red; tail covered with scattering reddish brown hairs.

Another style is yellowish brown and black above, and pale yellow and black on sides. Nose and sides of head with a large amount of white; under parts yellowish white from chin to tail, with the sides of the belly pale brown; tail yellowish; hands whitish; arms mixed black and yellowish white; legs and feet black to toes, which are whitish; ears blackish, edged with whitish hairs. The two styles are strikingly different.

Measurements. Total length, 777; tail, 220; hind foot, 83 (skin). Skull: total length, 96; greatest zygomatic width, 48; length of Hensel, 78; mastoid breadth, 34; palatal length, 22; upper tooth row, 20; height of lower jaw at condyle, 30.

408. melanurus (*Capromys*), Poey, Monatsb. K. Preuss. Akad.
    Wiss. Berl., 1864, p. 384.   Dobson, Proc. Zoöl. Soc., 1884,
    p. 233, pls. XVIII–XXI.
BLACK-TAILED HUTIA.

Type locality. Manzanillo, Cuba.

Geogr. Distr. Island of Cuba.

Genl. Char. Ears short, rounded, naked; eyes small; nostrils obliquely placed in end of muzzle; tail scaly, but clothed with long hairs; thumb rudimentary with small blunt claws; other digits with convex, acute claws; foot twice the size of the hand.

Color. General hue yellowish brown, darkest on head and palest beneath the body, mixed with long projecting black hairs; the hair on head much shorter than on the body; tail blackish brown.

Measurements. Total length, 595; tail, 266; ear, 25. Skull: occipito-nasal length, 75; edge of foramen magnum to alveoli of incisors, 57; length of nasals, 24; of frontal, 25; of parietal, 26; of upper molar series, 15.5.

409. prehensilis (*Capromys*), Poeppig, Journ. Acad. Nat. Scien.
    Phil., 1, 1824, p. 11.
PREHENSILE-TAILED HUTIA.

Type locality. Southern Cuba.

Geogr. Distr. Island of Cuba.

Genl. Char. Tail long, rather thickly covered with hair; pelage rather smooth; size moderate.

Color. Nose pale whitish brown; top and sides of head dark reddish brown mixed with white; upper parts dark reddish brown, being a mixture of dark brown, reddish brown, yellowish, and black, becoming more reddish on the rump; side paler, somewhat grayish; throat and breast grayish; rest of under parts yellowish, with a reddish tinge near the inguinal region; limbs, hands, and feet like back; tail covered with short reddish hair.

Measurements. Total length, 710; tail, 305; hind foot, 82; ear, 20. Skull: occipito-nasal length, 75; length of parietal, 27; of upper molar series, 18; width between upper premolars, 4; between posterior molars, 9.

a.—gundlachi (*Capromys*), Chapman, Bull. Am. Mus. Nat. Hist., 1901, p. 317.
GUNDLACH'S HUTIA.

Type locality. Nueva Gerona, Isle of Pines, near Cuba.

Genl. Char. Similar to *C. prehensilis*, but less rufous; zygomata heavier; postorbital process less produced; tail hairy throughout.

Color. Above mixed buff, black, and ferrugineous; rump brighter ferrugineous; crown and cheeks brown, nose buffy; lower part of cheek whitish; under parts and inner side of limbs buffy white; tail mixed rufous and brownish black.

Measurements. Total length, 695; tail, 300; hind foot, 80; ear, 23. Skull: greatest length, 80.5; greatest width, 40; width of postorbital processes, 24.5; length of nasals, 22.5; of frontal, 26.5; of parietal, 28; length of upper molar series, 11.5; height of lower jaw at condyle, 26.5.

410. elegans (*Capromys*), Latorre, Bol. Soc. Espan. Hist. Nat., Madrid, 1, 1901, p. 372.
LANCEOLATE-SPOT HUTIA.

Type locality. Cuba?

Geogr. Distr. Unique specimen in Madrid Museum, presumably from Cuba.

Genl. Char. Size smaller than *C. prehensilis*; large grayish red lanceolate spot on middle of back.

Color. General color brilliant yellowish red; head, tail, and feet chestnut brown; tawny yellow spot between the eyes; on the back is a large spot shaped like the head of a lance, the point directed towards the rump, of an intense grayish red graduating into black towards the broadest part. This spot is surrounded by an irregular white band, which extends on the right flank toward the under surface. On the shoulders, front of hind legs, and at base of tail are numerous white hairs mixed with the others; nails large, curved, and of a yellowish color; incisors large, and orange on the outer surface.

Measurements. Total length, 685; head with skull in skin, 95; tail 200; hind foot, 75.

B. Geocapromys.

Tail short, about equal to hind foot; claws shorter than in *Capromys*; inner toe of fore foot barely perceptible; ascending maxillary of zygomatic arch wider than in *Capromys*, the superior margin of squamosal narrower and without processes; occipital region lower.

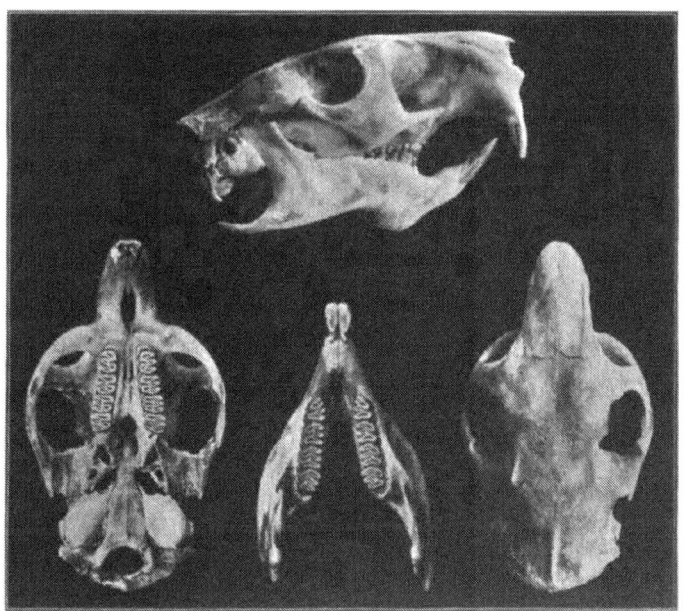

FIG. 74. CAPROMYS (GEOCAPROMYS) BROWNI.
No. 19147 Am. Mus. Nat. Hist. Coll. ⅓ nat. size.

411. browni (*Capromys*), Fisch., Syn. Mamm., Addend., 1830, p. 389.

brachyurus (*Capromys*), Hill, Gosse, Nat. Sojourn in Jamaica, 1851, p. 471.

SHORT-TAILED HUTIA.

Type locality. Jamaica.

Geogr. Distr. Island of Jamaica.

Genl. Char. Tail very short, stiff, scaly, naked at base, covered with short bristly hair above; fur dense, harsh; muzzle covered with down.

Color. Above mixed black and golden brown, beneath yellowish; tail black above, grayish brown beneath; hands and feet blackish; ears blackish gray.

Measurements. Total length, 450; tail, 35; hind foot, 60; ear, 6 (skin). Skull: occipito-nasal length, 101; length of Hensel, 81; length of nasals, 36; of frontals, 35; of upper molar series, 21; lower molar series, 21; width between rows of posterior molars, 10.

412. thoracatus (*Capromys*), True, Proc. U. S. Nat. Mus., 1888, p. 469.

WHITE-BANDED HUTIA.

Type locality. Little Swan Island, Gulf of Honduras.

Geogr. Distr. Type locality only.

Genl. Char. Tail very short, and in other characters also similar to *C. browni.*

Color. Above blackish brown; hairs plumbeous brown at base, and ringed in center with yellow; throat pale gray; breast crossed by a nearly pure white band; belly yellowish gray; tail dark brown; hands and feet ochraceous; fingers and toes blackish brown.

Measurements. Total length, 389; tail, 45; hind foot, 65; ear from occiput, 18. Skull: greatest length, 68.5; length of Hensel, 53.5; greatest breadth, 35.5; interorbital constriction, 17.7; length of nasals, 23; of frontals, 23; of upper molar series, 15.5; width between posterior upper molars, 5.3; length of lower molar series, 15.5.

413. ingrahami (*Capromys*), Allen, Bull. Am. Mus. Nat. Hist., 1891, p. 329.

INGRAHAM'S HUTIA.

Type locality. Plana Key, Bahama Islands.

Geogr. Distr. Bahama Islands.

Genl. Char. In size similar to *C. browni* from Jamaica; pelage coarse, harsh; tail very short, graduated, pointed, hairy.

Color. Above mixed yellowish brown, gray and black, or blackish brown, darkest on head and nape; sides similar to back, but paler; under parts pale yellowish brown; hands and feet reddish brown; tail rusty brown; bare on apical third beneath, which is black; ears black, fringed with reddish brown hairs.

Measurements. Total length, 335–375; tail vertebræ, 55; hind foot, 53–55; ear from crown, 16. Skull: total length, 63; greatest width, 32; interorbital constriction, 17.5; length of nasals, 21; of frontals, 21; of upper molar series, 15; width between rows of posterior molars, 5.5; length of lower jaw, 44; height of condyle, 18.5.

78. Plagiodontia.

$$I.\frac{1-1}{1-1};\ P.\frac{1-1}{1-1};\ M.\frac{3-3}{3-3} = 20.$$

Plagiodontia F. Cuv., Ann. Scien. Nat., 2me Sér., VI, 1836, p. 347, Zoöl.   Type *Plagiodontia aedium* F. Cuvier.

Skull broad; nasals broad, truncate posteriorly; infraorbital foramina very large, round; superior outline descending from parietal to end of nasals, and to occiput; zygomata moderately heavy, jugal broad, extended; thumb rudimentary with a flat nail; the four fingers with slender curved claws; toes larger, with strong compressed curved claws; the middle and two outer toes longest and nearly equal.   Teeth without roots, diminishing in upper row from last molar to premolar, which is smallest; each tooth has two oblique loops horizontal to the jaw, one internal and one external, parallel to each other.   Lower molars nearly equal in size, the premolar

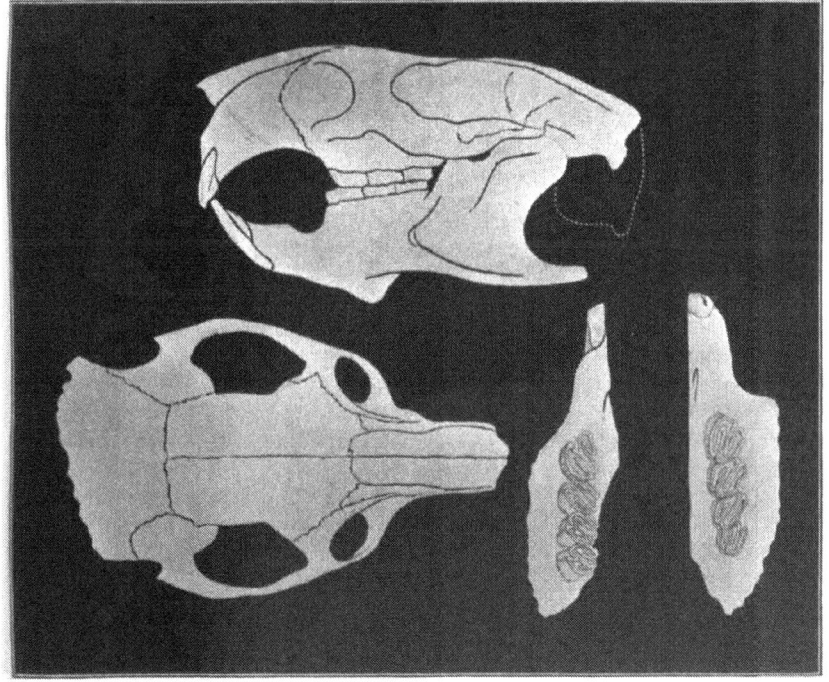

FIG. 75. PLAGIODONTIA AEDIUM.
ex Ann. Scien. Nat., Paris, 1836.   Plate 17, nat. size.

slightly smaller, with one external and two internal loops, the angle on latter much shallower than the others; lower incisors enter jaw to base of last molar.

414. aedium (*Plagiodontia*), F. Cuv., Ann. Scien. Nat., 2me Sér., vi, 1836, p. 347, pl. 17, Zoöl.

HAITIAN HUTIA.

Type locality. Island of Haiti.
Geogr. Distr. Island of Haiti.
Genl. Char. Those of the genus.

FIG. LIII. PLAGIODONTIA AEDIUM. HAITIAN HUTIA.
ex Ann. Scien. Nat., Paris, 1836, Pl. 17.

Color. General hue pale brown; hairs on upper parts gray for three-fourths their length, and fawn at the tips; long black hairs are intermingled with the others; hairs of under parts are paler than those of the back, and the long hairs are whitish; tail naked, scaly; incisors yellow.

Measurements. Total length, 425; tail, 125. Skull: occipital region imperfect; posterior border of frontal to end of nasals, 52; interorbital constriction, 19; zygomatic width, 42; length of nasals, 23.5; length of upper molar series, alveolar border, 21; length of mandible, angle to alveolus of incisor, 48; height at coronoid process, 23; length of lower molar series, alveolar border, 22.

Of the Family of "fretful porcupines," the semi-arboreal species of the genus ERETHIZON are found in North America. They prefer a cold climate, and if their lot is cast in tropical lands, such as Mexico, they endeavor to equalize matters by living in high altitudes in the mountains. While the American Porcupines are to a great extent arboreal, yet they are by no means restricted to a life in the trees, and the different species are frequently met with traveling on the ground, and in the western part of the United States it is not uncommon to find the Porcupine out on the prairie far from any timber. They are inoffensive animals when unmolested, but disagreeable creatures to handle or meddle with by either man or dog.

Fam. VII. **Erethizontidæ.**

Form stout; long acute spines loosely attached to skin. Skull with facial portion short, and the jugal without inferior angle; molars more or less completely rooted.

Subfam. I. **Erethizontinæ.**

79. Erethizon. Long-spined Porcupines.

$$ I.\frac{1-1}{1-1};\ P.\frac{1-1}{1-1};\ M.\frac{3-3}{3-3} = 20. $$

Erethizon F. Cuv., Mém. du Mus. Hist. Nat., Paris, ix, 1822, p. 426.
  Type *Hystrix dorsatus* Linnæus.
Eucritus Fisch., Mém. Soc. Imp. Moscow, v, 1817, pp. 372, 411.
Echinothrix Brookes, Cat. Anat. Zoöl., 1828, *Id.* Trans. Linn. Soc.
  Lond., xvi, 1829, pt. i, p. 97.
Echinoprocta Gray, Proc. Zoöl. Soc., 1865, pp. 321–322 desc.

Four toes on fore feet, five on hind feet, all with strong claws; limbs short, strong; no naked mesial line on upper lip, which is covered with hair and notched above the incisors; tail short, thick, non-prehensile, covered above with stiff hairs and spines, and on the sides and beneath with stiff bristles.

415. epixanthum (*Erethizon*), Brandt, Mém. Acad. Imp. Scien.,
    St. Petersb., 6th Ser., 1835, p. 390, pl. 1, 9. Elliot, Syn. N.
    Am. Mamm., 1901, p. 265.
  pilosus. Peale (nec Rich.), U. S. Expl. Exped., Mamm., 1848,
    p. 46.
WESTERN PORCUPINE.
  Type locality. California? Unalaska?
  Geogr. Distr. State of Sonora, Mexico, into New Mexico, eastward to Missouri, west to the Pacific, and north to Alaska and the limit of trees.
  Genl. Char. Light tips of long hairs of dorsal surface greenish yellow. Average length of nasals exceed interorbital breadth, or over

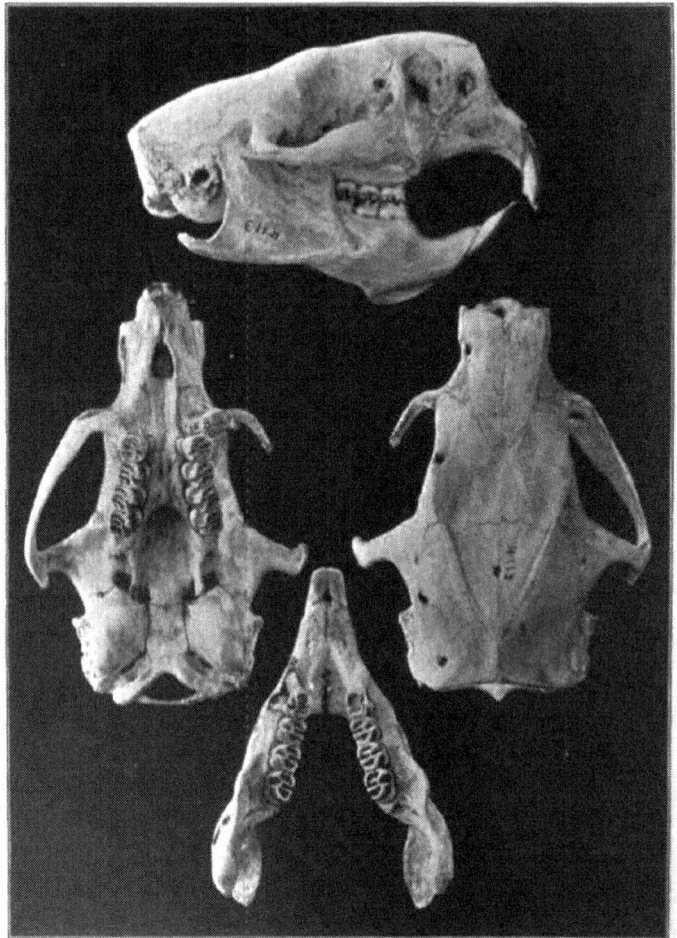

Fɪɢ. 76. Erethizon epixanthum.
No. 4113 Am. Mus. Nat. Hist. Coll. ⅓ nat. size.

one-third the skull's length; nasals extending backward to the orbits, or two-fifths length of skull.

Color. Similar to *E. dorsatum*, except tips of long hairs are green-ish yellow instead of yellowish white; central line of belly sooty brown; spines yellowish on the back, on the sides whitish, tipped with brown.

FIG. LIV. ERETHIZON EPIXANTHUM. WESTERN PORCUPINE.

Measurements. Total length, 825; tail vertebræ, 165. Skull: occipito-nasal length, 92; Hensel, 81; greatest zygomatic width, 70; mastoid breadth, 47; palatal length, 30; length of upper tooth row, 24.

The members of the genus COENDU are essentially Tree Porcupines having prehensile tails to aid them in their movements among the branches. They are more especially natives of South America, but one extends its range into Mexico. They are of a more slender form than the Ground Porcupines, and their quills are variously colored, and these are mixed among the hairs, exhibiting the transition stage, neither all hairs nor all quills.

80. Coendu. Short-spined Porcupines.

$$\text{I.}\frac{1-1}{1-1}; \ \text{P.}\frac{1-1}{1-1}; \ \text{M.}\frac{3-3}{3-3} = 20.$$

Coendu Lacép., Disc. d'ouvert. et de cloture du Cours Hist. Nat., Suppl., 1799, p. 11. Type *Hystrix prehensilis* Linnæus.

Senetheres (*sic*) F. Cuv., Mém. du Mus. Hist. Nat.. Paris, ix, 1822, p. 433.

Laboura Bilbberg. Syn. Faun. Scandinav., Mamm., I, 1828, Consp. A.

Cercolabes Brandt, Mamm. Exot. Nov., in Mém. Acad. Imp. St. Petersb., Sér. 3, III, 1835, p. 55.

Body covered with short, variously colored spines, close together, and mixed with hairs; hind feet with only four toes, hallux absent: fleshy pad on inner side of foot; tail prehensile.

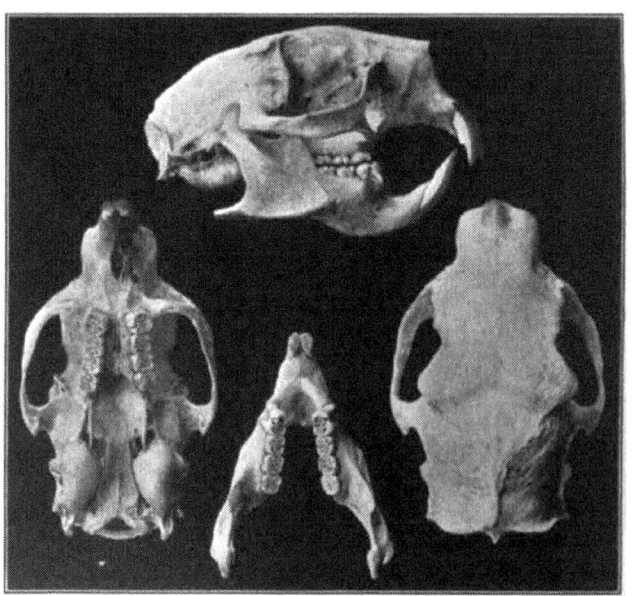

FIG. 77. COENDU MEXICANUM.
No. 102 Field Columbian Mus. Coll. ⅓ nat. size

KEY TO THE SPECIES AND SUBSPECIES.

A. Tail prehensile.

  a. Fur fulvous brown.   PAGE

    a.′ Hind foot, 52 mm.....................*C. pallidum*   401

  b. Fur speckled with white.

    a.′ Hind foot, 71 mm...................*C. rothschildi*   401

  c. Fur black.

    a.′ Hind foot, 88 mm................ ...*C. mexicanum*   402

    b.′ Hind foot, 74 mm.................*C. m. yucataniæ*   402

  d. Fur blackish brown; hind foot, 75 mm.......*C. lænatum*   402

416. pallidum (*Cercolabes*), Waterh., Mamm., II, 1848, p. 434.
LIGHT-COLORED PORCUPINE.

Type locality. "West Indies"?

Genl. Char. Fur soft; spines short; tail short, with slender spines on upper part of basal half.

Color. General color pale fulvous brown; limbs, lower part of flanks, and under parts darker brown; muzzle and feet dusky brown; tail brownish black; quills white, with black tips, hidden mostly in the fur.

Measurements. Total length, 513; tail, 193; hind foot, 53.3.

417. rothschildi (*Coendu*), Thomas, Ann. Mag. Nat. Hist., 7th Ser., x, 1902, p. 169.
ROTHSCHILD'S TREE PORCUPINE.

Type locality. Sevilla Island, off Chiriqui, Panama.

Genl. Char. Hair short; skull much inflated above orbits; nasal aperture large; fourth premolar not larger than molars.

Color. Profusely speckled with white; spines on back all tipped with white.

Measurements. Total length, 740; tail vertebræ, 330; hind foot, with claws, 68. Skull: basilar length, 71; length of upper molar series, 17.3.

FIG. LV. COENDU MEXICANUM. MEXICAN TREE PORCUPINE.

418. mexicanum (*Hystrix*), Kerr, Linn., Anim. King., 1792, p. 214.

novæ-hispaniæ Briss., Regn. Anim., 1756, p. 127.

prehensilis Schreb., Säugeth., IV, p. 603.

PREHENSILE-TAILED PORCUPINE.

Type locality. Mountains of Mexico.

Geogr. Distr. Mirador, State of Vera Cruz, Mexico, south to Costa Rica.

Genl. Char. Tail short, stout, bare of spines on apical half.

Color. Body covered with yellow and white spines, with black tips; the fur amid these on the body and limbs is black, with the tips of hairs white; basal half only of tail covered with spines, remainder with stiff black hairs.

Measurements. Total length, 750; tail, 320; hind foot, 88 (dried skin.) Skull: occipito-nasal length, 82; total length, 84; zygomatic width, 52; interorbital constriction, 27; palatal length (palatal arch to anterior edge of premolar), 16; length of upper molar series, alveolar border, 18; length of mandible, angle to alveolus of incisor, 54; length of lower tooth row, alveolar border, 22.

a.—yucatania (*Coendu*), Thomas, Ann. Mag. Nat. Hist., 7th Ser., x, 1902, p. 249.

YUCATAN TREE PORCUPINE.

Type locality. Izamal, Yucatan.

Genl. Char. Similar to *C. mexicanum*; black fur shorter. Skull with nasals parallel, not expanded anteriorly; forehead inflated; braincase narrow; anterior palate flat; bullæ high, narrow.

Color. Like that of *C. mexicanum*.

Measurements. Total length, 820; tail, 380; hind foot with claws, 74. Skull: greatest length, 88; basilar length, 75; zygomatic breadth, 48.5; nasals, 30.5 × 19; height of forehead above palate, 97; interorbital breadth, anteriorly, 30; posteriorly, 36; breadth of braincase behind zygomata, 33; palatal length, 38.5; palatal foramina, 8 × 3.2; length of upper tooth row, 18.1.

419. lænatum (*Coendu*), Thomas, Ann. Mag. Nat. Hist., 7th Ser., xi, 1903, p. 381.

CHIRIQUI PORCUPINE.   *Gato de Spinas* in Chiriqui.

Type locality. Boquete, Chiriqui, Panama.   Altitude, 5,000 feet.

Genl. Char. Spines not showing through fur on tail or limbs. Skull flat, not inflated on frontal region; anterior portion of premaxillæ projecting but slightly in front of nasals; nasals broad anteriorly, narrowing posteriorly; supraorbital edges square, ridges well defined; palatal foramina ending at premaxillo-maxillary suture; posterior edge of palate on line with posterior edge of second upper molar.

Color. Blackish brown above and beneath; spines of back yellowish white on basal three-fifths, remainder brownish black, tips horny; hands, feet, and tail black.

Measurements. Total length, 708; tail, 256; hind foot, with claws, 75. Skull: greatest length, 80; basilar length, 67; zygomatic breadth, 47; length of nasals, 24; breadth of nasals, anteriorly, 16; breadth of nasals at fronto-premaxillary suture, 12.8; interorbital constriction, 25.5; palatal length, 35.7; height of forehead, above premolars 27.7; length of upper tooth row, 19.

The Rodents with hoof-like claws of the family AGOUTIDÆ, resemble in their general outward appearance a ruminant, such as the little musk-deer, more than a rodent. Slender of form and limbs, they are small of stature with very short ears and tail. Two genera contain all the known species, distinguished from each other by the number of toes on the hind foot, the members of DASYPROCTA having three, and those of AGOUTI five. The former genus contains a number of species, distributed in Central and South America, and even on some of the West India islands; but on these last only two have been found as yet, one of which has a great range, for it is a native of South America as far south as Paraguay, but is not met with west of the Andean Chain of Mountains, being replaced in Ecuador by *A. taczanowski.* The absence of tail in the Agoutis is compensated for by the length of the hairs on the rump, which fall over so far that they would hide any moderately long tail. Agoutis are dwellers both of the woods and plains, agile in their movements, and swift runners. Nocturnal in habits, they remain hidden for the greater part of the day. Their food is chiefly vegetable.

Fam. VIII. **Agoutidæ. Agoutis.**

E. R. Alston. *The genus Dasyprocta, with Description of a New Species.* Proc. Zoöl. Soc. Lond., 1876, p. 347.

81. **Dasyprocta. Agoutis. Paca.**

$$I.\frac{1-1}{1-1}; \ P.\frac{1-1}{1-1}; \ M.\frac{3-3}{3-3} = 20.$$

Dasyprocta Illiger, Prodr. Syst. Mamm. et Av., 1811, p. 93. Type
———?

Cutia Liais, Climats. Géol. Faune et Geog. Bot. Brésil, 1872, p. 534.

Myoprocta Thomas, Ann. Mag. Nat. Hist., 7th Ser., XII, 1903, p. 464.

Molars semi-rooted, with external and internal enamel folds; claws hoof-like; tail obsolete; hind toes three.

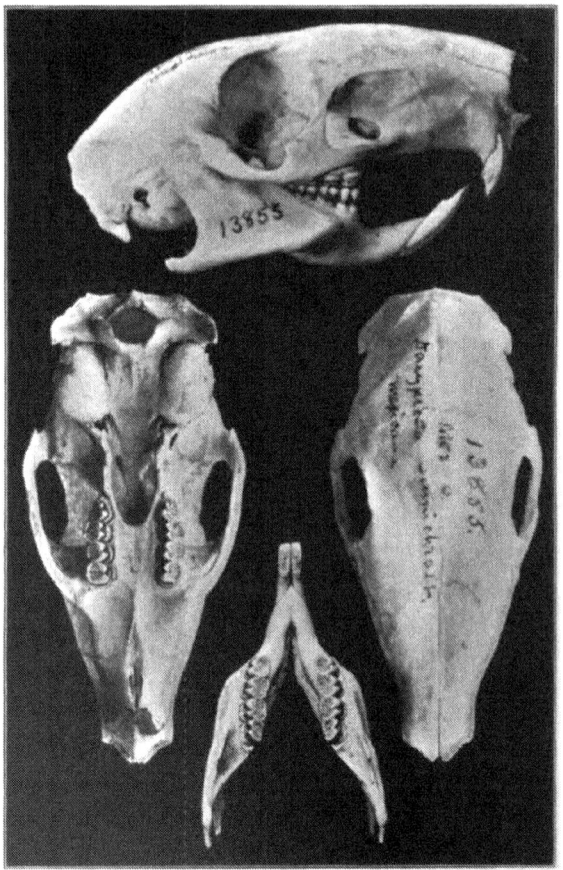

FIG. 78. DASYPROCTA MEXICANA.
No. 13855 U. S. Nat. Mus. Coll.   ¾ nat. size.

KEY TO THE SPECIES.

A. Without nuchal crest.
    a. Long hairs on rump not annulated.
        a.′ Above rufous, or hairs with yellow and black rings.

420. punctata (*Dasyprocta*), Gray, Ann. Mag. Nat. Hist., 1st Ser.,
    X, 1842, p. 264.
SPOTTED AGOUTI. *Cotusa* in Guatemala.
    Type locality. "South America."
    Geogr. Distr. Guatemala, Costa Rica, Central America.
    Genl. Char. Color variable, from a bright chestnut to a pale
yellow.
    Color. Uniform rufous, or with yellow and black rings on hairs
on all the body except a pale line on middle of abdomen; hairs of
rump only slightly elongate.
    Measurements. Total length about 550; hind foot, 100.

421. ruatanica (*Dasyprocta*), Thomas, Ann. Mag. Nat. Hist., 7th
    Ser., VIII, 1901, p. 272,
RUATAN ISLAND AGOUTI.
    Type locality. Island of Ruatan, Bay of Honduras.
    Geogr. Distr. Type locality only.
    Genl. Char. Similar to *D. punctata*, but smaller.
    Color. Above mixed black and yellow, like *D. punctata*; under
part like back, tinged with olivaceous; chin white; yellow patch on
belly; hands and feet grizzled or deep brown; ears blackish.
    Measurements. Length head and body, 435; hind foot, 101.
Skull: basal length, 70.5; zygomatic breadth, 18.7; interorbital
breadth, 27.5.

422. mexicana (*Dasyprocta*), Sauss., Rev. Mag. Zoöl., 2me Sér., XII,
    1860, p. 53.
MEXICAN AGOUTI.
    Type locality. "Hot zone of Mexico." State of Vera Cruz?
    Genl. Char. Long hairs black throughout their length; size
small.
    Color. Hairs on body above, and sides ringed with black and
pure white; rump black; throat and belly white.

Measurements. Total length about 430; hind foot, 90. Skull: occipito-nasal length, 108; greatest zygomatic width, 50; mastoid width, 46; length of Hensel, 79; palatal length, 14; length of upper tooth row (crown), 19; length of frontals, 40; height of lower jaw at condyle, 28.

423. callida (*Dasyprocta*), Bangs, Am. Nat., xxxv, 1901, p. 635.
CUNNING AGOUTI.

Type locality. San Miguel Island, Bay of Panama.

Genl. Char. Color pale; rump hairs white-tipped. Skull slender; rostrum long; nasals narrow.

Color. Upper parts with the hairs annulated with yellowish and black, giving a yellowish clay as the general color; more ochraceous in middle of back; rump hairs long, black with white tips; under parts soiled white, hairs annulated with drab; fore and hind feet brownish black.

Measurements. Total length, 420–510; tail vertebræ, 20–30; hind foot, 96–105; ear from notch, 33–38. Skull: basal length, 85.4; occipito-nasal length, 98.6; zygomatic width, 44; mastoid width, 32; interorbital constriction, 26.2; length of nasals, 38; width of nasals,

FIG. LVI. DASYPROCTA ISTHMICA. ISTHMIAN AGOUTI.

15.4; length of palate to palatal notch, 39; to end of pterygoid, 55.4; upper tooth row (four molar teeth), 17.2; greatest width of rostrum, 24; length of single half of mandible, 58; lower tooth row (four molar teeth), 18.8.

424. isthmicæ (*Dasyprocta*), Alston, Proc. Zoöl. Soc., 1876, p. 347.
ISTHMIAN AGOUTI.
Type locality. Colon, Columbia.
Geogr. Distr. Unknown, probably Costa Rica to Columbia.
Genl. Char. Long hairs of rump black, with broad pale tips.
Color. "Fur ringed with black and yellow; rump black more or less washed with orange or yellow, the long hairs being black at the base, scarcely annulated except close to the tips, which are broadly margined with the light color."
Measurements. "Total length about 22 inches; hind foot, 4.25 inches." (Alston, l. c.)

425. coibæ (*Dasyprocta*), Thomas, Novitat. Zoöl., IX, 1902, p. 136.
COIBA AGOUTI.
Type locality. Coiba Island, West Coast of Panama.
Genl. Char. Size as in *D. isthmicæ*; fur coarse, sparse; rump hairs about three inches in length; nasals parallel, not tapering.
Color. Upper parts grizzled brown; hairs ringed with brown and orange; rump hairs broadly tipped with orange; crown blackish brown; under parts soiled yellowish; upper surface of feet black; tail naked for about an inch; ears nearly naked, brown.
Measurements. Head and body, 570; tail, 25; hind foot, 105; with hoofs, 115; ear, 32. Skull: basilar length, 78; zygomatic breadth, 53; nasals, 40 × 24; interorbital breadth, 32; palatal length, 38.5; diastema, 26.5; length of palatal foramina, 4.2; length of bullæ, 15.2; length of upper molar series (crowns), 16.

426. cristata (*Cavia*), Desm., Nouv. Dict. Hist. Nat., I, 1816, p. 213.
antillensis, Sclat., Proc. Zoöl. Soc., 1874, p. 666.
CRESTED AGOUTI.
Type locality. "Surinam." Probably a West Indian form. (Alston.)
Geogr. Distr. Islands of St. Vincent, St. Lucia, and St. Thomas, West Indies.
Genl. Char. Nuchal crest present; colors dark.
Color. Hairs ringed with black and reddish or brownish yellow; nuchal tuft and rump black, with hairs ringed at base.
Measurements. Total length, 450; hind foot, 93.75.

The Paca is more robust than the Agouti, with coarse hair, and no tail worth mentioning, and the inner toe and the nails on each foot very small. Like the Agouti, it is nocturnal, hiding in underground retreats in the forest by day, coming out at night to feed. It excavates burrows several feet deep, mostly in the vicinity of rivers. The subspecies mentioned below, with its parent species, and a smaller one, *A. taczanowski*, from Ecuador, are the only representatives of the genus known, the *A. paca*, however, having a wide distribution. A remarkable character in the Paca is the unusual development of the cheek bone, the malar being greatly inflated and excavated, and its outer surface roughened in an extraordinary degree. The cavity in the cheek bone is lined with a mucous membrane and communicates with the mouth by a small opening. The *raison d'être* of this peculiar structure is unknown, and it can hardly be used as a pouch for food, like those of the Gophers and Chipmunks, for it would seem that any particles placed in this bony pouch would be apt to stay there, the animal having no means of extracting them.

82. Agouti. Paca.

$$I.\frac{1-1}{1-1}; \ P.M.\frac{1-1}{1-1}; \ M.\frac{3-3}{3-3} = 20.$$

Agouti Lacép., Tabl. Divis. Sous-divis. Ordres et Genres Mamm., Suppl. to Disc. d'ouvert et de cloture du Cours d'Hist. Nat., etc., 1799, p. 11. Type *Mus paca* Linnæus.

Coelogenus (sic) F. Cuv., Ann. du Mus., Hist. Nat. Paris, x, 1807, p. 203.

Paca Fischer, Zoogn., 1814, p. 85.

Osteopera Harlan, Faun. Amer., 1825, p. 126.

Genyscœlus Liais, Climats. Géol. Faune Brésil., 1872, p. 537.

Five toes on hind feet; zygomatic arches greatly expanded vertically, forming bony capsules on side of face, communicating with mouth by a small opening at bottom of inclosed cavity. Head large and broad; nose not pointed; tail a fleshy tubercle; inner toes and the nail of each foot very small.

paca virgata (*Agouti*), Bangs, Bull. Mus. Comp. Zoöl., xxxix, 1902, p. 47.

CENTRAL AMERICAN PACA.

Type locality. Divala, Chiriqui, Panama.

Geogr. Distr. Central America.

Genl. Char. Similar to the South American animal, but second stripes much less broken into spots; hind foot larger. Skull larger; palate narrower; audital bullæ flatter.

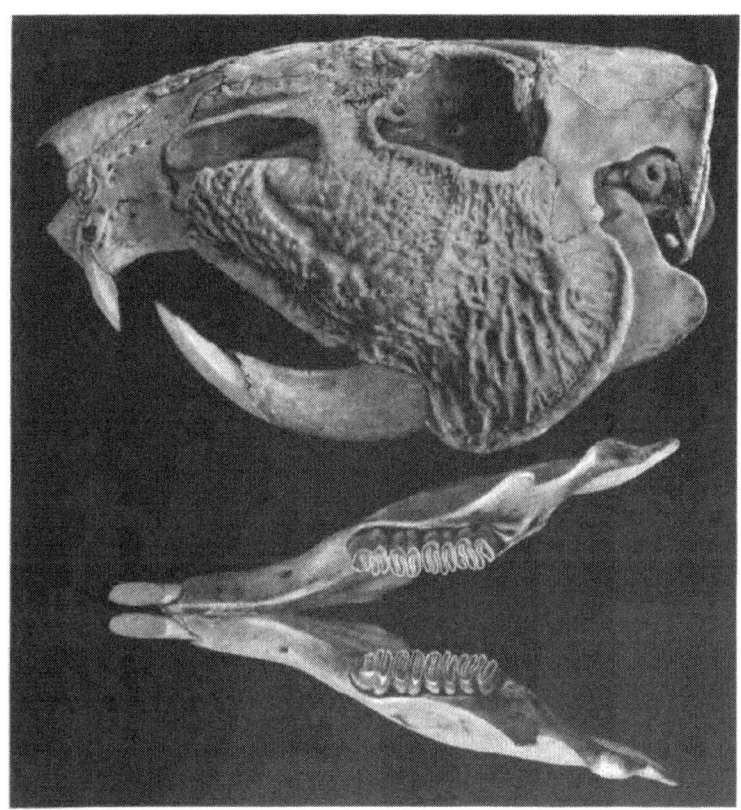

AGOUTI PACA VIRGATA.
No. 10079 Mus. Comp. Zoöl. Coll.   ¼ nat. size.

AGOUTI PACA VIRGATA.

No. 10070 Mus. Comp. Zoöl. Coll. ¾ nat. size.

Color. Variable; upper parts walnut brown; a stripe from shoulder to hip and a shorter one above it white; these break into a series of spots on the sides of neck and on flanks and are smaller on the latter. Two rows of small white spots above the others, the lower extending from sides of neck to flanks.

FIG. LVII. AGOUTI PACA VIRGATA. CENTRAL AMERICAN PACA.

Measurements. Total length, 740; tail vertebræ, 22; hind foot, 130; ear, 43. Skull: basal length, 139.6; occipito-nasal length, 151; zygomatic width, 104; mastoid width, 54.8; interorbital width, 47.2; length of nasals, 51.2; width of nasals, 26; length of palate to palatal notch, 76; width of palate at middle of second molariform tooth, 7; at middle of last molariform tooth, 10.2; upper molar series, 29.6; length of single half mandible, 107. (Bangs, l. c.)

No family of mammals is better known generally than that of the LEPORIDÆ, which includes the Hares and Rabbits. It is the one group of animals with which nearly all persons are familiar. The terms, Hares and Rabbits, although used indiscriminately by many persons, really indicate very important distinctions, viz.: Hares never make burrows in the earth, but construct loosely arranged nests or "forms," where they sit during the day, and in which they bring forth their young fully clothed with fur and able to take care of themselves. On the other hand, the Rabbit digs a deep burrow in the earth, with many connecting passages and openings, and the young are born underground, naked, blind, and helpless. These animals inhabit districts varying greatly in their conditions and situations. Some seek swamps, marshes, or dense thickets, like the southern canebrakes, and are partly aquatic; others delight in woods, bushy coverts, and tangled depths; while still others are at home only in the snow-covered northern wastes, or the wide wind-swept prairies or desert expanse. In the northern portion of the western hemisphere the greater portion of these animals are found, and the largest species occur in the extreme north and on the plains, and are represented by the Arctic Hares and Jack Rabbits. Certain species turn white in winter in districts where there are heavy falls of snow, the white coat assimilating with the snow, and affording concealment to the animal. This change, however, does not always occur throughout the range of every species, as witnessed by the Washington Hare, *L. washingtoni*, which is a white animal in winter in its northern range, while in the more southern part of its dispersion, about Puget Sound, it does not turn white in winter, the moderate snow fall in that section not making a white coat necessary for protection. On the contrary, an animal of such a color would be, probably, all the more conspicuous. Hares are remarkable for their lengthened ears and hind legs, and in some species these characters are carried to an extreme, but all members of the family have the hind legs considerably elongated, and it is by them that the great leaps made in flight are accomplished. Hares and Rabbits are absolutely without defense, flight (aided by a low order of strategy, illustrated by doubling on its tracks) being their only means of escape from their enemies. They are, however, always on the watch, their large eyes roving constantly over every object in range of their vision, and the long ears constantly in motion, attentive to every sound. Innumerable enemies of the earth and air are continually seeking their destruction, and it is only its amazing fecundity that enables the race to survive. The fore legs are very short, and are never used as hands, as is the case with many rodents, and although in the feeble combats indulged in by Hares, the fore feet may occasionally be used to strike an adversary, they are capable of inflicting only very slight injuries. Compared with many rodents, the teeth of Hares are weak, but they commit much damage with such as they have, gnawing trees, shrubs, etc., and are very destructive to growing crops, vegetables, and also to ornamental plants. The members of this family possess more teeth than those of any other among the rodents, and they are remarkable for having at birth three pairs of incisors in the upper jaw, the second pair small and placed behind the middle large pair. The second outer pair early becomes deciduous, but the inner small pair is retained through life. The food of these animals is strictly vegetable. Rabbits have been introduced into various parts of the world, and in some lands have multiplied to such an extent as to become very serious pests, and all kinds of methods for exterminating them have been tried in vain, illustrating in a very forcible and unpleasant manner the foolishness of man when he disturbs the harmony of Nature and interferes with her distribution of animal life upon the Globe. In sections of western North America Jack Rabbits, so-called, abound in such extraordinary numbers that great hunts are regularly organized and attended by all the ranchmen in the vicinity, and many thousands of these animals are killed in a single day, having been "rounded up" in a manner similar to that employed with the half-wild range cattle, except that the Hares are driven into a space inclosed with nets, from which there is no escape, and where they are speedily dispatched with clubs. In spite of these wholesale executions, and all other fatalities that overtake them, Hares still flourish.

One other family is comprised in this suborder, the LAGOMYIDÆ, containing the little Chief Hares, or Pikas. No species are found within the lands embraced in this work so far as known. Far up the mountain sides, sometimes at an elevation of many thousand feet, amid the ranges that form the "backbone" of the North American Continent, their fortress a hole amid the rocks, these little creatures, whose aspect is between that of a guinea-pig and a rabbit, live in colonies and betray their presence to the intruder on their domains by sharp, squeaking, querulous ventriloquial notes or cries, deceptive as to distance and locality. Very timid, the Pikas are shy and watchful, and survey an interloper from the farther side of some friendly stone. They lay up stores of provisions, such as grass and other herbage, against the long severe winter, and are very industrious. Four young are produced in the spring about May. Pikas are very small, tailless animals, about eight inches in length, with large, flat ears, small eyes, and a rudimentary thumb with claw.

Fam. IX.   Leporidæ.   Hares, Rabbits.

C. J. Forsyth-Major, *On Fossil and Recent Lagomorpha.* Trans. Zoöl. Soc., 1898, p. 433.

83.   Romerolagus.

Romerolagus Merr., Proc. Biol. Soc. Wash., x, 1896, p. 173.   Type *Romerolagus nelsoni* Merriam.

Small; ears, hind legs, and feet short. Skull similar to that of the subgenus *Sylvilagus*, but postorbital processes are lacking anteriorly, and jugals elongated posteriorly; clavicles articulating with sternum and scapula; prosternum broader than long before first pair of ribs; mesosternum of three segments; six pair of ribs articulating with sternum; fifth cervical vertebræ with transverse process directed outward, not backward; transverse process of lumbar vertebræ broadly expanded; small hypopophyses present on first three lumbar vertebræ; inferior crest of navicular bone short and not produced under base of metatarsal.   (ex Merr., l. c.)

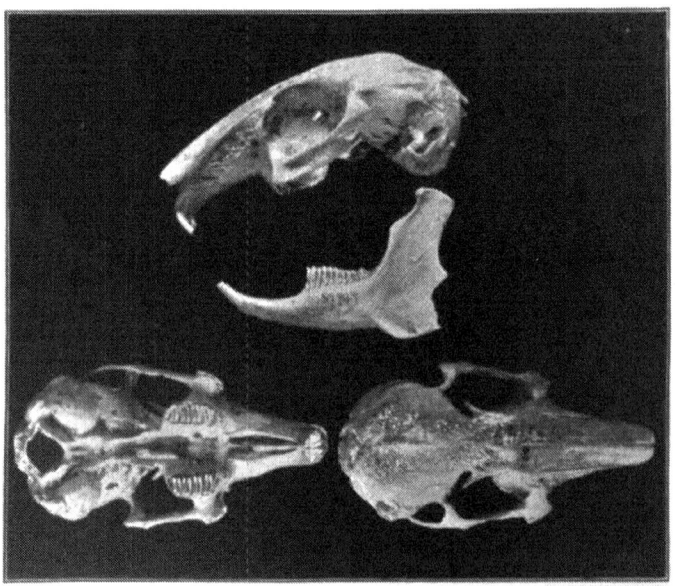

FIG. 79. ROMEROLAGUS NELSONI.
No. 57949 U. S. Nat. Mus. Coll. Nat. size.

427. *nelsoni (*Romerolagus*), Merr., Proc. Biol. Soc. Wash., x, 1896,
    p. 173.
    diazi. (*Lepus*), Ferrari-Perez, Cat. Comis. Geogr. Explor. Repub.
    Mexicana, 1893, pl. 42.

NELSON'S HARE.

    Type locality.   West slope of Mt. Popocatepetl, State of Mexico,
Mexico.   Altitude, 10,000–12,000 feet.

    * It is a moot question whether this species should not bear the name *diazi*
instead of *nelsoni*, the figure on plate 42 of the Catalogue above cited represent-
ing undoubtedly this species.   It is said that a description was also published
in a Mexican newspaper.   The figure in the plate gives a better idea of the animal than can be obtained from any description.

Geogr. Distr. Mt. Popocatepetl, State of Mexico, and Mt. Iztac-cihuatl ? Mexico.

Genl. Char. Those of genus.

Color. Upper parts, collar, and sides grayish brown and black, suffused with yellowish; chin and belly smoky gray washed with buff; feet buffy yellow.

Measurements. Total length, 311; tail vertebræ, 0; hind foot, 53; ear from notch (skin), 36. Skull: occipito-nasal length, 48; Hensel, 38; zygomatic width, 25; across orbital processes, 13; palatal length, 20; length of mandible, angle to symphysis, 31; height at coronoid process, 23.

84. Lepus.

$$I.\frac{2-2}{1-1}; \ P.\frac{3-3}{2-2}; \ M.\frac{3-3}{3-3} = 28.$$

Lepus Linn., Syst. Nat., I, 1758, p. 57; I, 1766, p. 79. Type *Lepus timidus* Linnæus.

Hydrolagus Gray, Ann. Mag. Nat. Hist., 3d Ser., xx, 1867, p. 221.

Silvilagus Gray, Ann. Mag. Nat. Hist., 3d Ser., xx, 1867, p. 222.

Tapeti Gray, Ann. Mag. Nat. Hist., 3d Ser., xx, 1867, p. 224.

Macrotolagus Mearns, Science, I, 1895, p. 698. *Id.* Proc. U. S. Nat. Mus., 1896, p. 552.

Microlagus Trouess., Cat. Mamm. vivent. quam fossil, 1897, fasc 3, p. 660.

Limnolagus Mearns, Science, N. S., v, 1897, p. 393.

Skull high, superior outline much curved, especially at occipital region; postorbital processes in the majority of species long, more or less divergent, flanking a deep wide notch, their posterior extremities not completely fused with skull; (exceptions to this are the swamp hares which have this process ankylosed to the cranium by its tip, or its internal margin); all the openings of the skull are large; facial surface of the maxilla reticulated; orbits very large, meeting in the mesial line of the cranium; teeth more numerous than in any other family of rodents; second pair of upper incisors small, situated behind the chief pair; the latter is grooved deeply in front, and all are deeply implanted in the skull and lower jaw; molars rootless; third upper molar minute; last lower molar larger, but still much the smallest of the lower series; palate a mere bridge between molars. The scapula ends in a process, which has near its termination a branch directed at right angles to the axis; tibia and fibula always ankylosed; fore feet with five toes, hind feet with four. A patch of hair covered skin on inner surface of cheeks extending backward from the angle of mouth. Hind legs elongate, in some species greatly so; ears very long; tail rudimentary.

KEY TO THE SUBGENERA.

A. Tail rudimentary.
  a. Interparietal persistent as a distinct bone in
    adults.
    a.' Ear shorter than hind foot.

A. Limnolagus.

"Interparietal present as a distinct bone in adults. Skull and teeth massive; rostral portion of skull wide as high; postorbital process of frontals ankylosed with the cranium for its entire length; frontals and parietals deeply pitted; skull rather straight above, about half as wide as long; pelage harsh; head small; ear, tail, and hind foot short, the latter scantily haired."

KEY TO THE SPECIES AND SUBSPECIES.

A. Skull large, heavy; frontals and parietals deeply
    pitted; pelage harsh.
  a. Basisphenoid and basioccipital forming an
    obtuse angle.   PAGE
    a.' Above pale buffy gray, lined with black..*L. a. attwateri* 414
    b.' Above yellowish brown, tinged with
      rufous and lined with black.............*L. palustris* 415
  b. Basisphenoid and basioccipital on same plane.....*L. truii* 415

aquaticus attwateri (*Lepus*), Allen, Bull. Am. Mus. Nat. Hist., 1895,
    p. 327. Elliot, Syn. N. Am. Mamm., 1901, p. 278.
ATTWATER'S SWAMP HARE.

Type locality. Medina River, 18 miles south of San Antonio, Bexar County, Texas.

Geogr. Distr. Yucatan, Mexico, to southeastern Texas.

Genl. Char. The general color paler than that of *L. aquaticus.* Size similar.

Color. Above pale buffy gray lined with black; sides whitish gray tinged with buff; dorsal region tinged with yellowish, darkest on the rump; belly and inside of legs white; back sometimes nearly all black;

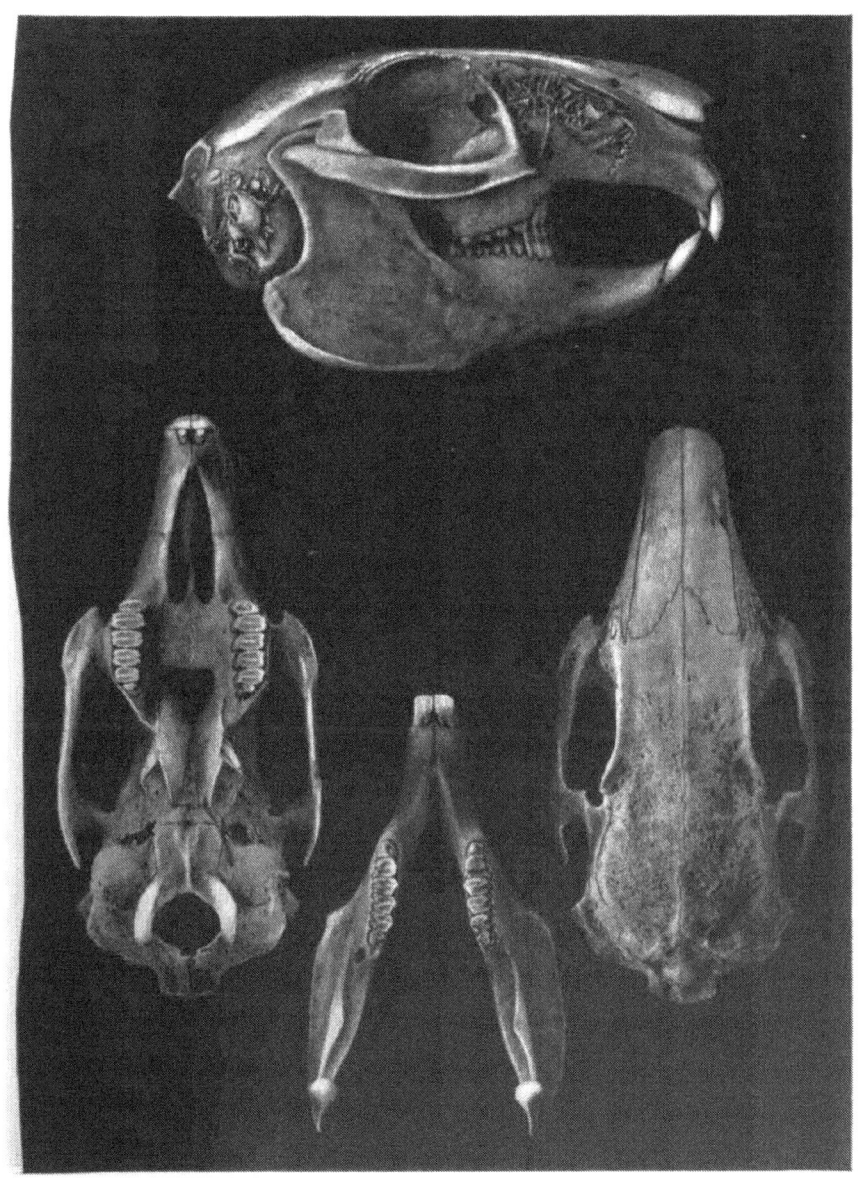

LEPUS AQUATICUS ATTWATERI.
No. 6131 Am. Mus. Nat. Hist. Coll.   Nat. size.

ears sparsely haired, reddish brown outside, with a narrow white border anteriorly; curved black line at corner of eye across cheek; tail reddish brown above, white below; feet fulvous above, soles dusky.

Measurements. Total length, 520; tail to end of hairs, 83; hind foot, 105; ear, 65. Skull: total length, 87; basal length, 79; zygomatic breadth, 40; mastoid breadth, 32; interorbital constriction, 32; length of nasals, 35; of lower jaw, 63; height of coronoid process, 37.

428. palustris (*Lepus*), Bachm., Journ. Acad. Nat. Scien. Phil., 1837, p. 194, pls. 15, 16. Elliot, Syn. N. Am. Mamm., 1901, p. 279.

douglasi Gray, Ann. Mag. Nat. Hist., 1837, p. 586.

SWAMP HARE. *Conejo* in Mexico.

Type locality. Eastern South Carolina?

Geogr. Distr. Yucatan, Mexico, north to Texas; Florida; and North Carolina.

Genl. Char. Size of *L. sylvaticus*, tail shorter; ears broader, more rounded; head larger; nails of toes exposed; incisors and molars, broader and heavier.

Color. Above yellowish brown tinged with rufous or rusty (the latter especially on the rump and outside of legs), and heavily shaded with black; beneath grayish white; fore neck, breast, and sides yellowish brown; chin grayish white; throat brownish gray; ear grayish rufous lined with black; tail above rufous and black, beneath grayish white; soles brownish.

Measurements. Total length, 444; tail vertebræ, 38; hind foot, 88; ear, 64. Skull: total length, 79; greatest width, 38; interorbital width, 15; lateral length of nasals, 32; posterior width of nasals, 19; length of upper molar series, 15; length of lower jaw, 60.

429 truii (*Lepus*), Allen, Bull. Amer. Mus. Nat. Hist., 1890, p. 192.

TRUE'S SWAMP HARE.

Type locality. Mirador, State of Vera Cruz, Mexico.

Geogr. Distr. State of Vera Cruz, Mexico.

Genl. Char. Similar to *L. palustris*, darker, smaller; skull with comparative differences.

Color. Similar to *L. palustris*, but smaller in size and more mixed with black on upper parts, especially on the dorsal region; beneath pale yellowish except a broad yellowish brown pectoral band; ears black and rufous; feet pale rufous.

Measurements. Total length, 335; hind foot, 75; ear from crown, 54. Skull: basal length, 57; greatest zygomatic breadth, 35; interorbital constriction, 25; mid-palatal length, 11; length of nasals, 27; length of upper molar series, 14; height of lower jaw at coronoid process, 32. (ex Type.)

B. Silvilagus.

"Interparietal persistent as a distinct bone in adults; rostrum wider than high; skull and teeth light; postorbital process united with cranium behind, inclosing a narrow foramen; upper surface of skull

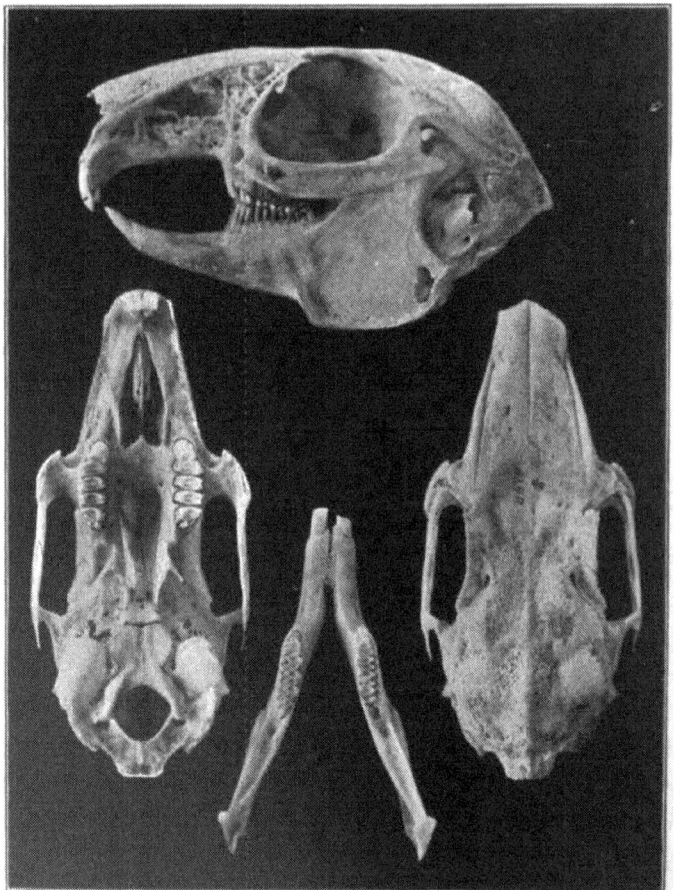

FIG. 80. LEPUS (SILVILAGUS) F. SUBCINCTUS.
No. 8678 Field Columbian Mus. Coll.   Nat. size.

less pitted; skull sometimes wider than half its length, much arched; pelage softer."

KEY TO THE SPECIES AND SUBSPECIES.

A. Ears shorter than hind foot.

   a. Ears equal to or shorter than head.   PAGE

floridanus subcinctus (*Lepus*), Miller, Proc. Acad. Nat. Scien. Phil.,
    1899, p. 386.

MICHOACAN HARE.

   Type locality. Hacienda El Molino, near Negrete, State of Michoacan, Mexico.

   Geogr. Distr. State of Michoacan, Mexico; range unknown.

   Genl. Char. Similar to *L. f. chapmani*; larger.

   Color. Above mixed black and buffy white; paler on rump and sides; nape light cinnamon; crown yellowish; orbital ring whitish; cheeks below and behind eyes blackish; lower throat and chest buffy cream sprinkled with black; rest of under parts whitish, with a buff band in front of hind legs; legs ochraceous cinnamon buff, palest on hind legs; tail mixed brown and whitish above, white beneath; ears gray exteriorly, buff interiorly; black along anterior border near tip, extreme anterior margin pale buff.

Measurements. Total length, 434; tail vertebræ, 47; hind foot, 86; ear from crown, 76. Skull: occipito-nasal length, 73; Hensel, 52; zygomatic width, 34; length of nasals (outer side), 30; (median), 22.5; palatal length, 7; length of incisive foramina, 17; upper tooth row, 7.

floridanus aztecus (Lepus), Allen, Bull. Amer. Mus. Nat. Hist.,
　　1890, p. 188.
TEHUANTEPEC HARE.

Type locality. Tehuantepec City, State of Oaxaca, Mexico.
Geogr. Distr. State of Oaxaca, Mexico.
Genl. Char. Smaller than *L. floridanus,* lighter in color and ears larger; palatal arch round.

Color. Above and sides of neck buffy cinnamon and black; sides gray, with buffy cinnamon band in front of thighs; nape and outer surface of limbs yellowish rufous; under parts white; chest yellowish white; eye stripe grayish white; cheeks gray; fore feet yellowish white; hind feet pure white; tail above cinnamon rufous, like rump; ears dark brownish gray, blackish at tip, anterior border white basally.

Measurements. Total length, 300; tail to end of hairs, 37; hind foot, 82; ear from notch, 64. Skull: occipito-nasal length, 71; median length of nasals, 25; lateral length of nasals, 33; posterior width of nasals, 16; anterior width, 9; distance from anterior premolar to alveolus of incisor, 19.5; length of palatine foramina, 16.5; least interorbital width anterior to postorbital processes, 18; length of palate, 7; width of palate, 9.

floridanus persultator (Lepus), Elliot, Pub. Field. Columb. Mus., III,
　　1903, p. 147. Zoölogy.
PUEBLA HARE.

Type locality. Puebla, State of Puebla, Mexico.
Genl. Char. In color very closely resembling *L. f. subcinctus,* but smaller in all its dimensions. Skull is distinguished for the straightness of its anterior superior outline, the nasals being flat and on a line with the frontals; posterior portion of skull from behind orbits curving sharply downward; nasals broad, abruptly compressed near anterior termination; ears short; palatal arch with azygos process in center.

Color. Top of head cinnamon rufous and black, rest of upper parts except rump mixed black and ochraceous buff; sides gray; rump gray and black; nape and outer surface of limbs yellowish rufous; ochraceous buff band in front of thighs; pectoral band pale yellowish rufous; under parts white; eye stripe buff; orbital ring cream buff; cheeks mixed gray buff and black; fore feet buff; hind feet white; tail above ochraceous buff, beneath white; ears dark brown sprinkled with buff, darkest at tip, anterior border for three-fourths its length from base, white.

Measurements. Hind foot, 72; ear from notch, 54; from head, 62 (skin). Skull: occipito-nasal length, 66; median length of nasals, 19; lateral length of nasals, 27; posterior width of nasals, 11; anterior width, 8; distance from anterior premolar to alveolus of incisor, 19; length of palatine foramina, 16; least interorbital width, anterior to postorbital processes, 11; length of palate, 5; width of palate, 8.

floridanus yucatanicus (*Lepus*), Miller, Proc. Acad. Nat. Scien. Phil., 1889, p. 384.

YUCATAN HARE.

Type locality. Merida, Yucatan, Mexico.

Geogr. Distr. Yucatan, Mexico; limits of range unknown.

Genl. Char. Similar to *L. f. aztecus*, but larger in size, and darker in color.

Color. Above mixed black and buff, lightest on rump and sides; nape cinnamon rufous, crown of head like back but darker; orbital ring whitish; breast ochraceous buff; rest of under parts white; thighs gray; legs cinnamon rufous with a white line on inner side, including the hind foot, but only reaching the wrist on fore legs; ears gray, fringed with pale buff on anterior margin, changing to black near tip; interior of ears whitish; tail mixed reddish brown and black above, white beneath.

Measurements. Total length, 430; hind foot, 198; ear from crown, 71. Skull: greatest length, 82; basal length, 64; posterior margin of palate to tip of hamular process, 17; zygomatic breadth, 39; interorbital constriction, 20; greatest length of nasals, 37; greatest breadth of nasals, 16; length of incisive foramina, 21; length of upper molar series, 14; length of mandible, 62; lower molar series, 14.6.

floridanus holzneri (*Lepus*), Mearns, Proc. U. S. Nat. Mus., 1896, p. 554. Elliot, Syn. N. Am. Mamm., 1901, p. 284.

rigidus. Mearns, Proc. U. S. Nat. Mus., 18, 1895, p. 555. (*Winter pelage.*)

HOLZNER'S HARE.

Type locality. Near the summit of Huachuca Mountains, southern Arizona.

Geogr. Distr. Southern Arizona and northern New Mexico to northern Mexico in States of Sonora and Chihuahua.

Genl. Char. Size large; ears rather short; hind feet long; colors dark; nasals extending beyond premaxillæ.

Color. Summer Pelage. Upper sides of head and back to rump vinaceous cinnamon mixed with gray and black; thighs and rump whitish gray, lined with black; beneath white, tinged with yellowish; pectoral band clay color; nape tawny; legs tawny, the inner side brownish white; ears reddish brown, gray, and black mixed, bordered anteriorly for basal two-thirds with white; tail above yellowish brown grizzled, beneath white.

Winter Pelage. Above gray, lined with black, washed with clay-color on back, hips, and ventral border; sides and thighs gray, lined with black; under parts grayish white on breast, tinged with clay-color, remainder pure white; orbital region whitish; nape russet; top and sides of head gray, washed with yellowish brown; tail grayish brown above, hairs tipped with hoary, beneath pure white; feet whitish; ears grayish white on lower part passing into gray mixed with black, with a narrow black band on terminal half and edged with white; inner side grayish white.

Measurements. Total length, 415; tail vertebræ, 64; ear from crown, 77; from notch, 65; hind foot, 99.5.

430. veræcrucis (*Lepus*), Thomas, Proc. Zoöl. Soc., 1890, p. 74, pl. VII.

VERA CRUZ HARE.

Type locality. Las Vigas, Jalapa, State of Vera Cruz, Mexico.

Geogr. Distr. State of Vera Cruz, Mexico; range unknown.

Genl. Char. Similar to *L. floridanus*, but larger, ears longer.

Color. Upper parts yellowish brown mixed with black; nape pale rufous; face grayish yellow and black; orbital ring pale cream color; ears pale gray, tips and outer edges blackish; inner surface yellowish; under parts yellowish gray; legs pale orange yellow or orange gray; tail above grayish brown; beneath white.

Measurements. Total length, 492; tail vertebræ, 32; hind foot, 104; ear from crown, 90. Skull: greatest length, 85; basal length, 69.5; length of nasals, 36; interorbital constriction, 19.4; length of interparietal, 5.3; length of palatine foramen, 20.4; of upper tooth row, crowns, 14; length of lower jaw to tips of incisors, 68.

431. russatus (*Lepus*), Allen, Bull. Am. Mus. Nat. Hist., 1904, p. 31. RUSSET HARE.

Type locality. Pasa Nueva, State of Vera Cruz, Mexico.

Genl. Char. Pelage coarse, harsh. Audital bullæ very large.

Color. Top of head and upper parts of body and tail pale brownish russet, varied with dark brown; sides and hips varied creamy white; nape, anterior surface of fore legs, and outer side of hind legs ferrugineous; cheeks and sides of neck like back, lined with black; pectoral band clay color; ventral surface yellowish white; upper surface of hind feet creamy white; ears externally grayish brown tinged with pale russet, blackish on apical third.

Measurements. Total length, 450; tail vertebræ, 42; hind foot, 80; ear from crown (dry skin), 62. Skull: occipital-nasal length, 78.5; basal length, 63; zygomatic width, 53.2; interorbital constriction, 17; mastoid breadth, 28; width of braincase, 25; length of nasals, 36; palatal bridge, 7.5; length of upper tooth row, 6.5; length of lower jaw, 55; height of condyle, 35.

432. parvulus (*Lepus*), Allen, Bull. Am. Mus. Nat. Hist., 1904, p. 34. LITTLE HARE.

Type locality. Apam, southern part of State of Hidalgo, Mexico.

Color. Above pale buff varied with black, slightly grayish on rump; chin, throat, and central ventral surface yellowish white; pectoral band broad, pale rusty brown; nape pale ferrugineous; sides of head buffy brown; upper surface of the feet pale rusty; hind feet deep buff; ears externally buffy grayish brown, internally pale yellowish brown with a deep buff edge; tail above blackish, tips of hairs buffy gray.

Measurements. Total length, 390; hind foot, 75; ear 65. Skull: occipito-nasal length, 65; basal length, 54.5; zygomatic width, 33; interorbital constriction, 18.4; mastoid breadth, 27; width of braincase, 25; length of nasals, 25; posterior width of nasals, 13.5; palatal bridge, 5.2; length of upper tooth row, 10; length of palatal foramina, 14.5; length of mandible, 43; height at angle, 25.5.

433 insolitus (*Lepus*), Allen, Bull. Amer. Mus. Nat. Hist., 1890, p. 189.

PLAINS HARE.

Type locality. Plains of Colima, State of Jalisco, Mexico.

Geogr. Distr. State of Jalisco, Mexico.

Genl. Char. Similar to *L. floridanus*, but larger and paler. Malar with a deep groove on the outer surface of the anterior half; postorbital processes not fused with the braincase, merely touching it.

Color. Upper parts sandy buff and black; sides grayish; nape and fore legs externally rufous; breast yellowish brown; rest of under parts white; hind leg externally yellowish brown; fore feet brownish yellow; hind feet white, as is also the anterior edge of leg; orbital ring buffy gray; tail and rump pale rusty brown and black; ears brownish gray, edge and tip blackish.

Measurements. Total length, 440; tail to end of hairs, 40; hind foot, 92; ear from crown, 78. Skull: total length, 83; basilar length, 66; greatest width, 39; interorbital constriction, 21; length of nasals,

34; of palatine foramen, 21; of upper molar series at crown, 22.5; length of lower jaw, 58; height at condyle, 40.

434. auduboni (*Lepus*), Baird, N. Am. Mamm., 1857, p. 608, pls. XIII, XLVIII, fig. 2. Elliot, Syn. N. Am. Mamm., 1901, p. 283. AUDUBON'S HARE.

Type locality. San Francisco, California.

Geogr. Distr. Lower California, Mexico, from Cape St. Lucas to vicinity of San Francisco, California.

Genl. Char. Size smaller than that of *L. f. mallurus*; ears longer than head; tail long.

Color. Above pale yellowish brown mixed with black; sides paler, with little or no black; nape pale rufous; fore feet above pale yellowish and rusty; hind feet whitish, sides rusty; pectoral band pale yellowish brown; under parts white; ears dark brown, the hairs with pale yellowish tips, so that this hue predominates, grading into black or brownish black at tips; tail above like back, beneath black.

Measurements. Total length, 457; tail vertebræ, 38; hind foot, 89; ear from notch, 70. Skull: occipito-nasal length, 69; Hensel, 53; zygomatic width, 32; lateral length of nasals, 30; median length of nasals, 23; posterior width of nasals, 17; anterior width of nasals, 8; length of lower jaw to end of incisors, 35; height at condyle, 31.

435. sanctidiegi (*Lepus*), Miller, Proc. Acad. Nat. Scien. Phil., 1899, p. 389. Elliot, Syn. N. Am. Mamm., 1901, p. 283. SAN DIEGO HARE.

Type locality. Mexican boundary line, monument No. 258, shore of Pacific Ocean, San Diego County, California.

Geogr. Distr. Northern Mexico to southwestern California.

Genl. Char. Similar to *L. auduboni*, but paler.

Color. Above grizzle of black and light cream buff; sides paler; rump whitish gray; nape ochraceous buff; tail dark brown above, white beneath; ears gray, lower half paler than crown of head, which is like the back; orbital ring whitish; chin and throat white, tinged with plumbeous; lower part of throat and chest cream buff; legs ochraceous buff; fore feet cream buff; hind feet white; white of belly reaching to wrists and back of hind feet.

Measurements. Total length, 385; tail vertebræ, 63; hind foot, 85; ear from crown, 78. Skull: greatest length, 69; basal length, 56; posterior margin of palate to tip of hamular process, 16.4; zygomatic breadth, 33; interorbital constriction, 19; length of nasals, 29; greatest width of nasals, 13.6; length of upper molar series, 12.8; length of incisive foramina, 6; length of mandible, 50; of lower molar series, 13.6.

436. arizonæ (*Lepus*), Allen, Mon. N. Am. Rod., 1877, p. 332.
Elliot, Syn. N. Am. Mamm., 1901, p. 285.
ARIZONA HARE.

Type locality. Beale's Springs, fifty miles west of Fort Whipple, Yavapai County, Arizona.

Geogr. Distr. Lower California and State of Sonora, Mexico, north to Deserts of Arizona, and the Chiricahua and Huachuca Mountains, (but not to the White Mountains,) up to 8,500 feet.

Genl. Char. Smaller than *L. nuttalli*, but similar; ears much longer and broader.

Color. Above pale yellowish gray, mixed sparingly with black; nape yellowish fulvous; sides pale gray, mixed sometimes with pale brown; chin white; pectoral band yellowish; rest of under parts white; feet pale yellowish brown; tail above darker than back, yellow brown, beneath white; ears pale grayish brown, outer edge whitish. Winter specimens are heavily lined with black above and on sides.

Measurements. Total length, 340-383; tail vertebræ, 35-54; hind foot, 76-94; ear, 69-78. Skull: total length, 65; greatest width, 35; interorbital constriction, 17; length of nasals, 28; upper molar series, 12; length of lower jaw, 46.

a.—major (*Lepus*), Mearns, Proc. U. S. Nat. Mus., 1896, p. 557.
Elliot, Syn. N. Am. Mamm., 1901, p. 286.
GREATER DESERT HARE.

Type locality. Calabasas, Pima County, Arizona.

Geogr. Distr. Poso de Luis, State of Sonora, to the basin of the Mimbres, State of Chihuahua, Mexico, and northward to Colorado Plateau of Arizona.

Genl. Char. "Similar to *L. arizonæ*, but larger, more reddish and darker."

Color. Above grayish drab, tinged with cinnamon, lined with black; sides paler; rump iron gray; nape and outer surface of limbs dull cinnamon; ears pale grayish on inner side, drab mixed with gray and black on outer side, tips black; pectoral band clay color; rest of under parts white.

Measurements. Total length to end of hairs of tail, 430; tail vertebræ, 42; hind foot, 92; ear from notch, 69 (skin.)

b.—minor (*Lepus*), Mearns, Proc. U. S. Nat. Mus., 1896, p. 557.
Elliot, Syn. N. Am. Mamm., 1901, p. 286.
LESSER DESERT HARE.

Type locality. El Paso, El Paso County, Texas.

Geogr. Distr. State of Chihuahua, Mexico. "Plains of Colorado, southward to the Rio Grande, and westward to the elevated central tract, where it intergrades with *L. a. major* in the pass between the southern end of the Rocky Mountains and northern extremity of the Sierra Madre.''

Genl. Char. Smaller than *L. arizonæ*; ears short, colors pale. Rostrum more elongate; mandible stouter and higher, and audital bullæ larger.

Color. Above yellowish brown, lined with black; rump grayish white, lined with black; sides yellowish gray, with a buff lateral line; head gray, tinged with yellowish brown on cheeks and crown; nape light cinnamon; fore legs wood brown; hind feet above white; pectoral band yellowish gray, rest of under parts white; tail above dusky, hairs tipped with yellowish brown and gray, beneath white; ears grayish white on dorsal surface behind, and gray mixed with yellowish brown and black in front, basal two-thirds of front edge white, tips black.

Measurements. Average total length, 345; tail vertebræ, 50.2; hind foot, 83.4; ear from notch, 65.8. Skull: occipito-nasal length, 29; Hensel, 46; zygomatic width, 32; lateral length of nasals, 34; medium length of nasals, 18; posterior width of nasals, 11; anterior width of nasals, 7; length of lower jaw to end of incisors, 46.5; height of condyle, 27.

c.—confinis (*Lepus*), Allen, Bull. Am. Mus. Nat. Hist., 1898, p. 146.

ALLIED HARE.

Type locality. Playa Maria, Lower California, Mexico.

Geogr. Distr. Lower California, Mexico.

Genl. Char. Similar to *L. arizonæ*, but darker.

Color. Upper parts dark grayish brown, mixed with black and pale brown; side pale grayish brown; top of nose and head to occiput like back; nape pale fulvous; side of head and nose mixed light gray and brown; pectoral band yellowish brown on sides, yellowish white in center; chin, upper part of throat, and rest of under parts white, the plumbeous base of the hairs showing; fore legs dark buff; hind legs gray; under sides of feet reddish brown; tail almost invisible, like back; ears on outside brownish black, reddish towards anterior edges, black inside.

Measurements. Total length, 310; hind foot, 61; ear from notch, 65. (ex Type.) Skull: total length, 60.5; basal length, 53; zygomatic breadth, 29; interorbital constriction, 9.5; length of nasals, 24.5; breadth of nasals posteriorly, 10.5.

437. durangæ (*Lepus*), Allen, Bull. Am. Mus. Nat. Hist., 1903, p. 609.

DURANGO HARE.

Type locality. Rancho Bailon, State of Durango, Mexico. Altitude, 7,800 feet.

Genl. Char. "Size of *Lepus insolitus*, but much less varied with black, and the general coloration much paler, except the nape patch, the legs, and feet, which are of the same deep rufous as in *L. insolitus*. In other respects the coloration is not distinctly different from that of *L. a. major*, collected at the same locality. From the latter it differs in being twice as large (in general bulk), and from both *L. insolitus* and *L. a. major* in important cranial characters. Skull similar in general contour to that of *L. a. major*, but very much larger, with actually smaller audital bullæ."

Measurements. Total length, 457; head and body, 406; tail vertebræ, 51; hind foot without claws, 95; ear from notch, 76. Skull: total length, 79; basilar length, 60; zygomatic breadth, 36.3; length of nasals, 35; width of nasals, posteriorly, 16; anteriorly, 9; alveolar length of upper tooth row, 13.3; length of lower jaw, 55; height of condyle, 30; alveolar length of lower tooth row, 13.6." (Allen, l. c.)

438. orizabæ (*Lepus*), Merr., Proc. Biol. Soc. Wash., VIII, 1893, p. 143.

ORIZABA HARE.

Type locality. Mt. Orizaba, State of Puebla, Mexico. Altitude about 9,500 feet.

Geogr. Distr. State of Puebla, Mexico.

Genl. Char. Similar to *L. arizonæ*, but darker; audital bullæ smaller.

Color. Upper parts except rump deep clay color and black; rump and flanks mixed gray and black; nape patch and feet fulvous; breast drab mixed with buffy; rest of under parts white; tail above grizzled drab gray, and buff, beneath white; ears grayish brown, edge near tip brown.

Measurements. Total length, 395; tail vertebræ, 51; hind foot, 90.

439. nuttalli (*Lepus*), Bachm., Jour. Acad. Nat. Scien. Phil., 1837, p. 345, pl. 22. Elliot, Syn. N. Am. Mamm., 1901, p. 284.

artemisia. Bachm., Jour. Acad. Nat. Scien. Phil., 1839, p. 94.

NUTTALL'S HARE.

Type locality. Plains of the Columbia near Walla Walla.

Geogr. Distr. State of Sonora, Mexico, north to 49th parallel.

Genl. Char. Size small; colors pale; ear short.

Color. Above yellowish gray mixed with black and brown; rump light gray and sides of body whitish yellow; nape pale cinnamon; pectoral band light buff, rest of under parts white; fore legs buffy white above; hind legs white; tail above like rump, beneath, white; ears pale yellowish brown and black, edged with white.

Measurements. Total length, 420; hind foot, 50; ear, 50. Skull: total length, 70; greatest width, 35; interorbital constriction, 18; length of nasals, 31; width of nasals posteriorly, 14; length of upper molar series, 12; length of lower jaw, 47.

440. graysoni (*Lepus*), Allen, Mon. N. Am. Rodent., 1877, p. 347. GRAYSON'S HARE.

Type locality. Maria Madre Island, Tres Marias Islands, State of Jalisco, Mexico.

Geogr. Distr. Tres Marias Islands, State of Jalisco, Mexico.

Genl. Char. Ear very short; colors pale.

Color. Upper parts pale cinnamon brown mixed with blackish brown; sides and rump paler; throat brown, rest of under parts white; outer surface of legs reddish brown, inner surface whitish; ears yellowish brown mixed with black at base, blackish brown toward tip; tail above blackish brown, beneath white; orbital ring pale brown.

Measurements. Total length, 388–433; tail vertebræ, 19–25; hind foot, 88; ear, 57. Skull: occipito-nasal length, 75; Hensel, 60; zygomatic width, 37; interorbital width, 18; median length of nasals, 24.5; lateral length of nasals, 30; posterior width of nasals, 14.5; anterior width of nasals, 9; palatal length, 30; length of upper tooth row, 13; length of mandible, angle to tip of incisors, 60; length of lower tooth row, 14.

O. Tapeti.

"Skull like that of *Lepus*, but the hinder supraorbital notch narrow, lobes short with a sharp inner edge; the front of the lower edge of the zygoma dilated, sharp-edged, porous above, hinder nasal opening rather narrow. Tail none. Ears short." (Gray, l. c.)

KEY TO THE SPECIES.

A. Ears very short; tail practically none.   PAGE
    a. Upper parts cinnamon brown and black..........*L. gabbi*   426
    b. Upper parts tawny ferrugineous and black.....*L. incitatus*   428

441. gabbi (*Lepus*), Allen, Mon. N. Am. Rodent., 1877, p. 349. GABB'S HARE.

Type locality. Talamanca, Costa Rica.

Geogr. Distr. Costa Rica.

Genl. Char. Ears very short; tail practically none.

Color. Above cinnamon brown and black; top of head yellow, ferrugineous, and black; sides of head yellowish brown varied with black; nape pale rufous; white spot bordered with brown behind nostril; breast yellowish brown; chin, throat, and rest of under parts white; outer side of legs rufous, inner whitish; tail above cinnamon brown and black, beneath yellowish brown; ears pale rufous and black, bordered narrowly with white.

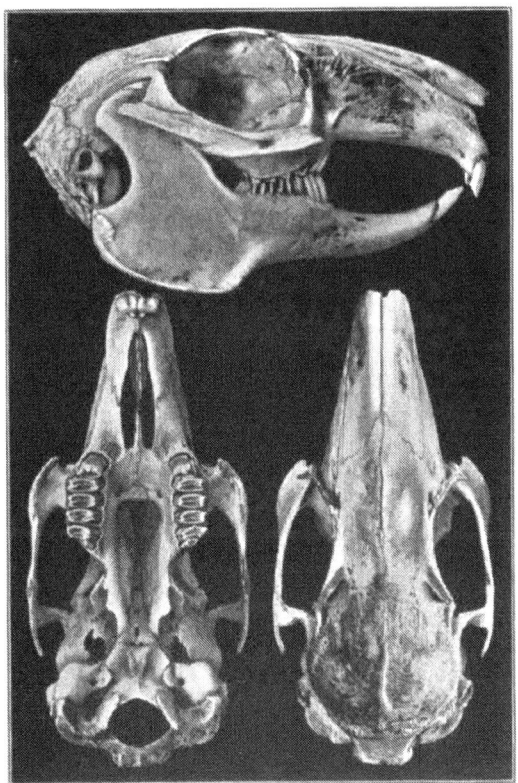

FIG. 81. LEPUS (TAPETI) GABBI.
Type U. S. Nat. Mus. Coll. Nat. size.

Measurements. Total length about 300; ear, 66; hind foot, 75. Skull: total length, 70; Hensel, 55; zygomatic width, 33; interorbital width, 15; median length of nasals, 20; lateral length of nasals, 28; palatal length, 27; length of upper tooth row, 13; length of mandible, 57; length of lower tooth row, 15.

442. incitatus (*Lepus*), Bangs, Am. Nat. xxxv., 1901, p. 633.
FLEET HARE.

Type locality. San Miguel Island, Bay of Panama.

Genl. Char. Similar to *L. gabbi*, but with a larger and heavier skull; rostrum wider and heavier, more rounded and arched; bony palate wider and longer; molar and incisor teeth heavier.

Color. Upper parts tawny ferrugineous; sides dull ochraceous; top of head and middle of back mixed with black; nuchal patch, arms, and outer side of legs bright tawny ferrugineous; superciliary stripe buffy white; outer side of ear like back, tip dusky, outer border yellowish white; under parts soiled white; sides of neck tawny ferrugineous; under side isabella.

Measurements. Total length, 420; tail vertebræ, 20; hind foot, 80; ear from notch, 45. Skull: basal length, 57; occipito-nasal length, 74.6; zygomatic width, 35.4; mastoid width, 23.8; interorbital constriction, 16.2; length of nasals, 30.4; width of nasals, 13; length of palatal bridge, (incisive foramina to palatal notch), 9; length of incisive foramina, 17; width of incisive foramina, 6.4; upper tooth row, alveolar border, 14.6; length of single half mandible, 56; lower tooth row, alveolar border, 15.4.

D. Microlagus.

"Ears longer than hind foot; tail short; skull narrow, low, and lightly ossified with postorbital process usually free, scarcely touching cranium behind." (Mearns.)

KEY TO THE SPECIES.

PAGE

A. Above light fulvous brown and black..........*L. cerrosensis*   428
B. Above yellowish, dark brown and black........*L. bachmani*   429
C. Above grayish brown and black.
  a. Small...................................*L. cinerascens*   429
  b. Large..................................*L. peninsularis*   430

443. cerrosensis (*Lepus*), Allen, Bull. Amer. Mus. Nat. Hist., 1898, p. 145.
CERROS ISLAND HARE.

Type locality. Cerros or Cedros Island, off west coast of Lower California, Mexico.

Geogr. Distr. Type locality only.

Color. Upper parts light fulvous brown, mixed with black; pectoral band pale yellowish brown; rest of under parts white; fore feet fawn; hind feet yellowish white; tail dark gray above, white beneath; ears grayish brown.

Measurements. Total length, 310; tail vertebræ, 25; hind foot, 54; ear from notch, 46. Skull: total length, 58; basal length, 49; greatest zygomatic breadth, 28.5; postorbital constriction, 10; length of nasals, 12; width posteriorly, 9.

444. bachmani (*Lepus*), Waterh., Proc. Zoöl. Soc., 1838, p. 103. Elliot, Syn. N. Am. Mamm., 1901, p. 281.

trowbridgii, Baird. Journ. Acad. Nat. Scien. Phil., 1855, p. 333.

BACHMAN'S WOOD HARE. *Conejo* in Mexico.

Type locality. Southwest coast of North America, probably California.

Geogr. Distr. Cape St. Lucas, Lower California, Mexico, to Fort Crook, California.

Genl. Char. Similar to *L. auduboni*, but smaller; ears equal to head in length; hind feet short; tail almost rudimentary; ears uniformly gray.

Color. Above yellowish brown, mixed with dark brown; throat, chest, and sides paler; beneath dusky gray, sometimes whitish; nape light rufous; legs and hind feet whitish, tinged with rufous; ears gray, at extreme base rusty, no black edging at tip; fur everywhere lead-color at base.

Measurements. Total length, 380; tail vertebræ, 30; hind foot, 75; ear, 61. Skull: occipito-nasal length, 72.5; Hensel, 54; inter-orbital constriction, 17; median length of nasals, 21; lateral length of nasals, 29; width of nasals posteriorly, 16.5; anterior width of nasals, 9; length of upper tooth row at alveolus, 19; length of lower jaw to tip of incisors, 55; height at condyle, 32.

445. cinerascens (*Lepus*), Allen, Bull. Am. Mus. Nat. Hist., 1890, p. 159. Elliot, Syn. N. Am. Mamm., 1901, p. 287.

ASH-COLORED HARE.

Type locality. San Fernando, Los Angeles County, California.

Geogr. Distr. Lower California, Mexico, and deserts of southern California.

Genl. Char. Similar to *L. bachmani*, but smaller and paler.

Color. Above gray and blackish brown mixed; nape pale rusty; dorsal region pale buffy gray; sides pure gray; under parts white; pectoral band brownish gray; fore legs yellowish brown; hind feet grayish brown; tail above dark gray, beneath white; ears brownish gray.

Measurements. Total length, 294; tail, 24; hind foot, 63; ear from notch, 58. Skull: total length, 62; basilar length, 48; width at postorbital constriction, 10; length of nasals, 25; posterior width of nasals, 12; length of upper molar series, at alveolar border, 6; length of lower jaw, 42; height at condyle, 27.

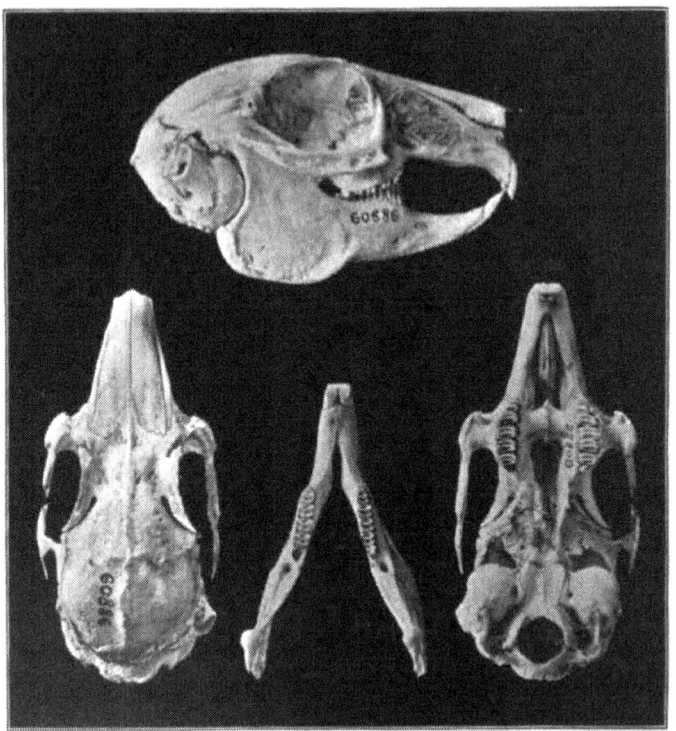

FIG. 82. LEPUS (MICROLAGUS) CINERASCENS.
No. 60886 U. S. Nat. Mus. Coll. Nat size.

446. peninsularis (*Lepus*), Allen, Bull. Amer. Mus. Nat. Hist., 1898,
    p. 144.

LOWER CALIFORNIA HARE.

Type locality. Santa Anita, Lower California, Mexico.

Geogr. Distr. Cape Region of Lower California, Mexico.

Genl. Char. Similar to *L. cinerascens*, but paler.

Color. Upper parts pale grayish brown and black, darkest on
dorsal region; sides grayer and paler; under parts and hind feet white;
fore feet brownish white.

Measurements. Total length, 324; tail vertebræ, 20; hind foot,
73; ear, 61. Skull: total length, 61.5; basal length, 52; greatest
zygomatic breadth, 30; width of postorbital constriction, 9.5; length
of nasals, 26; posterior width of nasals, 11; length of upper molar
series, 11.5; length of lower jaw, 43; height at condyle, 25.

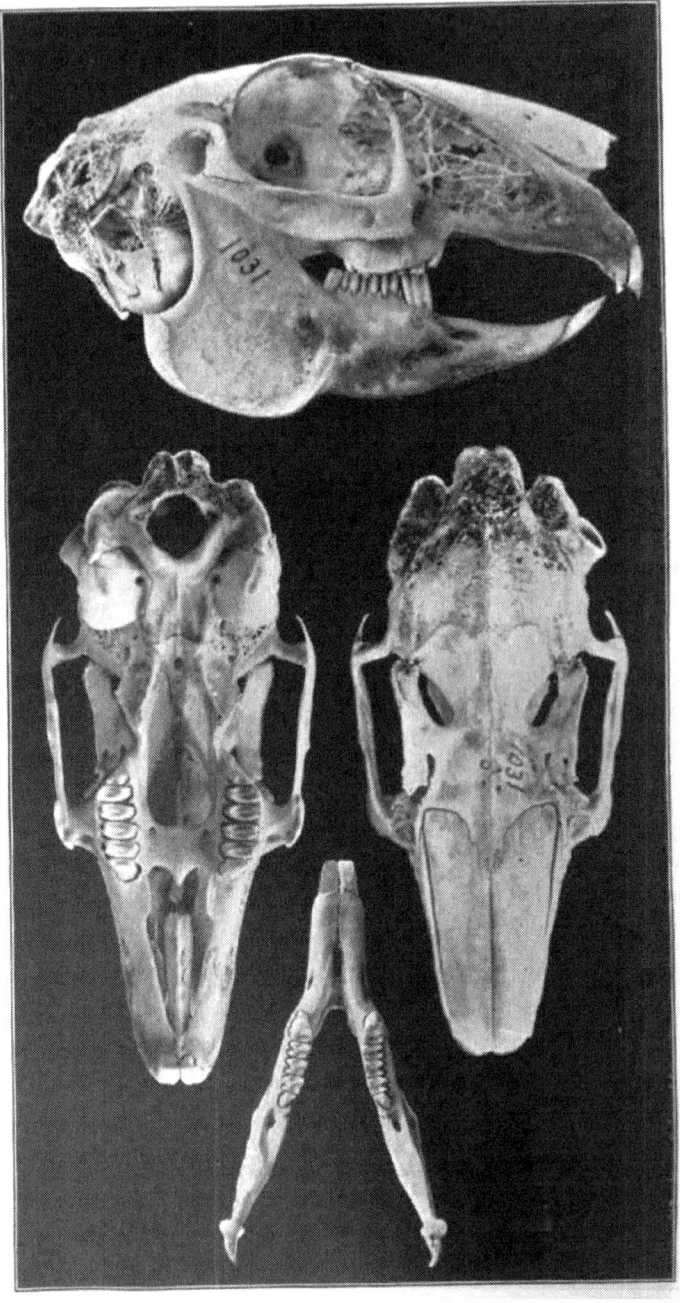

LEPUS T. EREMICUS.
No. 1031 Field Columbian Mus. Coll. Nat. size.

E. Macrotolagus.

Interparietal obliterated in adults; skull twice as long as wide; postorbital processes large, arching, and united to cranium by a suture, and inclosing a wide and long foramen; nasals lengthened; upper front incisors without distinct lateral groove; ear longer than hind foot.

KEY TO THE SPECIES AND SUBSPECIES.

447. callotis (*Lepus*), Wagl., Nat. Syst. Amph., 1830, p. 23. Elliot, Syn. N. Am. Mamm., 1901, p. 288.

nigricaudatus, Bennett, Proc. Zoöl. Soc., 1833, p. 41.

mexicanus (Licht.) Richards 6th Rep. Brit. Assoc., 1837, p. 150.

flavigularis, Wagl. Suppl. Schreib. Säugeth. IV, 1844, p. 107.

BEAUTIFUL-EARED JACK RABBIT. *Liebre* in Mexico.

Type locality. "Mexico."

Geogr. Distr. Through Mexico to Tehuantepec.

Genl. Char. Size large; similar to *L. texensis* Waterh., but the black on tips of ears almost obsolete, quite so in some specimens, the tips being pale yellowish or white.

Color. Above yellowish brown, mixed with black; sides paler; rump and thighs whitish ash, lined with black, and a black line in center of rump; nape black in summer; pectoral collar and throat pale brownish yellow; rest of under parts and hind feet white; outer surface of fore legs ashy gray, lined faintly with black; whitish spot on head; ears brownish yellow with yellowish white edging, usually white at tips; tail above black, beneath brownish gray.

Measurements. Total length, 560; tail vertebræ, 57; hind foot, 142; ear, height posteriorly, 137. Skull: total length, 47.5; greatest width, 44.5; interorbital constriction, 23; length of nasals, 45; posterior width of nasals, 46; length of upper molar series, 14; length of lower jaw, 67; height of lower jaw, 33.

448. merriami (*Lepus*), Mearns, Proc. U. S. Nat. Mus., 1896, p. 444. Elliot Syn. N. Am. Mamm., 1901, p. 289.

texensis, Aud. & Bachm., N. Am. Quad., III, 1853, p. 156, pl. CXXXIII.

MERRIAM'S JACK RABBIT.

Type locality. Fort Clark, Kinney County, Texas.

Geogr. Distr. Northern Mexico near boundary line and "Lower Gulf Coast to mouth of the Rio Grande, and up that stream to mouth of the Devil's River."

Genl. Char. Similar to *L. callotis*; ears shorter and tipped with black.

Color. Above grayish fawn mixed with black; nape black; sides of rump, thighs, and legs grayish white, lined with black; a black line on middle of rump; gular area clay color; rest of under parts white except a cream buff patch on sides of abdomen, such as are usually seen before the thighs; tail above black, beneath white; ears on anterior surface yellowish brown, mixed with black, the border buff, succeeded by a narrow black line; inner surface with base and tip black, intermediate space grayish white; inside ears buff with an elongated black patch near outer edge; the border is white at base, then buff, and black at tip.

Measurements. Total length, 570; tail vertebræ, 75; hind foot, 123; ear from crown, 142. Skull: occipito-nasal length, 90; Hensel, 72; zygomatic width, 41; interorbital constriction, 26; median length of nasals, 31; posterior width of nasals, 20.5; length of frontals, 37; of parietals, 18; palatal length, 9.5; length of lower jaw from tips of incisors, 71; height at condyle, 37.

449. gaillardi (*Lepus*), Mearns, Proc. U. S. Nat. Mus., 1896, p. 560.
Elliot, Syn. N. Am. Mamm., 1901, p. 289.
GAILLARD'S JACK RABBIT.

Type locality. Plagas Valley, near its west fork, near Monument No. 63, Mexican boundary line, Grant County, New Mexico.

Geogr. Distr. State of Sonora, Mexico, into plain east of the San Luis Mountains, at the head of the Rio Yaqui, and east of that river's watershed. Extent of range unknown.

Genl. Char. Similar to *L. callotis*, but more yellowish; ears smaller; no black patch on nape. Skull: nasals long and very wide; supraorbital processes elevated, massive.

Color. Head cream buff and black; whitish area about eye; nape ochraceous buff; back pale ochraceous cinnamon, mixed with black; rump and thighs white with a few black hairs; sides and under parts white; limbs white, washed with buff on outer side; gular patch buff; sides of neck and front of shoulders ochraceous; ears yellowish brown, mixed with black anteriorly, white posteriorly, tips white, fringe on anterior edge ochraceous buff, of tips and posterior edge white; tail above black with many white-tipped hairs, beneath white. But little difference between summer and winter pelages.

Measurements. Total length, 530; tail vertebræ, 77; ear from crown, 146; hind foot, 131.

a.—battyi (*Lepus*), Allen, Bull. Am. Mus. Nat. Hist., 1903, p. 607.
BATTY'S JACKASS RABBIT.

Type locality. Rancho Santuario, State of Durango, Mexico.

Genl. Char. "Similar to *L. galliardi*, but much smaller, the general coloration yellower and less rufescent, especially the under fur; prepectoral collar much paler, nearly white, or pale brownish white instead of buff; front of fore feet grayish white instead of buffy white, and upper surface of hind feet clearer or purer white; extreme terminal portion (about 25 mm.) of anterior border of ear blackish in both forms."

Measurements. "Type, total length, 511; head and body, 451; tail vertebræ, 60; hind foot without claws, 122; ear from notch, 127; from crown, 140. Skull: total length, 92; basal length, 82; zygomatic breadth, 41; greatest breadth across supraorbital processes, 31; postorbital constriction, 11; length of nasals, 40; anterior width of nasals, 11; posterior width of nasals, 20; palatal length, 9; length of premolar-molar series (alveolar border), 17." (Allen, l. c.)

450. alleni (*Lepus*), Mearns, Bull. Am. Mus. Nat. Hist., 1890, p. 294. Elliot, Syn. N. Am. Mamm., 1901, p. 288.

FIG. LVIII. LEPUS ALLENI. ALLEN'S JACK RABBIT.

ALLEN'S JACK RABBIT.

Type locality. Rillito Station, Southern Pacific Railroad, Pima County, Arizona.

Geogr. Distr. Desert region between Phœnix and Benson, Arizona. Mexico?

Genl. Char. Size large; ears large; fulvous gular patch.

Color. Above yellowish brown, mixed with black; nape fulvous; sides, hips, rump, and outer side of legs white, mixed with black, giving a gray effect; pectoral band fulvous, rest of under parts white; head pale yellowish gray; feet white above; tail above gray, with a line of plumbeous black extending onto the rump, beneath white; ears whitish, with fringe white.

Measurements. Total length, 643; tail vertebræ, 69; hind foot, 138; ear from notch, 156. Skull: occipito-nasal length, 108; Hensel, 86; zygomatic width, 48; median nasal length, 34; lateral length of nasals, 45; posterior width of nasals, 36; anterior width of nasals, 16; palatal length to tip of azygos termination, 11; length of upper tooth row, 19; length of frontals, 47; of parietals, 16; length of lower jaw from tips of incisors to angle, 84; height at condyle, 43.

a.—palitans (Lepus), Bangs, Proc. New Eng. Zoöl. Club, 1, 1900, p. 85. WANDERING JACK RABBIT.

Type locality. Agua Caliente, forty miles southeast of Mazatlan, State of Sinaloa, Mexico.

Geogr. Distr. State of Sinaloa, Mexico; range unknown.

Genl. Char. Similar to *L. alleni*, but smaller and darker; ear shorter; rostrum broader and shorter.

Color. Above yellowish brown; sides gray; flanks and rump white; chin, throat, and under parts, except neck, and inner side of legs white; under side of neck cinnamon; nape plumbeous; head grizzled gray; orbital ring whitish; tail above blackish, beneath white; ears naked except white fringe on edges.

Measurements. Ear from notch, 126; from crown, 150; hind foot, 129. Skull: occipito-nasal length, 95.4; zygomatic width, 46.2; inter-orbital constriction, 22; length of nasals, 42.2; greatest width of nasals, 23.6; length of upper tooth row, 16.4.

451. asellus (*Lepus*), Miller, Proc. Acad. Nat. Scien. Phil., 1899, p. 380.
DONKEY JACK RABBIT.

Type locality. City of San Luis Potosi, State of San Luis Potosi, Mexico.

Geogr. Distr. State of San Luis Potosi, Mexico; range unknown.

Genl. Char. Similar to *L. merriami*, with longer ears and shorter tail.

Color. Above grizzled black and white; dorsal line darkest; rump and sides paler; under parts white, tinged with bluish gray; collar buff; limbs smoky gray, tinged with broccoli brown on fore legs and feet; hind feet white; a white line on fore legs to wrist, and one on hind legs to feet; orbital ring whitish; nape grizzled like back; ears broccoli brown, edges buff, tip black; tail above black, below grayish white.

Measurements. Total length, 558; tail vertebræ, 62; hind foot, 120; ear from crown, 175. Skull: greatest length, 100; basal length, 84; zygomatic width, 44; interorbital constriction, 30; lateral length of palate, 7; posterior margin of palate to tip of hamular process, 21; length of nasals, 43; greatest width of nasals, 22; upper molar series, 16.8; length of incisive foramina, 10.4; length of mandible, 74; length of lower molar series, 17.

texensis eremicus, Elliot, Syn. N. Am. Mamm., 1901, p. 291.
    texianus (*sic*) *eremicus* (*Lepus*), Allen, Bull. Amer. Mus. Nat.
        Hist., 1894, p. 347.
DESERT JACK RABBIT.

Type locality. Fairbank, Cochise County, Arizona.

Geogr. Distr. State of Sonora, Mexico, north to the White Mountain region, southeastern Arizona.

Genl. Char. Similar to *L. texensis*, but smaller.

Color. Above pale yellowish brown and black, darker on head and tinged with rufous; nape pale yellowish brown; flanks yellowish white, grading into the pure white of under parts; spot on rump black; broad fulvous band on lower part of throat and breast; rest of under parts and inside of legs white; outer side of legs pale yellowish brown: tail above black, beneath whitish; ears outside yellowish brown, finely grizzled with black and fringed with white on edges; inside grayish, grizzled brown inside the white edge.

Measurements. Total length, 565; tail vertebræ, 74; hind foot, 123; ear from crown, 128. Skull: occipito-nasal length, 94; Hensel, 76; zygomatic width, 42; lateral nasal length, 37; median nasal length, 29; posterior nasal width, 20; anterior nasal width, 14; length of lower jaw to end of incisors, 72; height at condyle, 40.

texensis griseus, Elliot, Syn. N. Am. Mamm., 1901, p. 291.
  texianus (sic) griscus (Lepus), Mearns, Proc. U. S. Nat. Mus., 1896, p. 562.
GRAY DESERT JACK RABBIT.
  Type locality. Fort Hancock, El Paso County, Texas.
  Geogr. Distr. States of Coahuila and Chihuahua, Mexico, north to upper Rio Grande from Maverick and Kinney Counties, Texas, to Grant County, New Mexico.
  Genl. Char. Size about equal to that of *L. californicus*; ear larger.
  Color. Winter Pelage. Above brownish gray, lined with black; rump and thighs gray; sides gray, lined sparsely with black and tinged with yellowish brown; nape grayish white; top of head brownish gray mixed with black; sides of head and neck tinged with yellowish brown; gular patch grayish clay color; rest of under parts white; legs gray, tinged with clay color; tail above black, this color extending over and dividing the gray on the rump, beneath brownish; ears brownish gray on outside anteriorly, with brownish white fringes, posteriorly white, tipped with black and fringed with white.
  Measurements. Total length, 559.2; tail vertebræ, 91.5; hind foot, 127; ear from crown, 152.8.

texensis micropus.
  t ianus (sic) micropus (Lepus), Allen, Bull. Am. Mus. Nat. Hist., 1903, p. 605.
SMALL-FOOTED HARE.
  Type locality. Rio del Bocas, State of Durango, Mexico. Altitude, 6,800 feet.
  Genl. Char. "Similar to *L. t. eremicus* and *L. t. griseus*, but more brownish gray than the latter, and large-bodied, with shorter tail, smaller hind feet, and larger ears than either, and with less fulvous along the sides of the body; prepectoral area paler and more grayish."
  Measurements. "Type, total length, 535; head and body, 450; tail vertebræ, 76; hind foot, 114; ear from notch, 133; from crown,

175. Eight males. Total length, 564 (535–587); head and body, 493 (459–514); tail vertebræ, 71 (64–83); hind foot, 116 (108–127); ear from notch, 136.4 (133–146). Eleven females. Total length, 599 (559–626); head and body, 524 (483–546); tail vertebræ, 73.7 (64–89); hind foot, without claws, 118 (108–124); ear from notch, 137 (130–145)." (Allen, l. c.)

texensis deserticola, Elliot, Syn. N. Am. Mamm., 1901, p. 291.
   texianus (sic) deserticola (Lepus), Mearns, Proc. U. S. Nat. Mus., 18, 1896, p. 564.
WESTERN DESERT RABBIT.
   Type locality. Western edge of the Colorado Desert, at base of Coast Range Mountains, in San Diego County, California.
   Geogr. Distr. State of Sonora, Mexico, and the desert region between the Sonoyta Valley of Arizona and Sonora, and the Coast Range Mountains of California; Lower California, Mexico.
   Genl. Char. Size of *L. californicus*, with larger ears; colors pale.
   Color. Above clay color, mixed with gray and black; sides and gular patch ochraceous; beneath tinged with ochraceous buff; ear pale drab and white, as usually seen, tipped with black on under side.
   Measurements. Total length, 560; tail vertebræ, 110; hind foot, 125; ear from crown, 158. Skull: total length, 92.5; occipito-nasal length, 90; Hensel, 72; zygomatic width, 42; width between orbits, 26; median nasal length, 29; lateral nasal length, 38; width of palatal bridge at tip of azygos, 8; length of incisive foramina, 22; posterior width of incisive foramina, 10; length of upper tooth row, alveolar border, 16; mandibular length, angle to alveolus of incisor, 63; height at coronoid process, 37.

452. californicus *(Lepus)*, Gray, Charlesw. Mag. Nat. Hist., 1, 1837, p. 586. Bachm., Journ. Acad. Nat. Scien. Phil., 1839, p. 86. Elliot, Syn. N. Am. Mamm., 1901, p. 291.
   bennetti, Gray Zoöl. Sulphur, 1844, p. 36, pl. 14.
   richardsoni, Bachm. Journ. Acad. Nat. Scien. Phil., 1839, p. 88.
CALIFORNIA JACK RABBIT.
   Type locality. "St. Antoine," Santa Barbara County?, California.
   Geogr. Distr. Lower California, Mexico, from Cape St. Lucas to northern California.
   Genl. Char. Size large; ears and hind feet longer than the head; hind feet shorter than the ears.
   Color. Above yellowish brown, mixed with black; sides, rump, and thighs tinged with cinnamon; head like back, darkest on the crown; nape smoky gray; beneath white in center of belly and chin;

chest and fore legs pale yellowish brown; hind feet whitish; tail above black, this color extending on the rump and dividing the gray sides; beneath pale buff; ears dark brown, fringed anteriorly with white, and with a fulvous white border posteriorly, tips brownish black.

Measurements. Total length, 545; tail vertebræ, 107; hind foot, 161; ear from notch, 125; from crown, 155. Skull: occipito-nasal length, 80; Hensel, 54; zygomatic breadth, 41; interorbital constriction, 17; lateral length of nasals, 32; median length of nasals, 26; posterior width of nasals, 18; anterior width of nasals, 11; length of upper tooth row, 13; length of lower jaw to end of incisors, 61; height at coronoid process, 36.

a.—*xanti* (*Lepus*), Thomas, Ann. Mag. Nat. Hist., 7th Ser., I, 1898, p. 45.

LOWER CALIFORNIA JACK RABBIT.

Type locality. Santa Anita, Lower California, Mexico.

Geogr. Distr. Cape Region of Lower California, Mexico.

Genl. Char. Similar to *L. californicus*, but smaller and grayer.

Color. General color like pale *L. californicus*, with belly tinged with darkish buff; ears gray outside, white inside; anterior fringe white with black hairs intermixed; nape smoky gray or black.

Measurements. Total length, 540; tail, 63; hind foot, 120; ear from notch, 125. Skull: "greatest length, 88; basilar length, 69; greatest breadth, 41; diagonal length of nasals, 38; greatest breadth of nasals, 17.6; intertemporal breadth with ledges, 22.6; without ledges, 16.7; breadth of palatal bridge, 6; length of molar series, alveolar border, 14.7."

453. insularis (*Lepus*), Bryant, Proc. Calif. Acad. Scien., 2d Ser., III, 1891, p. 92.

edwardsi, St. Loup, Bull. Mus. Paris, 1895, p. 5.

ESPERITO SANTO ISLAND JACK RABBIT.

Type locality. Esperito Santo Island, coast of Lower California, Mexico.

Geogr. Distr. Known only from type locality.

Genl. Char. Size equal to that of *L. californicus*.

Color. Above black; under parts pale vinaceous cinnamon, darker on sides; cheeks gray; chin and orbital region grayish white; throat cinnamon rufous; limbs and fore feet cinnamon rufous; hind feet grayish white; toes brownish; tail black above, beneath cinnamon rufous; ears blackish gray, tips black, inferior margin white.

Measurements. Total length about 450; tail, 100; hind foot, 110 (dried skin).

454. martirensis (*Lepus*), Stowell, Proc. Calif. Acad. Scien., 2d Ser.,
v, 1895, p. 51.

SAN PEDRO MARTIR JACK RABBIT.

Type locality. La Grulla, San Pedro Martir Mountains, Lower California, Mexico.

Geogr. Distr. San Pedro Martir Range, Lower California, Mexico.

Genl. Char. Size of *L. californicus*; ears larger; color darker.

Color. Above mixed steel gray and black; sides lighter tinged with rufous; chin and throat yellowish white; neck beneath gray; breast and inner side of legs salmon; belly whitish, washed with salmon; ears gray, apical half black.

Measurements. Total length, 603; tail vertebræ, 95; hind foot, 126; ear from crown, 184. Skull: total length, 96; occipito-nasal length, 91; Hensel, 75; zygomatic width, 41; breadth between orbits, 24.5; median length of nasals, 30; lateral length of nasals, 34.5; posterior width of nasals, 17; width of palatal bridge to point of azygos, 10; length of incisive foramina, 24; posterior width of incisive foramina, 9; alveolar length of upper tooth row, 10; mandibular length, angle to alveolus of incisor, 61; height at coronoid process, 36.

INDEX OF LATIN NAMES.

VOL. IV, PART I.

i

*Lepus durangæ proves to be the same as L. holzneri, and therefore becomes a synonym of that species. See Allen, Bull. Am. Mus. Nat. Hist., 1904, p. 210.

INDEX OF COMMON NAMES.

VOL. IV, PART I.

www.ingramcontent.com/pod-product-compliance
Ingram Content Group UK Ltd.
Pitfield, Milton Keynes, MK11 3LW, UK
UKHW021654100125
4058UKWH00038B/533